CHEMISTRY
Imagination and Implication

A. TRUMAN SCHWARTZ

Macalester College

Academic Press
New York
London

ACADEMIC PRESS, INC.
111 Fifth Avenue, New York, New York 10003

United Kingdom Edition published by
ACADEMIC PRESS, INC. (LONDON) LTD.
24/28 Oval Road, London NW1

LIBRARY OF CONGRESS CATALOG CARD NUMBER: 72-84366

PRINTED IN THE UNITED STATES OF AMERICA

To my father and
the legacy of the past
To my son and
the promise of the future

Contents

Preface

To learn something new can be one of life's most exhilarating experiences; and to teach someone is often almost as rewarding. But students and teachers involved in a chemistry course designed primarily for nonscience majors are confronted by especially exciting opportunities. For there are few branches of inquiry which reveal so much of the beauty and wonder of nature and simultaneously exert so great a practical influence upon everyday existence. Of course, the correlation is hardly coincidental. It is precisely because chemists have succeeded in discovering so many profound and lovely things about our world that they have gained the power to change it for better or for worse.

The intellectual excitement of chemistry and its useful ubiquity are so intrinsic to the science that I have attempted to convey this complementarity in the subtitle: "Imagination and Implication." I hope it will be apparent in the pages that follow that chemistry has progressed through the exercise of creative imagination every bit as brilliant and perceptive as that which has produced mankind's greatest music, painting, and poetry. Moreover, I trust the many examples of applied (and misapplied) chemistry will make it manifestly clear that the implications of this science are as important as the consequences of political systems and economic policies.

In my attempt to concurrently develop the principles of chemistry and provide instances of their social significance, I am deviating somewhat from the approaches employed in most other texts. Some books written for students concentrating in the humanities, fine arts, and social sciences place their almost-exclusive emphasis upon the same concepts taught in general chemistry courses designed for science majors. On the other hand, several books have recently appeared which greatly minimize such traditional topics and devote most of their pages to the applications of chemistry. Still others treat both chemical science and chemical technology, but in separate chapters. Needless to say, I favor my own integrative approach. But I am also well aware that chemistry for philosophers, painters, political scientists, and poets will remain a highly idiosyncratic offering, reflecting the individual judgment and taste of the instructor. For example, although all teachers will probably agree that comprehensiveness is impossible in such a course, there will be many different opinions of what topics should be presented. My own selection includes more quantum mechanics and thermodynamics than is commonly taught in terminal courses. I justify this emphasis because these two subdisciplines have yielded fundamentally important observations about the organization and function of the universe. Those instructors desiring to reduce the coverage of either or both of these areas may do so by skipping parts of Chapters 9, 12, or 13.

There is an organizational logic to this book, and therefore I would prefer that it be read in its entirety in the order of presentation. However, I realize that the time limit imposed by a one quarter course may necessitate cuts. Under such circumstances, the last chapters are always the most vulnerable. And it is true that Chapters 15 and 16 are more-or-less independent and perhaps expendable. But I suspect that students will also find them among the most interesting. Professors who believe that stoichiometric calculations have no place in nonmajors' courses can deemphasize the quantitative aspects of Chapter 4.

On balance, I think that the heterogeneity which I foresee as characteristic of chemistry courses for nonscience majors is a good sign, for it reflects the humanity of the chemist/teacher and the humaneness of the science. In order to pump more life's blood into what the public often regards as the desiccated corpus of chemistry, I have adopted a phenomenological–historical approach which includes occasional quotations from the scientists involved, brief biographical vignettes, and references to contemporary events. If I appear to be wordy about the way in which I do this, it is, frankly, because I am long-winded. But my loquacity not withstanding, one equation is often worth a hundred words, and I have not often been able to rely upon that more concise form of symbolic communication. Instead, I have talked my way around equations, trying to be as rigorous as possible and to avoid condescending oversimplifications. Instructors who grow faint at seeing so much verbiage are respectfully reminded that many of the students in this course are accustomed to reading 2500 pages of English literature in one semester.

Finally, to further emphasize the fact that chemists are organic mixtures of flesh and blood (with feet of clay), I have deliberately written in a personal style. The occasional use of the first person, the outrageous puns, and the various stylistic warts clearly reveal that there is a human being, not a committee, behind this book. This is not to suggest that I produced the text unaided. Indeed, it would be difficult to overestimate the contributions of my colleagues in the Chemistry Department of Macalester College—Professors Phil Bryan, Earl Doomes, John McGrew, John Scott, Emil Slowinski, Fred Stocker, and Wayne Wolsey. All of them proved ready and willing to answer my questions and discuss selected portions of the manuscript. Other thoughtful suggestions came from two nonchemists who volunteered to read segments of the book—Professors Claude Welch of the Macalester Department of Biology and David White of our Philosophy Department. Professor Henry Bent of North Carolina State University read several chapters and provided insightful advice.

It does not minimize the very significant contributions of these gentlemen to observe that my most severe and most helpful critics were the students who studied the manuscript in an Interim Term course. I am certain that my book is much the better for their perceptive comments. Mr. William Norman was one of my first readers, and Mr. Bruce Williams, another student, checked the solutions to most of the problems.

This book owes much to the diligence, efficiency, cooperativeness, and enthusiasm of the staff of Academic Press who shepherded it through the intricacies of production. I sincerely thank all of these fine people and my wife Beverly, who typed the manuscript, corrected the novelties of my spelling, and gave me many valuable suggestions, unencumbered by a preexisting knowledge of chemistry—all with cheerful good humor and infinite patience. Finally, I am grateful to my many nonscientist friends who were always willing to comment on a passage or to offer their opinions on matters of content, style, or design, and who generously tolerated my preoccupation with this project. It is for people such as they and for students who share their interests that this book is intended.

A. Truman Schwartz

Never imagine yourself not to be other-
wise than what it might appear to others
that what you were or might have been
was not otherwise than what you had
been would have appeared to them to
be otherwise.

The Rev. Charles Lutwidge Dodgson

1

Introduction and apologia

Like it or not, you can't get away from
chemistry. And let's face it, there are some
things about chemistry that are not particularly likable—things you might very
well like to get away from. Take for example, a breath of air: 78.09% nitrogen,
20.93% oxygen, 0.03% carbon dioxide,
0.93% argon, 0.0025% other rare gases,
plus a few miscellaneous contributions of
modern chemistry such as 17.1 ppm (parts
per million) carbon monoxide, 0.13 ppm
sulfur dioxide, and 0.14 ppm nitric oxide
and nitrogen dioxide.* Admittedly, not a
very appetizing lungful. But before you ban-
ish chemistry, burn this book (more pollution!), and return to a clean-air utopia, it
might be well to remember that those miscellaneous gases came from the exhaust
pipe of the car or bus that may have
brought you to class this morning, the
smokestacks of the power plant providing
the electricity lighting the bulb by which
you may be reading these words, and the
chimneys of the factories which produced
the paper and ink before you. Understand,

* Data for Chicago, 1965, from "Air Quality Data from the National
Air Sampling Networks and Contributing State and Local Networks,
1964–1965," U. S. Department of Health, Education, and Welfare,
National Air Pollution Control Administration, Cincinnati, Ohio, 1966.

I make no claim for the unexpendability of this particular book, but I do not think too many of you will take issue with my premise that paper and printing and, for that matter, electric power and the internal combustion engine have enhanced civilization and the quality of existence. Chemistry has been a conspicuous contributor to such progress *and* to the pollution spawned by it.

There is no need for me to attempt to list even a small fraction of the thousands of instances of the ubiquitous (and sometimes iniquitous) utility of chemistry. If you are up to it, an evening of television commercials will provide more than enough examples: washday wonders, enzymatic enemies of dirt and stain, toothpastes promising fewer cavities and more sexual success, miracle fibers to put on or walk on, gasoline with more go, lotions and potions and notions. And, should all this television affect you adversely, there are pills and powders to sooth your aching head and neutralize your acid stomach.

Without doubt, chemistry has been spectacularly influential in its impact on daily living. Indeed, the social consequences of little pills containing estrogens, of sugar cubes coated with D-lysergic acid diethylamide, and of amphetamines, barbiturates, mescaline, and heroin have been nothing short of revolutionary. If relevance to modern life is the chief criterion for the existence of an academic discipline, chemistry is eminently suited for study.

As a matter of fact, I happen to believe that there are factors besides the pragmatic that add to a subject's reason for existence in the contemporary curriculum. I have the temerity to summarize these in what, I fear, may be a rather unfashionable term—intellectual excitement. I maintain that, even if chemistry had not contributed a single "practical" discovery to make everyday life better (or worse), it would still be worthy of serious scrutiny. Chemistry, the science of stuff, is devoted to increasing our understanding of the universe, what it is made of, how it is organized and structured, and how it works. These are audaciously bold aims, and in their efforts to attain them, chemists have developed approximations, models, and theories of great imagination, power, and beauty. The concepts of chemical identity and chemical combination, the principles of molecular structure, the Atomic Theory, thermodynamics, and quantum mechanics are among the outstanding intellectual achievements of man. I am convinced that these ideas are so important and so elegant that every educated man and woman should have some familiarity with them.

I also think that some knowledge of the methodology of science is essential. This would normally be the place for a paragraph on "the" scientific method. You know the sort of thing: several sentences on empiricism and the experimental basis of science; definitions of hypothesis, law, and theory; and a nod toward mathematics. You will be spared that for two reasons. First, the workings of science and scientists are too diverse and too individualistic to permit a simplistic summary. And second, this entire book is an attempt to describe the ways (not way) in which science is done.

Of course, the best way to learn what science is all about is to do it yourself—a standard rationale for required laboratories. Unfortunately, many so-called laboratory exercises are just that, chemical calisthenics. They emphasize obedience rather than independence, just as many texts teach memorization rather than understanding. Please do not infer that I disparage obedience or memorization; there probably never was a successful scientist who did not demonstrate at least to some extent both of these virtues. But though they may be necessary, they are definitely not sufficient. The point is that the counterpart of the cookbook laboratory manual is the *ex cathedra* text. Too often in our passion for neatness, we protect the students from some of the messier aspects of science. Believe it or not, most scientists are fascinated with ambiguities, often because they have a great urge to resolve them. This little known fact could hardly be inferred from books written with about as much uncertainty as the Decalogue.

It seems safe to say that, at present, physical science is more orderly than the social sciences, the humanities, or the fine arts. Procedures are better established, verification is more readily obtained, consensus is more easily achieved. But behind almost every law set in sacred boldface type was an encounter between conflicting ideas. To ignore these issues, to present science as a monolith rising majestically toward omniscience, is to do a great disservice to science and to the student. Anyone growing up in the last half of the twentieth century has probably developed some tolerance for uncertainty. Indeed, if the student is in art or literature or political science rather than a chemistry major, it may be because he loves the ambiguity of men and nature and fails to see it recognized by science. One aim of this book is to demonstrate that chemists too can be filled with wonder at the sometimes contradictory complexity of the universe. My approach in certain instances will be to introduce you to great issues in the development of chemistry. At other times I will lead you to the indistinct growing-edge of contemporary chemistry. I cannot promise to always suppress my professorial urge to provide pedantic comment or belabor the obvious. Ignore these outbursts if you please, but above all observe and judge for yourself the workings of this way of knowing.

I am neither so naive nor so provincial that I expect or even hope to transmute all of you into chemists. The economic consequences alone of such a mass conversion would be disastrous, especially to me. I realize full well that this may be the last text-book on chemistry many of you will read. Some of you have been dragooned against your will into the course served by this volume. Others have probably wound up in this particular classroom because the hour was open and you needed a science subject to satisfy distribution requirements. Perchance some of you are even enrolled because you like chemistry. I rather hope you all will by the time the course is over.

This book attempts to accomplish that goal by emphasizing the dual nature of chemistry. Its useful ubiquity and its intellectual excitement are inseparable. To concentrate upon either of these aspects at the expense of the other would create a badly biased view of the science. Therefore, in the pages which follow, I make every effort to present the essential concepts of chemistry along with numerous examples of their application or misapplication. In doing so, I emphasize certain topics to a greater extent than is typical of most introductory general chemistry texts. Of necessity, I slight others. I have made my selection with several ends in mind. First of all, I trust this encounter with chemistry will lead you to an increased awareness of the importance of the science to your everyday existence. Moreover, I hope you will acquire sufficient knowledge of the experimental facts, principles, and methods of chemistry to aid you in exercising intelligent and informed judgment in the growing number of instances where controversy surrounds the interaction of chemistry with society or the individual. But beyond the immediately and obviously practical aspects of chemistry, I hope you will come to appreciate the intellectual achievements of chemists in resolving the beautiful ambiguities of nature. You could do no better than to capture the awe of a famous physicist, the late J. Robert Oppenheimer:

We have learned that wonder resides, and paradox
and puzzlement and harmony and order, in many ordinary
things; in the stuff that matter is made of, in the flow of
the ocean's currents, in the migration of the birds, in
bubble and drop and clod. We have come to respect the
most pedestrian curiosity as a likely origin of learning
*unexpected and lovely things about our world.**

Finally then, is it too much to hope that all this might lead you to really rather like chemistry? If any or all of these ends are achieved, you and I will both be rewarded.

---

* J. Robert Oppenheimer, The Need for New Knowledge, *in* "Symposium on Basic Research" (Dael Wolfle, ed.), Publ. 56, p. 4, American Association for the Advancement of Science, Washington, D. C., © 1959.

For thought and action

1.1 Identify several inventions or "advances" based upon science and technology (the steam engine, radio and television, fertilizers and insecticides, computers, "miracle" drugs and surgical techniques, etc.), and prepare a "debit" and "credit" list for each, evaluating their relative contributions to the welfare or detriment of mankind.

1.2 At least three "scientific" revolutions—the Copernican, the Darwinian, and the Freudian—fired shots heard round the world. Discuss how these great theories of nature irreversibly altered life on this planet. Can you suggest any other intellectual advances, either within or without the realm of science, which were comparable in the magnitude of their impact? Defend your choices.

1.3 J. Robert Oppenheimer, watching the explosion of the first nuclear test device over the New Mexico desert in July, 1945, thought of these lines from the *Bhagavad Gita:*

"I am become Death, The shatterer of worlds."

The following words were spoken by other observers:

"Good God, I believe that the long-haired boys have lost control."
"The war's over. One or two of those things, and Japan will be finished."
"This was the nearest to doomsday one can possibly imagine."
"It was like being witness to the Second Coming of Christ."
"It worked, my God, the damn thing worked!"
"Now we're all sons of bitches!"
"My God, its beautiful!" "No, it's terrible!"

Consider and comment on these reactions in the context of the social responsibility of scientists.

1.4 Edgar Allen Poe's "Sonnet—To Science" begins with these lines:

Science! true daughter of Old Time thou art!
Who alterest all things with thy peering eyes.
Why preyest thou thus upon the poet's heart,
Vulture, whose wings are dull realities?

Evaluate, relative to your own present conception of science, the validity of Poe's vision.

Suggested readings

Bronowski, J., "Science and Human Values," Harper Torchbooks, New York, 1956.
Brown, Martin, ed., "The Social Responsibility of the Scientist," Free Press, New York, 1971.

Commoner, Barry, "Science and Survival," Ballantine Books, New York, 1963.

Kuhn, Thomas S., "The Structure of Scientific Revolutions," University of Chicago Press, Chicago, Illinois, 1962.

Russell, Bertrand, "The Scientific Outlook," Allen & Unwin, London, 1931.

Snow, C. P., "The Two Cultures: And A Second Look," The New American Library (Mentor Books), New York, 1963.

The matter matter

Some attic ideas about everything

This chapter is addressed to a rather fundamental question: "What is the world made of?" The problem is hardly a new one; in fact, it was asked and answered by the man usually considered to have been the first philosopher, Thales of Miletus (ca. 624–545 B.C.) Not to be outdone by the philosophers, we will also claim Thales as the first theoretical chemist. Back in those dear dead days of Western philosophy's infancy, its practitioners were still interested in things besides the meaning of words, and you couldn't really tell physics from metaphysics. This is not the place to evaluate Thales' contributions as a philosopher; but as a theoretical chemist, he was something short of an unqualified success. Put a bit more charitably, his conclusion that all matter is ultimately derived from water has undergone rather drastic revision in the past 2500 years. In fairness to Thales, we should note that his theory was perhaps a not unreasonable one for a man who was figuratively and at times, no doubt, literally drenched in the Homeric "wine-dark" waters of the Aegean. Maybe Thales' elementary inspiration came on that legendary day when, walking along, his eyes philosophically fixed on the heavens, he fell into a well. (Lest the reader draw any unwarranted inferences regarding a philosopher's relationship to reality, it should be noted that, according to another legend, Thales, on the basis of his meteorological observations, predicted a bumper crop of olives, and demonstrated his business accumen by renting all the olive presses in the area, only to sublet them at his own price when his prediction came true.)

Philosophers are not only lovers of wisdom, but lovers of argument as well. Therefore, it is not altogether surprising that certain of Thales' contemporaries and followers suggested other substances as the basic, elementary stuff of the universe. Anaximenes proposed air, Heraclitus favored fire, and Anaximander invented *apeiron*, an eternal and unlimited element unlike any of the others. Incidentally, what is probably the final phase of this debate appears in the pages of "Faust," Part II. Here Goethe introduces Thales and Anaxagoras, who argue the importance of water versus fire. (Thales wins.)

The first great synthesizer of ideas about the constitution of matter was Empedocles (ca. 490–430 b.c.). His universe was made of earth, water, air, and fire, combined by a sort of mutual love in various proportions to form all things. Not only did this theory have the virtue of comprehensiveness, a little reflection will show that it was amazingly consistent with certain observations. For example, when a log burns, fire obviously escapes, bubbles of sap and water may form along the log, and water can condense out of the vapors which rise, along with the various "airs" and smokes. And, of course, the ash which remains behind appears to be rich in the earthy element. The experimental results seem to be in qualitative agreement with the Empedoclean hypothesis.

Aristotle (384–322 b.c.) extended and embroidered the theory of Empedocles by emphasizing the properties of the four elements. There were also four of these properties: coldness, hotness, dryness, and wetness. Each of the elements in its purest form (not the forms encountered in common experience, but sort of Platonic ideals) contained a maximum of two complementary properties. Thus, earth was the cold and dry element, water was cold and wet, air was hot and wet, and fire was hot and dry. Any element could be changed into any other by altering its properties. For example, water could be changed to air by the addition of the property of heat, which then replaced cold, a procedure which appears to be operative in the boiling or evaporation of water. The explanation may be simplistic, but it is not without some correspondence to nature. Indeed, such early and perhaps naive models of matter are a good deal closer to the methodology of modern science than they might at first seem to be. Their significance in shaping the conception of the universe which per-

FIGURE 2.1   *A schematic representation of the four elements of Empedocles and Aristotle, and the four properties of these elements.*

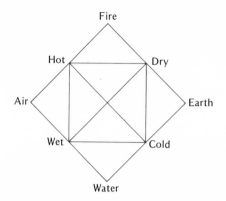

sisted for hundreds of years will be considered later. First, some words on ancient chemical technology.

The craft of the ancient and artful chemical engineer

The practical arts of metallurgy achieved a considerable degree of sophistication long before they were scientifically understood. Thus, the isolation and use of certain of the substances we now classify as elements predated the theories of Thales, Empedocles, and Aristotle by at least two millenia. That gold was one of the first metals to be used is a direct consequence of its chemical properties. Because of its inertness, that is, its low tendency to react chemically with other substances, gold is often found in an uncombined, pure metallic state. The beauty of the metal, its durability, its malleability, and its rarity soon lead to the use of gold as a material for ornamentation and a medium of exchange. The pursuit of it is hardly a new phenomenon.

In ancient times, silver was more valuable than gold—another consequence of chemistry. Because silver is more reactive than gold, nuggets of the former element are less common than pieces of pure gold, in spite of the greater total abundance of silver. The more abundant metal is usually found combined with other substances, so the isolation of significant supplies of silver required the development of the technique called cupelation, heating the impure sample in a porous cup of bone or ash which absorbs the dross.

Smelting, the heating of metallic ores with carbon, is necessary in order to win the more reactive metals from the chemical forms in which they commonly occur. The method was first applied to copper and tin in Asia Minor at least 3500 years before the birth of Christ. The presence of the tin imparted to the resulting metal a hardness which made it far superior to pure copper in many of its applications. This bit of applied chemistry launched the Bronze Age, during which Odysseus, armed in this alloy of 90% copper and 10% tin, took the long way home from Troy.

Although the nickel content of the earliest known iron objects suggests a meteoritic source of the metal, the technique of isolating iron from its ore was apparently discovered not long after the first smelting of copper. By 1500 B.C., the Hittites had developed a rather sophisticated ferrous metallurgy, and 300 years later the use of iron began to spread through the Middle East and, finally, Europe. The dawning of the Iron Age prompted a contemporary observer, Pliny the Elder (23–79 A.D.), to utter this ominous warning about the terrible new weaponry.

*It is by the aid of iron that we construct houses, cleave
rocks, and perform so many other useful offices of life.
But it is with iron also that wars, murders, and robberies
are effected, and this, not only hand to hand, but from a
distance even, by the aid of weapons and winged
weapons, now launched from engines, now hurled by
the human arm, and now furnished with feathery wings.
This last I regard as the most criminal artifice that has been
devised by the human mind; for, as if to bring death upon
man with still greater rapidity, we have given wings to
iron and taught it to fly. Let us, therefore, acquit Nature
of a charge that belongs to man himself.* *

There is evidence that independent discoveries of iron smelting also occurred in China, southeastern Africa, and among the Eskimos.

Two more metals, lead and mercury, completed the conveniently mystical seven known to the ancients. Unfortunately, not quite enough was known about the toxicity of the former, and certainly the decline and fall of some Roman citizens, if not the Empire itself, was attributable to the use of lead in cooking utensils and water pipes. (Similar effects have been observed more recently from ingesting the effluent of lead-soldered stills.)

The above few paragraphs have not begun to do justice to the fascinating and complex story of ancient chemical technology and its profound influence. Museums and contemporary accounts

---

* Pliny the Elder, "Natural History," Book XXXIV, Chapter 39, quoted in
M. E. Weeks, and H. M. Leicester, "Discovery of the Elements," p. 31,
Journal of Chemical Education, Easton, Pennsylvania, 1968.

(most conveniently, the Old Testament) are rich sources of information concerning this aspect of the interwoven history of civilization and of science. It is to be hoped that the survey was not so brief that the reader failed to note the fact that the application of certain chemical procedures far outstripped their accurate interpretation. What we will consider next is an attempt to effect an amalgamation of these practical arts with prevailing theories of matter in that quixotic quest called alchemy.

Making gold for fun and profit

Geoffrey Chaucer and Ben Jonson are responsible for giving *alchemy* and alchemists a bad reputation. The second canon of the "Canon Yeoman's Tale" in Chaucer's "Canterbury Tales" is as inventive as he is unscrupulous; and Subtle, the title character of Jonson's "The Alchemist," and his delightfully bawdy accomplices use alchemy as "a pretty kind of game, somewhat like the tricks o' the cards, to cheat a man." Admittedly, the search for a method of turning base metals into gold and preparing an Elixir of Life was subject to some debasement. But by and large, alchemists were serious; poor perhaps, but honest. Furthermore, they were, generally speaking, not would-be magicians or wizards, though their esoteric and enigmatic symbolism might suggest otherwise. Rather, the alchemists were seeking to achieve an end that was logically consistent with the working of the world as they understood it. In this respect, they were practicing a sort of science.

Apparently, alchemy emerged from the black soil of the Nile Delta sometime in the first century A.D. The origin is suggested by the name. Some scholars believe "alchemy" is derived from *Khem* or *chemia*, an Egyptian word meaning black, which was also a name for that country. The Arabic influence is indicated by the article *al*. Interestingly, the cognate of "chemistry" first appears in an edict of Emperor Diocletian, issued in 296 A.D., banning the art of making gold and ordering manuscripts on the subject to be burned. Not a very propitious start!

The earliest alchemical manuscripts extant include instructions for coloring or plating metals to give them the appearance of gold. In these papyri there is no suggestion that the priestly practitioners of these operations were under any illusions about actually bringing about the transmutation of one metal into another. Nevertheless, such high hopes manifested themselves in subsequent students of the Hermetic Art. The conclusion was not a surprising one; Aristotle and others had provided the theoretical basis. We have noted that the four basic elements could change into one another; consequently, the relative quantities of the four in, say, a metal could also change; *ergo,* the metal itself could change its identity. It even became an accepted belief that such metamorphoses were in the natural order of things. After all, change is a fact of life. Seeds become flowers, caterpillars turn into butterflies, and tadpoles grow up to be frogs. Confronted with such mind-boggling transmutations, it hardly strained the alchemists' faith to espouse the idea that base metals in the womb of the earth were ripening, gaining in perfection, tending towards gold. "Lead and other metals . . . would be gold if they had time," says Subtle in "The Alchemist," act II, scene 3.

The alchemists did not have the time, and so they launched their search for a catalyst to speed up the reaction—the *Philosopher's Stone.* Again, the idea of such an agent is consistent with the then prevailing conception of matter. Indeed, it is possible to argue that the Philosopher's Stone was thought to be something very like condensed *pneuma,* the spirit or breath of the universe. The idea is also akin to Aristotle's *quintaessentia,* an etherial fifth element which is eternal and unchangeable and comes close to being pure form. The Philosopher's Stone thus was the means by which changes in form and properties and, hence, identity could be accomplished. It was a purifying medium for the perfecting of men as well as metals.

Cryptic recipes for the preparation of the Stone make recent synthesis of nucleic acids and proteins seem trivially simple. For one thing, there was the problem of starting materials; some processes required such unusual reactants as the "blood of a red-

FIGURE 2.2   *"The Chemyst" by Cornelis Pietersz Bega (1620–1664), in the Fisher Collection of Alchemical and Historical Pictures.*

haired man" or the "urine of an ass." More often, gold and silver, the two most "perfect" metals, were chosen as the initial ingredients. These were converted to "philosopher's sulfur" and "philosopher's mercury," respectively, the components from which all metals were thought to be made. The reactants were hermetically sealed in a Philosopher's Egg or Hermetic Vessel and a whole series of steps hopefully ensued. The progress of the reaction could be followed by color changes. When the proper reaction time had elapsed, and providing the adept was "*homo frugi*, a Pious, Holy, and Religious Man, One free from mortal Sin, a very Virgin," the wondrous *Lapis Philosophorum* should have been virtue's reward. The proof of the powder was in the projection, and so, the final test involved mixing the Stone and base metal in a prescribed ratio and waiting for the transmutation into gold or silver. If the result was not the eagerly hoped for one, the alchemist presumably looked into his alembic, his ingredients, and/or his soul for contaminants.

It should not be too surprising that some of the first government-subsidized research projects were in alchemy. What may be unexpected is the fact that numerous accounts of supposedly successful transmutations do exist. So do some coins, which, according to their inscriptions, were minted from alchemical gold—an unusual but effective way to balance the budget. Explanation of these supposed successes include fraud, fiction, wishful thinking, and hallucinations. No doubt all of these influences were operative in one instance or another.

Yet, the fact that the alchemists spent most of the Middle Ages in a futile attempt to achieve an impossibility does not mean that they made no contribution. As I have previously suggested, their confidence in the operation of the regularities of nature (as they saw them) and their belief that these were amenable to man's study and mastery were in keeping with the basic tenets of modern science. Moreover, alchemy provided apparatus, observations, and even theories which gradually led to modern chemistry. But before we examine the great watershed that marks the beginnings of the science as you will study it, we must consider the ingenious though inadequate theory which prompted the Chemical Revolution. It was called the *Phlogiston Theory* and it grew out of alchemy.

What's burning?

The idea that a combustible principle is present in all flammable materials owes a debt to the ancient sulfur–mercury theory of composition. Geber and other Islamic alchemists held that all matter was composed of these two ingredients (though not the common forms of these substances). "Sulfur" was believed to impart hot and dry natures to metals and "mercury" provided the cold and moist properties. In the sixteenth century, Phillipus Theophrastus Aureolus Bombastus von Hohenheim—the amazing Paracelsus—added salt to sulfur and mercury to create an elementary trinity. Around 1670, these three were renamed *terra prima, terra secunda,* and *terra tertia,* respectively, by Johann Joachim Becher, a German physician. Becher also called the sulfurous combustible component *terra pinguis* and occasionally, *phlogiston,* after the Greek word for fire. It was the latter name which stuck.

Although Becher's contributions were chiefly in nomenclature, his writings did influence another medical doctor, Georg Ernest Stahl, who, in the later seventeenth century, did much to develop the first comprehensive chemical theory. *The outstanding feature of the Phlogiston Theory was that it explained combustion and calcination by the same mechanism.* Calcination had long been known as the conversion of a metal to its *calx,* a nonmetallic substance, often a dark, nonlustrous powder. Although examples such as the rusting of iron were familiar to everyone, the chemical nature of calxes was not known. At first glance, it would seem that the process of calcination bears little similarity to burning. Nevertheless, Stahl argued that all combustible substances and all calcinable metals contain phlogiston. Burning and calcination were both attributed to the loss of this flammable principle.

While it is true that the phlogistonists seldom if ever wrote equations, we can represent their interpretation of calcination and combustion in this convenient form. An arrow is used to indicate the occurrence of a change or transformation, and the rusting of iron thus becomes

Iron → Calx of iron + Phlogiston.

On the other hand, the burning of carbon does not yield a calx in the strict sense of the word. Experiment shows that a gas, not a solid, is formed. Nevertheless, the change was also explained as an escape of phlogiston. And therefore, we are reasonably justified in writing the phlogistonists' description of a charcoal fire as

Carbon → "Calx of carbon" + Phlogiston.

Certainly, any adherent of the theory in its declining days would understand the significance of the equation.

The Phlogiston Theory was easily extended to account for the conversion of calxes to metals. For a few calxes (the calx of mercury is a particularly important example), this transformation can be accomplished merely by the application of heat. Most calxes, however, require more complicated treatment. In spite of that fact, one very useful method for obtaining metals from their calxes has been practiced for thousands of years. You read about it a few pages back; it is smelting. Many metallic ores are calxes, and smelting involves heating these ores with charcoal or coke. A phlogistonist would remind you that carbon burns readily, and therefore must contain a good deal of phlogiston. In the smelter, this phlogiston is transferred to the metallic calx which thus becomes converted to the pure metal. Expressed in equation form, the reaction involving iron ore would be

Calx of iron + Carbon → Iron + "Calx of carbon."

Qualitatively, the Phlogiston Theory was a success, and correct in its discernment of the similarity between calcination and combustion. There were, of course, some difficulties with the concept of phlogiston; it had never been isolated, prepared, or characterized. Nevertheless, its existence was widely accepted by such competent eighteenth century chemists as Karl Wilhelm Scheele in Sweden, Joseph Black in Scotland, and Henry Cavendish and Joseph Priestley in England.

It would be a mistake to conclude that these men, all careful observers, were unaware of the fact that air was necessary for burning. To them, air was an essential receiver for the phlogiston released during combustion.

By 1770 it had been well established that air was not always and everywhere homogeneous and identical in its properties. Indeed, it was apparent that several kinds of "air" existed. Around 1750, Joseph Black, in his classical research on *magnesia alba* (magnesium carbonate), prepared a gas which extinguished lighted candles and, in high concentration, suffocated small animals. Black was able to show that his *fixed air* was the same as the gas found in the breath exhaled by animals and above burning charcoal. This, then, is what we called "calx of carbon" a moment ago.

Twenty years later, Daniel Rutherford, who had been a student of Black's at Edinburgh, isolated from ordinary air another colorless, odorless gas incapable of supporting combustion or life. This *noxious air* was obtained by allowing a mouse to expire in a closed container and chemically absorbing the fixed air exhaled by the late beastie. The gaseous residue remaining was clearly not fixed air because, unlike the latter, it did not turn lime water (calcium hydroxide solution) cloudy in appearance.

A revolution built on air

The history of the Chemical Revolution is very much the history of a third gas, prepared more-or-less independently, but by similar methods, by three men in three different countries in the span of three years. The "more-or-less" is inserted in the above sentence as a hedge to hold off historians of science, who have been arguing, almost since 1775, about priority and plagiarism in this case.

Quite clearly, the honor of having first prepared the gas belongs to the Swedish apothecary, Karl Wilhelm Scheele. It was in 1771 or 1772 that he obtained a gas, again colorless and odorless, but capable of supporting more vigorous combustion than ordinary air. His choice of a name, *fire air*, was a logical one. Unfortunately for himself, Scheele was a good deal less eager to rush into print than many modern chemists, and it was 1777 before a full account of his discovery was published. This was three years after Joseph Priestley's preparation of what he called *dephlogisticated air*, so named because the effect the gas had in promoting combustion suggested to its discoverer that it had a particularly high capacity for phlogiston.

An ideal title for a biography of Priestley would be "Sermons and Soda Water." As a nonconformist Unitarian minister he delivered quite a few of the former; and as an amateur chemist, he discovered the latter (he called it "mephetic julep"). In addition, Priestley had time to produce a prodigious corpus of writings on a wide range of subjects—philosophy, theology, education, political and social theory, history, and science. His extrachemical interests are evident in this comment on the unusually "good" air which he had prepared: "A moralist at least may say that the air which nature had provided for us is as good as we deserve." Nevertheless, Priestley does suggest that "pure dephlogisticated air might be very useful as a medicine" or "may become a fashionable article in luxury." His charmingly naive statement that "Hitherto only two mice and myself have had the privilege of breathing it" may have been true for the pure gas, but hardly for mixtures containing it.

Both Scheele and Priestley were convinced phlogistonists, and this colored their view of the gas they had prepared. Antoine Laurent Lavoisier, on the other hand, was not committed to the prevailing paradigm. Like Priestley, Lavoisier was a Renaissance man in the eighteenth century—an economist, an educational and agricultural reformer, a public servant and, unfortunately and fatally, a tax collector. Both men were in the intellectual vanguard of their day, and both suffered for their farsightedness. Priestley's books and apparatus were destroyed by a mob and he was finally forced to emigrate to Northumberland, Pennsylvania. Lavoisier paid more dearly; he gave his head to a society which smugly proclaimed, in sentencing him to the guillotine, *"La Republique n'a pas besoin des savants."* ("The Republic has no need of scholars.") In his understanding of *air eminemment pur,* or *vital air,* which he prepared in 1775, Lavoisier was far ahead of his English contemporary. In fact, although the nineteenth century chemist, Berthelot, may have exhibited an excess of Gallic chauvinism in his contention that "Chemistry is a French science. Its founder was Lavoisier of immortal memory," the evaluation was not too far from the truth.

If a single chemical reaction can be said to have launched *La Revolution Chemique,* it was surely the decomposition of *mercurius calcinatus per se,* the red calx of mercury. Scheele, Priestley, and Lavoisier all used this reaction to prepare the gas which we have thus far avoided identifying. There was agreement on the observables. Heating the red powder produced droplets of mercury and a colorless, odorless, tasteless gas which supported combustion and sustained life to a greater extent than ordinary air. But the ambiguity and the eventual clue to modern chemistry lay in the interpretation of these facts.

To Priestley and Scheele, the reaction could have been summarized

Heat + Calx of mercury + Phlogiston → Mercury +
Dephlogisticated air or Fire air.

The calx readily absorbed phlogiston and thus became converted to metallic mercury. Since heat was required for the reaction, the heat was presumably the source of the phlogiston. Indeed, Scheele, the most professional chemist of the three, argued that heat was a compound substance made up of fire air and phlogiston. Thus, the combination of the calx with the phlogiston of heat would leave fire air as well as yield mercury. Qualitatively, the undisputed facts were beautifully accounted for.

But there was a fly in the experimental ointment, which the Phlogiston Theory simply could not conveniently remove. The problem was one of mass, or its more familiar forceful manifestation, weight. As early as 1630, Jean Rey had noted that the mass of a metal increases on calcination, that is, the calx weighs more than the sample of metal from which it is prepared. This is rather peculiar behavior if the reaction is viewed as the loss of something called phlogiston. From time to time chemists worried about this, and several attempts were made to account for the observation. The best remembered *ad hoc* hypothesis introduced was the suggestion that phlogiston must have negative weight, and the more of it a substance contained, the less it weighed.

Not all phlogistonists, however, accepted this antigravity argument. Some suggested that heat had mass. Others simply concluded that mass was not a variable relevant to chemistry. To attempt to isolate the principle of flammability, they contended, was as futile as trying to bottle gravity or magnetism. Before the reader dismisses this antimaterialistic idea as mindlessly myopic, he should consider how important and sometimes how difficult the identification of significant variables is in the functioning of science. Discovering what depends on what is often more difficult than determining the functional, physical, or mathematical relationships involving the variables. One trivial example should suffice: we feel moderately secure in neglecting sunspot activity when analyzing chemicals or stock market trends, but to do so while studying cosmic ray intensity might be a serious oversight.

Lavoisier and the "new chemistry"

To Lavoisier, weight and mass were significant variables, and much of his great contribution of chemistry derives from his quantitative approach. He was the first to carry out the heating of the red calx of mercury under fully controlled so that everything could be weighed, that is, *quantitative* conditions. He found that the combined masses of mercury and vital air formed were exactly equal to the mass of calx consumed. Mass and matter were neither created nor destroyed. In the course of the reaction, the appearance and properties of the matter involved did change, but *the total mass remained invariant*. This has come to be accepted as one of the great generalizations about the working of our world and has been enshrined in textbooks as a law—the *Law of Conservation of Matter or Mass*.* As it relates to the specific reaction under consideration, the law suggests that the gas formed must be coming from the calx itself. Consequently, the calx must be composed of mercury plus the substance given off as a gas, and the reaction can be represented

Calx of mercury → Mercury + Vital air.

If this reasoning were right, it should be possible to reverse the process, to convert mercury to its calx by heating the metal in the presence of vital air. For that matter, if vital air were indeed a normal part of ordinary air, the mercury might be able to abstract this component from the atmosphere. Indeed, this is just what happened in another of Lavoisier's classic quantitative studies. Again, the mass of the calx formed was equal to the total masses of the combining mercury and vital air.

A pattern was emerging. *Calcination and combustion and even respiration were not due to a loss of phlogiston, but rather to the combination of substances with the most active gas in the atmos-*

---

* The Russian, Lomonosov, shares with Lavoisier the credit for formulating this law.

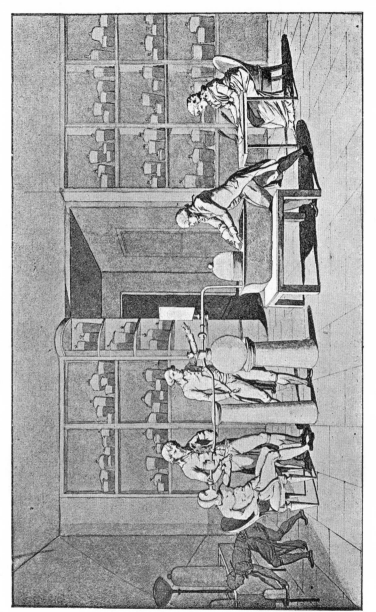

FIGURE 2.3 Lavoisier in his laboratory performing an experiment on respiration. (After a drawing by Mme. Lavoisier, pictured on the right.) (Reproduced from William A. Tilden, "Famous Chemists, The Men and Their Work," Dutton, New York, 1921.)

phere. This was fire air, dephlogisticated air, vital air, or, to at last arrive at Lavoisier's final name, *oxygene*. Thus, calxes were compound substances consisting of a metal plus oxygen; *oxides*, Lavoisier called them according to his new method of nomenclature. *Mercurius calcinatus per se* is mercuric oxide and the crucial reaction then is

Mercuric oxide → Mercury + Oxygen.

In modern terminology, we say that the oxide is *reduced* to the pure metal and oxygen is evolved.

Once the combining powers of oxygen were recognized, other phenomena could be explained equally well. The fixed air which Black had observed above burning carbon and in exhaled breath must be an oxide of carbon. We know it as *carbon dioxide*. (Why we specify carbon *di*oxide will become apparent later.) For the moment you should note that carbon dioxide is also formed during the smelting of metallic ores with charcoal or coke. The carbon "steals" the oxygen away from the metallic oxide, leaving the metal. This interpretation enables us to rewrite the smelting of iron calx as the reduction of iron oxide.

Iron oxide + Carbon → Iron + Carbon dioxide

Or consider this example. Henry Cavendish had shown that *inflammable air,* prepared as early as 1671 by Robert Boyle from the action of acid on iron, burned to form water. This behavior could be rationalized if inflammable air were made up of water plus phlogiston. Loss of the phlogiston during the burning process would leave behind the water, which, of course, is noncombustible.

Inflammable air → Water + Phlogiston

Even after the discoveries of Scheele and Priestley, the phlogistonist explanation could be stretched to include fire air or dephlogisticated air. Because the phlogiston content of this gas was held to be zero, or at least very low, it was able to absorb the phlogiston evolved during the combustion. Moreover, as we noted earlier, Scheele argued that the dephlogisticated air actually combined with phlogiston to yield heat. Thus, his equation for the reaction would be

Inflammable air (Water + Phlogiston) + Dephlogisticated air →
        Water + Heat (Dephlogisticated air + Phlogiston).

On the other hand, the application of Lavoisier's system led to the inescapable conclusion that water must be an oxide, a compound of oxygen and inflammable air, or *hydrogen*, as it was re-christened. Accordingly, the new representation for the burning of hydrogen was

Hydrogen + Oxygen → Water.

The formation of water was another instance of *oxidation,* the reaction of a substance with oxygen. To further test the hypothesis, Lavoisier ran the reaction backwards. He passed water through a hot iron gun barrel and found the two gases at the other end. Admittedly, this observation could also be accommodated by the old theory, but only with growing difficulty.

Suddenly there was no need to use the concept of phlogiston. Not only could the new theory explain all the qualitative aspects of calcination and combustion as well as the Phlogiston Theory, it could also account for quantitative relationships with far greater ease and elegance. There were naturally attempts to keep the older theory afloat; but the *ad hoc* hypotheses grew more and more clumsy, and the complex superstructure eventually sank from sheer weight and a few gaping holes. Some phlogistonists, like Joseph Black, lept overboard in time to be rescued by the "French Theory." Others, like Priestley, went bravely down with the ship. Ironically, it was his discovery of oxygen which had started the leak.

Lavoisier had literally turned chemistry upside down, and the implications of the inversion were monumental. Earth, air, fire, and water could no longer be maintained to be elementary substances. Air was obviously a mixture of gases—oxygen, nitrogen (as Lavoisier renamed Rutherford's noxious air), an oxide of carbon, and water vapor. Moreover, the relative ratio of these components could vary widely; the composition of air was definitely not exactly and specifically fixed. And finally, the constituent gases in air could be separated with relative ease. Surely, these individual com-

TABLE 2.1   *Gases which Played Important Roles in the Chemical Revolution*[a]

Phlogistonists' name	Modern name
Fixed air (Black)	Carbon dioxide
Noxious air (Rutherford)	Nitrogen
Dephlogisticated air (Priestley) ⎫ Fire air (Scheele) ⎬	Oxygen
Inflammable air (Cavendish)	Hydrogen

[a] The modern names are, in most cases, the same names suggested by Lavoisier.

ponents must be closer to being elementary substances than air itself. Similarly, matter as complex and as widely variable as earth could hardly be a fundamental sort of stuff. And even water could not be considered one of the basic forms of matter because it too could be broken down into simpler substances.

Elements, compounds, and mixtures

To us, as to Lavoisier, the true elements are not Aristotelian ideals somewhat resembling fire, air, water, and earth. Rather they are specific and very tangible substances—stuff like the seven ancient metals, like carbon and sulfur, hydrogen, oxygen, and nitrogen. Something very similar to the modern definition of an element had been enunciated by Robert Boyle in "The Sceptical Chymist" (1661):

I now mean by Elements . . . . certain Primitive and
Simple, or perfectly unmingled bodies; which not being
made of any other bodies, or of one another, are the
ingredients of which all those called perfectly mixed
Bodies are immediately compounded, and into which they
are ultimately resolved.

However, Boyle did not venture to identify any known substance as an element. As a matter of fact, this believer in alchemical transmutation actually questioned the existence of elements as he had defined them.

Lavoisier listed 33 in the first (1789) edition of "La traite elementaire de la chimie" (Table 2.2). Although certain of his selections, for example, light and heat, do not appear on modern lecture-room charts, most of his choices have retained their elementary status. Through the years they have been joined by over 70 others.

These one hundred-odd elements are the basic substances of the universe. Everything in existence, apart from some single particles with no real elementary identity, is made up of one or more of these sorts of stuff. Any object can, at least in principle, be resolved into its component elements. When elements are joined by chemical interaction, the resulting substance is termed a *compound*. As we have seen, it was the elucidation of the composition of a set of compounds, the oxides, which played a significant role in the history just recounted. In order to decompose a compound the procedures used, for example, the application of heat to water or mercuric oxide, must produce a *chemical reaction*. In general, a chemical reaction is a change which results in the conversion of one or more specific substances into one or more different distinct chemical entities. The emphasis here is upon chemical identity.

TABLE 2.2 *Reproduction of Lavoisier's Table of the Elements, from his "Traité élémentaire de chimie," 1789. (Reproduced from F. J. Moore, "A History of Chemistry," 3rd ed., p. 99, McGraw-Hill, New York, 1939).*

DES SUBSTANCES SIMPLES.

TABLEAU DES SUBSTANCES SIMPLES.

	Noms nouveaux.	Noms anciens correspondans.
Substances simples qui appartiennent aux trois règnes, & qu'on peut regarder comme les élémens des corps.	Lumière	Lumière.
	Calorique	Chaleur.
		Principe de la chaleur.
		Fluide igné.
		Feu.
		Matière du feu & de la chaleur.
	Oxygène	Air déphlogistiqué.
		Air empiréal.
		Air vital.
		Base de l'air vital.
	Azote	Gaz phlogistiqué.
		Mofète.
		Base de la mofète.
	Hydrogène	Gaz inflammable.
		Base du gaz inflammable.
Substances simples non métalliques oxidables & acidifiables.	Soufre	Soufre.
	Phosphore	Phosphore.
	Carbone	Charbon pur.
	Radical muriatique	Inconnu.
	Radical fluorique	Inconnu.
	Radical boracique	Inconnu.
Substances simples métalliques oxidables & acidifiables.	Antimoine	Antimoine.
	Argent	Argent.
	Arsenic	Arsenic.
	Bismuth	Bismuth.
	Cobalt	Cobalt.
	Cuivre	Cuivre.
	Etain	Etain.
	Fer	Fer.
	Manganèse	Manganèse.
	Mercure	Mercure.
	Molybdène	Molybdènes
	Nickel	Nickel.
	Or	Or.
	Platine	Platine.
	Plomb	Plomb.
	Tungstène	Tungstène.
	Zinc	Zinc.
Substances simples salifiables terreuses.	Chaux	Terre calcaire, chaux.
	Magnésie	Magnésie, base du sel d'epsom.
	Baryte	Barote, terre pesante.
	Alumine	Argile, terre de l'alun, base de l'alun.
	Silice	Terre siliceuse, terre vitrifiable.

Compounds, like elements, have, when they are purified, sharply defined chemical and physical properties. The melting point, boiling point, density, and crystal structure are among the physical properties which together serve to uniquely characterize a compound or element. Chapter 3 will present evidence that the constancy of properties for any pure compound is related to its fixed elementary composition.

In contrast, *mixtures* have neither constant composition nor invariant properties. Two deep breaths, one in downtown Los Angeles and the other in the High Sierra, should be enough to convince even the most skeptical reader of the variation in the mixture we live in. In fact, most of the matter we usually encounter is in the form of mixtures, some, like our own bodies, incredibly complex. Nevertheless, no matter how complicated the composition, a clever chemist with lots of time and patience should be able to separate any mixture into its component compounds and uncombined elements by physical, as distinct from chemical, means. This is not to imply that chemical reactions are never used to fractionate mixtures. For example, the separation of carbon dioxide from a mixture of gases can be effected by passing the mixture through a concentrated solution of potassium hydroxide. A chemical change occurs in which the carbon dioxide reacts with the potassium hydroxide to form potassium carbonate. The point remains that *mixtures can be separated by physical methods* (e.g., distillation) which take advantage of differences in physical properties (e.g., boiling points). By definition, *breaking down compounds into their constituent elements always requires chemical changes.*

A good deal of industrial time and effort is devoted to separating mixtures. For example, the fractional distillation of crude petroleum can be carried out to yield individual compounds with different boiling points. More commonly, however, it is sufficient to halt the process at a series of mixtures of similar compounds boiling over a relatively narrow range. These mixtures constitute gasoline, kerosene, diesel fuel, and so on. The same technique is commonly used to separate the various components in liquified air. The desalinization of salt water presents another important problem in the resolution of a mixture. Distillation is sometimes employed, but fractional crystallization can also be used on concentrated brine solutions. This procedure takes advantage of differences in the solubilities of various compounds in water. If the physical and chemical properties of the compounds in a mixture are very similar, their separation can be extremely difficult and may require specialized and sophisticated techniques. Fortunately, quite a number of very effective methods of fractionation have been devised.

In the final analysis (the choice of the word is intentional), the classification of matter into categories called elements, compounds, and mixtures is an experimental procedure. It illustrates the empirical basis of science, in general, and chemistry, in particular. But also note that this account of the slow evolution of our conceptions of matter and its composition has demonstrated that the scientific process is more than the mere collection of data or cataloging of observations. Scientific empiricism is tempered and organized by discriminating judgment, insight, and ingenuity. And even so fundamental a starting point as the nature of stuff shows the mark of man's mind.

For thought and action

2.1   A standard dictionary definition of science is "knowledge covering general truths or the operation of general laws." Does alchemy fit this definition? Why or why not?

2.2   Gold is oxidized only with very great difficulty; sodium is very readily oxidized. Gold is often found in an uncombined state; sodium never occurs naturally as the pure element. Relate these observations.

2.3   Explain the smelting of copper in terms of both the Phlogiston Theory and the Oxygen Theory.

2.4   Account for the rusting of iron according to the method of Priestley and the method of Lavoisier.

2.5   Provide modern, that is, post-Lavoisier, names for the calxes of lead, tin, and silver.

2.6   Sulfur burns to yield a gaseous oxide called sulfur dioxide. Devise and describe an experiment which would test this hypothesis.

2.7   Sulfur dioxide is a common air pollutant released from burning coal and fuel oils. Explain the source of the sulfur dioxide and suggest methods for combating this significant pollution problem.

2.8   Try your hand at classifying the following as elements, compounds, or mixtures. If you are in doubt, what experimental test or tests would you suggest to resolve the dilema? In the case of mixtures, suggest a suitable method for fractionation into pure substances: (a) sugar, (b) copper, (c) salt, (d) bronze, (e) beer, (f) aluminum, (g) magnesium carbonate, (h) carbon, (i) Ohio River water, (j) zinc oxide.

2.9   Suggest experiments which might be able to determine whether or not heat and light really belonged in Lavoisier's table of the elements.

2.10 Lavoisier was greatly influenced by the French philosopher,

Condillac, and took from his work the following motto: "The sciences have made progress, because philosophers have applied themselves with more attention to observe, and have communicated to their language that precision and accuracy which they have employed in their observations: in correcting their language they reason better." From what you know of Lavoisier's work and method, were they consistent with his motto?

2.11   On page 8 of this text, a statement is made that the four-element theory of Empedocles had the "virtue of comprehensiveness." Can an excess of comprehensiveness in a theory ever be a vice? Discuss.

2.12   Lavoisier was the author of the following statement: "Chemists have made a vague principle of phlogiston which is not strictly defined, and which in consequence accommodates itself to every explanation into which it is pressed. . . . It is a veritable Proteus which changes its form every moment." Comment on the implications of this statement, especially with regard to Question 2.11.

Suggested readings

Conant, James Bryant, ed., The Overthrow of the Phlogiston Theory, in "Harvard Case Histories in Experimental Science" (James Bryant Conant, general ed.), Vol. 1, pp. 65–115, Harvard Univ. Press, Cambridge, Massachusetts, 1966.

Holmyard, E. J., "Alchemy," Penguin Books, Harmondsworth, Middlesex, 1957.

Jonson, Ben, "The Alchemist," e.g., the edition edited by G. E. Bentley, Appleton, New York, 1947.

Lavoisier, Antoine, "Elements of Chemistry" (translated by Robert Kerr), Dover, New York, 1965.

Read, John, "Prelude to Chemistry, An Outline of Alchemy," MIT Press, Cambridge, Massachusetts, 1936, 1966.

Stillman, John Maxson, "The Story of Alchemy and Early Chemistry," Dover, New York, 1924, 1960.

Weeks, Mary Elvira, and Leicester, Henry M., "Discovery of the Elements," Journal of Chemical Education, Easton, Pennsylvania, 1968.

3

Atomism—the evidence of things not seen

The inferred atoms of the ancients

Newspapers used to declare with oracular assurance, "This is the Atomic Age!" But that was 25 years ago and the popular press had just discovered the atom in a mushroom-shaped cloud above the dying remains of a city called Hiroshima. The intervening years have seen first outer space and then our more immediate environment become the scientific issues which have captured popular imagination, interest, and support. Nevertheless, the atomic concept is so firmly fixed that it is taken either for granted or as an article of faith by most people. I would be willing to make a modest wager that few of the firm believers in the "absolute reality" of atoms could begin to satisfactorily cite the evidence for the discreteness of matter. The chief purpose of this chapter is to remedy that condition.

The atomic concept is not self-evident. Neither did it spring full-grown from the head of Zeus or Albert Einstein, though the former source is a good deal closer to the origin than the latter. Again, the Greeks can claim some priority, as they seemingly can for almost everything significant in Western civilization. Along with problems of the composition and origin of the universe, and change and permanency, the pre-Socratic philosophers also addressed their inquiring minds to the question of the structure and organization of matter. One of the key issues was whether matter was infinitely divisible. In practice it appeared to be. A lump of sulfur could be broken in two, each piece could in turn be divided, and so on until it was all ground into fine dust. Each speck of powdered sulfur was still sulfur; it had the same properties as the starting substance. Since no one had ever succeeded in arriving at the ultimate, indivisible limit, there was no particular profit in postulating its existence. At least so thought some philosophers.

Not all, however; Democritus of Abdera (ca. 460–370 B.C.) disagreed. And, since Democritus was ultimately proved right (or at least righter), his name is recorded in general chemistry texts. None of Democritus' writings has survived (the alchemical works attributed to him are obviously of a much later date). A chief source of information about his concept of the structure of matter is Aristotle, who goes into some length to present it in order to dispute it. Basically, Democritus seems to have held that somewhere or other, the dividing has to stop. There must be a lower limit of particles which are indivisible. And he called them just that, ατομος.

Democritus' *atoms*, to shed Hellenic hubris for the vulgar vernacular, were all made of the same stuff. They were all colorless, odorless, and tasteless; but differed in size and shape. Because all matter was made up of these minute particles, the properties of any substance apparently depended upon the combination of atoms in it. Moreover, Democritus argued that all atoms must be in constant motion—even in solids. And solids weren't really solid anyway, because there was always the "nothing" between the "something" of the atoms. The nothing was just that: empty space, a vacuum. And here is where the theory ran afoul of Aristotle, who abhorred a vacuum even more than nature does.

In his rejection of atomism, Aristotle showed a time-honored prerogative of students—he did not follow his teacher's teaching. Plato had developed a beautiful theory which used solid geometry to link the elementary and atomic concepts. Each of the elements—earth, water, air, and fire—was thought to be made up of atoms of a particular, characteristic shape: cubes, icosahedrons, octahedrons, and tetrahedrons, respectively. The packing of these atoms even predicted the general qualitative relationships among the densities of the four elements. But, as has been noted, Aristotle was not convinced.

Subsequent generations did not always show Aristotle's healthy skepticism of professorial authority, and for centuries atoms were more often out than in. It would, however, be a gross oversimplification to conclude that the natural philosophers who followed Aristotle all slavishly accepted the teachings of the master without thinking for themselves. There were, in fact, a number of convinced and outspoken atomists, notably the Roman poet, Lucretius (ca. 100–55 B.C.). The somewhat immodest title of his long poem, "De Rerum Natura" (On the Nature of Things) reflects its scope. Early in the work, Lucretius proposes his central thesis: "all nature as it is in itself consists of two things—bodies and the vacant space in which the bodies are situated and through which they move in different directions." Arguing from this premise, the poet goes on to consider the properties of atoms and the implications of this form of material organization for meteorology and geology, sensation and sex, cosmology and sociology, and even life and the mind. (Science, like most scientists, has become somewhat less ambitious as it has grown older.)

The atomistic concepts of Democritus and Lucretius were revived in the "corpuscular philosophy" of the Renaissance and the Enlightenment. The movement was well named, because initially it was a philosophically grounded rationale for certain physical phenomena. Or, more precisely, the phenomena were used to justify the philosophy. To be sure, among its adherents were eminent men of science: Galileo, Boyle, and Sir Isaac Newton. The latter, following the annunciation of his magnificent Theory of Gravitation, even went so far as to speculate about the nature of the attractive forces which atoms exert upon each other. Nevertheless, the language of Newton's famous statement is revealing: "It seems probable to me, that God in the Beginning form'd Matter in solid, massy, hard, impenetrable, moveable Particles." The adjectives are, of course, of utmost importance, but so is the implication of the theological underpinning of the systematic Newtonian universe. And the choice of the word "probable" reflects the status of atoms in 1700. There was no direct proof, even for the atomic hypotheses of the great English genius.

Indeed, the honest disagreement on the atomic issue illustrates how difficult the problem really is. Experiment and experience were and are subject to opposite interpretation. We have seen how the properties of solids seem to argue for the continuity of matter. Liquids also appear to be smooth, continuous, without breaks or graininess. On the other hand, fine granules or powders can be poured freely and seem to approximate fluidity. Therefore, the idea that a seemingly continuous liquid might be made up of minute particles does not necessarily contradict common sense. Moreover, the evaporation of after-shave lotion, the diffusion of perfume through a room, the solution of sugar in coffee all seem to suggest atomism—at least to the author. But admittedly, it is difficult to impartially interpret such phenomena because we have all been saturated with almost two centuries of atomic thinking. To be fair about it all, one should really hear an antiatomist on the issue, and they are even scarcer than flat-worlders.

Of course, the obvious problem is one of perception. It is rather like looking at a painting from across a gallery. You see masses of color and discern shapes and forms, but you cannot perceive individual brush strokes or bits of paint. Yet, the painting may be Pointillist, made up of tiny spots of color, points of pigment. From a distance, Seurat's famous "A Sunday Afternoon on the Grande Jatte" appears to be continuous; its artistic atomism is only visible at close range. But here the analogy breaks down, you cannot conveniently get that close to matter. How can you possibly talk intelligently about what you cannot see? No more than an instant's reflection should convince you that, for better or worse, we do it all the time. Lucretius recognized this at the end of the pre-Christian era, and his "De Rerum Natura" presents several examples of "bodies whose existence you must acknowledge though they cannot be seen." You do just that when you make inferences about the existence and nature of the absent inhabitant of a room. The basis for this knowledge is indirect evidence, and the method is as useful in science as it is elsewhere. In fact, some of the best support for atoms is indirectly but logically inferred from macroscopic relationships involving the masses of elements. This evidence is the subject of the next section. It formed the foundation of the Atomic Theory.

The matter of mass and the mass of matter

You will recall from your reading of Chapter 2 that the correct qualitative understanding of the elementary forms of matter grew out of quantitative information. In a somewhat similar way, our current conception of the structure of matter is also based upon numerical data. Incidentally, you may, with good justification, question the imprecision inherent in the undefined terms "forms of matter" and "structure of matter." If the words seem to overlap it is because the fundamental facts and assumptions underlying them are also interrelated. As this account of chemistry progresses, the relationship between the identity and properties of an element and its atomic structure will become more apparent.

In the paragraphs which immediately follow, we concentrate upon three fundamental laws of chemical combination. These are regularities about nature established not by a *priori* postulation, but by the painstaking accumulation of experimental evidence. Then, we will attempt to interpret these quantitative laws in much the same way as did a Manchester schoolmaster named John

Dalton. We will create a theoretical model consistent with the facts, a representation of reality. Subsequently, new facts and new revisions of the theory will be considered.

In presenting both the laws and the theory which explains them, two compounds will be used as prime examples: water and hydrogen peroxide. The choice is admittedly not completely justifiable from the standpoint of historical accuracy. Although water was much studied in this connection, hydrogen peroxide was not. Part of the difficulty lies in the fact that pure hydrogen peroxide is highly unstable stuff. Nevertheless, these substances illustrate the development of the Atomic Theory so well that they provide their own rationale.

EVERYTHING BALANCES: THE CONSERVATION OF MASS

Even so seemingly simple a regularity as the *Law of the Conservation of Mass or Matter* was not easily established. In order to measure mass changes during a chemical reaction, it is obviously helpful to know what substances are reacting and what substances are produced. For example, to weigh an unburned piece of charcoal and then to reweigh the ash remaining after combustion, and to assume that these were the only ponderables involved would lead to the conclusion that, at least in this instance, mass decreases during a chemical reaction. On the other hand, if the carbon dioxide formed were trapped and weighed and its weight totaled with that of the ash, the opposite conclusion would be indicated. As Lavoisier established, oxygen also must be included in the equation as one of the reactants.

Gradually the muddied water cleared, thanks particularly to reactions carried out in closed vessels. This was one of Lavoisier's techniques and it can also be applied to our example of hydrogen and oxygen. Samples of the two gases can be introduced into an evacuated bulb and the entire apparatus can be carefully weighed. Then, if the vessel has been provided with electrodes sealed through its walls, a spark can be set off which, within rather broad limits of the oxygen/hydrogen ratio, causes a fire or an explosion. The qualitative interpretation can be summarized as follows:

Hydrogen + Oxygen → Water (an oxide of hydrogen).

If the bulb with its watery contents is weighed after the combustion is over, the weight will be found to be identical (or very nearly so) to the initial weight.

It goes without saying that one experiment does not make a law. The chemist must always be wary of experimental error. He must try to minimize it as best he can and then be well informed of how much inaccuracy and uncertainty remain in his measurements. But repeated weighings indicate that within experimental error, no detectable change in weight occurs during the reaction.

Mass of hydrogen + Mass of oxygen = Mass of water

The total mass of the reactants equals the total mass of the products.

Through similar repeated verifications on other systems, a law emerges—the Law of Conservation of Mass. It can be stated in a variety of ways. *In an ordinary chemical reaction, the total mass of the substances reacting equals the total mass of the products formed; there is no detectable change in mass.* * Or, in a somewhat wordier version: matter, its exclusive property, mass, and the latter's measurable manifestation, weight, neither increase nor decrease in the course of a chemical reaction. Matter is neither created nor destroyed by chemical change; it is conserved during the process.

The consequences of this law for our environment are colossal. Madison Avenue and the advertising industry notwithstanding, there are no consumers, only converters. Because all matter is conserved, both a beer can and its contents are potential problems in pollution. In order to keep "spaceship Earth" inhabitable, both must be recycled. If we fail to learn the lesson of the Law of Conservation of Matter, we will surely suffocate in our own offal. And this beautiful blue-green ball will become a planetary garbage dump. To be sure, its mass will be unchanged, but its environment will be irrevocably altered.

WATER IS WATER: CONSTANT COMPOSITION

With the establishment of the idea that elements combine to form compounds and in the process follow the Law of Conservation of Mass, there came a great interest in the quantitative aspects of the composition of compounds. A number of careful and competent men devoted themselves to decomposing compounds and determining the mass ratios of the constituent elements. The aim was to discover whether or not a pair of elements could combine in only a single fixed mass ratio, in several ratios, or in an indefinite range of proportions. Again, the answer was not an obvious one. And even after much of the evidence was in, some of those careful and competent chemists differed in their conclusions.

The two chief adversaries were compatriots—Claude Louis

---

* Albert Einstein is indirectly responsible for the qualifying weasel words "ordinary chemical reaction" and "detectable." You will see why in Chapter 16.

Berthollet and Joseph Louis Proust. Because they were Frenchmen their controversy was carried on with the utmost brilliance and courtesy. Berthollet was a professor at the Ecole Normale, but his most famous pupil, Napoleon Bonaparte, was tutored privately. Proust's students at Madrid were somewhat less illustrious, which perhaps explains why, though the winner in this particular disagreement, he remains less well known than his intellectual antagonist.

Berthollet subscribed to the notion that elements could unite in a rather wide range of mass ratios and cited analytical results to support his position. Proust, on the other hand, held that the elementary composition by weight of any specific compound was exactly fixed, and he, too, presented experimental evidence for his thesis. The most convincing proof came when Proust was able to demonstrate that in all of Berthollet's examples of variable composition, the material analyzed was in fact a mixture of two compounds, each with a constant characteristic composition. The implication that two elements can combine in more than one mass ratio, but that each of these ratios is fixed and represents a different, individual compound is one we will return to shortly. At the moment, our chief emphasis is upon the idea of the constancy of composition which characterizes all compounds. This is the *Law of Constant Composition or Definite Proportions*. Proust summed it up in the following words:

*We are forced to recognize that the composition of true compounds is as invariable as their properties: between pole and pole they are identical in these two respects. The cinnabar of Japan has the same properties and composition as that of Spain; silver chloride is identically the same, whether obtained from Peru or Siberia; in all the world there is but one chloride of sodium, one saltpeter, one sulfate of lime, or one baryta. The native oxides have the same composition as the artificial. These are facts which analysis confirms at every step!**

Put a bit more briefly, the law states that *the mass ratio of the elements in any pure compound is fixed.*

Of course the law applies to water. Its decomposition can be effected by electrolysis, the passage of direct electric current through a sample. Hydrogen is evolved at the negative electrode, the cathode, and oxygen forms at the positive electrode, the anode. Repeated analyses show that the compound contains

---

* Joseph Louis Proust, quoted in L. H. Cragg and R. P. Graham, "An Introduction to the Principles of Chemistry," p. 133, Bobbs-Merrill, Indianapolis, Indiana, 1954.

FIGURE 3.1.   *The electrolysis of water to yield hydrogen and*
oxygen. Note that the volume of hydrogen is twice that
of oxygen.

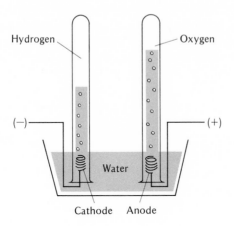

11.19% hydrogen and 88.81% oxygen by weight. Out of 100.00
grams (g) of water, 11.19 g are hydrogen and 88.81 g are oxygen.

This compositional information can also be expressed in a mass
ratio. Suppose we wish to find the weight of oxygen which is
associated with exactly one gram of hydrogen. We represent the
unknown weight with an x and write this simple equation:

$$\frac{x \text{ g oxygen}}{1.000 \text{ g hydrogen}} = \frac{88.81 \text{ g oxygen}}{11.19 \text{ g hydrogen}}.$$

Solving for x yields,

$$x = \frac{88.81 \text{ g oxygen}}{11.19 \text{ g hydrogen}} \times 1.000 \text{ g hydrogen}$$
$$= 7.937 \text{ g oxygen}.$$

It follows that for every gram of hydrogen found in water there are
7.937 g of oxygen. In other words, the mass ratio of oxygen to
hydrogen in water has a constant value of 7.931 to 1.000.

This ratio can be rounded off to the nearest whole numbers and
the mass relationship can be written in an equation,

9 g water → 1 g hydrogen + 8 g oxygen.

(Obviously, any unit of mass other than grams could also be
adopted, so long as the same unit was used throughout for all
products and reactants. Moreover, any multiple or submultiple of
the above equation would reflect the same composition ratio.)

The fact that each compound has a fixed quantitative elementary composition explains why each compound also has distinguishing physical and chemical properties. If the composition of

a compound were to vary, as does the composition of a mixture, its properties would also be inconstant. This is not to imply that the principle of constant composition–constant properties is not operative in mixtures. For example, a solution containing 44.0 volume percent ethylene glycol (the chief ingredient in commercial antifreeze) and 56.0% water freezes at a characteristic $-28.9°C$ ($-20°F$). However, such a solution is not a single substance. Its composition can readily be altered by the addition of either more water or more ethylene glycol. In the former case, the freezing point goes up; in the latter, it goes down (within certain interesting compositional limits). Moreover, a mixture of ethylene glycol and water can be separated by fractional distillation into its two constituent compounds, each of which individually illustrates the Law of Constant Composition.

Seven paragraphs have been devoted to "proving" this law, and what is written in them is basically correct, but nature is a little more complicated than this brief summary suggests. There are instances of compounds (for example, many compounds of carbon and hydrogen) which have identical elementary composition by weight, yet differ so markedly in physical and chemical properties that they obviously must be classified as distinct substances. On the other hand, there are compounds with identical properties but slightly different elementary mass proportions. And finally, there are some so-called nonstoichiometric compounds which show small variations in both properties and composition. Happily, all these exceptions have been explained by the Atomic Theory and its extensions, but more of that later.

MORE WAYS THAN ONE: MULTIPLE PROPORTIONS

Water may be water, but water is not the only compound which is composed exclusively of hydrogen and oxygen. The other is hydrogen peroxide, a liquid which looks somewhat like water but behaves rather differently. Its freezing point is $-0.89°C$, very near that of water, but its normal boiling point is $151.4°C$ compared to $100.0°C$ for water. It is about one and one-half times as dense as its common cousin. Chemically, the two oxides are quite different. Hydrogen peroxide readily decomposes to yield water and oxygen. The oxygen in turn can react with brown hair to make it blonde or with rocket fuel to make it burn. It is this oxygen-releasing property that leads to the occasional use of a 3% aqueous solution of "peroxide" as a disinfectant.

As we have seen, the fact that a pair of elements can form more than one compound was recognized by Proust and used to explain Berthollet's analytical data. Proust was also able to show that each

of the binary compounds had a unique, fixed composition. However, he failed to note the importance of the elementary mass
relationships *between* two different compounds of the same elements, probably because of faulty analyses. As a result, he missed
the generalization which has come to be known as the *Law of
Multiple Proportions.*

The law is easier to illustrate than to state. Returning to hydrogen peroxide, we find experimentally that the compound contains
5.89% hydrogen and 94.11% oxygen by weight. This result does
not seem particularly informative or interesting until it is converted to an oxygen/hydrogen mass ratio of 15.98 to 1.000. If you
recall that for water the same ratio has the numerical value of
7.937 to 1.000, it will become immediately obvious that 15.98 is
almost exactly two times 7.937. Of course, the percentages quoted
are the averages of many repeated experiments. And many sets of
compounds had to be analyzed to confirm the regularity. To be
sure, the ratio of the masses of one element combined, in two
compounds, with a fixed mass of another element is not always 2
to 1, as it is for the oxides of hydrogen and carbon. For the oxides
of iron the ratio is 3 to 2, and the situation is even more complicated for the five different oxides of nitrogen. But the ratio can
always be expressed in integers, that is, whole numbers. Thus, the
Law of Multiple Proportions holds: *when two elements, A and B,
combine in different mass ratios to form two different compounds,
the mass of element A combined with a fixed mass of element B
in one compound stands in a simple, whole number ratio to the
mass of A combined with the same fixed mass of B in the other
compound.* If all this still sounds a little confusing perhaps the
symbolic representation of the law in Figure 3.2 will help.

FIGURE 3.2.   *A summary representation of the three mass laws
of chemical combination. A and B are elements, and C and C'
are compounds.*

Law of Conservation of Mass

$$x \text{ g } A + y \text{ g } B \rightarrow (x + y) \text{ g } C$$
$$z \text{ g } A + y \text{ g } B \rightarrow (z + y) \text{ g } C'$$

Law of Constant Composition

For pure C $x/y = k$ (a constant)
For pure C' $z/y = k'$ (a constant)

Law of Multiple Proportions

$x/z = m/n$ or $k/k' = m/n$ (*m* and *n* are integers)

Quite clearly, it makes no difference which element is regarded as the constant mass reference. For example, the question can be phrased, how many grams of hydrogen combine with 8 g of oxygen? Approximating our results to better reflect the experimental accuracy of the year 1800, we find that 0.5 g of hydrogen is combined with 8.0 g of oxygen in hydrogen peroxide as compared with a corresponding value of 1.0 g of hydrogen in water. The ratio of the former mass of hydrogen to the latter is 1:2.

Early in the nineteenth century, tables of chemical equivalences began to be established and one of the standards adopted was the one just used, 8 g of oxygen. The *gram equivalent weight (GEW) of an element was defined as the weight (in grams) of an element which would exactly and completely combine with 8 g of oxygen.* As in the water–hydrogen peroxide example, the gram equivalent weight of an element is not fixed, but can often have several values, depending upon the compound analyzed. Thus, the gram equivalent weights of carbon in its two oxides are 3 and 6 g. And nitrogen, in its five oxides, demonstrates five different values, 2.8, 3.5, 4.7, 7.0, and 14.0 g. Things seem to be getting more complicated instead of clearer.

But there is order in this confusion and it was first glimpsed through the myopic, colorblind eyes of John Dalton. The Law of Multiple Proportions is generally attributed to Dalton, and it served as a cornerstone of his famous theory. So did the Laws of Conservation of Mass and Constant Composition. Therefore, before you move on to a study of the Atomic Theory, it might be worthwhile to have a last look at Figure 3.2, a symbolic summary of the three mass laws of chemical composition. Their importance in the establishment of the atomic concept is considerable.

Dalton's billiard balls

The previous section was about experiment and the laws generalized from laborious laboratory data. The laws, as summary statements about the organization of matter, are more intrinsically interesting and more useful than individual pieces of data. Nevertheless, these codified regularities are not very satisfactory aesthetically. They raise more questions than they answer, all sorts of intriguing enigmas: Why do the facts correspond to these particular laws? What if anything underlies the phenomena? How can experiment be explained? And so we come to the search for a model of reality—an approximation which will help us to understand and rationalize how this small aspect of the universe functions. Our only limitation is our imagination; our only constraint, fidelity to fact. Perhaps it is possible to create more than one theoretical system consistent with the observations. While you speculate on your own model, you would do well to consider John Dalton's.

There is reason to suspect that Dalton's interest in atomism was an outgrowth of his hobby of meteorology. Every day for 57 years, from 1787 until the day before he died in 1844, Dalton observed and recorded the weather. While making over 200,000 observations and entries, he had ample opportunity to note the properties of air and wind. He was undoubtedly aware of the fact that at constant temperature, the volume of a gas decreases with increasing pressure (scientists refer to this as *Boyle's Law*). And similarly, he must have known that the volume of a gas maintained at constant pressure increases with increasing temperature (*Charles'* or *Gay-Lussac's Law*). Moreover, Dalton was himself the formulator of the *Law of Partial Pressures*, the recognition that the total pressure exerted by a mixture of gases in a container is equal to the sum of the partial pressures of the component gases. Thus, he was

FIGURE 3.3.   *A portrait of John Dalton. (Reproduced from F. J. Moore, "A History of Chemistry," 3rd ed., p. 114, McGraw-Hill, New York, 1939.)*

familiar with the compression, expansion, and diffusion of gases.*
All three of these phenomena can be explained if a gas is regarded
as composed of tiny particles, sort of submicroscopic billiard balls.
When a gas is compressed the particles get closer together; when
it expands they get further apart; and when two gases diffuse into
one another, their respective particles become mixed.

But the idea that matter is composed of miniature billiard balls
can account for much more than the properties of gases. Dalton
realized the chemical implications of atomism and, in "A New System of Chemical Philosophy" (1808), he published the first full
exposition of his theory and its application to chemical combination. "The ultimate particles of all homogeneous bodies," he
wrote, "are perfectly alike in weight, figure, &c. In other words,
every particle of water is like every other particle of water, every
particle of hydrogen is like every other particle of hydrogen,
&c. . . ." Each element was thus assumed to be made up of its
own, unique, identical atoms. Compounds, on the other hand,
must consist of compound particles—the associated atoms of the
constituent elements. Each of these *molecules*† was held to have
a fixed, characteristic composition, an integral number of atoms
stuck together by an ill-defined affinity. Because Dalton's atoms
were immutable and indivisible, they were conserved in the course
of a chemical reaction. Indeed, such a change was regarded as a
rearrangement of atoms to form new molecules and therefore new
substances.

Clearly, the conservation of mass could be considered a consequence of the conservation of atoms. The Law of Constant Composition could be attributed to the constancy of the atomic ratio
characteristic of the molecules of a specific compound. And
finally, the Law of Multiple Proportions could be rationalized as
the result of integral differences in atomic ratios. The correspondence between fact and theory seemed very high indeed. Experiment had provided evidence of the unseen and atoms had offered
an explanation of the observable.

Nevertheless, two gnawing and closely related questions remained unanswered. What is the atomic ratio characteristic of any
particular compound? And what is the mass characteristic of any
particular elementary atom? Mass ratios of the elements in a compound had been and could be determined with relative ease, but

---

* These properties of gases are treated in some detail in Chapter 10.

† Dalton did not use the term "molecule"; he called even the compound particles "atoms." We will find it semantically less confusing
to use the modern nomenclature, reserving the word atom for the
smallest particle of an element.

there was no apparent way of discovering how many atoms there were in a given mass of matter, or how many atoms there were in any molecule. Without knowing atomic ratios, it was impossible to determine even relative atomic weights, to say nothing of absolute values.

Faced with this dilemma, Dalton guessed:

*If there are two bodies, A and B, which are disposed
to combine, the following is the order in which the
combination may take place, beginning with the most
simple: namely,*
   1 atom of A + 1 atom of B = 1 atom of C, binary
   1 atom of A + 2 atoms of B = 1 atom of C, tertiary

     . . . .

*. . . When only one combination of two bodies can be
obtained, it must be presumed to be a binary one, unless
some cause appears to the contrary.**

Dalton stated that his rules of chemical synthesis, of which the above quotation is just a part, were based upon the principle of "greatest simplicity." And, as long as one is hypothesizing, it perhaps is not a bad idea to keep Ockham's razor handy, and opt for simplicity. Nevertheless, the rules are purely arbitrary.

Even arbitrary rules have specific consequences and can lead to agreement with at least some experimental facts. Consider Dalton's equation for the formation of water:

$$\odot \quad + \quad \bigcirc \quad \rightarrow \quad \odot\bigcirc$$

3.1   Hydrogen   Oxygen   Water.

If one accepts the assumption that water exists in molecules ("binary atoms") of one atom of hydrogen and one of oxygen, the weight composition of the compound immediately dictates the relative masses of hydrogen and oxygen atoms. Since the weight ratio of oxygen to hydrogen in water is approximately 8 to 1, this must also represent the atomic weight ratio. (Actually, Dalton. who as an experimentalist demonstrated more patience than precision, determined the relative weights of the atoms of these two elements to be 7 to 1. We will use the more exact analytical data previously cited.)

Because we are dealing here with relative, not absolute atomic masses, it is necessary to establish a reference. Dalton's selection was a sound one; he set the atomic weight of hydrogen, the least dense element he knew, equal to unity. By analyzing a num-

* John Dalton, "A New System of Chemical Philosophy," Part I, Russell, Manchester, 1808, reprinted in "Foundations of the Atomic Theory," Alembic Club Reprints, No. 2, Edinburgh, 1899.

ber of compounds and applying his rules, he was able to draw up
a table of atomic weights, which was unfortunately marred by the
quality of both the analyses and the rules. Also, the values were
all rounded to the nearest whole number, though there was no
sound reason for this action.

Once the apparent relative atomic weights of a pair of elements
had been established, the results could be coupled with weight
analyses of other compounds to infer formulas, that is, atomic
ratios. For example, if the relative masses of hydrogen and oxygen
are taken as 1 and 8, respectively, hydrogen peroxide, which exhibits a hydrogen/oxygen weight ratio of 1 to 16, must exist in
molecules containing twice as many atoms of oxygen as does a
water molecule. It follows that the symbol of a hydrogen peroxide
molecule must be $\bigcirc\!\odot\!\bigcirc$, two atoms of oxygen and one of hydrogen. Thus, the Daltonian model as applied to our examples can
be summarized in the following tabulation.

Substance	Symbol	Relative mass
Hydrogen	$\odot$	1
Oxygen	$\bigcirc$	8
Water	$\odot\bigcirc$	9
Hydrogen peroxide	$\bigcirc\odot\bigcirc$	17

It should be emphasized that the above formulas and relative
weights are completely consistent with the experimentally determined weight composition of the two compounds. In that respect,
the model is perfectly adequate. However, it must also be remembered that the entire argument and assignment is predicated on
the arbitrary assumption that water is a diatomic, binary molecule.
In the very year in which Dalton's "New System" appeared, a discovery was made which threw doubt upon the validity of that
assumption.

Atoms A.D. (After Dalton)

Thus far, our emphasis has been upon the mass relationships of
combining elements, generalizations beautifully clarified by the
Daltonian Atomic Theory. But another property was to prove of
utmost importance in revising and extending that theory. Starting
in 1805, Joseph Louis Gay-Lussac launched a series of investiga-

tions on the combining volumes of reacting gases. The first reaction he studied was a familiar one, the combination of oxygen and hydrogen to form water. After many repeated determinations Gay-Lussac concluded that the volume (vol) of hydrogen reacting is always exactly twice as great as the volume of oxygen involved. In the form of an equation:

2 vol hydrogen + 1 vol oxygen → Water.

Of course, since the volume of any gas depends upon temperature and pressure, it was necessary to measure the volumes of both gases at the same values of these two variables. Gay-Lussac kept this carefully in mind as he went on to perform similar experiments on other gaseous reactions. In every instance, he found that *the volumes of the reacting gases always stood in simple, integral ratios to each other*. This is the *Law of Combining Volumes*, formulated by Gay-Lussac in 1808.

Gay-Lussac knew of Dalton's theory and recognized that his own discovery must have some bearing on it. Nevertheless, he did not specifically suggest a possible connection between volume ratios and atomic ratios. That insight was due to Jöns Jakob Berzelius, one of the great organizers of chemistry. The inference was that equal volumes of gases (at constant temperature and pressure) must contain equal numbers of particles, or at least the numbers of particles in equal volumes must be related to each other by an integral multiplier. In the absence of any evidence to the contrary, the former, simpler hypothesis was tentatively accepted.

Application of this idea to the formation of water immediately requires an alteration of the molecular formula of water. (At this point it seems appropriate to abandon Dalton's symbols for the elements and adopt those of Berzelius. The former thought the alphabetical symbols to be "unscientific" and would have regarded the following equation in the same light.)

3.2
$$2H + O \rightarrow H_2O$$
Hydrogen   Oxygen   Water

Here the number of hydrogen and oxygen atoms combining has been made to conform to the same ratio as the combining volumes.

The final emergence of the familiar formula for water, after so long a struggle, may prompt you to throw down your book with a sigh of relief—or disbelief. Hopefully, the disbelief will win out and induce you to read on. Thus far in this chapter, neither the above equation nor the hypothesis underlying it has been established firmly enough to convince a twentieth century student skeptic. The equation has some implications which cry for investigation.

In the first place, either the equation or our previously accepted relative atomic weights must be incorrect. They cannot both be right and maintain fidelity to the experimentally determined weight composition of water. If the atomic weights of oxygen and hydrogen are maintained at 8 and 1, respectively, the formula H_2O predicts an oxygen/hydrogen mass ratio of 8 to 2, instead of the repeatedly observed value of 8 to 1. If we play chemistry according to the generally accepted rules, we simply cannot disregard the well-established experimental result. A fact is a fact! At this stage in our argument, atomic weights and molecular formulas do not have quite that status. One of these must give, and since we have at least tentatively accepted the idea of equal volumes—equal number of particles, it must be the atomic weights. After all, there was nothing sacred about our previous selection.

It should be apparent that the model represented by the Equation 3.2 requires either halving the relative mass of hydrogen or doubling that of oxygen. In order to retain whole number atomic weights, we will do the latter, thus making 16 the new value for oxygen. The relative weight of a water molecule, H_2O, thus becomes 18. With this change, theory is again reconciled with experiment—at least until more information comes along.

Dalton himself pointed out the contradictory evidence. Although the example he chose was not the one we have been using, water lends itself equally well to illustrating the problem. The difficulty arises from a simple extension of Gay-Lussac's method. If the volume of water vapor formed is measured at the same temperature and pressure used in determining the volumes of the combining gases, the volume of the water vapor is found to be equal to that of the hydrogen and twice that of the oxygen, that is,

2 vol hydrogen + 1 vol oxygen → 2 vol water vapor.

But Equation 3.2, coupled with a consistent application of the idea that equal volumes of gases contain equal numbers of particles, clearly predicts

2 vol hydrogen + 1 vol oxygen → 1 vol water vapor.

To Dalton, evidence such as this was a clear vindication of his original thesis that atoms and molecules of different substances (and consequently different masses) also have different volumes. Indeed, the idea of elementary variations in atomic size had preceded Dalton's conclusions about differences in atomic weight. True, one might argue to the contrary that a gas is mostly empty space, and consequently the "effective" volume occupied by any particle (again, at a fixed temperature and pressure) must be independent of the actual mass or size of the atom or molecule, and thus equal for all gaseous substances. Nevertheless, application of this hypothesis to the oxidation of hydrogen led to the irrefutable inconsistency just cited.

The way out of this jungle of apparent confusion was shown by a professor of physics at the University of Turino, Amadeo Avogadro. The famous hypothesis which bears his name was first propounded in 1811 in a journal of physics. It is not exactly to the glory of chemistry that the idea was not widely accepted until 1860, when Stanislao Cannizzaro, one of Avogadro's former students, reintroduced it in a masterful paper distributed at a chemical conference, called to consider the problem of assigning atomic weights.

Avogadro's great contribution to chemistry was the suggestion that *elements as well as compounds can exist as compound atoms, that is, molecules.* In fact, the Italian physicist carefully differentiated among "elementary molecules"—the individual atoms of an element, "integral molecules"—the molecules of compounds, and "constituent molecules"—the molecules of an element. The distinction is a significant one, and in our modern usage of the word "molecule" for the molecules of both elements and compounds we have lost an organizationally useful semantic device.

Using the example of water, Avogadro pointed out that the volume data could be rationalized if he assumed that elementary hydrogen and oxygen both exist in molecules (returning to the current nomenclature) of two atoms each. Writing these particles as H_2 and O_2, respectively, their combination becomes

3.3 $2H_2 + O_2 \rightarrow 2H_2O.$

If it is the number of these molecules, these compound particles, which is constant for equal volumes of all gases, the hydrogen: oxygen:water vapor volume ratio of $2:1:2$ is at once explained. This distinction between atoms and molecules underlies *Avogadro's Hypothesis: equal volumes of gases at a fixed temperature and pressure contain equal numbers of molecules.* Moreover, the concept of molecules of an element differentiates this statement from that attributed to Berzelius on p. 42.

Returning briefly to our touchstone of mass relationships, it is clear that the latest (and last!) version of our equation is experimentally honest. Using, as before, 1 and 16 as the atomic weights of hydrogen and oxygen, their molecular weights become 2 and 32, and the oxygen/hydrogen weight ratio in water is the old familiar 8 to 1.

Avogadro realized that the same mass and volume ratios would also follow if the elements existed in molecules of 4, 6, or other even numbers of atoms. However, this would require modification of the molecular form of water in order to account for the observed volume change. Thus, one would have to write

$$2H_4 + O_4 \rightarrow 2H_4O_2.$$

Avogadro had at his disposal no ready way of ascertaining the number of atoms in a molecule, and hence the above equation could not be ruled out. Today we take a molecular formula to represent not only a mass relationship, but a distinct physical entity as well. And experiments can demonstrate that the molecules of gaseous hydrogen, oxygen, and water are indeed H_2, O_2, and H_2O. We also know that the molecules of some elements are made up of more than two atoms. For example, under certain conditions, molecules of gaseous phosphorus consist of four atoms each, and sulfur vapor can contain molecules of formulas S_8 and S_6. On the other hand, mercury and helium are examples of elements which exist, in the gaseous state, as individual atoms. There are no molecules of gaseous helium. Or, if you prefer, the molecules of the element are identical to its atoms. All this implies the ability to determine atomic and molecular weights, a capacity which, in no small measure, derives from Avogadro's Hypothesis.

But before we move on to the challenge of weighing the unweighable, you are invited to consider another challenge—hydrogen peroxide. Buried somewhere in the previous paragraphs are the weight composition data for this compound. Another potentially useful fact is that the hydrogen/oxygen/hydrogen peroxide vapor volume ratio is 1:1:1. As mathematics textbooks are wont to state, this is left as an exercise for the interested reader: The student, armed as he is with atoms and information, should devise and defend a logical formula and molecular weight for hydrogen peroxide. The proof of the pedagogy is in the pupil's performance.

Weighing the unweighable

Even after the eventual acceptance of Avogadro's Hypothesis, the task of determining relative atomic weights remained a challenging one. The necessity of making careful, quantitative measurements of weight composition or combining weights was by no means obviated by the theoretical advances. Nevertheless, Avogadro's generalization did provide a means for measuring molecular weights. And from such information, atomic weights could be inferred.

The procedure originally outlined by Cannizzaro is an exercise in logic applied to experimental evidence. One begins by weighing identical volumes of pure gaseous compounds and elements, all at the same convenient constant temperature and pressure. According to Avogadro, these volumes contain equal numbers of molecules. To be sure, neither Cannizzaro nor his famous teacher knew how many molecules of each substance were involved. But they were confident that when they compared the masses of equal volumes of gases, each containing thousands of particles, they were also comparing the masses of individual molecules. In other words, the ratio of the observed weights must be identical to the ratio of the molecular weights of the compounds and elements under investigation.

Just such a method can be used to find the relative molecular weights of the five oxides of nitrogen. Of course, some agreement upon a standard of reference is necessary if meaningful comparisons are to be made. Therefore, in Table 3.1 we adopt 32.0 as the molecular weight of oxygen. After our lengthy aquatic experience of the previous section, the choice is not altogether accidental. Relative to this reference, the oxides of nitrogen assume their indicated molecular weights; oxide A exhibits a value of 44.0, B is 30.0, and so on.

But note that since we are ignorant of the formulas of these compounds, that is, their molecular composition, we still do not know the atomic weight of nitrogen. As the next step towards attaining this information, Cannizzaro suggested that the elementary composition (by weight percent) be determined for each of the compounds. Here is where the chemical analysis comes in. And the results of such analyses are summarized in the table. Compound A contains 63.6% nitrogen and 36.4% oxygen, B consists of 46.7% nitrogen and 53.3% oxygen, and so on. These compositional data are then coupled with the molecular weights to

TABLE 3.1   *Illustration of Cannizzaro's Method of Estimating Atomic Weights from Molecular Weights and Composition Data*

| Compound | Molecular weight[a] | Weight percent | | Relative weight | | Formula |
		Nitrogen	Oxygen	Nitrogen	Oxygen	
A	44.0	63.6	36.4	28.0	16.0	N_2O
B	30.0	46.7	53.3	14.0	16.0	NO
C	76.0	36.8	63.2	28.0	48.0	N_2O_3
D	46.0	30.4	69.6	14.0	32.0	NO_2
E	108.0	25.9	74.1	28.0	80.0	N_2O_5

[a] On a scale where the molecular weight of oxygen is 32.0. Note also that on this same scale the molecular weight of nitrogen is 28.0.

compute the relative weights of the elements in each molecule. Let us use oxide A as an illustration. Since the compound is 63.6% nitrogen, that same percentage of a single molecule will be nitrogen. Thus, we can obtain the relative weight of nitrogen in a molecule of A by multiplying the molecular weight by the percent of nitrogen divided by 100.0,

$$\text{Relative weight of nitrogen} = 44.0 \times \frac{63.6}{100.0}$$
$$= 28.0.$$

In other words, out of every 44.0 g of oxide A, 28.0 g is nitrogen. The remaining 16.0 g must be oxygen. You will note that the ratio of the mass of nitrogen to the chemically corresponding mass of oxygen is identical to that expressed by the percentage composition. The latter stipulates that 63.6 g of nitrogen are combined with 36.4 g of oxygen in each 100.0 g of oxide A. Therefore, we can write

$$\frac{\text{Mass of nitrogen}}{\text{Mass of oxygen}} = \frac{63.6}{36.4} = \frac{28.0}{16.0} = 1.75.$$

Comparable calculations for the other oxides of nitrogen yield the tabulated relative weights of the two constituent elements. You will note that the values vary depending upon the compound. Cannizzaro argued that in such a list, the smallest value for the relative weight of an element is probably its correct atomic weight. Hence, we tentatively assign 14.0 as the atomic weight of nitrogen and 16.0 as the atomic weight of oxygen. A moment's reflection, however, should convince you that this logic works only if at least one of the compounds in the series is made up of molecules which contain only a single atom of the element in question. But the *minimum* value of the relative weight unquestionably places a *maximum* limit on the atomic weight. For example, the atomic weight of nitrogen could not be 28.0, since compounds exist which contain, relatively speaking, half that much nitrogen. There is no room for half atoms in Dalton's theory. A relative weight of 28.0, as in compound A, means that the molecule must contain at least two atoms of the element. And, if we are lucky and the atomic weight of nitrogen is indeed 14.0, such a molecule will contain *only* two atoms of nitrogen. Moreover, because the molecular weight of pure gaseous nitrogen is known to be 28.0, each molecule of the element would then consist of two atoms, N_2. Similarly, choosing 16.0 as the atomic weight of oxygen implies that this element, with a molecular weight of 32.0, must also exist in diatomic molecules, O_2. As we have seen, this conclusion is in harmony with Avogadro's arguments.

Returning now to oxide A, we see that our atomic weights permit the prediction of the atomic ratio characteristic of the compound. The relative mass values of 28.0 for nitrogen and 16.0 for oxygen suggest that each molecule or A consists of two atoms of nitrogen and one of oxygen. Thus, its formula would be N_2O. And, because nitrogen and oxygen are both made up of diatomic molecules, the equation for the direct formation of oxide A from its constituent elements could be written

$$2N_2 + O_2 \rightarrow 2N_2O.$$

Quite obviously, the equation is directly analogous to that summarizing the formation of water from hydrogen and oxygen. In conformity with that example, one would predict that two volumes of nitrogen gas should combine with one volume of oxygen to yield two volumes of A. And this, in fact, is what is observed.

2 vol nitrogen + 1 vol oxygen $\rightarrow$ 2 vol A

The same reasoning, buttressed by volume data, applies to the case of compound B. If our estimated atomic weights are correct, each molecule of this oxide contains a single atom of nitrogen and one of oxygen. Symbolically, its formula must be NO. Together, the atomic weights of 14.0 and 16.0 add up to yield 30.0, the molecular weight of the product. Moreover, volume measurements show that, at constant temperature and pressure,

1 vol nitrogen + 1 vol oxygen $\rightarrow$ 2 vol B.

This result is, again, in agreement with the existence of the diatomic molecules N_2 and O_2. Their combination can therefore be represented as

$$N_2 + O_2 \rightarrow 2NO.$$

It turns out that this is the most common direct reaction of these two elements. You will learn later why it is also one of the most significant steps in air pollution.

A skeptical student may object with some justification that the argument just advanced does not absolutely and unambiguously prove the correctness of the conclusions. But, for that matter, no scientific theory is probably ever proved to be true with the certitude of a mathematical identity. The point is that the relative weights of 16.0 for an atom of oxygen and 14.0 for nitrogen are fully consistent with experimental evidence, including both mass and volume data. On the other hand, any alternative values would require a cumbersome set of ad hoc hypotheses to justify their validity. Moreover, as additional compounds were investigated, fresh evidence was amassed for the correctness of these and other atomic weights. Because the elements, through their compounds, form a quantitatively related network, each new atomic weight determined made values for other elements easier to ascertain. With increased refinement in experimental and analytical techniques, it has become possible to report atomic weights to as many as six significant figures. But even at much lower precision, it is apparent that atomic weights are not the simple whole numbers Dalton envisaged and which we have been writing. To be sure, there are whole number aspects about atomic masses and atomic structure, but they are not obvious in the average atomic weights tabulated in texts and wall charts.

It is essential to keep in mind that for all their precision, atomic weights are relative and not absolute quantities. One element must be designated as the reference and assigned an arbitrary (but, hopefully, convenient) atomic weight. All other atomic weights are then reported relative to this standard. As a matter of fact, different elementary standards have been used at different times. But some elements are more satisfactory choices than others. And oxygen, because it reacts with so many different elements, served as the reference substance over most of the history of modern chemistry. It was variously given the atomic weights of 1, 8, 16, and 100. However, the value which persisted the longest was 16.0000. For a time, it was thought that relative to this standard, hydrogen had an atomic weight of exactly one. This was attractive because both Dalton and Cannizzaro had assigned an atomic weight of unity to the lightest element. But around the turn of this century, data appeared which called for a revision of this value to 1.0080.

Shortly thereafter, it was discovered that it is common for a single element to consist of atoms of more than one atomic weight (isotopes). This was one of a whole series of developments which revolutionized our concept of atoms and which will be treated in considerable depth in Chapter 6. For the purpose of the present argument, it is sufficient to note that normal atomic weights are really averages of allowable individual isotopic weights, weighted by distribution factors. In the 1930s, physicists established a new standard which took this atomic heterogeneity into account when they assigned an atomic weight of 16.0000 to a specific isotope of oxygen. On this scale, the average atomic weight of the mixture of isotopes found in ordinary samples of oxygen has a value of 16.0044. Chemists, on the other hand, continued to assign a reference atomic weight of 16.0000 to ordinary oxygen. The 0.0275% disagreement between chemistry and physics was finally resolved in 1961 by an interdisciplinary compromise which settled on the isotope carbon-12 as the universal standard, with an atomic weight of exactly twelve. On this basis, the former chemical standard, ordinary oxygen, has an atomic weight of 15.9994, and the other elements assume the values listed in Table 3.2. All of this might

TABLE 3.2

	Symbol	Atomic number	Atomic weight°
Actinium	Ac	89	(227)
Aluminum	Al	13	26.9815
Americium	Am	95	(243)
Antimony	Sb	51	121.75
Argon	Ar	18	39.948
Arsenic	As	33	74.9216
Astatine	At	85	(210)
Barium	Ba	56	137.34
Berkelium	Bk	97	(247)
Beryllium	Be	4	9.01218
Bismuth	Bi	83	208.9806
Boron	B	5	10.81
Bromine	Br	35	79.904
Cadmium	Cd	48	112.40
Calcium	Ca	20	40.08
Californium	Cf	98	(251)
Carbon	C	6	12.011
Cerium	Ce	58	140.12
Cesium	Cs	55	132.9055
Chlorine	Cl	17	35.453
Chromium	Cr	24	51.996
Cobalt	Co	27	58.9332
Copper	Cu	29	63.546
Curium	Cm	96	(247)
Dysprosium	Dy	66	162.50
Einsteinium	Es	99	(254)
Erbium	Er	68	167.26
Europium	Eu	63	151.96
Fermium	Fm	100	(257)
Fluorine	F	9	18.9984
Francium	Fr	87	(223)
Gadolinium	Gd	64	157.25
Gallium	Ga	31	69.72
Germanium	Ge	32	72.59
Gold	Au	79	196.9665
Hafnium	Hf	72	178.49
Hahnium (?)	Ha (?)	105	(260)
Helium	He	2	4.00260
Holmium	Ho	67	164.9303
Hydrogen	H	1	1.00807
Indium	In	49	114.82
Iodine	I	53	126.9045
Iridium	Ir	77	192.22
Iron	Fe	26	55.847
Krypton	Kr	36	83.80
Kurchatovium (?)	Ku (?)	104	(261)
Lanthanum	La	57	138.9055
Lawrencium	Lw	103	(257)
Lead	Pb	82	207.2
Lithium	Li	3	6.941
Lutetium	Lu	71	174.97
Magnesium	Mg	12	24.305
Manganese	Mn	25	54.9380
Mendelevium	Md	101	(256)
Mercury	Hg	80	200.59
Molybdenum	Mo	42	95.94
Neodymium	Nd	60	144.24

The Chemical Elements

	Symbol	Atomic number	Atomic weight°
Neon	Ne	10	20.179
Neptunium	Np	93	237.0482
Nickel	Ni	28	58.71
Niobium	Nb	41	92.9064
Nitrogen	N	7	14.0067
Nobelium	No	102	(254)
Osmium	Os	76	190.2
Oxygen	O	8	15.9994
Palladium	Pd	46	106.4
Phosphorus	P	15	30.9738
Platinum	Pt	78	195.09
Plutonium	Pu	94	(244)
Polonium	Po	84	(210)
Potassium	K	19	39.102
Praseodymium	Pr	59	140.9077
Promethium	Pm	61	(145)
Protactinium	Pa	91	231.0359
Radium	Ra	88	226.0254
Radon	Rn	86	(222)
Rhenium	Re	75	186.2
Rhodium	Rh	45	102.9055
Rubidium	Rb	37	85.4678
Ruthenium	Ru	44	101.07
Samarium	Sm	62	150.4
Scandium	Sc	21	44.9559
Selenium	Se	34	78.96
Silicon	Si	14	28.086
Silver	Ag	47	107.868
Sodium	Na	11	22.9898
Strontium	Sr	38	87.62
Sulfur	S	16	32.06
Tantalum	Ta	73	180.9479
Technetium	Tc	43	98.9062
Tellurium	Te	52	127.60
Terbium	Tb	65	158.9254
Thallium	Tl	81	204.37
Thorium	Th	90	232.0381
Thulium	Tm	69	168.9342
Tin	Sn	50	118.69
Titanium	Ti	22	47.90
Tungsten	W	74	183.85
Uranium	U	92	238.029
Vanadium	V	23	50.9414
Xenon	Xe	54	131.30
Ytterbium	Yb	70	173.04
Yttrium	Y	39	88.9059
Zinc	Zn	30	65.37
Zirconium	Zr	40	91.22

°The atomic weights are based on carbon-12 = 12.0000.
Numbers in parentheses are the mass numbers (number
of protons + neutrons) of the most stable isotopes
of these radioactive elements. For most other elements,
the values listed reflect the natural distribution of the
various isotopes. The number of significant figures
reported is an indication of the accuracy with which
the atomic weight is known.

seem to be much ado about next-to-nothing. But, it does reflect that modern science has the technique to pick some pretty tiny nits.

The perceptive reader will, by now, have observed what might appear to be a significant omission. I have been writing molecular and atomic weights without specifying any units. But the even more insightful student has probably realized that this is quite proper. Strictly speaking, atomic and molecular weights are not weights at all. They are really ratios—relative mass values expressed with respect to some standard. This explains why there are no "grams" or "pounds" or "tons" attached to these pure numbers. Nevertheless, it is often useful to speak of atomic and molecular weights in terms of some definite, weighable quantity of matter. Since scientists universally employ the metric system, many of our measurements and calculations are done in grams. Therefore, we frequently speak of *gram atomic weights (GAW's)* and *gram molecular weights (GMW's).** These are simply the corresponding atomic and molecular weights with the designation "grams" appended. The numerical values are the same. Thus, expressed to the nearest tenth of a gram, the gram atomic weight of oxygen is 16.0 g and the GAW of nitrogen is 14.0 g. These masses of the pure elements contain the same number of atoms. Similarly, the gram molecular weight of O_2 is 32.0 g, and the GMW's of N_2, N_2O, and H_2O are 28.0 g, 44.0 g, and 18.0 g, respectively.

The usefulness of gram atomic weights is well illustrated by the process which permits the determination of atomic weights for metals. Most of the metallic elements—well-known substances like iron, copper, and aluminum, and less familiar examples such as cadmium, molybdenum and titanium—have very high boiling points. In other words, it is extremely difficult to vaporize them. As a result, their molecular weights cannot be conveniently determined by measuring the masses of equal volumes of the gaseous metals.† Moreover, these elements form few, if any, volatile compounds. This means that Canizzaro's method of estimating atomic weights from molecular weights and composition data cannot be employed.

Fortunately, however, there is an alternative approach which applies especially to metallic elements. In 1819, two French professors, Pierre Louis Dulong and Alexis Thérèse Petit, called attention to an unusual relationship between gram atomic weights and

* The term *gram formula weight (GFW)* is sometimes used in place of gram molecular weight, particularly for compounds which do not exist as molecules (see Chapter 9).

† You will learn in Chapter 10 that the concept of molecules does not really apply to solid metals.

specific heats. The *specific heat* of a substance is the amount of heat necessary to raise the temperature of 1 gram of the material 1 degree Centigrade. The heat is measured in *calories,* a unit defined as the amount of heat required to increase the temperature of 1 gram of water 1 degree Centigrade.* The defining temperature interval is set between 14.5° and 15.5°C, because the specific heat of liquid water does vary somewhat with temperature. Nevertheless, its value is always near 1 cal/g °C, read "one calorie per gram per degree Centigrade."

The long and short of the generalization proposed by Dulong and Petit was the fact that *the product of the gram atomic weight and the specific heat is approximately 6 cal/°C for many elements,* particularly metals. Representing the specific heat by its conventional symbol, c_p, we can write

3.4   (GAW) $c_p \dfrac{cal}{g\ °C} = \dfrac{6\ cal}{°C}$.

Since the idea of gram atomic weight implies equal numbers of atoms, the inference is that the specific heat of a solid element depends chiefly upon the number of atoms present in a gram, and not their elementary identity. Accurate measurements have demonstrated that this is not an exact relationship. Nevertheless, for many elements it has provided a useful means of approximating the atomic weight.

For example, experiment indicates that the specific heat of copper is 0.0921 cal/g °C. This quantity of heat is required to raise the temperature of exactly 1 g of this metal 1 degree Centigrade. Substituting this value of c_p in Equation 3.4 and solving for GAW we find the following:

$$GAW = \frac{6\ cal/°C}{0.0921\ cal/g\ °C} = 65\ g.$$

As we have already noted, this value is not exact. But it is close enough to guide our subsequent calculations. These computations must involve compositional data obtained in accurate experiments. In this specific instance, we need a value for the gram equivalent weight (GEW) of copper, the weight of the element which reacts completely with 8.000 g of oxygen. Once we find this GEW, we can take advantage of the fact that the GAW of an element is a whole number multiple of its GEW.

3.5   GAW = n(GEW)   where $n = 1, 2, 3, \ldots$

---

* Note that the calories which weight-watchers are continually counting are really *kilo*calories, each 1000 times larger than the calorie defined above.

To illustrate the experimental and mathematical procedure involved, let me invent and introduce some data. Experiments of
this kind are often carried out in introductory undergraduate laboratory courses. Suppose a student seeks the gram equivalent and
gram atomic weights of copper. He starts by carefully weighing a
quantity of copper oxide into a clean, dry test tube. Let us say that
the black powder weighs 1.538 g. The student does not know the
composition or the formula of the oxide (or he's willing to pretend
he doesn't). That is, he is ignorant of the relative weights of copper and oxygen present in the compound. Therefore, he is also
unaware of how much of the mass of the 1.538 g sample is due to
copper and how much is due to oxygen. But he can find out by
reducing the oxide to pure metallic copper. In other words, he
removes the oxygen from the compound. You recall that Lavoisier
(and others) did a similar sort of thing by heating the oxide of
mercury. The oxygen escaped, leaving the metal behind. But in
copper oxide, the oxygen is more tightly held. Heating alone is
insufficient to cause the compound to break up into its constituent
elements. Consequently, the student must imitate the copper
smelters of 5000 years ago. But instead of heating the oxide with
carbon, as they did, the contemporary chemist passes hydrogen
over the sample. Hydrogen has even more affinity for oxygen than
does copper. Alas, the latter cannot compete with the attractive
interloper! And so, hydrogen breaks up a beautiful couple to form
a new but familiar union, H_2O. The water escapes, and the de-
serted copper is left in elementary isolation.

The student weighs this metallic residue and finds that it
amounts to 1.229 g. The difference between the original mass of
1.538 g and this mass must obviously be due to oxygen.

1.538 g copper oxide − 1.229 g copper = 0.309 g oxygen.

This result provides all the information necessary for calculating
the gram equivalent weight of copper in the oxide. For this particular compound, the copper/oxygen mass ratio is 1.229:0.309.
The message of the Law of Constant Composition is that the mass
ratio of the elements in a pure compound is fixed. Therefore, the ratio just determined should apply to all samples of the
oxide. We are thus fully justified in expressing this ratio relative
to a reference quantity of oxygen, specifically, 8.000 g. And the
weight of copper thus related to 8.000 g of oxygen will be its gram
equivalent weight.

$$\frac{1.229 \text{ g copper}}{0.309 \text{ g oxygen}} = \frac{GEW \text{ g copper}}{8.000 \text{ g oxygen}}$$

$$GEW = \frac{(1.229 \text{ g copper})(8.000 \text{ g oxygen})}{(0.309 \text{ g oxygen})}$$

$$= 31.82 \text{ g copper}$$

We need only consult the approximate atomic weight, estimated as 65 g, in the specific heat experiment, to observe that the GAW is about twice the GEW of 31.82. Consequently, n, the whole number multiplier in Equation 3.5, must be 2. We therefore can write

$$GAW = 2(31.82 \text{ g}) = 63.64 \text{ g}.$$

The gram equivalent weight, based as it is upon careful analysis, is a more accurate quantity than the gram atomic weight calculated from the Law of Dulong and Petit. It follows that 63.64 must be a more reliable atomic weight for copper than is 65. And more exact measurements of a similar sort can yield atomic weights of even higher refinement. To be sure, the task is arduous and at times, no doubt tedious. But there are few elementary properties of more fundamental importance to chemistry than accurate atomic weights. Their existence is a tribute to the logical and experimental skills of scientists who dared to weigh the unweighable.

Counting the uncountable

There is, unquestionably, tremendous utility and considerable satisfaction in knowing the relative masses of the atoms of the various elements. But you could hardly expect anyone with a normal sense of curiosity to be satisfied at stopping there! Obviously, the next challenge is to find out the absolute weight of an atom, which of course requires a knowledge of how many atoms there might be in a given mass of matter. To discuss such a subject may seem as profitless as disputation over the angelic capacity of a pinhead ballroom. Yet, as formidable as the task sounds (and is!), it too has been accomplished.

Unfortunately, a detailed survey of the various procedures which have been employed to evaluate the number of atoms or molecules in a given volume or mass of matter would require more pages than one could responsibly devote to the subject in a book of this sort. However, a brief mention of the physical phenomena forming the basis for such determinations and the names of some of the scientists involved might lead to more library, and possibly laboratory, research by interested readers.

One of the first scientific papers to suggest a means of calculating the number of molecules in a definite volume of a gas under specified physical conditions was published in 1865 by Joseph Loschmidt. The properties of gaseous viscosity and diffusibility were proposed as the experimental bases for such determinations, and the theoretical framework for the calculations was provided by the Kinetic Theory of Gases, which treats statistically the behavior of the molecules making up the gas. By these means, a number of scientists, including the great mathematical physicist, James Clerk Maxwell, made some of the first estimates of the *Loschmidt number*, the number of gaseous molecules contained in a 1 cubic centimeter (cc) volume at 0°C and 1 atmosphere pressure.

In 1899, Lord Rayleigh pointed out that his explanation of why the sky was blue—a consequence of the scattering of sunlight by atmospheric molecules and dust particles—had implications for the determination of Loschmidt's number. Lord Kelvin went on to elaborate these ideas and came up with a value of 2.47×10^{19} for the number. The research was good, but one could hardly say it was peerless.

Perhaps the most accurate of the early molecular nosecounts was based on Brownian motion—the incessant, erratic movement of small particles suspended in air or liquid. This microscopic motion is due to the bombardment of the particles by molecules of the suspending medium. A theoretical and mathematical investigation of this phenomenon was done in 1905, by an examiner in the Swiss patent office. If this work is sometimes forgotten, it may be because its author, Albert Einstein, produced in the same year a paper introducing the Special Theory of Relativity. Three years later Jean Baptiste Perrin began a series of experiments in which he observed the Brownian motion of gamboge particles of very nearly uniform size. Sometimes science profits from some rather unlikely stuff, and this study of suspended sludge is a case in point. Gamboge is an orange gum resin obtained from certain Asiatic trees and has diverse uses as a pigment and a laxative. But in this particular investigation, it aided the determination of what Perrin called *Avogadro's number—the number of molecules in one gram molecular weight of a substance*. Following the theory of the Italian physicist honored in the name, all substances should show the same value for Avogadro's number, since gram molecular weights and gram atomic weights of all elements and compounds contain equal numbers of particles. The values obtained by Perrin for that constant ranged from 6.5×10^{23} to 7.2×10^{23}.

Three other methods which have been employed for determining Avogadro's number also deserve brief mention. They involve soap, crystals, and electrons. The first of these was used by Lecomte du Noüy, a French nobleman who wrote semipopular books about science, philosophy, and theology when he was not investigating surface phenomena. Du Noüy's technique involved measuring the area of soap films. By estimating the size of an individual soap molecule and making some guesses about the way in which they packed to form a film, he was able to calculate a value for the elusive number.

The second method is somewhat similar in basic principle. The distance of interatomic separation and the arrangement of atoms in a crystal of an element (often a metal) is determined by bouncing X rays off the atoms. This corresponds to a knowledge of the number of atoms in a given volume of the solid. This information, coupled with the density and atomic weight of the element, yields Avogadro's number.

Finally, the third method recognizes that it is considerably easier to count electrons than atoms. It turns out that the quantity of electricity passing through a circuit, a readily measurable property, is directly proportional to the number of electrons flowing. Faraday's Law and its extensions conveniently relate the quantity of electricity to the weight of an element electroplated by the current.* By this means, it is possible to arrive at the number of atoms present in 1 gram atomic weight. Details of this method will have to wait for a more propitious moment.

The value that these various methods have agreed upon for Avogadro's number is 6.023×10^{23}. The magnitude of such a number is almost beyond comprehension, but possibly a few "gee-whiz" examples will help make it more meaningful. For example, the gram molecular weight of water is 18.0154 g. Since the density of water is very nearly 1 g/cc, the volume of a gram molecular weight's worth of water is about 18 cc or 0.61 fluid ounces, a small swallow for 6.023×10^{23} molecules! Or, consider what you get for a penny: A freshly minted United States one cent piece weighs about 3.0 g. Since 1 gram atomic weight of copper weighs 63.5 g and contains one Avogadro number of atoms, a penny must contain $3.0 \text{ g} \times \dfrac{6.023 \times 10^{23} \text{ atoms/GAW}}{63.5 \text{ g/GAW}}$ or 2.8×10^{22} atoms. Each one of these weighs $\dfrac{63.5 \text{ g/GAW}}{6.023 \times 10^{23} \text{ atoms/GAW}}$ or 1.05×10^{22} g. This means that you could rub about 500,000,000,000,000,000 atoms off Honest Abe's beard before you could even detect the difference on a standard laboratory analytical balance capable of weighing to the nearest 0.0001 g. Such magnitudes dwarf even the national debt!

To introduce yet another example, if each atom had the size and weight of a grain of rice [about 6×3 millimeters (mm) and 20 milligrams (mg), respectively], 6.023×10^{23} of them, packed as rice grains normally do, would fill a cubic volume 146 miles on each side and weigh 13.2 quadrillion tons. And finally, we can even suggest an answer to the theological problem stated earlier. Coupling what we know about the arrangements of atoms in iron and the size of a typical pin head with the not unreasonable as-

---

* Chapter 6 provides more information on this subject.

sumption of one angel per atom, we arrive at 3.8×10^{13} as an upper limit for the heavenly host capacity of a standard straight pin.

The word "mole" is often used to designate an Avogadro's number of atoms, molecules, or even elementary particles such as electrons. The name suggests gram molecular weight and, indeed, it is proper to say that a mole of water weighs 18.0154 g. One can, however, also speak of a mole of oxygen molecules (31.9988 g) or a mole of oxygen atoms (15.9994 g). Although changes in atomic weight scales quite obviously do not influence the weight of any individual atom, they do change the number of atoms in a mole, that is, Avogadro's number. Thus, the switch from the ordinary oxygen standard to the carbon-12 reference could be interpreted as having decreased the value of Avogadro's number by 2.25×10^{19}. As a percentage of the total, this is an almost negligible difference, but in absolute magnitude, it is staggering.

The mole concept is one of considerable utility in chemistry and we will apply it in the next chapter. But before we move on to that, a few words should be devoted to a modern technique which permits the determination of the masses of individual atoms or molecules: *mass spectrometry.* Gaseous atoms (or molecules) of the substance under investigation are given a positive electric charge. The ions (charged atoms or molecules) thus formed are passed simultaneously through electric and magnetic fields. The extent to which the fields deflect the beam of ions depends upon the ratio of charge to mass for the particles involved. The deflection is measured in the experiment; the electric and magnetic field strengths are known; and the ionic charge can be independently determined. Thus, the only unknown is the mass, which can be readily computed—if you have a multithousand dollar black box providing you with data. Happily, the results are consistent with those obtained somewhat less precisely (albeit less expensively) from dancing particles of a laxitive pigment called gamboge.

It is a long way from the speculations of Democritus to the absolute, individual atomic masses determined by modern mass spectrometry. Atoms have now been "seen" by special electron microscopes and a technique called field-ion microscopy. Seeing is believing, and Figure 3.4 is an example of the evidence. So perhaps we are safe in concluding that atoms, though quite different from Dalton's concept, are real after all—or at least as real as anything else we perceive. But it would be well to keep in mind that in all this long search for structure, the sharpest insight came not through complex electronic focusing devices, but through the thick lenses and colorblind eyes of a Manchester schoolmaster. For his was a creative vision, and his Atomic Theory was a meta-

FIGURE 3.4. *A field ion photomicrograph of a platinum point.*
Under the magnification of over one million diameters, the
individual atoms appear as spots of light. Courtesy of R. W.
Newman, University of Florida.

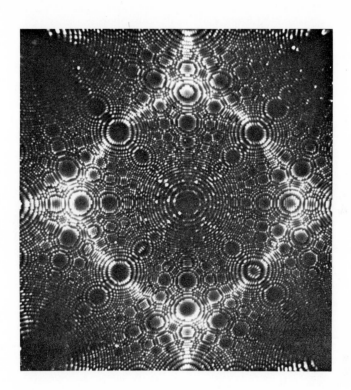

phoric model of the awesome, yet incredibly miniscule order and
elegance of matter.

For thought and action

3.1   Use the information in this chapter to determine the formula
      of hydrogen peroxide and its molecular weight. Also write an
      equation for the direct formation of hydrogen peroxide from
      hydrogen and oxygen.

3.2   Dalton used his principle of "maximum simplicity" to assign the
      formula NH to ammonia, a compound which is 82.4% nitrogen
      and 17.6% hydrogen (by weight). (a) Calculate the atomic weight
      which would be predicted for nitrogen if this formula were cor-
      rect. (b) Experiments show that at constant temperature and
      pressure

1 vol nitrogen + 3 vol hydrogen → 2 vol ammonia.   .

Translate this information into an equation reflecting the true symbols and formulas of the reactants and the product.

3.3   Recall the example of nitric oxide, NO. What is the gram equivalent weight of nitrogen in this compound?

3.4   Write a balanced chemical equation for the formation of compound D in Table 3.1 by the direct combination of nitrogen and oxygen. Also predict the volume ratio of nitrogen, oxygen, and oxide D (all measured at the same temperature and pressure) for this reaction.

3.5   Demonstrate that the formula N_2O_3 is consistent with the mass composition data for oxide C in Table 3.1 if the atomic weights of nitrogen and oxygen are 14.0 and 16.0, respectively.

3.6   Two different oxides of iron are analyzed. Compound A is found to consist of 77.8% iron and 22.2% oxygen (by weight); compound B contains 70.0% iron and 30.0% oxygen. Calculate the iron/oxygen weight ratios in the two compounds and demonstrate how they illustrate the Law of Multiple Proportions.

3.7   In one of Lavoisier's famous experiments he heated 45 grains of the red calx of mercury and obtained 41.5 grains of mercury and 8 cubic inches of oxygen. Calculate the gram (or grain) equivalent weight of mercury in the compound. Also attempt to determine the formula of the compound, assuming mercury to have an atomic weight of 201.

3.8   Various standards for the atomic weight scale have been proposed from time to time. One of these sets the atomic weight of oxygen arbitrarily equal to 100. Using this scale, calculate (a) the atomic weight of hydrogen and (b) the molecular weight of water.

3.9   Calculate the number of atoms in (a) 1.008 g hydrogen, (b) 63.546 g copper, and (c) 207.19 g lead. (Consult the atomic weights given on the end papers of this book.)

3.10   Calculate the number of atoms in (a) 10.0 g hydrogen, (b) 10.0 g copper, and (c) 10.0 g lead.

3.11   A glass contains 250 g of water. Calculate the following: (a) the number of moles of H_2O present, (b) the number of molecules of H_2O present, and (c) the number of hydrogen and oxygen atoms present. (Of course, the atoms are combined to form molecules.)

3.12   Calculate the mass of a single atom of gold (atomic weight = 197.0).

3.13   Calculate the mass of one billion atoms of iron (atomic weight = 55.85).

3.14   The specific heat of a metallic element is 0.029 cal/g °C. Estimate its atomic weight.

3.15 The relative weights of equal volumes of gaseous oxygen, methane, and chlorine (at the same temperature and pressure) are 1.00, 0.50, and 2.21, respectively. What are the molecular weights of (a) methane and (b) chlorine? The atomic weight of chlorine is 35.45. What does this fact, coupled with your above answer, imply about the usual structural form of gaseous chlorine?

3.16 In a series of experiments, certain chemical properties of a metallic element are investigated. 1.90 cal of heat are required to raise the temperature of a 3.07 g sample of the unknown element 10°C. The 3.07 g are then oxidized to yield 4.61 g of the corresponding metal oxide.
(a) Calculate an approximate atomic weight for the element.
(b) Calculate the gram equivalent weight of the metal in the oxide formed.
(c) Combine the results of parts (a) and (b) to calculate an exact value for the atomic weight of the element.

3.17 Carbon forms two different oxides. Analysis shows that oxide A contains 42.8% carbon and oxide B, 27.3% carbon. In another experiment, the weights of equal volumes of oxygen, oxide A, oxide B are weighed at identical conditions of temperature and pressure. The oxygen is found to weigh 0.253 g, oxide A weighs 0.221 g, and oxide B weighs 0.348 g. Assuming oxygen to have a molecular weight of 32.0, use Cannizzaro's method to estimate the atomic weights of carbon and oxygen. Also suggest formulas for the two oxides which are consistent with these atomic weights. Which compound, A or B, would be formed from the incomplete oxidation of carbon and other fuels? Finally, try naming the two oxides.

3.18 A theory may be internally self-consistent and in agreement with some experimental data and still be incorrect. Dalton's formulas for "olefiant gas" and "carbonic oxide" provide a case in point. The following composition data apply to these compounds.

Compound	Dalton's formula	Composition by weight	
"Olefiant gas"	CH	75.0% C	25.0% H
"Carbonic oxide"	CO	27.3% C	72.7% O

Show that these formulas are compatible with the experimental data if the atomic weights of hydrogen, carbon, and oxygen are 1, 3, and 8, respectively. Predict the formulas of these two compounds using the modern atomic weights of 1, 12, and 16.

3.19 Check the accuracy of the estimate of the capacity of the heavenly ball room reported on page 58. The atoms of iron are arranged in a square grid with 2.87 Å separation between adjacent atoms. (1 Å $= 10^{-8}$ cm.) The pinhead was assumed to be 2 mm in diameter and the area of a circle is πr^2.

3.20   What differences, if any, do you perceive between atomism as a philosophical tenet (Democritus and Lucretius) and atomism as a scientific theory (Dalton)?

Suggested readings

Alembic Club Reprints, No. 2, "Foundations of the Atomic Theory: Comprising Papers and Extracts by John Dalton, William Hyde Wollaston, M.D., and Thomas Thomson, M.D. (1802–1808)," Alembic Club, Edinburgh, 1899.

Freund, Ida, "The Study of Chemical Composition, An Account of Its Method and Historical Development," Dover, New York, 1968.

Lucretius, "On the Nature of Things" (translated by H. A. J. Munro), Washington Square Press, New York.

Nash, Leonard K., The Atomic-Molecular Theory, in "Harvard Case Histories in Experimental Science" (James Bryant Conant, general ed.), Vol. 1, pp. 215–321, Harvard Univ. Press, Cambridge, Massachusetts, 1966.

Toulmin, Stephen, and Goodfield, June, "The Architecture of Matter," Harper Torchbooks, New York, 1962.

4

The chemical code

The elementary alphabet

It does not take very long for any branch of knowledge to develop a specialized vocabulary. And there are, of course, sound reasons for establishing a symbolic system within a discipline. If there is agreement on and acceptance of the terms applied to the concepts and objects germane to the field, internal communication is greatly facilitated. However, what is an undoubted internal asset, may be a considerable impediment to understanding by the wider society. Sometimes one suspects that this latter effect might be the desired one. Certainly, the poor layman, bleeding either literally or figuratively, is often mystified, awed, or enraged (or any combination thereof) when confronted with medical or legal Latin. If something of the miraculous aura of the witch doctor or medicine man still lingers about the modern physician, the chemist, too occasionally enjoys the esoteric and occult image fostered by medieval alchemists and modern nomenclature. There is, after all, something rather exciting about being mistaken for Merlin or Mandrake. And, to most nonscientists, the list of ingredients in a soft drink— carbonated water, citric acid, saccharin, sodium citrate, gum arabic, brominated vegetable oil, and benzoate of soda—is only slightly more comprehensible than a recipe calling for alkanna, orpiment, realgar, and verdigris. "Eye of newt and toe of frog, wool of bat and tongue of dog" are at least intelligible!

In point of fact, some alchemical symbolism was intentionally obscure. The lucrative nature of the goal which the alchemists so assiduously sought did not exactly encourage intellectual openness. On the other hand, certain symbols were widely accepted at an early date. A case in point is provided by the seven metals we have already indicated as being familiar to the ancients. The sun and moon and the five then-known planets provided a convenient set of celestial counterparts to the more mundane metals. These heavenly bodies lent their astrological symbols, and presumably their astrological influences, to the elements.

Table 4.1 includes, in addition to these alchemical–astrological symbols, the Latin names of the seven ancient elements and the symbols derived from them. The symbols of the great majority of elements, however, correspond closely to their modern English names. This is evident from the alphabetical listing of elementary names, symbols, and atomic weights in Table 3.2.

The origins and meanings of the names summarize a fascinating amount of information about the discovery, occurrence, or properties of the elements. Some elements are named after compounds in which they commonly occur; for example, sodium from soda and potassium from potash.*

The names francium (Fr), polonium (Po), americium (Am), californium (Cf), and berkelium (Bk) celebrate the romantic and exotic

TABLE 4.1.   *Names and symbols of some elements*

Modern name	Modern symbol	Latin name	Alchemical symbol
Gold	Au	Aurum	☉
Silver	Ag	Argentum	☾
Copper	Cu	Cuprum	♀
Iron	Fe	Ferrum	♂
Tin	Sn	Stannum	♃
Lead	Pb	Plumbum	♄
Mercury	Hg	Hydrargyrum	☿

* The symbols of these elements are abbreviations of their German names, *natrium* (Na) and *kalium* (K).

places in which these elements or their discoverers were first found. The undisputed prizewinner in this category is Ytterby, a mining town near Stockholm, which has provided names for no less than four elements isolated from its ores: ytterbium (Yb), yttrium (Y), erbium (Er), and terbium (Tm). A more remote and inaccessible place of discovery, albeit a more familiar one, is commemorated in the name helium (He) from the Greek word *helios*, "the sun." Other heavenly bodies lending their names to earthly elements include Uranus (U), Neptune (Np), and Pluto (Pu). Among the scientific stars who have been honored by having an *ium* appended to their names are the Curies (Cu), Einstein (Es), Fermi (Fm), Mendeleev (Md), Nobel (No), and Lawrence (Lr).

Perhaps the names describing properties are the most ingenious of all. Lavoisier used Greek to express the chemical characteristics of hydrogen (H), "water-former." But the property implied by the name "oxygen" (O), "acid-former," is misleading. Contrary to what the great Frenchman thought, this element is not an essential component of all acids. Appropriately, radioactive radium (Ra) is named after the Latin *radius*, "ray." A rose by any other name might be rhodium (Rh), an element whose compounds dissolve to form pink solutions. And the name thallium (Tl) is derived from *thallus*, the Latin word for a budding twig. The not-altogether-obvious reason: a characteristic green color which appears when thallium compounds are heated in a flame.

Finally, there is the story of mean old Tantalus and his tearful daughter, Niobe. You may recall that Tantalus, one of Zeus' numerous mortal offspring, invited the gods to a banquet at which he served them his only son, Pelops. Apalled by this horrible menu, the Olympians devised a cunningly cruel punishment. Tantalus was placed in a pool in Hades which receded beyond reach whenever he stooped to drink. The metallic element discovered in 1802 by Ekeberg was equally unaffected by water and most acids. Moreover, its isolation proved tantalizingly difficult. Hence the name, tantalum (Ta). As for the ambitious and arrogant daughter, she was remembered when it came to naming the element always found in association with tantalum. For a time, some American scientists chose chauvinism over classicism and columbium over niobium (Nb). But mythology is not dead—even among chemists. Niobe has joined her tormented father in the elementary Olympus, though the underworld seems a more suitable place for such unsavory characters.

Formulas: a chemical spelling lesson

The 100-odd symbols of the elements provide the "alphabet" of chemistry, just as the substances which they represent constitute the basic stuff of the science and of the universe. Carrying our alphabetical analogy a bit further, combinations of these elementary "letters" are the "words" of chemistry. These formulas, of course, symbolize compounds; and like ordinary words, they can be misspelled. For that matter, one can invent nonsense compounds as well as nonsense words. When conceived by the wit of a Lewis Carroll or an Ogden Nash, orthographical invention can be effective and amusing. " 'Twas brillig and the slythy toves/Did gyre and gimble in the wabe" even makes a perverse sort of sense in Looking-Glass Land. But chemists tend to be a bit stuffy about their symbolism, and any hilarity engendered by a formula like HO_2 would undoubtedly be at the expense of its author.

The obvious point is that the formulas of compounds must bear some correspondence to reality—or at least reality as it is conventionally conceived by chemists. The "word" HO_2 implies a compound in which the atomic ratio of oxygen to hydrogen is 2 to 1, and it so happens that such a compound is unknown. Of course, if someone were able to unequivocally demonstrate the existence of a substance corresponding to such a formula, it would quickly become incorporated into the chemical vocabulary. But certainly, just writing HO_2 doesn't make it so.

The formula of a compound, like everything else in chemistry, is ultimately based upon experiment. Perhaps a rather lengthy example, involving a real compound of considerable importance, will illustrate the procedure by which such a representation is determined. Suppose that eminent scientist, Al Chymist, is investigating a compound found in petroleum—one which is a major component in gasoline. He has isolated the pure substance from the original mixture by distillation or some other suitable separatory technique. And, after a series of qualitative tests, he has discovered that the compound is composed of two elements, hydrogen and carbon. It is, in common terminology, a "hydrocarbon," and it must have the general formula $C_m H_n$. But the chemist would like to be able to assign numbers—correct numbers— to m and n. His task is to determine the formula of the compound. In other words, he seeks to find the atomic ratio of carbon and hydrogen which characterizes this specific hydrocarbon. Unfortunately, he cannot measure this ratio directly. But he can calculate it from a knowledge of the weight composition of the compound, that is, the mass ratio of carbon to hydrogen. Therefore, Al designs an experiment to yield this information.

The analytical procedure involves making use of the same chemical change which occurs when the hydrocarbon is employed as a fuel. A weighed sample of the compound is burned in plenty of air or oxygen. As a result of this combustion reaction, all the carbon originally present in the compound will be converted to

FIGURE 4.1. *Schematic diagram of an apparatus for the combustion analysis of compounds containing hydrogen and carbon.*

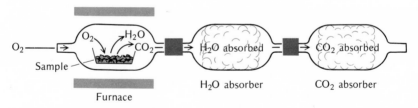

carbon dioxide, CO_2, and all the hydrogen initially present will appear as water vapor, H_2O. The CO_2 and H_2O thus formed are trapped in two different chemical absorbers in an apparatus pictured in Figure 4.1. The traps are weighed both before and after the burning, and the increase in weight is attributed to the water and carbon dioxide generated. Al has taken precautions to prevent the absorption of these two compounds from the atmosphere or any source other than the burning hydrocarbon. Therefore, he can be reasonably certain that the H_2O and CO_2 collected contain hydrogen and carbon from the original sample *only*. And, since he thoroughly burned the sample, he can also assume that the absorbed H_2O and CO_2 contain *all* of the hydrogen and carbon initially present in the compound of unknown formula.

In order to determine the composition of the compound, we turn to data gathered by our hypothetical friend:

Initial weight (wt) of hydrocarbon	10.0 g
Wt of water absorbed	14.3 g
Wt of carbon dioxide absorbed	30.8 g

What we and Dr. Chymist are after is the way in which that 10.0 g of hydrocarbon is divided up among carbon and hydrogen. And the other two weights, or even one of them, will provide the answer. Al argues that since all the hydrogen orginally found in the unknown substance is now part of the absorbed water, the weight of hydrogen in 14.3 g H_2O should also be the weight of hydrogen in 10.0 g of the starting sample. So he sets out to calculate this quantity. He has read Chapter 3 and is convinced that H_2O is indeed the correct formula for water. Moreover, he is trusting enough to accept 1.0 as the atomic weight of hydrogen and 16.0 as the atomic weight of oxygen. (There are limits to his skepticism.) This gives H_2O a molecular weight of $2 \times 1.0 + 1 \times 16.0$ or 18.0. In other words, 2.0 g out of every 18.0 g of water are hydrogen. If the Law of Constant Composition is anything to go by (and it is), this same ratio should apply to all samples of pure water. Therefore, Al Chymist assigns an x to the weight of hydrogen in 14.3 g of H_2O and confidently writes

$$\frac{x \text{ g H}}{14.3 \text{ g H}_2\text{O}} = \frac{2.0 \text{ g H}}{18.0 \text{ g H}_2\text{O}}$$

$$x = 14.3 \text{ g H}_2\text{O} \times \frac{2.0 \text{ g H}}{18.0 \text{ g H}_2\text{O}}$$

$$= 1.59 \text{ g H.}$$

The chemist thus concludes that the 10.0 g sample of hydrocarbon which he burned contained 1.59 g of hydrogen.

Of course, the calculation could also have been done by recalling that water contains 11.19% hydrogen by weight. This simply means that hydrogen accounts for 11.19 g out of every 100.0 g of water. Then the weight of hydrogen in 14.3 g of H_2O is obtained by a simple multiplication:

$$x = 14.3 \text{ g H}_2\text{O} \times \frac{11.19 \text{ g H}}{100.0 \text{ g H}_2\text{O}} = 1.60 \text{ g H.}$$

The fact that the two answers are not identical is a consequence of rounding off atomic weights and experimental data to the nearest tenth of a gram. In practice, much more accurate weighings are possible. The point is, the values are close enough to prove the equivalence of the two methods.

It turns out that this experimental result is sufficient to yield the weight composition of the unknown compound and subsequently permit the calculation of the formula. After all, the remainder of the 10.0 g sample must be carbon, since it contained only two elements.

Wt of carbon = 10.0 g hydrocarbon − 1.59 g hydrogen
= 8.41 g

But Dr. Chymist is a cautious experimenter, and so he also calculates the carbon content from the weight of the absorbed carbon dioxide. In order to do this, he must first determine the elementary mass composition of carbon dioxide. The general procedure is illustrated in Example 4.1.

EXAMPLE 4.1.   *Procedure for calculating the weight percent of elements present in a compound of known formula.*

1.   Look up the atomic weights of all the participating elements.
2.   Calculate the molecular or formula weight of the compound by multiplying each atomic weight by the corresponding subscript in the correct formula for the compound and adding, for example,

   GMW of $A_l B_m C_n = l(\text{GAW})_A + m(\text{GAW})_B + n(\text{GAW})_C$.

3.   Calculate the weight percent of each element by dividing the weight of each element in one molecular weight by the molecular weight itself and multiplying by 100.

$$\%A = \frac{l(GAW)_A}{GMW} \times 100 \quad \%B = \frac{m(GAW)_B}{GMW} \times 100$$

$$\%C = \frac{n(GAW)_C}{GMW} \times 100$$

In this specific instance, our experimenter assumes the formula for carbon dioxide is CO_2. Moreover, he knows that the atomic weight of carbon is 12.0 and that of oxygen is 16.0. Combining this information he finds that the molecular weight of CO_2 is $12.0 + 2 \times 16.0$ or 44.0. Therefore, the fraction of CO_2 which is carbon is represented by the mass ratio 12.0:44.0. Or, expressed relative to 100.0 g of CO_2, the percentage composition is

$$\frac{12.0 \text{ g C}}{44.0 \text{ g } CO_2} \times 100 = 27.3\% \text{ C.}$$

Either of these expressions of elementary composition by weight could be used to calculate the mass of carbon in 30.8 g of CO_2 and hence in 10.0 g of the gasoline compound. According to my slide rule

$$30.8 \text{ g } CO_2 \times \frac{12.0 \text{ g C}}{44.0 \text{ g } CO_2} = 8.40 \text{ g C}$$

or

$$30.8 \text{ g } CO_2 \times \frac{27.3 \text{ g C}}{100.0 \text{ g } CO_2} = 8.41 \text{ g C.}$$

The agreement between the weight of carbon calculated earlier by difference and this more direct determination is good, and Albert is encouraged. But before proceeding to the determination of the formula, he pauses briefly to calculate the weight percent composition of the hydrocarbon. After all, he and we have seen that such information can be useful.

$$\text{Wt percent H in } C_mH_n = \frac{1.59 \text{ g H}}{10.0 \text{ g } C_mH_n} \times 100 = 15.9\%$$

It follows at once that the weight percent of carbon in the compound must be 84.1%.

In order to finally arrive at the correct formula, Al Chymist must now convert the rather awkward mass ratio of the two elements into a simpler atomic ratio. The crucial question can be simply phrased: How many gram atomic weights of hydrogen make up 1.59 g of the element, and how many GAW's of carbon constitute 8.41 g? The answer is perhaps easiest to see in ratio form. For instance, we know that 1 GAW of hydrogen weighs 1.0 g or, a bit more exactly, 1.01 g. Therefore, 1.59 g must amount to about 1½ GAW's. Our chemist represents this unknown as y in the following expression, and it soon becomes a known:

$$\frac{y \text{ GAW's H}}{1.59 \text{ g H}} = \frac{1 \text{ GAW H}}{1.01 \text{ g H}}$$

$$y = 1.59 \text{ g H} \times \frac{1 \text{ GAW H}}{1.01 \text{ g H}}$$

$$= 1.58 \text{ GAW's H.}$$

The calculation of the number of GAW's of carbon is equally easy. One merely divides the weight of the element, 8.41 g, by its gram atomic weight, 12.0 g.

$$8.41 \text{ g C} \times \frac{1 \text{ GAW C}}{12.0 \text{ g C}} = 0.70 \text{ GAW C}$$

We thus see that the initial 10.0 g sample of the compound under investigation contained 1.58 GAW's or moles of hydrogen and 0.70 mole of carbon.

Our fictional friend could incorporate these results into a formula and write $C_{0.70}H_{1.58}$. The correct atomic ratio would indeed be reflected by the subscripts. However, chemists are simple folk and like the simplicity of whole numbers and whole atoms. Fortunately, it is not difficult to convert these decimals into integers. First, the larger number is divided by the smaller: $1.58/0.70 = 2.26$. This makes the C/H atomic ratio 1/2.26. The numerical value of this ratio is exactly that which was first obtained. But it is still not expressed in the whole numbers which the chemist seeks for m and n. So he attempts to convert the fraction to an equivalent ratio of integers by multiplying both the numerator and the denominator by the same whole number. He first tries 2 as the multiplier,

$$\frac{C}{H} = \frac{1}{2.26} \times \frac{2}{2} = \frac{2}{4.52}.$$

The value of the ratio remains unchanged, but you see at once that this is still not the desired result. It is also readily apparent that 4 would have been a better choice,

$$\frac{C}{H} = \frac{1}{2.26} \times \frac{4}{4} = \frac{4}{9.04}.$$

This is very close to being the ratio of two whole numbers, and AI is justified in rounding off the 9.04 as 9. It follows that the simplest possible formula of the hydrocarbon must be C_4H_9. The compound contains nine atoms of hydrogen for every four of carbon. The experimental data for the composition of the hydrocarbon, by weight, have thus been converted to a whole-number ratio of gram atomic weights. And any accurate set of elementary mass values for the compound in question would yield the identical result. For example, you can readily ascertain that the same formula emerges if the GAW ratio is calculated from the weight percent values for this hydrocarbon. After all, although the weight of a sample of a compound may vary widely, its weight composition and its atomic ratio are constant.

But it turns out that the simplest formula is not necessarily an accurate representation of the true molecular formula. Recall that the formula of a compound should reflect the composition of the physical entity called a molecule, not just the atomic ratio. Therefore, Dr. Chymist needs one more piece of important information, the molecular weight. He would probably obtain this by determining the number of gram molecular weights in a known weight of the vaporized hydrocarbon. The details of such an experiment are discussed in Chapter 10. For the moment we will merely report the result, 114 g. A little bit of arithmetic indicates that the gram molecular weight of C_4H_9 would be about one-half that value.

$$4 \text{ GAW's C} \times \frac{12.0 \text{ g}}{\text{GAW}} + 9 \text{ GAW's H} \times \frac{1.0 \text{ g}}{\text{GAW}} = 57.0 \text{ g}$$

Therefore, our hero triumphantly concludes that the true formula must involve twice as much of these elements as indicated in the simplest formula. The unknown must be C_8H_{18}, and each of its molecules must consist of 8 carbon atoms and 18 hydrogen atoms. And, once we know there are eight carbons, we can even recognize that fact in a name—octane.

The general procedure just followed is frequently employed in the determination of formulas, even if the compound includes more than two elements. Example 4.2 provides a summary of the steps involved.

EXAMPLE 4.2 *Procedure for determining the formula of a compound from analytical data.*

1. Determine the composition, by weight, of all the elements present in the compound.

    $$x \text{ g A} + y \text{ g B} + z \text{ g C} = w \text{ g } A_l B_m C_n \ (l, m, n, \text{ unknown})$$

    a. Data are obtained experimentally, often based upon the known composition and mass of other compounds.
    b. If the total mass of the compound is known, the mass of one of the constituent elements need not be directly determined, but can be found by difference, for example,

        $$z \text{ g C} = w \text{ g } A_l B_m C_n - (x \text{ g A} + y \text{ g B}).$$

    c. The data can be in terms of weight percents.

2. Determine the number of GAW's of each of the elements present in a given mass of compound.

    $$l' = \frac{x}{(GAW)_A} \qquad m' = \frac{y}{(GAW)_B} \qquad n' = \frac{z}{(GAW)_C}$$

    (l', m', and n' will most likely not be integers)

3. Convert numbers of GAW's to integers by dividing each of them by the smallest value, for example, if m' is the smallest number of GAW's,

$$l = \frac{l'}{m'} \qquad m = \frac{m'}{m'} = 1 \qquad n = \frac{n'}{m'}.$$

4. If nonintegral numbers of GAW's still remain, multiply all values by the smallest factor necessary to convert all of them to whole numbers.

What one does is to experimentally measure the elementary composition of the compound by weight. The resulting weight ratio is then converted to a ratio of gram atomic weights, and this ratio is reexpressed in terms of whole numbers, which finally appear in the simplest formula. In some instances, it may be necessary to change this expression by a whole number multiple in order to bring it into conformity with the true molecular structure. Of course, the fact that the mass data underlying the formula calculation must be experimentally obtained means that there is room for error. Human and mechanical frailties and failings are undeniable facts of chemistry and life. Not all investigators are as accurate as Albert Chymist. A relatively small loss of material or a mistake in weighing could easily lead to an incorrect formula. Therefore, a careful researcher repeats his analyses, attempts to minimize his errors, and estimates the unavoidable uncertainties in his measurements. Indeed, physical scientists can often be reasonably sure of just how good or how bad their data actually are—a degree of self-knowledge which seems more difficult to attain in the social sciences. Ideally, this should keep chemists and their colleagues from basing conclusions on questionable evidence. In practice it doesn't always work that way.

By this time it may have occurred to you that in spite of the fact that you know quite a bit about finding the formula of compounds, you have learned little about naming them. There are rules and conventions, all right. But part of the problem is that we have not yet considered all the concepts required to adequately explain the issues. Some of these will be introduced later in the text. Nonetheless, chemicals did have names before these ideas were fully developed. Unfortunately, the names were sometimes not very precise and usually were not related in any rational scheme. It would be carrying the historical emphasis to an absurd extreme to ask you to memorize such names as butter of antimony or corrosive sublimate or muriatic acid or marsh gas. At the moment, antimony trichloride, mercuric chloride, hydrochloric acid, and methane may not seem much better. But when you are presented with the corresponding formulas, $SbCl_3$ $HgCl_2$, HCl, and CH_4 hopefully you begin to see the connection; and certainly you acquire a good deal of information about the composition of these compounds.

Because it would be somewhat premature to discuss the naming of compounds at this point, the author has resorted to the time-honored repository of important facts that don't precisely fit any-where—an appendix. Some rules of nomenclature and some examples, for both inorganic and organic (carbon-containing) sub-stances, are summarized there. Whenever new compounds are introduced into the body of the text, their names and formulas will be indicated. Beyond that, you are warmly encouraged to make frequent use of Appendix 3. Hopefully, it will prove more valu-able than its intestinal namesake.

Equations and chemical syntax

If the symbols of the elements are the letters of the chemical alphabet, and formulas of compounds constitute the vocabulary of chemistry, chemical equations must surely be the science's sentences, for equations are concise and meaningful statements about chemical facts and have both qualitative and quantitative significance. You will recall that equations were originally intro-duced in Chapter 3, but this is the place for a more systematic study. As in the previous section, we will carefully consider an example—the same example. First we will write a chemical sen-tence describing the burning of octane, the reaction of that com-pound with oxygen. When the equation is completed, we will in-vestigate its meanings and applications to the use of gasoline as a fuel.

We already know the nouns we need to construct the appro-priate chemical sentence. We just learned how to spell one, C_8H_{18}, and the others are O_2, H_2O, and CO_2. The first two formulas represent *reactants*, compounds which react with each other to form products. In this particular chemical change, H_2O and CO_2 are the *products*. We write down the reactants on the left and the products on the right:

$$C_8H_{18} + O_2 \rightarrow H_2O + CO_2.$$

The plus sign and the arrow are operative indicators, verbs if you're still keeping the analogy going. Their meaning should be self-evident.

Something is undoubtedly missing from this equation. As a matter of fact, it is not even an equation because it does not balance. Mass, matter, and atoms are conserved in chemical reac-tions. The total mass of all the reactants must equal the total mass of all the products. Therefore, the number and identity of atoms must also be the same on both sides of the arrow. The above un-equation annihilates 7 atoms of carbon and 16 of hydrogen while manufacturing one oxygen atom.

In an attempt to rectify the situation, we first note that one molecule of C_8H_{18} will yield 8 atoms of carbon and 18 atoms of hydrogen. The former reappear on the right side of the arrow as 8 molecules of CO_2. And the hydrogen atoms become incorporated into 9 molecules of H_2O. Thus, we can write

$$C_8H_{18} + ?O_2 \rightarrow 9H_2O + 8CO_2.$$

Only the number of oxygen molecules remains unspecified. The answer is easily found, however, by counting the atoms of oxygen on the product side of the equation. The 9 molecules of H_2O contribute one oxygen atom each, and the 8 molecules of CO_2 account for 16 more. Thus, the total number of oxygen atoms is 25, and the number of binary O_2 molecules in the balanced equation must be $^{25}/_2$.

4.1 $$C_8H_{18} + {}^{25}/_2\, O_2 \rightarrow 9H_2O + 8CO_2$$

A count and comparison of atoms on either side of the arrow indicate that the sentence obeys a cardinal rule of chemical syntax, the Law of Conservation of Mass. If the fractional coefficient for O_2 bothers you, just multiply the entire equation by 2:

4.2 $$2C_8H_{18} + 25O_2 \rightarrow 18H_2O + 16CO_2.$$

Its meaning remains essentially unaltered.

Part of that meaning is a qualitative description of the chemistry which takes place. Thus, Equation 4.2 states the fact that octane reacts with oxygen to yield water and carbon dioxide. If we wish, we can include additional descriptive information by specifying the physical states of the participants in the reaction. This is commonly done by using (s), (l), and (g), respectively, to designate solid, liquid, and gas. Since C_8H_{18} and H_2O are liquids at room temperature, we can write the following equation, which would be accurate at 25°C:

$$2C_8H_{18}(l) + 25O_2(g) \rightarrow 18H_2O(l) + 16CO_2(g).$$

But at the considerably higher temperature of an internal combustion engine, the octane and water are both vaporized. We recognize this by writing

$$2C_8H_{18}(g) + 25O_2(g) \rightarrow 18H_2O(g) + 16CO_2(g).$$

Sometimes an upward arrow ($\uparrow$) is employed to indicate the escape of a gaseous product, while a downward arrow ($\downarrow$) identifies a precipitating solid, that is, a solid which comes out of solution during the reaction.

It turns out that the physical states of reactants and products can also have quantitative implications, particularly with respect to energy changes associated with the reaction. And certainly it is as a source of energy that the burning of gasoline has its principal application. However, the mathematical manipulations of matter which follow do not depend upon physical states. Therefore, we will largely ignore these factors.

The primary quantitative significance of a balanced equation arises from the fact that each correct formula communicates specific information about the chemical composition of the corresponding substance. And the equation relates the reactants and products in a uniquely characteristic ratio. On the most microscopic level, Equation 4.2 can be interpreted as a statement about molecules.

$$2C_2H_{18} \quad + \quad 25O_2 \quad \rightarrow \quad 18H_2O \quad + \quad 16CO_2$$
2 molecules + 25 molecules → 18 molecules + 16 molecules

Note, first of all, that there are 27 molecules on the left side and 34 on the right. This is no cause for alarm; the number of product molecules need not equal the number of reactant molecules. Nor is it necessary for the volume of the products to equal the volume of the reactants. Indeed, the fact that it often does not was used to good advantage in the last chapter. However, *it is always true of any chemical reaction and the balanced equation describing it, that mass and the number and identity of the atoms involved must remain unaltered.* A chemical change is merely a rearrangement of atoms.

Because a chemical equation is basically a ratio, we can multiply the relative number of molecules participating in the reaction by a constant number without altering that ratio. You will recognize that this is just what we did in going from Equation 4.1 to Equation 4.2. Now suppose we select as the multiplier Avogadro's number, 6.023×10^{23}.

$$2C_8H_{18} \quad + \quad 25O_2 \quad \rightarrow$$
$$18H_2O \quad + \quad 16O_2$$
$6.023 \times 10^{23} (2 + 25)$ molecules →
$$6.023 \times 10^{23} (18 + 16) \text{ molecules}$$

But 6.023×10^{23} molecules constitute 1 mole, $2 \times 6.023 \times 10^{23}$ molecules represent 2 moles, and so on. Therefore, the above equation also makes a statement about the relative number of moles of the various chemical substances participating in the reaction.

4.3

$$2C_8H_{18} + 25O_2 \rightarrow 18H_2O + 16O_2$$

2 moles + 25 moles → 18 moles + 16 moles

The numerical coefficients in the balanced equation indicate the molar ratio of the reactants and products in this process.

Every reaction has a plan or system. The participants in it bear a unique quantitative relationship to each other, and this relationship is specified by the balanced equation. In this particular instance, 2 moles of C_8H_{18} react with 25 moles of O_2. The two quantities are said to be *chemically equivalent*. This does not mean that 2 moles of octane weigh as much as 25 moles of oxygen, or that they contain the same number of atoms, or that they are equal in any way. But it does indicate that 2 moles of C_8H_{18} and 25 moles of O_2 are chemically linked in this molar (or molecular) ratio during the combustion of octane. Furthermore, one can also look to the products and see that 18 moles of H_2O and 16 moles of CO_2 are chemically equivalent to 2 moles of C_8H_{18}. All these quantities are associated by the chemistry summarized in Equation 4.2. A chemical equation is, among other things, a statement about chemical equivalence.

Sometimes chemically equivalent quantities are referred to as *stoichiometric quantities*. The name is from the Greek words for "element," *stoicheion*, and "measure," *metron*. But as we just observed, it is "elementary measure" of a very special chemical sort. Since 2 moles of C_8H_{18} do not really equal 25 moles of O_2 we cannot very well use an equal sign between them. So we introduce a new symbol, $\simeq$, which is read "is equivalent to." Thus, we can write the following:

4.4   2 moles $C_8H_{18} \simeq$ 25 moles $O_2 \simeq$ 18 moles $H_2O \simeq$ 16 moles CO_2.

This notation will be useful in doing some *stoichiometric calculations*, that is, calculations based on balanced equations.

To cite a simple example, we can readily determine the number of moles of carbon dioxide which could be formed from any given number of moles of octane in the reaction summarized by Equation 4.2. Let us arbitrarily select 10 moles of C_8H_{18} as the starting quantity. What we are looking for is the number of moles of CO_2 which are chemically equivalent to this amount of C_8H_{18}. Since the answer is temporarily unknown, we give it the standard symbol for the mysterious, x. And we represent the equivalence as follows:

10 moles $C_8H_{18} \simeq$ x moles CO_2.

Since the stoichiometric relationship between two compounds is specified by the balanced equation, we simply record this ratio as

2 moles $C_8H_{18} \simeq$ 16 moles CO_2.

The same ratio must hold for any equivalent molar quantities of

C_8H_{18} and CO_2, including 10 moles of C_8H_{18} and x moles of CO_2. We indicate this fact by equating the two expressions for the ratio, that is,

$$\frac{10 \text{ moles } C_8H_{18}}{x \text{ moles } CO_2} = \frac{2 \text{ moles } C_8H_{18}}{16 \text{ moles } CO_2}.$$

Cross multiplying yields

$$(2 \text{ moles } C_8H_{18})(x \text{ moles } CO_2) = (10 \text{ moles } C_8H_{18})(16 \text{ moles } CO_2).$$

And solving for x provides the desired answer:

$$x = \frac{(10 \text{ moles } C_8H_{18})(16 \text{ moles } CO_2)}{(2 \text{ moles } C_8H_{18})} = 80 \text{ moles } CO_2.$$

Of course, the ratio of the coefficients of the balanced equation applies equally well to the other participants in the reaction. Therefore, we can use essentially the same method to calculate that 125 moles of O_2 and 90 moles of H_2O are also chemically equivalent to 10 moles of C_8H_{18}. Or, put in somewhat more phenomenological terms, 125 moles of oxygen are required to completely burn 10 moles of octane to form 80 moles of carbon dioxide and 90 moles of water.

Computations of this sort are far more than mere exercises in applied algebra. They are of great practical importance to chemistry. We have just seen how they permit the determination of the maximum possible yield of product expected from a given quantity of reactants—useful information if you happen to be manufacturing soap or plastic or birth control pills or anything else. Moreover, you will soon discover that stoichiometric calculations can be of assistance in understanding problems of environmental pollution. But it is considerably more convenient to weigh bunches of atoms and molecules than to count them. Therefore, such applications often involve the masses of the participants in the reaction, not the number of moles. For that reason, we move to the next section and a survey of the weighty significance of chemical equations.

Weighty sentences

You have seen that a balanced equation represents the specific molar ratio of reactants and products characteristic of a particular ·chemical change. But because atomic and molecular weights are known, this ratio can readily be converted into a mass ratio. At the risk of belaboring topics already touched on in Chapter 3, it might be well to comment further on our example. First we calculate the gram molecular weights (gram formula weights) of each of the substances involved in the reaction. Rounding off atomic weights to the nearest tenth of a gram, we obtain the following results:

C_8H_{18}	114.0 g	O_2	32.0 g
H_2O	18.0 g	CO_2	44.0 g.

Now each of these GMW's is multiplied by the appropriate coefficient from the balanced equation. We thus obtain an expression analagous to Equation 4.3, but using mass instead of moles.

4.5
$$2C_8H_{18} + 25O_2 \rightarrow 18H_2O + 16CO_2$$
$$2(114.0 \text{ g}) + 25(32.0 \text{ g}) \rightarrow 18(18.0 \text{ g}) + 16(44.0 \text{ g}) \quad \text{or}$$
$$228 \text{ g} + 800 \text{ g} \rightarrow 324 \text{ g} + 704 \text{ g}$$

When the indicated additions are actually carried out it becomes clear that the mass of the reactants and the mass of the products both equal 1028 g. The equation conforms to the conservation of matter. Moreover, implicit in the above expression is another way of writing the chemical equivalencies which characterize this particular reaction. Thus, we can translate the moles of Equation 4.4 into the grams of Equation 4.6:

4.6 $228 \text{ g } C_8H_{18} \simeq 800 \text{ g } O_2 \simeq 324 \text{ g } H_2O \simeq 704 \text{ g } CO_2$.

It requires only a moment's inspection to see that the mass ratios of Equation 4.6 are not numerically the same as the molar ratios of Equation 4.4. Nor should one expect them to be. This could happen only if all the participating compounds had identical molecular weights. Generally speaking, the relationship between chemically equivalent weights of compounds is not a simple whole number ratio of the sort which characterizes their molar equivalence. Perhaps a trivial example from applied economics will help further clarify the issue. Since it is reasonably convenient to count coins and bills, we express our monetary wealth in integral units of a common denominator—dollars for some people, cents for others. For example, one nickel is equivalent in purchasing power to five 1 cent pieces, that is,

1 nickel $\simeq$ 5 pennies.

But we could choose to use weight as a basis for comparison. It turns out that one new penny weighs about 3.13 g and one freshly minted nickel weighs 4.94 g. Therefore, 5 $\times$ 3.13 or 15.65 g of pennies are equal in exchange value to 4.94 g of nickels. Or, in our notation,

4.94 g nickels $\simeq$ 15.65 g pennies.

Restated relative to 1 g of nickels, the ratio becomes

1.00 g nickels $\simeq$ 3.17 g pennies.

Quite clearly, the expression of equivalence in terms of coins is not the same as the ratio of equivalent masses. The latter is not the neat whole number relationship of the former. But in spite of the fact that the ratios are different, they do represent the same basic information. The analogy is obvious: molecules, like coins, come in various weights. And at the moment, we are concerned with chemical computations using units of weight.

Let me illustrate the procedure with a specific example. Suppose I wish to predict the mass of oxygen required to react completely with 1 g of octane (about 15 drops). Now most likely this is not something you have been longing to know. But consider the implications of the problem. If less than a chemically equivalent quantity of oxygen is available, the reaction will not go to completion. Not all the C_8H_{18} will be converted into H_2O and CO_2. Some of the octane may remain unburned, as it frequently does in a car with an incorrectly adjusted carburetor (the gadget for mixing gasoline and air). This means not only a waste of gasoline, but also the emission of air pollutants. The foul black flatus of unburned fuel which emanates from a jet airplane as it lands or takes off is another instance of just this phenomenon. Sometimes, when insufficient oxygen is present, the fuel is partially broken down into smaller hydrocarbon molecules containing fewer atoms of carbon and hydrogen. These escape in the exhaust and can contribute significantly to smog. And the formation of deadly carbon monoxide, CO, is yet another consequence of operating an engine with less than a stoichiometric quantity of oxygen. There simply is not enough of the element to completely oxidize the carbon of the fuel to CO_2 and the process stops short of this point.

Obviously, the problems of pollution identified in the previous paragraph are of sufficient magnitude and complexity to warrant a considerable amount of attention. And we will return to these and other issues of environmental quality in Chapter 11. For the moment, however, let us restrict ourselves to the fundamental problem posed earlier—that of calculating the minimum amount of oxygen necessary to completely convert 1 g of octane to carbon dioxide and water. Our guide will be the balanced equation, 4.2, and its various extensions. Thus, expression 4.6 includes the chemical equivalence which relates octane and oxygen,

$$228 \text{ g } C_8H_{18} \simeq 800 \text{ g } O_2.$$

We seek the mass of O_2 chemically equivalent to 1.00 g of C_8H_{18}. So we represent the unknown with a z and write an analogous expression,

$$1.00 \text{ g } C_8H_{18} \simeq z \text{ g } O_2.$$

The mass ratio stated in the first of these two expressions must apply to any stoichiometric quantities of octane and oxygen participating in this reaction. Therefore, it also applies to the specific instance represented by the second expression. We can thus equate the two ratios:

$$\frac{228 \text{ g C}_8\text{H}_{18}}{800 \text{ g O}_2} = \frac{1.00 \text{ g C}_8\text{H}_{18}}{z \text{ g O}_2}.$$

Cross-multiplying and solving for z yields the desired answer:

$$(228 \text{ g C}_8\text{H}_{18})(z \text{ g O}_2) = (1.00 \text{ g C}_8\text{H}_{18})(800 \text{ g O}_2)$$

$$z = \frac{(1.00 \text{ g C}_8\text{H}_{18})(800 \text{ g O}_2)}{(228 \text{ g C}_8\text{H}_{18})}$$

$$= 3.51 \text{ g O}_2.$$

About 3½ g of oxygen are required to burn each gram of octane. This weight ratio, of course, is exactly equivalent to the corresponding molar ratio stated in the balanced equation, that is, 2 moles $C_8H_{18} \simeq 25$ moles O_2.

Information of this sort is essential to an automotive engineer designing a carburetor. Naturally, he must also take other variables into consideration—the fraction of oxygen in the air, the rate at which the gasoline and the oxygen are delivered to the carburetor, the thoroughness of mixing, and so on. But the stoichiometric relationship of the reaction, summarized in its balanced equation, provides a definite limiting condition. No matter what the design of the carburetor, if the mass ratio of O_2 to C_8H_{18} is less than 3.51 to 1.00, the reaction cannot go to completion. Thus, if only 3.00 g of O_2 are present for every 1.00 g of C_8H_{18}, some of the fuel will escape unburned or only partially converted to CO_2 and H_2O. Obviously, 3.00 g of oxygen weigh more than 1.00 g of octane. So in an absolute sense, there is more O_2 than C_8H_{18} in the mixture. But in a reaction, it is chemically equivalent quantities which count. And the balanced equation stipulates a stoichiometric ratio of 2 moles $C_8H_{18} \simeq 25$ moles O_2 or 1.00 g $C_8H_{18} \simeq 3.51$ g O_2. Therefore, a 3 to 1 mass ratio represents a chemical excess of octane and a deficiency of oxygen. In automobile jargon, the mixture is said to be "too rich" (with respect to gasoline). The oxygen is the reactant present in the relatively lower or limiting quantity. After it has all been consumed, there can be no further reaction. If there were, it would be tantamount to making something from nothing at all, a process not likely to succeed in this particular universe.

Of course, it is also possible to have an excess of oxygen. Suppose the mass ratio of O_2 to C_8H_{18} is 4.00 to 1.00. Now the former element is present in excess while the latter compound limits the reaction. If the reactants are well mixed and sufficient time is allowed for combustion, all the C_8H_{18} should emerge from the exhaust as CO_2 and H_2O. To be sure, there will be some unreacted oxygen left over, but this will certainly not pollute the air. Unfortunately, however, there are other complications. The combustion of gasoline is not the only chemical reaction which occurs in an automobile engine. The oxygen comes mixed with nitrogen in the air which is drawn into the carburetor. During the explosion in the cylinder, these two elements combine to form nitric oxide, NO,

$$N_2 + O_2 \rightarrow 2NO.$$

And as the relative amount of air in the cylinder increases, so does the extent of this reaction. The NO thus formed is responsible for some very undesirable atmospheric effects. But that story will also have to wait until the chapter on solutions. It is all tied up with a devilishly complicated set of simultaneous and competing reactions occurring in the solution called air. Since scientists stand little chance of making sense of such a complex situation without knowing how to handle considerably simpler single reactions, we return to the problem at hand.

Really, there are not too many more observations which need to be made. The procedure just used to calculate the amount of oxygen required to burn 1 g of octane can easily be extended to any weight of the compound. In fact, you are not even limited by the metric system. After all, there is nothing sacred about grams. The mass ratios of a balanced chemical equation apply equally well to any mass units. Thus, a calculation of the number of pounds of O_2 required to react completely with 1.00 lb C_8H_{18} (about one-sixth of a gallon) would be identical to the example just completed, except, of course, that the units would be different. The numerical answers would be the same, 3.51 lb of oxygen would be needed. If you do mix your units, the usual admonition about not adding apples and oranges applies. For example, if you try to determine the number of pounds of O_2 chemically equivalent to 1.00 g C_8H_{18}, make certain you remain consistent within your calculation and use the proper conversion factors.

If it is convenient, you can even make use of units like "pound atomic weights" and "pound molecular weights." These are simply atomic and molecular weights expressed in pounds and are numerically identical to the corresponding gram atomic and gram molecular weights. For example, the pound atomic weight of oxygen is 16.00 lb, just as the gram atomic weight of the element is 16.00 g. Note, however, that 16.00 lb of oxygen contain considerably more matter and more atoms than 16.00 g. A problem at the end of the chapter asks you to calculate how many. Some day, when this country finally joins the other nations of the world in adopting the metric system, we can eliminate such tedious exercises along with the archaic system which makes them necessary.

But whatever the mass units employed, there is obviously no need to restrict yourself to calculating the oxygen/octane mass ratio. That would be of limited utility and even less fun. Because a balanced equation specifies the mole and, indirectly, the mass ratios of *all* participants, the same method can be used to determine the weight of any reactant or product chemically equivalent to a given weight of any other. Thus, in our example, you can find out how many grams of CO_2 and how many grams of H_2O will be formed when you burn 1.00 g of C_8H_{18} with at least 3.51 g of O_2. Just remember that the correct mass ratio is that represented by Equation 4.6. The problem is restated at the end of the chapter and the answers are given in Appendix 4. And once you have mastered the method for this particular reaction, you can apply it to any other one. Just follow the general procedure outlined in Example 4.3. If you understand it, you understand equations.

EXAMPLE 4.3.   *Procedure for stoichiometric calculations.*

1. Make sure the equation is balanced.
2. Write down the molar ratio of all the products and reactants.
3. If the problem calls for an answer in numbers of moles, calculate the unknown relative to a given known number of moles of a participant in the reaction, making use of the corresponding molar equivalences stated in the balanced equation.
4. If the problem calls for an answer in mass units, convert numbers of moles to mass by multiplying chemically equivalent numbers of moles by the appropriate molecular or formula weights. Calculate the unknown mass relative to a given known mass of a participant in the reaction, making use of the corresponding mass equivalences implied by the balanced equation.
5. If the reactants are not present in chemically equivalent quantities, make sure the calculation is based upon the reactant present in the limiting lower relative amount.

Hopefully, your efforts will indicate that you have acquired a reasonable fluency with the letters, words, and sentences which make up the language of chemistry.

For thought and action

4.1   Compute the molecular weights of the following compounds, using atomic weights from Table 3.2 or the end papers of this book.
      (a)   Nitrogen pentoxide, N_2O_5
      (b)   Carbon tetrachloride, CCl_4
      (c)   Methyl ("wood") alcohol, CH_4O
      (d)   Sucrose (sugar), $C_{12}H_{22}O_{11}$
      (e)   Mescaline, $C_{11}H_{17}O_3N$

4.2   Calculate the weight percent of the elements found in each of the compounds listed in Problem 4.1.

4.3   Given the following weight percent data, calculate the simplest formulas of the following compounds:
(a)   Nitrous oxide ("laughing gas") 63.6% N, 36.4% O
(b)   Sodium chloride (salt) 39.3% Na, 60.7% Cl
(c)   Sulfuric acid 32.7% S, 65.3% O, 2.0% H
(d)   Urea (excreted in urine) 20.0% C, 6.7% H, 46.7% N, 26.6 % O

4.4   On page 70, the text reads, "you can readily ascertain that the same formula emerges if the GAW ratio is calculated from the weight percent values for this hydrocarbon." See for yourself by actually doing it. Recall that the compound in question contains 84.1% carbon and 15.9% hydrogen.

4.5   If you experienced difficulty in determining the formula of the red calx of mercury from Lavoisier's data (Problem 3.7), try again, using the procedure outlined in this chapter. Also write the balanced equation for the decomposition of this compound to yield mercury and oxygen.

4.6   Recall your earlier experience with hydrogen peroxide, a compound consisting of 5.89% hydrogen and 94.11% oxygen by weight. Use the method of this chapter and the accepted atomic weights to calculate its simplest formula. The molecular weight of hydrogen peroxide is 34.0. Show that the molecular formula obtained from these data is in agreement with your answer to Problem 3.1 and consistent with the observed hydrogen/oxygen/hydrogen peroxide volume ratio of 1:1:1.

4.7   Benzene is a common hydrocarbon compound widely used as a solvent. When 5.00 g of this substance are burned with an excess of oxygen in an analysis apparatus such as that pictured in Figure 4.1, 3.46 g of H_2O and 16.90 g of CO_2 are formed. Calculate the simplest formula for benzene. Another experiment indicates that the molecular weight of this compound is 78.0. Determine its true molecular formula. Using this information, write a balanced equation for the combustion of benzene.

4.8   The combustion method described in the text for the analysis of C_8H_{18} can be applied to compounds which contain other elements in addition to carbon and hydrogen. For example, it can be used to determine the composition of compounds containing carbon, hydrogen and oxygen. But in such a case, the weight of oxygen in the original sample must be obtained by difference, that is,

Wt oxygen = Wt compound − (Wt carbon + Wt hydrogen).

Explain why the weight of oxygen cannot be determined directly from the weights of water and carbon dioxide formed.

4.9   Use the reasoning and method of Problem 4.8 to determine the simplest formula for ethyl alcohol (the drinking kind) from the following data obtained in a combustion analysis apparatus. The compound contains three elements: carbon, hydrogen, and oxygen.

Initial wt ethyl alcohol	10.0 g
Wt H_2O absorbed	11.7 g
Wt CO_2 absorbed	19.1 g

4.10 Balance the following equations:

(a) The formation of ammonia by the direct combination of nitrogen and hydrogen:

$$N_2 + H_2 \rightarrow NH_3$$

(b) The reaction of hydrochloric acid and barium hydroxide:

$$HCl + Ba(OH)_2 \rightarrow BaCl_2 + H_2O$$

(c) The generation of hydrogen by the action of nitric acid on zinc:

$$HNO_3 + Zn \rightarrow Zn(NO_3)_2 + H_2$$

(d) The oxidation of glucose (a sugar):

$$C_6H_{12}O_6 + O_2 \rightarrow CO_2 + H_2O$$

4.11 Determine how many grams of CO_2 and H_2O are formed when 1.00 g of C_8H_{18} is burned with at least 3.51 g of O_2. The relevant balanced equation is given in the body of this chapter.

4.12 The bottled gas frequently used as a fuel for stoves and heaters and occasionally employed as an automobile fuel is propane, C_3H_8.

(a) Write the balanced equation for the combustion of propane to yield carbon dioxide and water. When you have satisfied yourself that the equation is correct (check the answer appendix) try the following questions.

(b) How many moles of CO_2 are chemically equivalent to 1 mole of C_3H_8? to 0.5 mole? to 5 moles?

(c) How many moles of H_2O could be formed from burning 2 moles of C_3H_8? How many H_2O molecules would this be?

(d) Calculate the number of grams of oxygen required to react completely with 1.00 g of propane and compare your result with the corresponding calculation for octane in the text. Describe what would happen if less than that quantity of oxygen were present.

4.13 That staple of the American diet, aspirin (acetylsalicylic acid) can be made by reacting salicylic acid and acetic acid according to the equation,

$$C_7H_6O_3 \quad + \quad C_2H_4O_2 \quad \rightarrow C_9H_8O_4 + H_2O.$$
salicylic acid   acetic acid   aspirin

(a) Determine the minimum quantities (in grams) of salicyclic acid and acetic acid required to yield 100.0 g of aspirin.

(b) If, by any chance, these problems have been a bit painful, you might find it informative to calculate the number of

5 grain aspirin tablets which could be made from 100.0 g of the pure compound (1 grain = 0.0648 g).

4.14 The reduction of iron oxide (ferric oxide) with coke (carbon) can be represented by the following equation:

$$2Fe_2O_3 + 3C \rightarrow 4Fe + 3CO_2.$$

(a) Suppose the operator of a smelter mixes 1.00 ton of iron oxide with 0.50 ton of coke. How many tons of pure iron can he hope to obtain?

(b) What yield can he expect if he attempts to save money by using only 0.20 ton of coke with 1.00 ton of iron oxide?

(c) Finally, how much iron could be formed from 1.00 ton of iron oxide and 0.10 ton of coke?

(d) Explain the significance of your answers.

4.15 Earlier you were promised this opportunity to calculate the number of atoms of oxygen in 1 pound atomic weight of the element (1 lb = 454 g).

4.16 A certain type of coal contains 2.0% sulfur (by weight). When the coal is burned, the sulfur is converted to sulfur dioxide, SO_2, a dangerous air pollutant. Assuming that all the sulfur in the coal reacts in this manner, calculate the number of pounds of SO_2 formed per ton of coal burned.

4.17 The "ethyl" added to gasoline as an antiknock agent is tetraethyl lead, with the formula $Pb(C_2H_5)_4$.

(a) Calculate the percent of lead (by weight) in this compound.

(b) On the average, leaded American gasolines contain 3.8 g of tetraethyl lead per gallon. About 67% of the lead is exhausted into the atmosphere when the gasoline is burned. Use this information and the answer to part (a) to calculate the weight of lead (in kilograms) which is added to the air in one year by a car which, in that period of time, is driven 15,000 miles at an average of 18 miles per gallon.

4.18 Analysis shows that hemoglobin, the blood protein responsible for oxygen transport and exchange, contains 0.333% iron. Each molecule contains four iron atoms. From these data calculate the molecular weight of hemoglobin.

Suggested readings

Kieffer, William F., "The Mole Concept in Chemistry," Van Nostrand-Reinhold, Princeton, New Jersey, 1962.

Walker, R. A., and Johnston, H., "The Language of Chemistry," Prentice-Hall, Englewood Cliffs, New Jersey, 1967.

Additional examples and problems of the sort included in this chapter can be found in any introductory general chemistry textbook and in a number of special books on problem solving in general chemistry.

*The most general definition of beauty,
therefore, is—that I may fulfil my threat
of plaguing my readers with hard words—
Multëity in Unity.*

Samuel Taylor Coleridge (1814)

Ordering the elements: properties and periodicity

About order

Much of man's intellectual endeavor (and I use the adjective in its broadest sense) has been a search for order. A universe without order would be inconceivable; if order did not exist, man would probably have to invent it. Too often scientists have presumed this quest to be uniquely characteristic of their way of appraising the world. Such parochial myopia fails to recognize the insight of the quotation which introduces this chapter. Note that Coleridge was not speaking about science, or even specifically about poetry, but about beauty. To be sure, as Jacob Bronowski has observed, "Science is nothing else than the search to discover unity in the wild variety of nature—or more exactly, in the variety of experience."* But this search does not isolate science from the rest of man's striving for beauty and truth; it integrates it.

This chapter is primarily about such a quest, the effort to make systematic sense out of the one hundred-odd substances from which the universe is fashioned. Be-

* J. Bronowski, "Science and Human Values," p. 27, Harper Torchbooks, New York, 1956.

fore we consider the organizational efforts, however, it might be advisable to study some of the evidence. Unfortunately, even a superficial summary of the physical and chemical properties of all the elements is beyond the scope of this book. Therefore, we will be selective and concentrate upon several groups, each one consisting of elements with certain basic similarities. The approach will be the hybrid phenomenological-historical one which has by now established itself as dear to the author's heart (if not the reader's). Hopefully, it will lead to a familiarity with the properties of at least a few elements, an acknowledgment of the existence of order among these fundamental substances, an appreciation of how chemistry and chemists work, and a preparedness for the concept of elementary periodicity.

The overactive elements

The most reactive metallic elements known are members of two groups called the *alkali metals* and the *alkaline earths*. Included in the former family are lithium (Li), sodium (Na), potassium (K), rubidium (Rb), cesium (Cs), and francium (Fr). Members of the set of similar elements identified as alkaline earths are beryllium (Be), magnesium (Mg), calcium (Ca), strontium (Sr), barium (Ba), and radium (Ra). The compounds of the alkali metals and alkaline earths are among the most common of all substances; they have been known and used for thousands of years. Yet, the pure elements themselves were not isolated until early in the nineteenth century. The difficulty in obtaining the uncombined elements is, as you might expect, a direct consequence of their chemistry—the fact that they are always tightly tied up with other elements in very stable compounds. Once the method to free the metals was developed, one man, Sir Humphry Davy, isolated six of them in two years. But before we study his famous experiments, we should devote some attention to his starting materials.

Few compounds have had as much influence on the course of human history as common salt: sodium chloride, NaCl. Supplies and shortages of the substance dictated commerce and trade routes, international alliances and enmities. Among the first technologies mastered by many societies were the arts of mining salt and obtaining it by the evaporation of sea water. The frequency with which salt appears in ancient literature testifies to its ubiquity. You will recall, for example, that Lot's wife underwent a salty transmutation in punishment for her last nostalgic look at Sodom. On the other hand, the use of the substance in the Sermon on the Mount is more metaphoric: "Ye are the salt of the earth." Salt mines, those saline synonyms for hard labor, are distributed from

Krakow to Detroit. And even today, sodium chloride remains one of the most widely used raw materials of the chemical industry.

Of comparable commercial importance is sodium carbonate, Na_2CO_3. The ancients (and some moderns) called it soda and knew its properties well. It imparted a bitter taste to water and made it feel slippery—characteristics which also came to be associated with other "alkalis." Five thousand years ago, sodium carbonate had already found uses which persist to the present day: as an aid in washing and, along with sand (chiefly silica or silicon dioxide, SiO_2), as an ingredient of glass.

Often, in ancient writings, soda was confused with a very similar compound called potash. This latter substance, potassium carbonate, K_2CO_3, in modern nomenclature, was also a mild alkali. It's origin was vegetable, as distinct from the mineral sources of sodium carbonate. The name suggests the means of preparation, by leaching wood ashes with water. Because of the comparable characteristics of the two carbonates, they were sometimes used interchangeably.

On the other hand, there was little chance of confusing limestone with soda and potash, though chemically all three are closely related. Many minerals are mixtures containing considerable quantities of this compound, chemically, calcium carbonate, $CaCO_3$. Limestone was a well-known building material long before it was used in the local courthouse. And before the Parthenon, there were the mountains. Hannibal had to contend with them when he brought his elephants over the Alps in 218 B.C. Legend has it that he used chemistry to make the way easier. Strong vinegar (acetic acid, $HC_2H_3O_2$) was poured on the rocks, which reacted in a way we succinctly summarize as

$$CaCO_3 + 2HC_2H_3O_2 \rightarrow Ca(C_2H_3O_2)_2 + H_2O + CO_2.$$

Apparently, Hannibal did not carry enough vinegar to induce the Alps to fizz away into so many mole hills, but at least the rocks were said to have cracked.

The great sculpture of classical Greece was done in marble and a lamp found in the famous tomb of King Tutankhamen (ca. 1358 B.C.) is carved of calcite, another mineral form of $CaCO_3$. Furthermore, the tomb itself was plastered with a calcium compound, plaster of Paris, still made by heating gypsum, calcium sulfate dihydrate, $CaCO_4 \cdot 2H_2O$. (Incidentally, the ancients used gypsum to "plaster" wine as well as walls. The mineral helped clarify the wine, improved its color, and increased its alcoholic content by absorbing water.)

A more durable plaster was prepared when the Romans heated limestone. This reaction is particularly important because when carried out long enough it yields lime, which is calcium oxide, CaO, plus carbon dioxide:

$$CaCO_3 \rightarrow CaO + CO_2.$$

Calcium oxide is a strong alkali, a caustic substance which can badly burn skin. Indeed, lime has long been used to promote the destruction of animal tissue. Moreover, calcium oxide reacts with water to yield calcium hydroxide, $Ca(OH)_2$, and in the process a considerable quantity of heat is evolved.

$$CaO + H_2O \rightarrow Ca(OH)_2$$

Calcium oxide can also convert sodium and potassium carbonates to the corresponding oxides. This reaction, for soda, is written

$$Na_2CO_3 + CaO \rightarrow Na_2O + CaCO_3.$$

Calcium carbonate is quite insoluble in water (a property promoting the durability of mountains and courthouses). Therefore, if the above reaction is carried out in solution, the chief product remaining dissolved will be sodium oxide or, thanks to this reaction with water, sodium hydroxide, NaOH.

$$Na_2O + H_2O \rightarrow 2NaOH$$

Similar equations can be written for the corresponding potassium compounds. Both NaOH and KOH are, like $Ca(OH)_2$, strong alkalis or bases. You will see in a later chapter how such compounds form a class which is in many ways the opposite of the acids.

The elucidation of much of the chemistry discussed in the past few paragraphs and the equations they contain was the work of Joseph Black. Although Black began his career as a phlogistonist, he became a convert to Lavoisier's new French system of chemistry and contributed greatly to the critical formative years of the science. Much of Black's work was with *magnesia alba* or magnesium carbonate, $MgCO_3$, a compound with many properties similar to those of calcium carbonate. Magnesium carbonate, however, was hardly the best known compound of magnesium. That distinction must belong to the mineral contaminant of a spring in Epsom, Surrey. The water tasted so bitter that thirsty cattle refused to drink it, even in the drought of 1618. Human beings showed rather less discriminating palates; and, following the old rule that if it tastes bad it must be good for you, they turned Epsom into a fashionable spa by the middle of that century. Epsom salts, magnesium sulfate, $MgSO_4$, remains a standard item in many modern medicine cabinets. Interestingly, most of the magnesium-containing minerals are considerably softer than the corresponding calcium compounds. They include serpentine, soapstone, asbestos, and meerschaum, that rich and rare pipe dream.

To complete the cast of compounds, we must add minerals from two rather widely separated sources, Bologna in Italy and Strontian in Scotland. Ultimately, the former yielded barium, the latter, strontium.

We are at last almost ready to consider the process used by Sir Humphry Davy to isolate these two elements and sodium, potassium, calcium, and magnesium from some of the raw materials just described. Your position is, of course, a little unfair. You already know more than Davy did at the outset of his experiments. Nevertheless, Davy did suspect that potash and soda must be compounds containing elements which had not yet been isolated.

The method the young Cornishman chose was to apply that exciting new source of energy—electricity. It is more than just a bad pun to say that Benjamin Franklin's kite flying sparked a good deal of interest in electricity. Indeed, to a modern reader, it may be difficult to conceive of the great popular appeal of this and other scientific discoveries of the late eighteenth and early nineteenth centuries. Davy himself was, at the time of the experiments we are about to recount, lecturer in chemistry and director of the laboratory at the Royal Institution in London. In this capacity, he often performed lecture-demonstrations, if not for the masses of that city, certainly for packed houses of the intellectually and socially elite. This was also a period during which the newly discovered gases provoked considerable inquisitive attention, and a favorite and fashionable way of turning on was at "laughing gas" parties. The results of an excess of nitrous oxide, N_2O, were sometimes disastrous, as is attested to by the print shown in Figure 5.1.

Davy's own investigations of gases were begun in Bristol at Dr. Thomas Beddoes' Pneumatic Institution. It was while Davy was serving as superintendent of this research center for the study of the therapeutic value of gases that he came to know the poets Coleridge, William Wordsworth, and Robert Southey. Whether the chemical knowledge of these three literary lights ever approached their fascination with the science is not clear. But in a letter to Davy, dated February 3, 1801, Coleridge asks for assistance in a scheme to study chemistry, along with Wordsworth and a mutual friend, William Calvert. Specifically, the poet requests a list of books, directions for setting up a laboratory, help in acquiring apparatus, and "your advice how to *begin.*" "You know how long, how ardently I have wished to initiate myself in Chemical science." The author of this text has not discovered if the project ever materialized. At any rate, in another (earlier) letter, Coleridge does say of himself, "I am a so-so chemist, & I love chemistry."

Davy could have similarly said, "I am a so-so poet, & I love

FIGURE 5.1. A "laughing gas" demonstration at the Royal Institution. Humphry Davy is behind the bench with a bellows in his hand. (Engraving by Gillray.) (Reproduced from William A. Tilden, "Famous Chemists, The Men and Their Work," facing p. 88, **Dutton**, New York, 1921.)

poetry." And, whether or not Davy was able to do the scientific favor asked of him, he was of unquestioned literary assistance to Wordsworth and Coleridge. The chemist corrected punctuation, proofread, and generally shepherded the second edition of Wordsworth's "Lyrical Ballads" through the press. But such contributions are all but forgotten because of the magnitude of Davy's achievements in experimental science.

Sir Humphry's first attempts to electrically break down potash and soda into their constituent elements failed. He electrified aqueous solutions of the compounds and succeeded only in decomposing the water to hydrogen and oxygen, a reaction he had already carefully studied. Then, on October 6, 1807, he changed his plan of attack and placed a small lump of moist potash in contact with the platinum electrodes from his battery. In Davy's own words:

*Under these circumstances a vivid action was soon
observed to take place. The potash began to fuse at both
its points of electrization. There was a violent effervescence
at the upper surface; at the lower, or negative, surface,
there was no liberation of elastic fluid; but small globules
having a high metallic lustre, and being precisely similar
in visible characters to quicksilver, appeared, some of
which burnt with explosion and bright flame, as soon as
they were formed, and others remained, and were merely
tarnished, and finally covered by a white film which
formed on their surfaces. These globules, numerous
experiments soon shewed to be the substance I was in
search of, and a peculiar inflammable principle the basis
of potash.**

Edmund Davy, who was serving as his cousin's assistant at the time, wrote that when Humphry saw the globules of the metal burst into flame, "he could not contain his joy—he actually bounded about the room in ecstatic delight; and some little time was required for him to compose himself sufficiently to continue the experiment."

Davy named the new element potassium, and several days later successfully applied the same process to soda. A similar reaction occurred: globules of a bright, lustrous metal, this time sodium, appeared at the cathode, or negative electrode, while oxygen gas

* Humphry Davy, On Some New Phenomena of Chemical Changes
Produced by Electricity . . . , *Philosophical Transactions* 98, 1ff.
(1808). Also reprinted in "A Source Book in Chemistry, 1400–1900"
by Henry M. Leicester and Herbert S. Klickstein, pp. 244–246,
Harvard Univ. Press, Cambridge, Massachusetts, 1952.

FIGURE 5.2. *Electrolysis of potassium oxide to produce potassium and oxygen.*

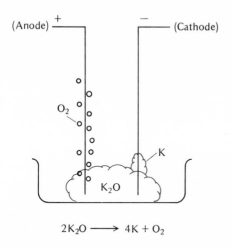

$$2K_2O \longrightarrow 4K + O_2$$

was evolved at the positive anode. Most likely, Davy was working not with the pure carbonates of potassium and sodium, but with the oxides or hydroxides of these elements. If so, the reactions which so delighted him can be represented by the following equations:

$$2M_2O \rightarrow 4M + O_2$$

or

$$2M(OH)_2 \rightarrow 2M + O_2 + 2H_2O,$$

where M is K or Na.

Davy soon demonstrated that potassium and sodium share many common properties. In their reaction with oxygen, described in the above quotation, these two elements dramatically illustrate the fundamental identity between calcination and combustion. Both processes, you will recall, are oxidations. And here are two metals which sometimes "rust" so vigorously that they evolve the light and heat associated with burning.

$$4M + O_2 \rightarrow 2M_2O$$

It is also possible to burn less reactive metals like iron, but this requires very high temperatures and/or supplies of pure oxygen. But potassium and sodium react readily with the oxygen of the air to form the corresponding oxide, the "white film" which Davy observed. This behavior explains, in part, why no unreacted samples of these elements are found in nature. The metals are just too reactive. In fact, sodium and potassium, still isolated electrolytically, must be protected from the air by a layer of kerosene or other hydrocarbon.

Water would definitely not do as a protective liquid because both elements react violently with H_2O:

$$2M + 2H_2O \rightarrow 2MOH + H_2.$$

The hydrogen which is evolved is highly flammable, and there is enough heat generated in the reaction that the gas is often explosively ignited. Clearly, this property is another example of the extreme reactivity of these alkali metals. The name is appropriate, since the metals are prepared from alkalis and readily form alkalis (e.g., K_2O and KOH) in their reactions with air and water.

Flushed with his success on the alkali metals, Sir Humphry turned his attention and his Voltaic piles to the alkaline earths: lime, magnesia, baryta, and strontia. Lavoisier had included the first three in his list of 33 elements. Nevertheless, he states elsewhere: "It is probable that we know only part of the metallic substances which exist in Nature. . . . It is possible, simply speaking, that all substances which we call earths (and Lavoisier specifically mentions baryta) may be simple metallic oxides irreducible by the methods we employ."

The methods Lavoisier was referring to were chiefly heating with carbon. We encountered this process as smelting early in the second chapter, but now would seem an appropriate place to write an equation,

$$2CuO + C \rightarrow 2Cu + CO_2.$$

The above expression signifies the reduction of copper oxide, the conversion of the calx of copper to the metallic element by the action of a reducing agent, carbon. You will recall that mercuric oxide, HgO, can be reduced simply by heating it.

$$2HgO \rightarrow 2Hg + O_2$$

However, copper and iron exhibit stronger affinity for oxygen and require the presence of a reducing agent like carbon to effect the comparable change. The oxygen of the alkaline earths is still more tightly bound and consequently more extreme methods are required. Again electricity provided the necessary means.

Davy's notebooks describe a number of unsuccessful efforts before he finally arrived at a procedure involving the electrolysis of a mixture of moist lime and mercuric oxide. The product was an amalgam, a solution of the new metal in mercury. He then distilled off the mercury to obtain, for the first time, a sample of a silvery white metal he named calcium after the Latin word for lime, *calcis*. The reaction brought about by the action of the applied electric field was

$$2CaO \rightarrow 2Ca + O_2.$$

Similar procedures yielded, in the same year of 1808, barium from baryta, strontium from strontia, and magnesium from magnesia.

All of these alkaline earth metals demonstrate the same high reactivity toward oxygen and water already reported for potassium and sodium. Of the four, magnesium is the least reactive, and samples of the pure metal can exist in air without spontaneously burning. Consequently, magnesium can be used as a material for construction, usually mixed with other metals to form lightweight alloys. The element does burn with a brilliant white light when heated in air, a property which leads to its use in flashbulbs.

One significant difference between the alkali metals and the alkaline earth metals is implicit in the equations you have been reading. Sodium and potassium form oxides with the general formula M_2O. In contrast, the corresponding compounds of magnesium, calcium, barium, and strontium have the formulas MO. This difference in atomic ratios proved to be useful in the classification of these elements into two different families.

For completeness sake, it should be noted that, after Davy's pioneering research, four more alkali metals and two more alkaline earth elements were isolated. The former include lithium, which was prepared by Sir Humphry in 1818; rubidium and cesium, both the discoveries of Robert Bunsen of burner fame; and francium, an element found in 1939 as a product of radioactive decay. The other alkaline earth elements are beryllium, a component of beryl and emeralds, and radium. It was the chemical similarity of the latter element to calcium and barium which aided the Curies in their isolation of that highly radioactive substance. We will return to their romantic story in the next chapter. At this moment, we must consider a very different set of elements, the rare gases.

The new, lazy, hidden strangers

The great reactivity of the alkali metals and the alkaline earths kept them locked in some of the most common compounds for centuries until Sir Humphry Davy and electricity provided the means of liberating them. Quite the opposite holds true for the *rare* or *noble gases:* helium (He), neon (Ne), argon (Ar), krypton (Kr), xenon (Xe), and radon (Rn). Their extreme resistance to reaction, coupled with their low natural abundance, kept these elements hidden until the turn of this century.

As seems so often the case in science, the facts were faintly glimpsed long before they were understood. In 1785, Henry Cavendish passed electric sparks through air mixed with extra oxygen in a closed glass tube. By this means, he caused all the nitrogen in the tube to combine with oxygen to form nitrogen dioxide, NO_2. This gas and the small quantity of carbon dioxide originally present in the air were chemically absorbed, a process which should have removed all the then-known gases. Nevertheless, a small portion of the air remained unreacted. Cavendish estimated that the residue was "certainly not more than 1/120 of the bulk of the phlogisticated air (nitrogen) let up into the tube." If the conditions of the experiment were accurate, this gas could not be nitrogen, oxygen, nitrogen dioxide, or carbon dioxide. It was, in fact, a mixture of the rare gas elements found in all samples of common air.

Cavendish did not pursue his discovery, and it went largely unnoticed for a century. Then, appropriately enough at the Cavendish Laboratory of Cambridge University, it suddenly acquired new significance. The Professor of Physics at that prestigious establishment, John William Strutt, Lord Rayleigh, had set out to determine whether or not the ratio of the atomic weight of oxygen to that of hydrogen is exactly 16 to 1. His method was to make very careful measurements of gas *density*, that is, *weight per unit volume*. Avogadro's Principle—by 1882 so widely accepted that it was really more than a hypothesis—predicted that equal volumes of gases contain equal numbers of molecules under identical conditions of temperature and pressure. Therefore, comparing the masses of some convenient volume, say 1 liter of each gas, should be equivalent to comparing the masses of individual molecules. It follows that the ratio of the densities of oxygen and hydrogen should be equal to the ratio of the molecular weights of these elements. Moreover, since both oxygen and hydrogen exist in diatomic molecules, their atomic weights should also stand in the same ratio as their densities.

In the course of time Rayleigh found his answer: The oxygen-to-hydrogen atomic weight ratio is, in fact, slightly smaller than 16 to 1, 15.882 to 1 to be exact. But, because he was a painstaking experimenter who kept his eyes open for serendipitous surprises, he found something else, too. In conjunction with this study, he also determined the densities of several other gases, including nitrogen. And here Lord Rayleigh made a puzzling observation. The nitrogen prepared chemically from ammonia, NH_3, and other nitrogen-containing compounds was always less dense than nitrogen prepared from air by absorbing out the oxygen, carbon dioxide, and water vapor. To be sure, the difference was not great, about 5 parts in 1000, but nevertheless, it was real and reproducible.

Lord Rayleigh reported the results in the famous British journal, *Nature*, and asked for explanations. None were offered and the

Professor prepared to test some of his own hypotheses. The experiment seemed to suggest that the molecular weight of "natural" nitrogen was somehow greater than that of "chemical" nitrogen. Certainly, the two samples could not exhibit different densities and still be identical. As a matter of fact, even within a single sample of gas, there might be molecules of differing weights. And Strutt knew perfectly well that when he compared densities, he might be comparing *average* molecular weights. Although each gas was supposedly pure nitrogen, there was no guarantee that every molecule in a sample was just like every other. One (or both) of the gases could be a mixture of substances, each with its own molecular weight. Either the sample obtained by chemical means was contaminated with a gas having a molecular weight lower than that of nitrogen, or the atmospheric sample contained some molecules of a higher weight. There was no way of knowing which without additional experimentation.

Specifically, Rayleigh conjectured that the chemically produced nitrogen might contain some hydrogen. Since the approximate molecular weights of the two elements are 28 and 2, respectively, the net average molecular weight of such a mixture would be less than 28. However, no trace of hydrogen could be found. This negative result suggested a heavyweight contaminant in atmospheric nitrogen. Rayleigh tested his hypothesis that this substance might be oxygen (molecular weight about 32) and again found none.

It was William Ramsay of University College, Bristol, who finally solved the riddle in August of 1894. His experimental method consisted of passing atmospheric nitrogen back and forth over red-hot magnesium. Such treatment results in the formation of magnesium nitride, Mg_3N_2, according to the equation

$$3Mg + N_2 \rightarrow Mg_3N_2.$$

After prolonged exposure, no further decrease in volume was observed. A portion of gas remained unreacted, a gas considerably denser than nitrogen. At first Rayleigh and Ramsay believed it to be a form of nitrogen with a higher-than-average molecular weight but with lower-than-average reactivity. Specifically, they thought it might consist of N_3 molecules rather than the normal N_2 species.

In putting this explanation to an experimental test, the British scientists employed a very important technique called *spectroscopy*. Chapters 6 and 7 will provide a good deal more information about this method and its meaning. Right now, the point to grasp is that if an element gets excited enough, it gives off light. Moreover, each energized element always emits a unique set of colors called its *spectrum*. For example, the light coming from hot vapors of sodium is always yellow. And the spectrum of electrically or thermally excited hydrogen contains red, green, and blue components or "lines." It makes no difference whether the element is in pure form or combined with others in a compound; the colors remain constant and characteristic. Furthermore, they can be measured with great precision in an instrument called a *spectroscope, spectrograph,* or *spectrometer.* In brief, the spectrum of an element is sort of a technicolor fingerprint which provides a means of unambiguous identification.

For this reason, Rayleigh and Ramsay probably expected to see only the spectrum of nitrogen when they placed a sample of the newly isolated gas in their spectroscope. Instead they saw groups of red and green lines which had never before been observed for any other substance. Here was proof that a new element had been found! Efforts were at once concentrated upon obtaining enough of it in pure form to permit chemical and physical characterization. But the element said very little for itself, except to state its molecular and atomic weight as 39.95. In fact, the gas is so indifferent to all chemical reactions that Ramsay and Rayleigh commemorated its laziness in the name argon (from the Greek *argos,* lazy).

The co-discoverers of argon worked in a scientific society whose elements had been organized by Mendeleev's periodic system. You will see in the following section that one of the main messages of this comprehensive scheme is that the properties of the elements tend to repeat themselves. Elements come in families, the members of which exhibit similar characteristics. Therefore, Rayleigh and Ramsay immediately concluded that their new element must be one of at least six unreactive gases. And so, the search for the missing five began at once.

Ironically, one of them had been discovered a quarter of a century earlier, not in an earthly laboratory, but in a solar prominence. The story began in 1868 in India where Jules Janssen, a French astronomer, had gone to observe a total eclipse of the sun. For the first time in history, he focused a spectroscope on the chromosphere, the luminescent region at the sun's surface. He was rewarded by the familiar spectroscopic lines characteristic of excited hydrogen. But in addition, Janssen saw a new yellow color—one that did not coincide with the spectrum of any known element. It required courage, confidence, and chutzpah, but the English astronomer Sir Norman Lockyer not only regarded the elementary origin of the new line as real, he even gave it a name, helium. Perhaps an earthbound chemist would not have dared to thus aspire to the stars.

For 25 years helium remained, at best, a hypothetical heleotropic element. Then, in 1888, an American mineralogist, William Hillebrand, noticed that an inert gas was evolved when he treated the mineral uranite with sulfuric acid. Unfortunately, Hillebrand did not pursue the investigation. However, when Sir William Ramsay read the American's paper after his own discovery of argon, he repeated the experiment with a related uranium mineral. Ramsay collected a sample of the gas and sent it, by happy co-incidence, to Sir Norman Lockyer for spectroscopic analysis. Enraptured, Lockyer wrote, "The glorious yellow effulgence of the capillary, while the current was passing, was a sight to see." His daring prediction had been vindicated. Helium had finally made its earthly debut!

Almost simultaneously, a sample of helium was isolated in Sweden, and searches for the element were undertaken in the natural gases of Italy. But the first major terrestrial source of the element was initially regarded as a disappointment by the citizens of Dexter, Kansas. Sometime in the spring of 1903 these good people gathered for the dedication of the new natural gas well which they hoped would mean prosperity for their little commu-nity. No doubt they were filled with anticipation as the torch was brought near the jet for the lighting ceremony. But their eager expectation was just as surely replaced by embarrased consterna-tion when the jet blew out the torch! But all's well that ends well, and this well probably made more money producing helium than it ever would have issuing natural gas. Certainly, it is just the prop-erty of incombustibility which, coupled with low density, makes helium an ideal replacement for dangerously flammable hydrogen in lighter-than-air craft. Had it been used more widely, the spec-tacular Hindenberg disaster of 1937 would have been averted.

Morris William Travers collaborated with Ramsay in the dis-covery of three more noble gases. All of these were obtained by the fractional distillation of liquefied air. The first isolated—on May 30, 1898—was named krypton, meaning hidden.* The atomic weight of the new element was found to be 83.80.

Ramsay and Travers had expected a lighter gas and less than a month later they found it. A fraction was isolated and a spectrum was run in an electric discharge tube. The enthusiastic words of Travers eloquently refute those who maintain that science is a cold and passionless pursuit.

The blaze of crimson light from the tube told its own story, and it was a sight to dwell upon and never to forget. It was worth the struggle of the previous two years; and all the

---

* Several students, collectors of memorable trivia, have identified Krypton as the planet which produced Earth's most famous extra-gallactic guest, Superman. As far as the author has been able to ascer-tain, the rare gas in question was named well before mild-mannered Clark Kent first changed clothes in a phone booth.

difficulties yet to be overcome before the research was
finished. The undiscovered gas had come to light in a
manner which was no less than dramatic. For the moment,
the actual spectrum of the gas did not matter in the least,
for nothing in the world gave a glow such as we had seen.*

You have seen that glow, that "blaze of crimson light," hundreds of times, and possibly with a good deal less joy than it engendered in its discoverer. Travers was writing about neon.

Hot on the heels of neon (named after the Greek word for new), atomic weight 20.18, came a much heavier rare gas. This element, which gives a blue glow in a vacuum tube, was named xenon, "the stranger." Its atomic weight was found to be 131.30. A still heavier element of this family, radon, remained undetected until 1900, when it was found as a radioactive decay product of radium.

We have come full circle; radium is where the last section ended. But before moving on, you should know that the inert gases are not completely inert after all. The big scientific news story of 1962 was the preparation of the first compounds containing xenon. The partner of this reticent stranger was the most reactive of all elements, fluorine (F).

Fervid Fluorine, though just Nine,
Knows her aim in life: combine!
In fact, of things that like to mingle,
None's less likely to stay single.†

So seductive is fluorine that even the stern and stalwart stuff of the rare gases finally succumbed to her electric charms. The fruits of these unions now include XeF_2, XeF_4, XeF_6, KrF_2, KrF_4, and RnF_4. Oxides of xenon have also been prepared.

The Siberian and his system

The two accounts of chemical sleuthing which you have just read differ in one significant respect. Davy's isolation of the alkali and alkaline earth metals was done before the major patterns of order among the elements had been discerned; the rare gases were discovered after that watershed. The former findings were instrumental in the establishment of the periodic table of the elements. This organizational scheme, in turn, informed the search

* M. W. Travers, "The Discovery of the Rare Gases," pp. 95–97, Arnold, London, 1928.
† This quatrain is one of a very clever series from "Adam's Atoms: Making Light of the Elements," by Vernon Newton: Copyright © 1965 by Vernon C. J. Newton, Jr. All rights reserved. (Reprinted by permission of The Viking Press, Inc., New York.)

for the noble gases. Between the literal tabular extremes of the alkali metals and the rare gases lies the unifying concept of *elementary periodicity.*

The first tentative steps toward this great generalization were limited in scope. Around 1829, Johann Wolfgang Döbereiner noted that certain similar elements can be grouped in triads. When the members of any one of these elementary triumverates are arranged in order of increasing atomic weight, many properties of the middle element lie approximately midway between the corresponding properties of the other two. Indeed, the atomic weight of the middle element is close to the mean of the atomic weights of the flanking substances. The selected examples in Table 5.1 indicate that although the concept provides a rough approximation of property values, it is by no means exact.

TABLE 5.1. *Selected Properties of an Elementary Triad*

| | Calcium | Strontium | | Barium |
		Mean[a]	Measured	
Atomic wt.	40.0	88.7	87.6	137.3
Density (g/cm^3)	1.6	2.6	2.6	3.5
Melting point (°C)	850	777	770	704
Boiling point (°C)	1490	1564	1370	1638

In 1866, J. A. R. Newlands reported a more general scheme to the Chemical Society of London. He arranged the then-known elements in order of increasing atomic weights and assigned each of the elements in this sequence a number. Thus, the first seven were

1	2	3	4	5	6	7
H	Li	Be	B	C	N	O

Next in order came fluorine, and Newlands argued that this corrosive, gaseous element was enough like hydrogen in its properties to justify grouping the two together. Certainly, sodium, element number 9 in his system, belonged with lithium; and magnesium, number 10, was similar to beryllium. Reasoning in this fashion, Newlands set down a second series of seven elements:

8	9	10	11	12	13	14
F	Na	Mg	Al	Si	P	S

There were six more of these seven-membered series in New-lands' arrangement. Potassium, rubidium, and cesium appeared in the same group with lithium and sodium, and the alkaline earth elements fell in line. Fluorine, chlorine (Cl), bromine (Br), and iodine (I), all similar nonmetals, could conveniently be classed to-gether. In most cases, Newlands could justifiably claim that "the eighth element starting from a given one is kind of a repetition of the first. This peculiar relationship," he wrote, "I propose to provisionally term the *Law of Octaves.*"

The term was an unfortunate one. Most likely Newlands meant the reference to the musical scale as no more than an analogy. But to the members of the Chemical Society, the phrase must have smacked of numerology or supposed celestial harmony. At any rate, one auditor waggishly inquired whether Newlands had thought of examining the elements in order of their initial letters. Other reactions were equally skeptical and the paper was rejected by the *Journal of the Chemical Society.* It was not until 1887, well after the work of Mendeleev and Lothar Meyer, that Newlands finally received belated recognition for his pioneering work.

Granting that the episode does provide an unfortunate example of chemical conservatism, it should also be noted that Newlands' system of elementary organization had a number of major short-comings. He had failed to leave blank spaces for undiscovered elements when they seemed to be indicated; and on the basis of their properties, some elements were clearly misplaced. Later schemes were able to avoid these errors.

The honor of enunciating the most successful of these schemes is shared by a German and a Russian. Although thus forever linked in the history of science, it would be difficult to find two chemists more different than Julius Lothar Meyer and Dmitri Ivanovich Mendeleev. The former, a son of a physician, spent part of his youth on the grounds of the summer palace of the Grand Duke of Oldenburg. After earning a medical degree, Meyer did further study in chemistry, physics, and mathematics. Most of his career was spent in academe and he ultimately rose to Professor of Chem-istry and Rector of the University of Tübingen. It is by no means a disparaging reflection upon the unquestioned abilities of Lothar Meyer to observe that his life was a success story well within the comfortable confines of the bourgeois establishment. The fact that Mendeleev did not fit this familiar mold makes him a more in-teresting character, if not a better scientist, and justifies a few paragraphs of biography.

To begin with, Dmitri Ivanovich Mendeleev was born in Siberia, of Russian and Mongolian ancestry, the youngest child of a family of at least 14. When his father, director of a high school, went blind, Maria Kornileva Mendeleeva established and ran a glass works to support her numerous offspring. But after about ten years of operation, the factory burned down and the elder Mendeleev died of tuberculosis. At this time Dmitri was 16 and had just completed his secondary education. His excellent mind clearly called for further training; and so his remarkable mother obtained horses and set off, with Dmitri and his sister, for Moscow, over a thousand miles away.

But Maria Mendeleeva's efforts to enroll her son in the University at Moscow proved unsuccessful. Histories state that she lacked sufficient political influence. The fact that her son-in-law (Dmitri's first science teacher) was an exiled revolutionary certainly could not have helped matters. Nor, perhaps, did young Mendeleev's disdain of Latin and the classical curriculum. Finally, aided by a friend, Dmitri Ivanovich was admitted, with financial aid, to the Central Pedagogic Institute of St. Petersburg. A few months later his mother died, ending a life which could have formed a plot for Dostoevsky.

At the Institute, Mendeleev performed brilliantly in chemistry, mathematics, and physics. After several years of teaching in secondary schools he was permitted to continue his study and research in Paris and Heidelberg. Finally, in 1861, he returned to St. Petersburg, obtained his doctorate, and soon after was appointed Professor of Chemistry at the Technological Institute there. Several years later he became a professor at the University of St. Petersburg and spent most of his subsequent career in that capacity.

Even in the loose and liberal academic atmosphere of the 1970s, Dmitri Mendeleev might provoke a few stares were he to walk across a contemporary campus. The lower half of his large face was covered with a great untrimmed beard which appears, from photographs, to have had the consistency of shredded wheat. His wiry hair hung down to his shoulders. It was cut once a year, in the spring, and even an audience with Czar Alexander III was not enough to induce him to depart from that hirsute habit.

For Czarist Russia of a century ago, Mendeleev's opinions were probably as unorthodox as his appearance—and potentially a good deal more dangerous. He was an outspoken advocate of educational reform and called for making the curriculum more relevant to the needs of society. No doubt because of these ideas and his popularity as an excellent teacher, Mendeleev, in 1890, agreed to present a student petition to the Minister of Education. The response was a sharp reminder that such meddling was not suitable conduct for a professor of chemistry. As a consequence of the controversy which followed, Mendeleev resigned his chair

FIGURE 5.3.   *A photographic portrait of Dmitri Mendeleev.*
(Reproduced from Mary Elvira Weeks and Henry M. Leicester,
"Discovery of the Elements," 7th ed., p. 642, Journal of
Chemical Education, Easton, Pennsylvania, 1968.)

and assumed the directorship of the Bureau of Weights and Measures. In retrospect, it seems likely that the specific incident which
precipitated the resignation was a pretext to get rid of a trouble-
maker. With characteristic candor and courage, Mendeleev often
identified himself with causes unpopular with officialdom. For example, he was a supporter, though not without some reservations,
of the feminist movement. On other political issues he considered
himself a peaceful evolutionist. "I am not afraid," he wrote, "of
the admission of foreign, even of socialistic ideas into Russia, because I have faith in the Russian people." Reading these words one
suspects that had it not been for Mendeleev's great international
scientific reputation and prestige, the authorities would not have
tolerated his personal, pedogogical, and political peculiarities as
long as they did.
    The chief source of that reputation, though Mendeleev's con-

tributions to chemistry were manifold, was his periodic system of the elements. This great scheme grew out of its author's efforts to organize the chapters of his famous textbook, "Principles of Chemistry." His method was to group similar elements together and discuss their properties. Early in 1869 he was working on chapters covering the alkali metals and the *halogens*. You have already studied the former elements in some detail; the latter are almost as well known. They include fluorine, that hyperactive greenish yellow gas; chlorine, the greenish yellow gas used to disinfect water; bromine, a corrosive brown liquid with a choking odor; and iodine, the violet crystals whose most familiar use is in an alcoholic solution as a treatment for wounds. The halogens are all nonmetals and all react readily with alkali metals to form compounds of the formula MX, where M represents a metal and X a halogen. Sodium chloride, NaCl, is, of course, the most common example of an alkali metal halide.

February 17, 1869, found Mendeleev at his desk, arranging slips of paper, each of which carried the name of an element and its approximate atomic weight. One of the sequences he tried was this:

$$F\ (19)\quad Cl(35.5)\quad Br\ (80)\quad I\ (127)$$
$$Li(7)\quad Na(23)\quad K\ (39)\quad Rb(88.4)\quad Cs(133)$$

Immediately he saw a pattern. The increases in atomic weights were roughly the same in both families of elements. *There appeared to be a regular, periodic repetition of elementary properties which recurred as a function of atomic weight.* Moreover, there were definite trends within each group. As the atomic weight increased, the melting and boiling points of the halogens also increased. (Recall the transition from gaseous fluorine to solid iodine.) Also, the members of each of the two families became somewhat less reactive with increasing atomic weight.

Mendeleev wasted no time in generalizing these observations into a comprehensive system for all the then-known elements. Less than a month after his discovery, his paper on "The Relation of the Properties to the Atomic Weights of the Elements" was read. The communication emphasized the repeating periodicity of properties when elements are arranged in order of increasing atomic weight. And the regularities evident in this organizational scheme seemed to justify Mendeleev's conclusion that *"the size of the atomic weight determines the nature of the elements."* The fact that some 40 years later it was demonstrated that the atomic weight is not the unique identity-determining property of an element should in no way detract from Mendeleev's contribution.

TABLE 5.2.   Mendeleev's Periodic Table (from Annalen der Chemie, supplemental vol. 8, 1872). Note the blanks left for undiscovered elements. (Reproduced from William F. Kieffer, "Chemistry: A Cultural Approach," p. 30, Harper, New York, 1971.)

Tabelle II.

Reihen	Gruppe I. — R²O	Gruppe II. — RO	Gruppe III. — R²O³	Gruppe IV. RH⁴ RO²	Gruppe V. RH³ R²O⁵	Gruppe VI. RH² RO³	Gruppe VII. RH R²O⁷	Gruppe VIII. — RO⁴
1	H = 1							
2	Li = 7	Be = 9,4	B = 11	C = 12	N = 14	O = 16	F = 19	
3	Na = 23	Mg = 24	Al = 27,3	Si = 28	P = 31	S = 32	Cl = 35,5	
4	K = 39	Ca = 40	— = 44	Ti = 48	V = 51	Cr = 52	Mn = 55	Fe = 56, Co = 59, Ni = 59, Cu = 63.
5	(Cu = 63)	Zn = 65	— = 68	— = 72	As = 75	Se = 78	Br = 80	
6	Rb = 85	Sr = 87	?Yt = 88	Zr = 90	Nb = 94	Mo = 96	— = 100	Ru = 104, Rh = 104, Pd = 106, Ag = 108.
7	(Ag = 108)	Cd = 112	In = 113	Sn = 118	Sb = 122	Te = 125	J = 127	
8	Cs = 133	Ba = 137	?Di = 138	?Ce = 140	—	—	—	— — — —
9	(—)	—	—	—	—	—	—	
10	—	—	?Er = 178	?La = 180	Ta = 182	W = 184	—	Os = 195, Ir = 197, Pt = 198, Au = 199.
11	(Au = 199)	Hg = 200	Tl = 204	Pb = 207	Bi = 208	—	—	
12	—	—	—	Th = 231	—	U = 240	—	— — — —

The basic form of the *periodic table* of 1869 is still evident in Table 5.4 on pp. 110–111. The chief differences between the two lie in the number of elements included. Mendeleev terminated his table with uranium (U). It was the heaviest element he knew of and, indeed, is today regarded as the heaviest element which occurs naturally in any significant abundance. The manmade transuranium elements are the strange offspring of the atomic age. Obviously, the rare gases were also missing from Mendeleev's list. Because they form a complete family, no member of which was known in Mendeleev's day, there was no *a priori* reason to postulate their existence. On the other hand, Mendeleev did have sound scientific reason to leave gaps within his table for other undiscovered elements.

When the elements known a century ago were ranked in order of increasing atomic weight, arsenic (As), with an atomic weight of 74.9, followed zinc (Zn), atomic weight 65.4. Had Mendeleev slavishly followed his system, he would have placed arsenic immediately after zinc and right under aluminum. But the properties of arsenic and aluminum are not at all alike. Arsenic has much more in common with phosphorus, and therefore, Mendeleev was justified in grouping these elements in the same column. Moreover, this placement put the next two elements in order of atomic weight, selenium (Se) and bromine, directly below sulfur and chlorine, respectively. On the basis of their chemical and physical behavior, this is clearly where they belong.

All this was to the good, but a two-element gap now remained between zinc and arsenic.

Al (27.0) Si (28.1) P (31.0) S (32.1) Cl(35.5)
Zn (65.4) —————— —————— As (74.9) Se (79.0) Br(79.0)
Cd(112.4) In(114.8) Sn(118.7) Sb(121.8) Te(127.6) I (126.9)

Mendeleev had no recourse but to conclude that lost sheep were missing from two fine old flocks. In one case, the absent element must have properties (including atomic weight) somewhere intermediate between the corresponding characteristics of aluminum (Al) and indium (In). This, to use Mendeleev's temporary title for the undiscovered element, was "ekaaluminum." The other, tentatively designated "ekasilicon," lay somewhere between silicon (Si) and tin (Sn). In each instance, Mendeleev argued, family resemblances would aid in identifying the missing member. And so, with supreme confidence, he proceeded to predict the properties of ekasilicon and ekaaluminum. From what he knew of silicon and tin, he drew up the list of characteristics for ekasilicon, reproduced in Table 5.3. The agreement between these predictions and the properties actually observed after the discovery of germanium (Ge) in 1885 is astounding. And similar, albeit somewhat less spectacular, successes were recorded for ekaaluminum, which proved to be gallium (Ga), and ekaboron, which was later identified as scandium (Sc). The accuracy of the prognostications was undisputed proof of the power of the periodic system of the elements.

In fact, it is probably because of these successful predictions that Mendeleev's name is more closely associated with the periodic table than that of Lothar Meyer. The latter's classification was based on curves of a variety of elementary properties—atomic volumes, melting and boiling points, malleability, brittleness, and so on—plotted versus atomic weight. The periodic fluctuation of these properties with increasing atomic weight is obvious from the curves; and Meyer's table of the elements, which appeared in

TABLE 5.3. *Predicted Properties of Ekasilicon and Observed Properties of Germanium*

	Ekasilicon (Es)	Germanium (Ge)
Atomic weight	72	72.6
Color of element	gray	gray
Density of element (g/cm^3)	5.5	5.36
Formula of oxide	EsO$_2$	GeO$_2$
Density of oxide (g/cm^3)	4.7	4.703
Formula of chloride	EsCl$_4$	GeCl$_4$
Density of chloride (g/cm^3)	1.9	1.887
Boiling point of chloride (°C)	under 100	86

1870, was fundamentally the same as Mendeleev's. Nevertheless, by his own admission, Lothar Meyer "did not have the boldness for such further assumptions as Mendeleev later correctly proposed." Perhaps the clue to the differences in risk-taking exhibited by these two men is not unrelated to the differences between a ducal summer palace and the Siberian steppes. Certainly, Mendeleev's actions were fully consistent with his philosophy of science:

Each law of natural science is of particular value scientifically only when it is possible to draw from it practical consequences, if I may so express it; that is, such logical conclusions as explain what had not previously been explained; which show phenomena not known up to its time; and especially, which permit making predictions that can be confirmed by experiment. Then the value of the law becomes evident and it is possible to test its truth.*

Tabular trends

It would be a bit immodest and highly imprudent to undertake, in this final section, a complete survey of the one hundred-odd inhabitants of the periodic table. That is the aim of all of chemistry. However, a brief discussion of trends within the periodic table should help make you more familiar with the stuff of this science and prepare you for your future study of the principles of atomic structure which underlie elementary periodicity. Remember that Mendeleev and Meyer knew nothing of these latter concepts. To scientists of their generation, atoms were still solid, indivisible, and indestructable particles with no fine structure. Nevertheless, by studying solely the properties of the elements and their compounds, these men were able to develop, independently, a classification scheme essentially identical to that used today.

The modern periodic table, reproduced as Table 5.4, begins with a short horizontal *row* or *period* of only two elements, hydrogen and helium. Hydrogen defies easy classification and sometimes is grouped with the alkali metals, sometimes with the halogens, and sometimes with both. Helium is clearly an inert, rare gas and hence heads the *column* (*group* or *family*) which also includes neon, argon, and other related elements.

Lithium, the third element in order of increasing atomic weight

* D. I. Mendeleev, quoted in Yu. I. Solov'ev, Chemistry, Chapter 5 of "History of Science in Russia," Vol. II, pp. 508–509, Academy of Sciences Press, Moscow, 1960.

starts a new period. It is immediately recognizable as one of the alkali metals, a group designated 1A. Beryllium is an alkaline earth element which is important in a number of alloys. Boron, the fifth element, heads a column usually marked 3A. This element does not show the high luster, malleability, and electrical and thermal conductivity associated with metals. It is often classified as a *metalloid*, an element with properties intermediate between the metallic and nonmetallic. Carbon is decidedly nonmetallic in most of its properties and exists in two forms or *allotropes*. Both of them are familiar, but one form, graphite, is considerably commoner than the other, diamond. The affluent allotrope and its lowly counterpart are both exclusively made up of identical carbon atoms. But, like the allotropes of any element, they differ in the arrangement of these atoms. Chapter 9 will provide an opportunity to study the nature of these differences for carbon.

Concluding this rapid survey of the second period, we encounter four gases. Nitrogen, which makes up the bulk of our atmosphere, is relatively inert. It does not easily combine with many other elements. If it did, our air would probably contain a much higher concentration of the oxides of nitrogen. Nevertheless, there are many vitally important nitrogen-containing compounds, including quite a few found in you and me. Our continued existence is, of course, also very much dependent upon the next element in the periodic table. In fact, much of the chemistry on this planet involves oxygen; breathing, burning, and rusting are only the most obvious examples. It is no accident that the properties and reactions of this gas provided the key to understanding elements and thus unlocked the laboratory of modern chemical science. As we have observed, fluorine is even more reactive than oxygen, and thus it is sort of a nonmetallic counterpart of lithium. In marked contrast is the inertness of neon, which brings to an end the second row of the periodic table.

The third period also contains eight members, all of them quite similar to the elements immediately above them in the table. Again, the trend is from a reactive metal in group 1A, sodium, to a reactive nonmetal in group 7A, chlorine. Another noble gas, argon, completes the row.

The fourth period includes 18 elements, all of which are strung out in a row across a modern table. As we have noted, potassium and calcium present no problems in classification. However, the next element, scandium (Sc), was never very comfortable as a member of the family headed by boron. Although early tables did attempt to make this association, scandium remained somewhat in the status of a bastard at a family reunion. Now this element is more happily grouped with its heavier brothers, yttrium (Y) and lanthanum (La) in a subfamily which still bears the notation 3B. By

TABLE 5.4.   *The Periodic Table of the Elements*

1A								8
1 H 1.008	2A							
3 Li 6.941	4 Be 9.012							
11 Na 22.99	12 Mg 24.31	3B	4B	5B	6B	7B		
19 K 39.10	20 Ca 40.08	21 Sc 44.96	22 Ti 47.90	23 V 50.94	24 Cr 52.00	25 Mn 54.94	26 Fe 55.85	27 Co 58.93
37 Rb 85.47	38 Sr 87.62	39 Y 88.91	40 Zr 91.22	41 Nb 92.91	42 Mo 95.94	43 Tc 98.91	44 Ru 101.07	45 Rh 102.91
55 Cs 132.91	56 Ba 137.34	57* La 138.91	72 Hf 178.49	73 Ta 180.95	74 W 183.85	75 Re 186.2	76 Os 190.2	77 Ir 192.22
87 Fr (223)	88 Ra 226.03	89† Ac (227)	104 (Ku) (261)	105 (Ha) (260)				

	58	59	60	61	62
*Lanthanide series:	Ce 140.12	Pr 140.91	Nd 144.24	Pm (145)	Sm 150.4

	90	91	92	93	94
† Actinide series:	Th 232.04	Pa 231.04	U 238.03	Np 237.05	Pu (244)

The atomic weights are based on carbon-12 = 12.0000.
More exact values are given on the back end paper.
Numbers in parentheses are the mass numbers (number
of protons + neutrons) of the most stable isotopes of
these radioactive elements. Most of the elements are
metallic. Nonmetals are grouped in the shaded area.

much the same means titanium (Ti), vanadium (V), chromium (Cr),
and manganese (Mn) also became the heads of new elementary
households. Iron (Fe), cobalt (Co), and nickel (Ni), next in line,
all have somewhat similar properties. Together they start group 8.

							7A	0
							1	2
							H	He
			3A	4A	5A	6A	1.008	4.003
			5	6	7	8	9	10
			B	C	N	O	F	Ne
			10.811	12.011	14.007	15.999	18.998	20.183
			13	14	15	16	17	18
			Al	Si	P	S	Cl	Ar
	1B	2B	26.98	28.09	30.97	32.06	35.45	39.95
28	29	30	31	32	33	34	35	36
Ni	Cu	Zn	Ga	Ge	As	Se	Br	Kr
58.71	63.55	65.37	69.72	72.59	74.92	78.96	79.90	83.80
46	47	48	49	50	51	52	53	54
Pd	Ag	Cd	In	Sn	Sb	Te	I	Xe
106.4	107.87	112.40	114.82	118.69	121.75	127.60	126.90	131.30
78	79	80	81	82	83	84	85	86
Pt	Au	Hg	Tl	Pb	Bi	Po	At	Rn
195.09	196.97	200.59	204.37	207.2	208.98	(210)	(210)	(222)

63	64	65	66	67	68	69	70	71
Eu	Gd	Tb	Dy	Ho	Er	Tm	Yb	Lu
151.96	157.25	158.93	162.50	164.93	167.26	168.93	173.04	174.97

95	96	97	98	99	100	101	102	103
Am	Cm	Bk	Cf	Es	Fm	Md	No	Lr
(243)	(247)	(247)	(251)	(254)	(257)	(256)	(254)	(257)

Copper (Cu) and zinc (Zn) complete this series of ten metallic elements. Although there are readily identifiable differences among these substances, they have enough in common to be collectively known as the *first transition series*. The transition referred to is back to the pattern established in period two, group 3A through the inert gases.

Period five is essentially a repeat of four, but period six adds 14 more elements. These *rare earths* are all so much alike that they are not placed in any of the existing families. This series is considered as starting with lanthanum (La), and it runs through lutetium (Lu). The *lanthanides*, as these elements are commonly called, are followed by ten more transition metals and then by members of the familiar final six groups. The seventh and last period begins, as do the others, with an alkali metal, francium (Fr) and an alkaline earth, radium (Ra). Actinium (Ac) and the *actinides* (corresponding to the lanthanides) follow.

One of the greatest advantages of the periodic table is the way it succinctly relates the chemical and physical properties of the elements. Because of family resemblances, if information is known about one of the members of a group, an intelligent and informed guess can often be made about the characteristics of other members. Moreover, the trends within the table can further simplify the process. We have noted that elements to the left of the table are metals and have a strong tendency to combine with the nonmetals on the right. Furthermore, the atomic ratios exhibited in these compounds are characteristic of the families. For example, the first four elements of the second period form fluorides with the formulas LiF, BeF_2, BF_3, and CF_4. Nitrogen, oxygen, and fluorine generally have a greater tendency to react with metals than with other nonmetals. Thus, they form the following compounds with lithium: Li_3N, Li_2O, and LiF. With some exceptions, the same atomic ratios generally apply to corresponding compounds of other elements in the respective groups. Moreover, other information about the chemical properties of the elements of the second period provides at least a first approximation to the behavior of the heavier elements below them.

In the periodic table (Table 5.4) on pages 110–111, nonmetals are grouped in the shaded area. A quick glance is enough to inform you that there are many more metallic than nonmetallic elements. To be sure, not all metals have the structural strength we associate with iron and certain other transition elements. The alkali metals and alkaline earths are, naturally, too reactive and too soft for such applications. The rare earths are neither very rare nor earthy. Because of their chemical similarities, they are very difficult to separate and are often used in a mixture called Misch metal. One familiar application of lanthanide mixtures is in cigarette lighter "flints," where it is alloyed with iron. Some lanthanides also find use as phosphorescent coatings for television picture tubes. All members of the actinide series are radioactive. The metalloids, which mark the stair-step separation of the two major classes of elements, have recently found important uses in transistors and other semiconductor devices. Some of these elements, for example, arsenic (As), antimony (Sb), and tin (Sn) can actually exist in both metallic and nonmetallic allotropes.

All of this is, of course, very well and good as an invaluable aid in organizing and classifying the basic substances of our universe. But hopefully this excursion into historical and descriptive chemistry has raised as many questions as it has answered. For the periodic table, useful as it is, does not really explain anything! The order in diversity, the "Multëity in Unity" is, to be sure, beautiful. But the *why* is equally beautiful. And the experimental basis for that explanation is the subject of the next chapter.

For thought and action

5.1   If you exhale your breath through a tube into a solution of calcium oxide (lime water) the solution turns cloudy. This does not happen, or happens only very slowly, if air is bubbled through lime water. Explain the chemistry which is taking place and write an equation for the reaction.

5.2   Recall your introduction, in Chapter 2, to Joseph Black. One of Black's contributions was preparing "fixed air" by heating *magnesia alba* (magnesium carbonate). Write a balanced equation for this reaction.

5.3   Black also observed that although limestone reacts effervescently with acids to give off a gas, lime does not evolve a gas when similarly treated. Explain these observations in a manner consistent with the chemical composition of these compounds. If possible, support your answer with equations.

5.4   Write the formulas of the following compounds: (a) lithium hydroxide, (b) sodium bromide, (c) cesium chloride, (d) barium oxide, (e) strontium carbonate.

5.5   Name the compounds with the following formulas: (a) K_2O, (b) $BaCl_2$, (c) Rb_2CO_3, (d) NaI, (e) $Ba(OH)_2$.

5.6   Write a balanced equation for each of the following:
(a)   The reaction of sodium with water
(b)   The reaction of potassium with oxygen
(c)   The reaction of calcium with water
(d)   The preparation of barium from its oxide.

5.7   Flushing chunks of sodium down the toilet is an ill-advised method of disposal. Explain why.

5.8   Suppose the popular press reports the discovery of a large deposit of pure potassium in northern Canada. What is your reaction as an informed citizen?

5.9   Magnesium chloride, $MgCl_2$, is present in large amounts in sea water. Devise a method or methods for isolating magnesium from this source. (It may be helpful for you to know that the process employed commercially takes advantage of the fact that the solubility of magnesium hydroxide is very low.)

5.10   Can you think of any other explanations which might account for the difference in density between "chemical" and "atmospheric" nitrogen noted by Lord Rayleigh?

5.11   You have most likely noticed that balloons filled with helium lose their gas faster than balloons filled with air. Provide a reason.

5.12   Argon is frequently used to fill tungsten-filament electric light bulbs, as distinct from neon-type tubes. The argon does not glow, nevertheless, it does serve a function. What is it?

5.13   If you were filling a time capsule with artifacts of the 1970s to be opened 500 years from now, what characteristics would you seek for the materials from which the cannister was built and for the gas with which it was filled? Do you have any specific suggestions?

5.14   Explain why Mendeleev left empty spaces in his table for the elements we today know as gallium and germanium, but did not predict the rare gases. Then try your own hand at prophesy by speculating on whether or not any undiscovered elements or families of elements exist now. If there are such elements, where in the periodic table would they most likely belong?

5.15   Below are some selected properties of lithium and potassium. Estimate values for the corresponding properties of sodium.

	Lithium	*Potassium*
Atomic weight	6.9	39.1
Density (g/cm³)	0.53	0.86
Melting point (°C)	180	63.4
Boiling point (°C)	1330	757

5.16   Classify the following elements as metals or nonmetals: (a) sodium (Na), (b) sulfur (S), (c) iodine (I), (d) chromium (Cr), (e) tantalum (Ta).

5.17   Match the elements with the proper classification:

(a)	chlorine (Cl)	(1)	alkali metal
(b)	iron (Fe)	(2)	lanthanide
(c)	uranium (U)	(3)	alkaline earth
(d)	radium (Ra)	(4)	halogen
(e)	radon (Rn)	(5)	actinide
(f)	lithium (Li)	(6)	transition metal (first series)

(g) palladium (Pd)   (7) rare gas

(h) gadolinium (Gd)   (8) transition metal (second series)

5.18 This chapter has included instances of both conservative traditionalism and bold innovation in the workings of science. Comment on the virtues and drawbacks of both characteristics, using specific examples.

Suggested readings

Bigelow, M. Jerome, "The Representative Elements," Bogden & Quigley, Croton-on-Hudson, New York, 1970.

Davis, Helen Miles, "The Chemical Elements," Science Service/Ballantine Books, Washington/New York, 1952, 1959.

Laidler, Keith J., and Ford-Smith, Michael, "The Chemical Elements," Bogden & Quigley, Tarrytown-on-Hudson, New York, 1970.

Latimer, Wendell M., and Hildebrand, Joel H., "Reference Book of Inorganic Chemistry," Macmillan, New York, 1951.

Newton, Vernon, "Adam's Atoms: Making Light of the Elements," Viking Press, New York, 1965.

Seaborg, Glenn T., and Valens, Evans G., "Elements of the Universe," Dutton, New York, 1965.

Weeks, Mary Elvira, and Leicester, Henry M., "Discovery of the Elements," 7th ed., Journal of Chemical Education, Easton, Pennsylvania, 1968.

Messages from inner space

Let there be light

In the previous chapter you encountered several references to spectroscopic results, but no explanation of spectroscopy. You will recall, for example, that a yellow spectroscopic line provided the first evidence of solar helium and a red light signaled the discovery of neon. Most likely, you are still in the dark about the source of the light. This may be disconcerting, but it should be of some consolation for you to know that early spectroscopists shared your ignorance. Like you, they used the results before they understood them. Happily and hopefully, you will emerge from this chapter and the next knowing far more about spectroscopy (and cathode rays and X rays and radioactivity) than the discoverers and early investigators of these phenomena.

The men credited with the discovery that forms the subject of this section were Robert Wilhelm Bunsen and Gustav Robert Kirchhoff, both of the University of Heidelberg. Of course, the phenomena underlying the spectroscope were known long before 1859 when the two German scientists built their first instrument. The fact that ordinary sunlight can be split into many colors is often painted in a bold bow across the sky. And it was discovered

at least three and a half centuries ago that a glass prism can duplicate the action of water droplets. Today we know that this effect
arises because the beams of the various colors present in ordinary
light are all bent to different extents in a raindrop or a prism. This
bending process separates the beams and makes them individually
visible. Together they form the familiar red-to-violet rainbow of
colors known as the *continuous visible spectrum.*

Stated with obvious simplicity, the world comes to us in living
color because light itself is multicolored and because objects of
different "color" react differently with the various colored components of light. Commonly, this interaction of matter with light
involves the absorption of light of certain colors and the reflection
of other parts of the spectrum. Thus, for example, a lemon is colored as it is because it absorbs blue, green, and red light and
reflects yellow. On the other hand, sodium chloride, common salt,
does not selectively absorb any part of the visible spectrum, but
rather reflects it all. Therefore, it is white or "colorless."

However, when a small pinch of sodium chloride is heated in a
flame, a somewhat different phenomenon occurs. The flame, although it may be light blue or nearly colorless in the absence of
salt, becomes bright yellow. Repeated observations with many
other compounds indicated, long ago, that the yellow color is due
to the presence of sodium. No matter what the substance, if it
contains sodium, it will impart this characteristic color to a flame
in which it is vaporized. In contrast, potassium and its compounds
give off a reddish violet light when heated in a flame. This difference in behavior suggests an easy way to distinguish between
compounds of these two similar alkali metals. Indeed, the test was
used as early as 1758.

Over the years, other elements were found to have their own
distinctive flame colors: calcium gives an orange color; strontium
is red; barium, light green; copper, a darker green; and so on.
These colors are familiar in fireworks and fireplaces, but they contribute far more significantly to science. For the flame color associated with any particular element is always the same shade. It is
a uniquely identifying fingerprint which can be used with great
sensitivity to ascertain the presence and concentration of a tiny bit
of mercury in fish, strontium in milk, or lead in bread. But for precise and exact results, the technique requires instrumental assistance; and here is where the spectrograph enters the scene.

In order to facilitate viewing the colors imparted by a variety of
elements to the flame of his newly designed gas burner, Bunsen,
a chemist, collaborating with the physicist Kirchhoff, passed the
light beam through a narrow slit and then through a prism. The
prism split the light into its various colored components, which
appeared not as a continuum, but as sharp lines on a viewing

screen or photographic plate. For example, the yellow flame of sodium was seen to arise from two bright lines in the yellow region of the spectrum. In the nomenclature of the times, these two lines were designated D_1 and D_2. Since the light seemed to come from the elements themselves, the flame colors of sodium and other elements came to be known as *emission spectra*.

It was soon discovered that the emission of light could also be induced by applying an electric field between electrodes sealed in a tube containing vapor of the element in question. For example, such a tube filled with a small quantity of mercury vapor exhibits a unique pattern of yellow, green, and violet lines pictured in Figure 6.1. Together, these combine to give the familiar bluish color of mercury vapor street lights.

Kirchhoff and Bunsen also shed new light on another fairly familiar phenomenon—dark line spectra. The experiment required a source of light containing a continuum of all the colors of the spectrum, as distinct from the sharp, separated light lines of a flame or a gas discharge tube. Such "white" light is given off by glowing, incandescent filaments of carbon, calcium oxide, and a variety of other substances. When the Heidelberg scientists passed the beam from one of these sources through a sample of sodium vapor and then through their spectroscope, two dark lines appeared in the yellow region. These lines were missing from the otherwise continuous spectrum of the source because the sodium was absorbing light of these particular colors. Kirchhoff recognized that the absorbed lines were in the same position as the D doublet emitted by a sodium-containing flame.

The conclusion was clear: an element can either absorb or emit light of a particular, characteristic color (or colors), depending upon the experimental arrangement. The absorption and emission phenomena must therefore be physically related. Moreover, the *dark lines* which had been observed in the solar spectrum in 1814 by Fraunhofer (and which still bear his name) must be due to the presence of various vaporized elements in the relatively cooler solar atmosphere above the incandescent surface of the sun. A reinvestigation of these spectra disclosed Fraunhofer absorption lines corresponding to the emission lines of sodium, hydrogen, calcium, and many other elements. The analytical applications of spectroscopy had literally reached astronomical proportions.

Once the elementary specificity of spectroscopic emission and absorption lines had been ascertained, it became particularly im-

FIGURE 6.1.   *The visible emission spectra of hydrogen, helium, mercury, and uranium. The numbers refer to wavelengths in Ångstroms. Used, by permission, from "Chemistry" by M. J. Sienko and R. A. Plane (2nd ed.). Copyright 1961, McGraw-Hill Book Company, New York.*

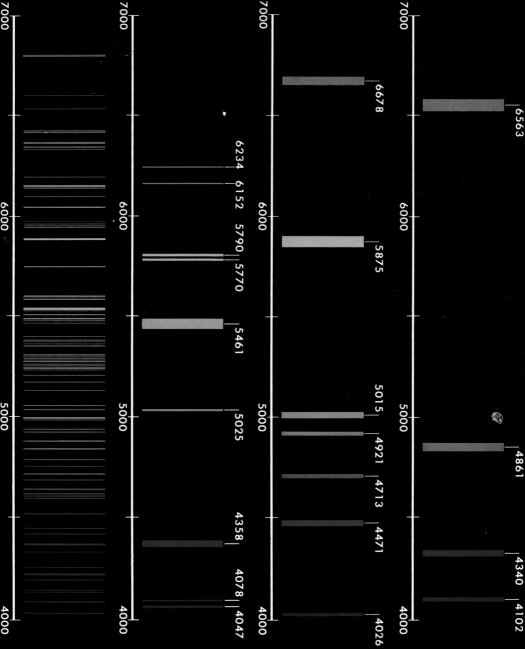

FIGURE 6.2. *Schematic representation of wave motion.*

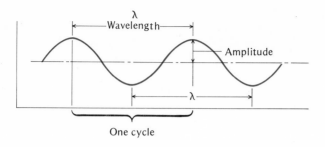

One cycle

portant to be able to precisely and quantitatively designate colors. Poets and publicists tend to be qualitative, comparative, and subjective in their colorful descriptions. And, while one can appreciate and apply phrases like "baby blue eyes" or "hot pink lipstick," there are times when it is helpful to be a bit more exact. This was one of those times. Understand, I am not suggesting that Shakespeare's standard for redness, "Coral is far more red than her lip's red," should be sharpened to read, "6500 ångstroms is far more red than her lip's red." But the point is, it could be done, because colors can be identified with extreme precision.

The method depends upon the fact that in many respects light acts as though it were made up of waves, an idea advanced by Huygens about 300 years ago. (It is well worth noting that it is in the beautiful but cussed ambiguity of nature, that light in many other respects acts as though it were made up of little energetic bullets, a concept championed by Newton, also about three centuries ago. Somewhat surprisingly, both men were right, but more of that later.) A wave is cyclic or periodic; it repeats itself. And the shortest repeat distance is called the *wavelength*. For example, the wavelength indicated in Figure 6.2 is the distance between successive crests. The principle is the same, whether you are dealing with an ocean, a guitar string, or a light wave.

The idea, as far as light goes, is that different colors have different wavelengths. Thus, the two yellow lines of the sodium doublet have wavelengths of 0.00005896 cm for D_1 and 0.00005890 cm for D_2. In exponential notation, these values become 5.896×10^{-5} and 5.890×10^{-5} cm. Perhaps some appreciation of how small these numbers really are can be gained by realizing that the diameter of a period on this page is about ten million times greater than the wavelengths of the sodium D lines. Because these dimensions are so miniscule compared to the centimeter, other smaller standard units are often employed to designate wavelengths. Most common are the Ångstrom (Å) and the nanometer (nm) or millimicron (mμ). The equivalences relating these units can be summarized as follows:

$1 \text{ Å} = 10^{-8}$ cm
$1 \text{ nm} = 1 \text{ m}\mu = 10^{-7}$ cm $= 10$ Å.

Consequently,

$$5.896 \times 10^{-5} \text{ cm} \times \frac{1 \text{ Å}}{10^{-8} \text{ cm}} = 5.896 \times 10^{3} \text{ Å} \qquad \text{or} \qquad 5896 \text{ Å}$$

Similarly,

$$5.896 \times 10^{-5} \text{ cm} \times \frac{1 \text{ nm}}{10^{-7} \text{ cm}} = 5.896 \times 10^{2} \text{ nm}$$

$$\text{or} \qquad 589.6 \text{ nm} = 589.6 \text{ m}\mu.$$

Naturally, the actual wavelength is the same, though the units used and the number of these units required to make up the length are different.

Small as these lengths are, they can, nonetheless, be measured with extreme accuracy. In fact, the international standard upon which the meter is now based is the wavelength of a particular orange-red line in the krypton spectrum. The meter is defined as 1,650,763.73 times this wavelength.

The color of light can also be specified in terms of its *frequency*. As the name implies, frequency is a measure of "how often?" To find the frequency of light, you simply count how many wave crests pass a certain point in space over a given period of time. Each crest represents a complete cycle, so the frequency of a wave is counted in cycles per second (cps). Incidentally, you'd better be a pretty fast counter, since the frequencies of the yellow lines of sodium are 5.085×10^{14} and 5.090×10^{14} cps. In expressing the dimensions of frequency, scientists omit the "cycles" and simply write "per second," that is, 1/sec or sec^{-1}. Sometimes this unit is also called a herz (Hz).

Wavelength and frequency are related to each other through an important general property of light—its speed. The equation is an example of Greco-Roman cooperation:

$$\lambda = \frac{c}{\nu}$$

The Greek letter lambda (λ) traditionally designates wavelength and the letter nu (ν) symbolizes frequency. The speed of light is represented by the letter c. Although you may know this fundamental constant of nature as 186,000 miles/sec, in scientific work it is commonly expressed as its metric equivalent, 2.998×10^{10} cm/sec. You will notice that when these units are used for c and the frequency is expressed as sec^{-1}, the wavelength is calculated in centimeters. That is,

$$\lambda \text{ (cm)} = \frac{c \text{ (cm/sec)}}{\nu \text{ (1/sec)}} \qquad \text{or} \qquad \frac{c \text{ (cm sec}^{-1})}{\nu \text{ (sec}^{-1})}.$$

Clearly, if you know the frequency of a certain color of light you can easily calculate its wavelength from this equation. Similarly, you can readily obtain the frequency corresponding to any known wavelength. Just to check the correctness of this relationship, we can calculate a value for the speed of light by multiplying the wavelength and frequency of the sodium D_1 line.*

$$c = \lambda\nu = 5.896 \times 10^{-5} \text{ cm} \times 5.085 \times 10^{14} \text{ sec}^{-1}$$
$$= 29.98 \times 10^9 \quad \text{or} \quad 2.998 \times 10^{10} \text{ cm sec}^{-1}$$

The above relationship involving frequency, wavelength, and velocity applies equally well to all sorts of wave phenomena. Of course, not all waves travel with the speed of light, and the equation must be modified by replacing c with the proper speed. For example, the velocity of sound in air is about 35,000 cm/sec, the exact value depending upon pressure and temperature. The frequencies of audible sound are much lower than those of visible light, between 20 and 20,000 cps, while the wavelengths are correspondingly longer, between 1.7×10^3 and 1.7 cm.

The visible spectrum, which is so important in identifying elements and illuminating their atomic structure (as well as our daily lives), is only a small part of the much broader *electromagnetic spectrum*. The rods and cones which form the physiological and biochemical light-detection apparatus of our eyes are sensitive only to radiation within the relatively narrow band between 4000 and 7000 Å (4.0×10^{-5} and 7.0×10^{-5} cm). The lower wavelength is seen as what we have come to call violet; the higher represents red. On either side of these limits of our personal visual perception stretches the great expanse of the entire spectrum. Wavelengths range from as small as 10^{-12} cm to at least 10^8 cm—a fantastic multiplicative factor of 10^{20}! Fortunately, however, we are not blind to these vast regions of radiation. For an array of instruments far more sensitive than human eyes—devices with names like Geiger counters, bubble chambers, phototubes, thermocouples, and radio receivers—have enabled men to "see" and use the entire electromagnetic spectrum.

At wavelengths greater than 7000 Å lies the region of *infrared* radiation, invisible, yet perceptible as heat. Around a wavelength of 1 centimeter, one encounters a part of the spectrum known as the *microwave* region. It is the realm of radar and the source of energy for recently developed high-speed ovens. And at still longer wavelengths of about 10^2 to 10^5 cm, between a meter and a kilometer, is the *radio* region. Of course, frequency decreases with increasing wavelength (all electromagnetic radiation travels at the

---

* Do not make the mistake of assuming that this is all there is to determining the speed of light!

FIGURE 6.3.   *The electromagnetic spectrum.*

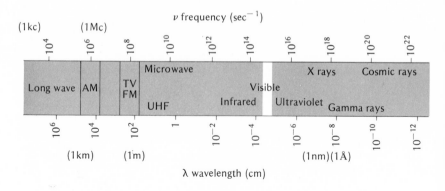

speed of light). The range of frequencies on the FM and AM dials of a radio, 88 to 108 megacycles (88×10^6 to 108×10^6 cps) and 55 to 160 kilocycles (55×10^3 to 160×10^3 cps), respectively, are thus within this wavelength interval. Shortwave radio and ultra-high-frequency (UHF) television transmission occur at somewhat higher frequencies.

On the other side of the visible portion of the spectrum, at wavelengths shorter than 4000 Å, is the *ultraviolet* region. Between $\lambda = 10^{-6}$ cm and $\lambda = 10^{-9}$ cm are found the once-mysterious X rays, an indispensable aid to modern medicine and to our understanding of the atom. *Gamma rays,* potentially lethal, but another vital clue in the search we are undertaking, occur at still higher frequencies and shorter wavelengths of 1.4×10^{-8} to 5×10^{-11} cm. And finally, at the extreme end of the spectrum, with frequencies as high as 10^{22} cps and wavelengths as short as 10^{-12} cm, are the highly energetic *cosmic rays* which continually bombard our planet.

The interaction of matter with radiation occurs not just within the narrow confines of visible light, but over almost the entire electromagnetic spectrum. For example, many colorless compounds absorb in the ultraviolet, infrared, and microwave regions. In the century since Bunsen and Kirchhoff perfected their spectroscope, characteristic spectroscopic patterns over wide frequency ranges have been collected and catalogued for all the elements and thousands of compounds. The importance of such information to qualitative and quantitative analysis should be obvious. But there is another application of transcendent significance. For spectra are messages from the microcosm of matter. It is through radiation that we communicate with atoms and molecules.

As history is measured, it required only a remarkably short time to learn the language. In the latter half of the nineteenth century scientists began to realize that spectra provided energetic evidence of atomic and molecular events. The distinctive yellow spectroscopic lines emitted by sodium must signal an occurrence at the lowest and most fundamental level of material organization. The simple Daltonian model of miniscule billiard balls could not possibly account for such phenomena. Stuff just had to be more complicated.

A video view of the electron

In the same year that Bunsen and Kirchhoff built their first spectroscope, 1859, a device of comparable importance was being constructed about one hundred miles to the north by another university scientist. The scene was Bonn, a sleepy little Rhineland city which 89 years earlier had heard the first cacaphonic sounds produced by a baby named Ludwig von Beethoven, and which 90 years later was to become the seat of government for a defeated and divided Germany. The scientist was Julius Plücker, mathematician and physicist. The gadget was the *gas discharge tube*.

The design of the tube was simple enough (Figure 6.4 is a more complicated version). Metallic electrodes were sealed in either end of a glass envelope, and the air was pumped out until what little remained had a pressure of less than one-millionth of an atmosphere. The electrodes were then connected to a big battery or other source of high electrical potential. Immediately, the walls of the tube opposite the negative electrode or cathode began to give off an eerie bluish glow. It is not likely that Plücker glimpsed in that glow the possibility of transmitting pictures over hundreds and thousands of miles. What he did see was this: something seemed to be striking the glass, and that something appeared to come from the cathode.

Fortunately, the hypothesis was easy to test. Other materials besides glass glowed brightly in the otherwise-invisible beam. One of these substances, zinc sulfide (ZnS), was used to coat a screen which was placed between the two electrodes. When an object opaque to the radiation was positioned between the cathode and the screen it cast a shadow on the screen. But there was no shadow if the object was located between the anode and the screen. The rays were indeed *cathode rays*, but what else were they?

The above question attracted a number of answers. A predictable one placed the radiation in the electromagnetic spectrum. However, the correct explanation came in 1897 from the Cavendish Professor of Experimental Physics at Cambridge University, J. J. Thomson. Thanks to his experimental ingenuity, Thomson was able to demonstrate that cathode rays are bent when they are passed between the poles of a magnet or between the oppositely charged plates of an electrical condenser. Electromagnetic rays are not influenced in this manner by magnetic or electric fields, but a stream of charged particles would show just such deflection. Moreover, the fact that the beam of cathode rays was attracted by the positively charged plate and repelled by the negative one indicated clearly that the charge borne by the particles must be negative. Thomson could summarize his results concisely, but by his own admission they introduced some new problems.

As the cathode rays carry a charge of negative electricity, are deflected by an electrostatic force as if they were negatively electrified, and are acted on by a magnetic force in just the way in which this force would act on a negatively electrified body moving along the path of these rays, I can see no escape from the conclusion that they are charges of negative electricity carried by particles of matter. The question next arises, What are these particles? are they atoms, or molecules, or matter in a still finer state of subdivision?*

An obvious way to answer this second round of riddles would be to determine the mass of a cathode ray particle. But that is far easier said than done, so Thomson initially had to be content with measuring the charge-to-mass ratio of such a particle. The quantity of electrical charge (usually abbreviated q) is measured in coulombs and, as you know, mass (m) is conventionally reported in grams. The Cavendish Professor was thus in quest of q/m in coulombs per gram.

One of the two techniques he employed capitalized upon the previously observed interaction of cathode rays with magnetic and electric fields. A special discharge tube apparatus was built with the poles of an electromagnet oriented at right angles to two electrically chargeable plates (see Figure 6.4). The beam of cathode rays passed through the regions of both the electric and magnetic fields. When neither field was applied, the rays went undeflected. However, when the two parallel plates were charged at different electrical potentials, the beam bent toward the positive one. The position of the beam was signaled by the glowing spot at which the rays struck the end of the tube, and consequently the extent of displacement could be measured accurately. This electrically induced curvature was then counteracted by turning on the magnetic field and adjusting its strength so that the beam was returned

* J. J. Thomson, Cathode Rays, *Philosophical Magazine* **44**, 302 (1897).

FIGURE 6.4. *Schematic diagram of a gas discharge tube equipped for measuring q/m for cathode ray particles. T$_1$ is the undeflected trajectory of the cathode ray when the electrostatic and magnetic fields either balance or are both off; T$_2$ is the trajectory with only the electric field on; T$_3$ is the trajectory with only the magnetic field on.*

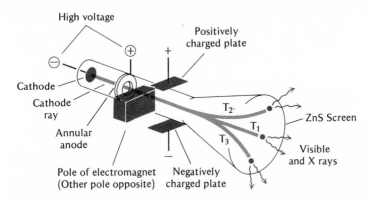

to its original, undeflected position. As a third step, the electric field was turned off and the amount of bending caused by the magnetic field alone was determined. The experimental data thus consisted of the observed beam displacements, the measured values for the magnitudes of the counterbalancing electric and magnetic fields, and certain dimensions of the apparatus. When related by some relatively simple and well-known (at least to physicists) equations, these data yielded the desired ratio.

A numerical value of about 2.0×10^8 coulombs/g was calculated for q/m of a cathode ray particle. Moreover, the ratio was independent of the gaseous elements present in the discharge tube. It appeared to be a constant characteristic of the rays. Unfortunately, there was no way of knowing what the charge per particle was, since there was no way of knowing how many particles were present in a gram. Nevertheless, the result did say something very significant about cathode rays. It did so in comparison with known q/m values for electrically charged atoms or ions.

Ions, as you will learn in Chapter 9, account for a number of familiar phenomena. For example, the conduction of electricity by molten sodium chloride is attributed to the movement of positively charged sodium ions and negatively charged chloride ions.* The same mechanism also explains the electrical conductivity of aqueous solutions of salt. And, of particular importance, the elec-

---

* Appropriately, ion means "wanderer" in the original Greek.

trical properties of ions in solution underlie the generalizations proposed in 1833 by Michael Faraday, a protégé of Sir Humphry Davy's and his successor at the Royal Institution. It is these laws of electrolysis, still associated with Faraday's name, which permit the determination of charge-to-mass ratios for ions in solution.

Faraday found that the mass of an element liberated from solution by the influence of electricity is directly proportional to the quantity of electricity passed through the solution. This is not to suggest that a given quantity of electricity always liberates the same weight of any and all elements. Quite the contrary, the quantity of electricity required to liberate 1.0 g of hydrogen in the electrolysis of water, 96,500 coulombs, will liberate 107.9 g of silver, 31.8 g of copper, and 9.0 g of aluminum. But 96,500 coulombs of electricity (1 faraday) always result in the electrolytic evolution of 1.0 g of hydrogen, 2 faradays yield 2.0 g, and so on.

The quantity of electric charge required to displace hydrogen is presumably associated with the hydrogen atoms, or better, hydrogen ions, since neutral atoms carry no charge. Hydrogen ions, on the other hand, are attracted by negatively charged electrodes. Therefore, each of these ions must bear a positive charge. Moreover, the experiments of Faraday's successors indicated that the magnitude of this charge must be 96,500 coulombs per gram of hydrogen. Again, this value does not specifically represent the absolute charge on each individual ion; it is instead a charge-to-mass ratio. Other elements were found to exhibit their own characteristic q/m values, all of them somewhat smaller than the hydrogen ratio, but of roughly comparable magnitude. No particles were discovered with higher values—that is, not until the cathode ray particle. Its q/m was a whopping 2000 times greater than that of the hydrogen ion. In symbols,

$$(q/m)_{\text{cathode ray}} \simeq 2000(q/m)_{\text{hydrogen ion}}$$

and in numbers,

$$2.0 \times 10^8 \text{ coulombs/g} \simeq 2000 \times 9.65 \times 10^4 \text{ coulombs/g}.$$

There are two ways, really three ways, to explain this relationship of ratios.

(1) $q_{\text{cathode ray}} \simeq 2000 \, q_{\text{hydrogen ion}}$,

(2) $m_{\text{cathode ray}} \simeq \dfrac{1}{2000} \, m_{\text{hydrogen ion}}$, or

(3)   some mathematically equivalent combination of
$q_{\text{cathode ray}} > q_{\text{hydrogen ion}}$ and
$m_{\text{cathode ray}} < m_{\text{hydrogen ion}}$.

Remember that Thomson did not know the charge of a single hydrogen ion any more than he knew the charge of a single cathode ray particle. Nor did he know with certainty the mass of an individual hydrogen ion or atom. Nevertheless, he could make an educated guess (though accurate determinations of Avogadro's number were not made until several years later). At any rate, Thomson chose the second option and in so doing, made the staggering suggestion that there exists in nature a particle with a mass only about 1/2000 that of the lightest atom.

*Thus on this view we have in the cathode rays matter in a new state, a state in which the subdivision of matter is carried very much further than in the ordinary gaseous state: a state in which all matter—that is, matter derived from different sources such as hydrogen, oxygen, &c.—is of one and the same kind; this matter being the substance from which all the chemical elements are built up.**

The statement was simple and straightforward, but its implications were momentous. A basic form of stuff, if not *the* basic form, had been found. The atom could no longer be considered the smallest unit of matter!

In dwelling on the structural significance of the minute mass of the cathode ray particle, we should be careful not to overlook the other term in the q/m ratio. For Thomson's beautiful research also identified the basic carrier of electrical current. Faraday's laws had codified the relationship between electricity and matter—mass is proportional to the quantity of electricity interacting with it. But, as Dalton demonstrated, matter behaves as if it were discrete; it comes in little chunks called atoms. Therefore, it is logical to conclude, on the basis of Faraday's observation, that electricity also comes in little packages. A particular number of these packages are required to evolve a given weight of an element from solution. Using exactly this line of reasoning, G. J. Stoney, in 1891, had suggested that the "natural unit of electricity" should be called an *electron*. Stoney was speaking about a fundamental unit of charge; he did not specifically associate it with a unit of matter. J. J. Thomson did that. His cathode ray particles, "corpuscles" he called them, are at the same time basic units of both electricity and matter. They are, in fact, particles of electricity. And they have acquired Stoney's term—electrons.

After his brilliant determination of q/m, or e/m, to switch to a common notation that represents the electrical charge on an electron with an e, Thomson turned his instruments and his attention to the direct measurement of that charge. He obtained approximate results which suggested that his hypothesis had been

* J. J. Thomson, Cathode Rays, *Philosophical Magazine* **44**, 312 (1897).

correct, *the charge on an electron is equal in magnitude to that on a hydrogen ion, though opposite in sign.* A more exact determination had to wait until 1909 and the elegant oil-drop experiment of Robert A. Millikan. The University of Chicago scientist introduced tiny droplets of oil between the parallel, horizontal plates of a condenser. When the plates were uncharged, gravity simply caused the drops to fall in the same downward direction exhibited earlier by Newton's apocryphal apple. However, the motion of the oil droplets could be reversed if they were charged with negative electricity and the upper condenser plate bore a positive charge. The electrostatic attraction could be made to equal or even exceed the pull on the drop due to the force of gravity.

Through a microscope, Millikan observed the oil droplets and measured their speed and direction of movement under various electric fields. Moreover, he repeated the experiment with many different charges on the drops. Thanks to the mathematical relationships of physics, he was able to calculate these charges from the magnitude (voltage) of the electric field applied across the plates and the velocity of the oil droplets. What physics could not tell him directly was how many units of electricity—how many electrons—were responsible for these charges. Nevertheless, an investigation of the various experimental values disclosed a common numerical factor; all charges were an integral multiple of 1.60×10^{-19} coulomb. Logically, this value must be the long-sought e, the charge on a single electron, the fundamental unit of electricity.

Once e was determined, it became trivially simple to calculate m. Using modern values of both e and e/m to illustrate the accuracy with which these constants are now known, we find the following:

$$m = \frac{1.6021 \times 10^{-19} \text{ coulomb}}{1.7588 \times 10^{8} \text{ coulombs/g}} = 9.1091 \times 10^{-28} \text{ g}.$$

A quick consultation of Chapter 3 will disclose that the mass of a hydrogen atom is the gram atomic weight divided by Avogadro's number,

$$\text{Mass H atom} = \frac{1.008 \text{ g/GAW}}{6.023 \times 10^{23} \text{ atoms/GAW}}$$
$$= 1.674 \times 10^{-24} \text{ g}.$$

Only a simple division is required to demonstrate that the mass of a hydrogen atom is 1838 times greater than that of an electron. This, of course, is a refinement of the original estimate of 2000.

A knowledge of the size of the basic bundle of electricity also makes possible the calculation of the number of electrons involved in any electrochemical reaction. Recall, for example, that

96,500 coulombs are required to divorce 1.0 g of hydrogen (or, to be more exact, 1.008 g) from its aqueous union with oxygen. The number of electrons necessary to carry this quantity of electricity must be

$$\frac{9.6500 \times 10^4 \text{ coulombs}}{1.6021 \times 10^{-19} \text{ coulomb/electron}} = 6.023 \times 10^{23} \text{ electrons.}$$

The number should be familiar, we used it only a few lines ago. The unit of electrical charge called a faraday must correspond to one Avogadro number (one mole) of electrons. Indeed, calculations of this sort were used to confirm Perrin's value of Avogadro's number. This many electrons are chemically equivalent (not equal in mass, of course) to 1.008 g of hydrogen. But 1.008 g of hydrogen is an atomic weight's worth; it consists of 6.023×10^{23} atoms. Therefore, the number of electrons involved in this particular reaction is equal to the number of hydrogen atoms; there is a direct one-to-one correspondence.

Although it may be getting ahead of the game a little, this is a convenient place to translate the previous sentence into an equation:

$$H^+ + e^- \rightarrow \tfrac{1}{2} H_2.$$

The state of hydrogen in water can be regarded as corresponding formally (though not physically) to an ion carrying a charge equal to that of an electron, but opposite in sign. When water is electrolyzed, these H^+ ions acquire one electron, e^-, each and are converted into neutral H atoms. Since atoms of hydrogen do not exist singly under normal conditions, but rather combine to form diatomic molecules, $\tfrac{1}{2} H_2$ has been used in the above equation.

Evidence of positively charged particles also comes from gas discharge tubes, but from the anode, not the cathode. Thomson determined the charge-to-mass ratio of these *anode* or *canal rays* by essentially the same technique he had employed earlier on cathode rays. Unlike the results for electrons, the larger q/m ratios for the positive ray particles were found to depend upon the specific gas present in the tube. The actual values corresponded closely to those calculated from electrolysis experiments. Presumably, the potential applied across the electrodes of the gas discharge tube removed some electrons from the atoms of the gas, thus producing positively charged ions. The former particles, attracted by the oppositely charged anode, formed the cathode rays; the latter made up the anode rays. The q/m values found for the positive particles thus reflected the ionic charges and atomic weights of the various elements examined. The mass spectrograph used today to determine atomic weights operates on the same basic principle as Thomson's apparatus.

There was also another result of this investigation of anode rays, and it must have been almost unbelievable. Even when the tube contained a pure element, several distinct values of q/m could occasionally be measured. The differences were slight, but reproducible. They could not be attributed to differences in the charge, which should vary in electronic increments of 1.60×10^{-19} coulomb. Apparently and inexplicably, the atoms of a single element were exhibiting more than one mass.

This was rank heresy! To suggest that the atomic weight of any element was not a unique identity-determining property of that substance seemed to contradict the basic principles of the Daltonian atomic theory. But at the turn of the century, all of physics was in ferment. There were almost daily reminders that none of the laws written for nature by scientists have the status of sacred, absolute, and infallible TRUTH. Apparently, even atoms had lost their indestructibility, individuality, and integrity. They were giving off light and negatively charged particles and, if the reports from Paris were true, even more!

But before we move on to the City of Light, let us cast one last nostalgic look at the device which started the section—the cathode ray tube. To be sure, the average American probably spends far too much time looking at the one in his television set. And the typical video viewer is no doubt more enthralled by the image than by the electrons producing it. But at least to this chemist's biased eyes, the little bits of atoms which bombard the phosphorescent screen of the picture tube are far more fascinating than commercials, soap operas, situation comedies, or even professional football.

Of serendipity and searching

It was Wednesday, February 26, 1896. The sky over Paris was leaden. Heavy banks of clouds hung low over the gargoyled Gothic towers of Notre Dame, the smooth, bald dome of Sacre Coeur, and the steel spire built seven years earlier by the engineer Eiffel for the World's Fair of 1889. Henri Becquerel, 36, like his father and grandfather before him Professor of Physics at the Museum of Natural History, eyed the sky with impatience. The sun was necessary for his experiments and, in the competitive race to gain scientific priority and the prestige that went with it, there was little time to lose. Becquerel was wasting none.

It was only about a month earlier that he had first heard of the rays discovered late in the previous year by the German, Wilhelm Röntgen. The rays were apparently given off when the beam from the cathode of a gas discharge tube struck the glass walls of the bulb. The most remarkable thing about these rays was their fantastic power of penetration; they could expose a photographic plate through several layers of lightproof paper. In fact, Röntgen had reported that they easily passed through a thousand-page book or a double pack of cards. Wood was also found to be transparent to the rays; but metals, particularly lead, absorbed them with fairly high opacity. Little else was known about the rays, and Röntgen's appellation, "X," seemed indeed appropriate.

Röntgen had been quick to grasp the practical importance of his discovery, and along with the copies of the research report which he mailed to physicists and friends on New Year's Day, 1896, he enclosed several photographs taken with the aid of X rays. One of the pictures clearly showed the shadows cast by the bones of his wife's hand. Two months later, physicians were already making medical use of similar photographs. Soon the popular press would be filled with the story of the mysterious Röntgen rays, and lead-lined underwear would become a popular garment for modest maidens and matrons who wished to keep their privates private and chastely protected from prying eyes.

The scientific repercussions of Röntgen's discovery were no less revolutionary than its proposed applications. Members of the close-knit international scientific fraternity quickly sensed that this was something big. Becquerel must have known that across the English Channel, J. J. Thomson was already deeply involved in an investigation of the nature of the cathode rays which appeared to induce the formation of X rays. The French scientist was more interested in the mechanism by which these latter rays originated. Therefore, he focused his attention upon the glowing, fluorescent spot on the tube—the region from which the X rays emerged.

The phenomenon of *fluorescence* was familiar to Becquerel. He knew of many minerals which exhibited it. They glowed bright red or green or yellow when exposed, not to ordinary visible light, but to ultraviolet or "black" light. The intense colors provided powerful evidence that the fluorescent substances were absorbing light in one wavelength region (in this instance, the ultraviolet) and emitting it in another (the visible). When Becquerel looked at the report from Röntgen's laboratory, he concluded that invisible radiation also formed part of fluorescent emission. Certainly, X rays appeared to be associated with the fluorescence of the glass cathode ray tube. It was, therefore, reasonable to expect that ordinary fluorescence, induced by ultraviolet light instead of cathode rays, might also give off highly penetrating radiation.

Becquerel required only a simple experimental arrangement to test his hypothesis. His procedure was to wrap unexposed photographic plates in several layers of heavy black paper and then to place upon them crystals of the fluorescent substances under investigation. The plates and crystals were next transferred into a cheap and convenient source of ultraviolet radiation—sunlight. After a predetermined exposure interval, Becquerel would take the plates into his darkroom, upwrap them, and develop them. His first results were disappointing. The photographic plates remained unaffected. None of the compounds and minerals investigated emitted rays capable of penetrating the paper.

But toward the end of February, Becquerel's luck took a change for the better. He repeated his experiment with a compound called potassium uranyl sulfate, ($K_2SO_4 \cdot UO_2SO_4 \cdot 2H_2O$). When he developed the plates after several hours in the sun, he found what he was hoping for, unmistakable dark images where the crystals had lain. He immediately tried again, separating the crystals from the wrapped film with a coin in one case and a plate of glass in another. The outline of the coin appeared on the plate, surrounded by a blackened region; the glass was largely transparent to the rays as they exposed the film beneath. The conclusion seemed as clear and unambiguous as the images in the photographic emulsion. On February 24, Becquerel proudly reported to the Academie des Sciences that his hypothesis had been borne out by experiment: penetrating rays, presumably X rays, could be induced by sunlight and emitted as part of normal fluorescence.

But Becquerel did not stop his investigations with this single positive result. He realized that more evidence would be necessary to reconfirm his ideas and that there probably were many other interesting things to learn about X-ray fluorescence. In brief, Becquerel was compulsive and curious. And because he was, he glowered back at the dark clouds which hid the sun he needed for his next studies. His photographic plates were already prepared and wrapped and the crystals of potassium uranyl sulfate were in place. But without the sun there was no point in continuing the experiment. After all, Becquerel knew that ultraviolet light was essential for the fluorescence of the compound, and he was almost as certain that the penetrating rays always accompanied fluorescence. So, without otherwise disturbing his equipment, he put the plates and crystals into a dark drawer to await the sunshine.

They had to wait three full days, February 27, 28, and 29 (1896 was a leap year). By March 1, Becquerel was eager enough to work on a Sunday. But even then, the plates did not see the sun, because on that day, Henri Becquerel developed them. Why he did is still a mystery. Perhaps it was his compulsiveness, maybe professorial forgetfulness. Certainly, there was no logical reason to do so. After all, he was saving the plates for an experiment and they were still presumably as good as new. But when Becquerel removed the plates from the developing solution he found that they had been far from that. Where he had expected to see nothing or next to nothing, he saw the images of the crystals, darker than in any previous experiment. A massive dose of penetrating radiation had fallen upon the photographic emulsion and exposed it heavily.

If Becquerel was astonished and dumbfounded by what he saw (and it is difficult to imagine how he could have been otherwise), he recovered quickly. First, he did what any scientist confronted with an unbelievable result would do, he carefully repeated the experiment, this time in absolute darkness. It confirmed the accidental one; even in the dark, the uranium compound took its own photograph. Faced with such irrefutable evidence, Becquerel was forced to reject his former hypothesis. There was no way that the emission of the rays in total darkness could be reconciled with a mechanism that linked their production to ultraviolet light-induced fluorescence. A new conclusion was necessary and Becquerel provided it:

*The radiations of uranium salts are emitted not only when the substances are exposed to light but when they are kept in the dark, and for more than two months the same pieces of different salts, kept protected from all known exciting radiations, continued to emit, almost without perceptible enfeeblement, the new radiations.**

Chance, in the form of a cloudy sky, had favored Henri Becquerel, and his mind had been prepared to make the most of it.

Carefully, he continued to explore the consequences of his discovery. He studied a variety of compounds, both fluorescent and nonfluorescent ones. There was no correlation between this property and that of giving off rays, that is, being *radioactive*. To be sure, some of the fluorescent substances he tested did exhibit the new phenomenon, but not all of them. What these radioactive materials did have in common was the presence of uranium. And, even nonfluorescent compounds of this element fogged photographic plates. Uranium, not fluorescence, was the source of the rays. To further corroborate this conclusion, Becquerel needed to examine the properties of uranium metal. Here he was again aided by his extraordinary good fortune and good timing. In another Paris laboratory, Henri Moissan was just finishing the first preparation of the pure element. As soon as a sample was obtained, Becquerel tested it. The rays it emitted were more intense than any he had previously seen. Obviously they came from uranium,

* Henri Becquerel, Sur les Radiations Émises par Phosphorescence, *Comptes Rendus* **122**, 420 (1896). Reprinted in translation in "A Source Book in Physics" by William Francis Magie, p. 610, McGraw-Hill, New York, 1935.

but by what mechanism? And what were they? These were the next questions eloquently calling for answers.

Fortunately, Becquerel had a capable and energetic young graduate student who thought the problem would make a good topic for her doctoral dissertation. Marie Sklodowska had left her native Warsaw five years earlier to study at the Sorbonne. In Paris she had acquired her preliminary degrees, a brilliant physicist husband named Pierre Curie, and a daughter, Irène. It occurred to Madame Curie that perhaps uranium was not the only element which was a ray-giver, and so, with characteristic thoroughness, she began to systematically search for others. She investigated dozens of elements, compounds, and minerals. In the course of this study she found that thorium (Th), another heavy element, was also radioactive. However, she made another, more curious observation. Pitchblende, an ore of uranium, always emitted far more radiation than would be predicted or could be explained on the basis of its relatively low uranium content. Although it was unlikely that the enhanced radioactivity was due to some peculiar crystalline structure or atomic arrangement in the mineral, the idea had to be tested. Marie Curie did this by mixing up a sample of artificial pitchblende from purified laboratory chemicals or reagents. The radioactivity of the product was no more than expected from its uranium content. The answer had to lie elsewhere. Natural pitchblende must contain a contaminating trace element of far greater radioactivity than uranium.

The search for this needle in a huge haystack of ore was long and arduous. Marie Curie was joined by her husband and together they began the work which was to make them the most famous scientific spouses in history. Pitchblende is a complex mixture of many elements. The problem of separating them was primarily a chemical one and the Curies, as physicists, had to learn the techniques of analytical chemistry. At the time, these were chiefly based upon the different solubilities of various compounds. First the pitchblende had to be ground and dissolved in large quantities of acid. Then chemicals were added to precipitate the numerous elements present in the ore, that is, to convert them into insoluble compounds. These solids were separated by filtration and then often redissolved and reprecipitated to further purify them. By such tedious and repetitious processes, the Curies were able to capitalize on differences in the chemical properties of the elements in the mineral mixture.

For one phase of their work, the starting material consisted of several tons of pitchblende residues from the Joachimsthal mine in Bohemia. Their laboratory was "a cross between a horse stable and a potato cellar." Nevertheless, the Curies had a signal to follow. The rays given off by the hidden element led them on.

FIGURE 6.5. *The laboratory in which the Curies discovered radium. (Reproduced from Mary Elvira Weeks and Henry M. Leicester, "Discovery of the Elements," 7th ed., p. 782, Journal of Chemical Education, Easton, Pennsylvania, 1968.)*

Finally, in July 1898, they succeeded in isolating a small sample of a new sulfide. The compound was quite similar to bismuth sulfide, but unlike Bi_2S_3, it was radioactive. The source of those rays must be the metal with which the sulfur was combined. It was, in fact, a new element, and Marie Sklodowska Curie, always a staunch Polish nationalist, named it polonium (Po).

But the search was not yet over. The instruments with which the Curies measured radioactivity indicated that there was another, even more radioactive element in pitchblende. So they returned to their piles of ore and their liters of solutions and their uninviting shed. This time the radioactivity was closely associated with barium. Thus, when barium sulfate ($BaSO_4$) was precipitated from solution by adding sulfuric acid (H_2SO_4) or some soluble sulfate, the radioactivity went with it. The barium was not the source of the rays, but another element very much like it was. Finally, in December 1898, the Curies managed to obtain a chloride sample rich enough in the chloride of the new element to permit spectroscopic analysis. The lines did not conform to those of any known substance. Even more convincing was the level of radioactivity, 900 times greater than that of uranium metal. Only a speck of the element had been isolated, but it was enough to identify and enough to name—radium, "the ray-giver."

The choice of name was an apt one. For radium not only emits an extremely high level of radioactivity, its compounds glow in the dark. "One of our joys," Marie Curie wrote much later, "was to go into our workroom at night; we then perceived on all sides the feebly luminous silhouettes of the bottles containing our products. It was really a lovely sight and always new to us." The Curies' hours of work were at last rewarded.

In 1903 they received another sort of reward—the Nobel Prize in Physics, which they shared with Becquerel. Madame Curie, the first woman to be so honored, went on, with her 1911 Prize in Chemistry, to become the first person to be twice recognized by the Nobel committee. Other honors were showered on her by many nations, particularly when the medical uses of radium became known. She herself was instrumental and influential in furthering the applications of her element in the treatment of cancer. The rays which it emits kill cells, particularly fast-growing, cancerous ones. Unfortunately, they also take a heavy toll of healthy cells. Ironically, Madame Curie's own death, in 1934, was most likely caused in part by the large quantities of radiation her body had absorbed during her years of research. It was not the first tragedy in a life filled with triumphs and accolades. Three years after his Nobel Prize, Pierre Curie had been struck and killed by a heavy horse-drawn cart on a rainy Paris street.

The α, β, γ of radioactivity

While the Curies were locked in their epic struggle with pitchblende, other scientists were investigating the nature of the rays emitted by uranium and other radioactive elements. One of the physicists who became fascinated by the phenomenon was a young New Zealander named Ernest Rutherford. Rutherford arrived at Cambridge's Cavendish laboratory late in 1895, only a few months before Röntgen's discovery of X rays. Stimulated by the scientific *Zeitgeist* and by the leadership of J. J. Thomson, the young man wasted no time in brilliantly fulfilling the great promise and potential he had evidenced as an undergraduate in New Zealand. His contributions were to constitute a succession of scientific achievements unequaled in the era which rewrote man's concepts of the atom. Indeed, Rutherford's researches on the nature of radiation, radioactive decay, and atomic structure were of such significance that to call him "the father of the modern atom" is an almost indisputable assignment of paternity. In recognition of these accomplishments, Ernest Rutherford was awarded with professorships at McGill, Manchester, and Cambridge Universities; the Nobel Prize in Chemistry for 1908; a knighthood; and a baronetcy.

But these honors were still far in the future and perhaps even undreamed of when Rutherford began his investigations of the effect of thin metal foils upon uranium radiation. He successively placed layers of aluminum foil (each only 5.0×10^{-4} cm thick) over a sample of uranium oxide (UO_2). With the addition of each layer, he measured the level of radioactivity penetrating the foil. The intensity of the radiation with one thickness of aluminum between the source and the detector was about 40% of that coming directly from the unshielded uranium oxide. With two pieces of foil in place, the radiation level was about 40% of that penetrating only one foil. And so it continued for four layers of foil, each one in succession reducing the previous intensity by about 60%. After the fourth thickness, however, the effect of the foil diminished greatly. Eight more layers were added and the intensity of the radiation was only slightly lowered. Some of the uranium rays were apparently being stopped by four thin sheets of aluminum metal (a total thickness of 0.002 cm). On the other hand, there seemed to be another sort of rays which could pass through 0.006 cm of aluminum with only minor absorption. Rutherford summed up the results and added a Greek touch.

These experiments show that the uranium radiation is complex, and that there are present at least two distinct types of radiation—one that is very readily absorbed, which will be termed for convenience the α radiation, and the other of a more penetrative character, which will be termed the β radiation.*

By now you have observed enough about the way science grows to know that a paragraph like that just quoted is an open invitation to more research. In this case, the challenge was to determine the nature of the two types of radiation reported by Rutherford. An obvious starting point was to pass radioactive emission between the poles of a magnet and between plates bearing opposite electrical charges. Both alpha (α) and beta (β) rays were deflected by the magnetic and electrical fields. As you recall from your study of cathode rays, such behavior indicated that both forms of radiation consisted of electrically charged particles. The fact that the beam of α particles was attracted by a negatively charged plate was evidence that its charge was positive. In contrast, β particles were attracted to the positive electrode and must therefore be negatively charged.

During investigations of this sort, yet another kind of radioactive emission manifested itself on the photographic plates used as detectors. The new rays were not influenced by either magnetic or electrical fields; they passed through both undeflected. There-

---

* E. Rutherford, Uranium Radiation and the Electrical Conduction Produced by It, *Philosophical Magazine* **47**, 116 (1899).

FIGURE 6.6.   *Representation of passage of α, β, and γ rays through an electric field.*

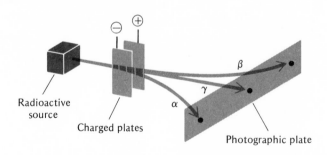

fore, unlike α and β radiation, the beam of gamma (γ) rays could not be a stream of charged particles. In fact, it wasn't a stream of particles at all! The ease with which the γ rays penetrated metals and other materials exceeded the penetrating power of α or β particles. These characteristics were much like those of X rays. Consequently, the newly discovered form of radioactive emission was also assigned to the electromagnetic spectrum. The wavelengths of γ rays proved to be far shorter than those of the visible spectrum, shorter even than those of X rays.

Quite clearly, one could not measure the charge-to-mass ratio of γ rays, since this radiation possessed neither property. But such information was potentially very useful in characterizing α and β particles. The deflection of the rays by magnetic and electrical fields again provided the method. Becquerel at once went to work on β radiation. Soon he reported a value for the charge-to-mass ratio. It was a familiar one—1.76 × 10⁸ coulombs/g. The French physicist had not only used J. J. Thomson's experimental technique, he had obtained the same numerical result Thomson had found for cathode ray particles. But the value was not an artifact of the experiment, it was a property of nature—specifically, of the part of nature called an electron. Subsequent research has confirmed the fact that the particles streaming from the cathode of a gas discharge tube and the penetrating negative rays emitted by radioactive elements are identical to each other and to the negatively charged particles which emerge from a battery or generator. However, physical identity does not necessarily imply common atomic origin, an idea you might well ponder until you again encounter electrons within the atom.

Rutherford himself undertook the task of characterizing α radiation. He too began by measuring the charge-to-mass ratio of the particles making up the rays. The value he found was 50,700 coulombs/g, much closer to that of ions than electrons. If the charge were of similar magnitude to that on an electron, the mass of an α particle must be much greater. In order to eliminate the "if" Rutherford set about to measure the charge. When he succeeded, in 1908, Millikan had not yet done his definitive determination of the charge on an electron. Nevertheless, Rutherford established that his own value of 3.12×10^{-19} coulomb for the α particle was about twice that for an electron and, of course, opposite in sign. Coupling his values for q/m and q, Rutherford calculated that the mass of an α particle must be approximately 6.15×10^{-24} g. Although he was a physicist, Rutherford was not unaware of the periodic table. The mass of an α particle was about four times that of a hydrogen atom and within experimental error of that of the recently rediscovered helium atom. Therefore, he could conclude that *"the α particle, after it has lost its positive charge, is a helium atom."*

The conclusion was based on more than a mere numerical coincidence. It had been known, since the work of the Curies, that a gas is slowly formed over substances which contain radium. This "radium emanation" had all the properties (or lack of them) of the rare gases isolated by Ramsay and Rayleigh. It, too, was inert to chemical reactions, much like helium. However, radium emanation was definitely not helium, nor apparently any other known noble gas. Its atomic weight was too great for that. Moreover, it emitted α particles and appeared to give off helium. If a sample of radium emanation were left undisturbed over a period of time, a growing concentration of helium could be spectroscopically detected in the gas.

In order to ascertain whether there was a connection between the α emission and the appearance of helium, Rutherford designed the simple yet elegant apparatus pictured in Figure 6.7. He enclosed a sample of radium emanation in a glass tube with walls thin enough to permit the passage of α particles but thick enough to prevent the diffusion of helium atoms. The enclosed space around the tube was pumped down to a high vacuum and periodically examined with a spectroscope. Twenty-four hours passed and there was no sign of helium; a day later a faint yellow line was detected in the spectrum; after two more days, the lines were brighter and others appeared. Finally, after a total of six days, the evidence was unambiguous and unassailable. The spectrum of helium was clearly visible in the gas outside the emanation tube. Alpha particles were passing through its walls and being converted into helium atoms. The association suggested earlier by Rutherford was experimentally verified.

This was an exceedingly strange way for stuff to behave! Radium was giving rise to radium emanation which looked suspiciously like an unknown rare gas element. And the emanation was in turn

FIGURE 6.7.   *Rutherford's apparatus for demonstrating that α particles are helium nuclei. Radium emanation was placed in the thin-walled inner tube, and the particles emitted passed through these walls into the heavier-walled outer tube. Mercury was used to compress the resulting helium gas into the capillary tube where its spectrum was measured.*

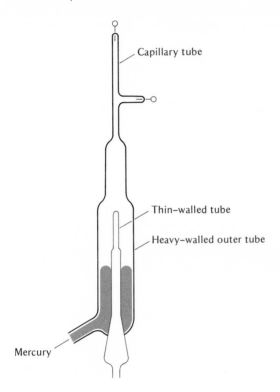

giving off helium. The elements themselves, immutable since Dalton, if not since time began, appeared to be acting capriciously to say the least. In Alice's apt albeit ungrammatical phrase, things were getting "curiouser and curiouser!" The most sacred tenants of physical science were being sorely threatened. Doctrines were disintegrating along with atoms. It was not a time for the faint of heart, and no doubt many men faced a crisis of scientific faith. But Ernest Rutherford was hardly a timid man. If he appeared to be a heretic it was because he had a deeper vision of the order of nature.

Alchemy revisited

In part, Rutherford's vision was conditioned by research he and Frederick Soddy had done in 1902 in Montreal at McGill. In this work they had seen other evidence which led them to question chemical orthodoxy. Their study was part of an international investigation of radioactive emission from three elements—uranium, thorium, and actinium. To put it simply, things were in a hell of a mess. Many samples studied gave off α, β, and γ radiation simultaneously. The intensity of radiation from what appeared to be a single element would sometimes decrease in a regular fashion, then suddenly increase again, often to fall for a second time.

Fortunately, chemistry came to the rescue of physics. The problem was that, in most cases, the physicists were not dealing with single elements, but with mixtures. By applying the methods of chemical analysis, chemists were able to separate the various elementary components. Thus it became possible to measure the radioactive properties of individual elements. Studies of this sort indicated that the intensity of radiation for any single radioactive element was directly proportional to the concentration of the element. Stated with the economy of symbols,

6.1 $I = kc,$

where I is intensity, c concentration, and k a constant of proportionality. Doubling the concentration doubles the intensity or rate of radioactive emission. The natural process is such that when one starts with a fixed amount of a radioactive element, the intensity of radiation decreases with time. This implies that the concentration of the radioactive element is also decreasing proportionately. And indeed, the conclusion is borne out by experiment.

It soon became evident that each radioactive element exhibits a characteristic rate of emission or decay. Put in a slightly different way, each element has its own unique constant of proportionality, k. A high value means the element will give off radiation at a great rate, losing its radioactivity very rapidly. On the other hand, elements with low numerical values for k emit relatively low intensities of radiation that persist for a long time. Another way of expressing such behavior is in terms of half-life, the time required for the intensity of radiation from a sample of a radioactive element to fall to one-half of its original value. Depending on the substance, half-lives can range from less-than-seconds to more-than-centuries. Of course, the equation and the principles governing this phenomenon also suggest that in a half-life, the concentration of the element drops to one-half of its original value. No matter what the starting concentration might be, after a half-life has passed, its own value will be halved.

There were some things about the sort of behavior described above that were comfortably familiar. Chemists could cite all sorts of chemical reactions which fit the same general mathematical form. After all, Equation 6.1 is essentially the rate expression for a reaction which follows something known as "first-order kinetics."

Kinetics is a general term for the rate of chemical (and sometimes physical) changes, and *first order* refers to the fact that the rate is proportional to the first power of the concentration, as distinct from the concentration squared, cubed, and so on.

However, there were other aspects of radioactive emission that made a chemist's mind boggle. For one thing, the rate of emission of the α, β, or γ rays by any given element was independent of temperature, while every well-behaved chemical reaction was sure to speed up as the temperature was raised. Moreover, the intensity of an element's radiation also did not vary with state of chemical combination; chemical combination had no influence whatsoever. For example, the radioactivity of uranium was the same (on a per gram of pure uranium basis) whether the element was present as the pure metal or combined in compounds like potassium uranyl sulfate or uranium oxide. This sort of indifference to chemical combination clearly put radioactivity beyond the ken of conventional chemistry and chemists. Furthermore, it indicated that *radioactivity must have an atomic and not a molecular origin*. The particular molecule in which a radioactive atom happened to find itself had no effect on the chances that the atom might break down.

The breakdown was the third and most perplexing problem. And it was here that the collaboration of Soddy and Rutherford (another example of the peaceful and productive coexistence of a chemist and a physicist) bore fruit. The specific system under investigation was thorium, an element exhibiting what appeared to be a rather high level of radioactivity. However, Soddy's chemical separation showed that the actual source of much of this radiation was something called thorium X, an α-emitter with a half-life of about 4 days. This something had all the properties of an element—but not the element thorium! Moreover, thorium X appeared at the same time thorium disappeared, again attended by the loss of α particles. If the Law of Conservation of Matter still held, the inescapable conclusion was that the thorium was changing into thorium X, another genuine element. If the accepted ideas of immutable Daltonian elements and atoms held, this could not be the explanation. These two representations of nature could not both survive this dilemma unscathed. One would have to yield.

Rutherford and Soddy repeated their experiments and reevaluated their data. There was no question about the accuracy of the results. Experiment was ruthlessly forcing a decision. And so they made their choice and wrote a paper describing the apparent "transformation" of one element, thorium, to another, thorium X. With intentional care they avoided the obvious word, "transmutation." After all, the paper was to be submitted to the *Transactions of the Chemical Society*, not the Alchemical Society.

The paper was accepted by the editors and more communications from Montreal followed. Each successive one introduced additional evidence for the correctness of the startling hypothesis that *one element could spontaneously change into another by altering its atomic structure.* By 1902, Rutherford and Soddy could confidently conclude:

*Since, therefore, radioactivity is at once an atomic phenomenon and accompanied by chemical changes in which new types of matter are produced, these changes must be occurring within the atom, and the radioactive elements must be undergoing spontaneous transformation.**

Or, in somewhat sexier symbolism, whenever an elementary radioactive "mother" gives parthenogenic birth to a "daughter" element, the happy atomic event is signaled by the emission of an α or β particle.†

Within only a few years, the long geneologies of the uranium, thorium, and actinium families were worked out. Typically, daughters were named for their mothers, though there might be precious little family resemblance. Thus, thorium spawned, in succession, mesothorium 1 and 2, radiothorium, thorium X, thorium emanation, and thorium A, B, C, C' or C", and D. Similarly, uranium, actinium, and radium each had a long list of namesake elements which they begat as radiation occurred. In fact, at least 37 substances with different radioactive properties were identified among the decay products of the three series. If all of these were unique elements, the periodic table would soon get completely out of hand and Mendeleev's beautifully neat system might have to be scrapped altogether. Fortunately, however, the table again demonstrated its great elasticity. Or perhaps it would be more correct to say that nature cooperated by again exhibiting her extraordinary orderliness.

The point is that chemical characterization soon showed many of the various "elements" found in the three decay series had identical chemical and physical properties—all, that is, except atomic weight and radioactivity. For example, radioactinium (with an atomic weight of approximately 227), radiothorium (228), ionium (230), uranium Y (231), and uranium X_1 all proved to be chemically indistinguishable from thorium (232). Similarly, the final, nonradioactive end products of the three series were orig-

* E. Rutherford and F. Soddy, The Cause and Nature of Radioactivity. I, *Philosophical Magazine* 4, 395 (1902).
† This is not exclusively correct; other mechanisms of spontaneous elementary transmutation have since been discovered.

FIGURE 6.8.   *Thorium decay series. Symbols in parentheses reflect the approximate isotope weight and elementary identity of the various substances involved.*

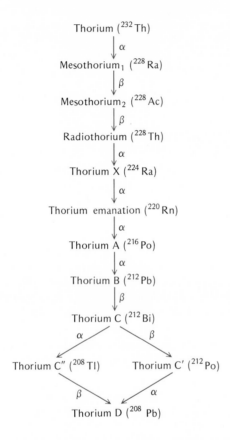

Thorium (^{232}Th)

$\downarrow \alpha$

Mesothorium$_1$ (^{228}Ra)

$\downarrow \beta$

Mesothorium$_2$ (^{228}Ac)

$\downarrow \beta$

Radiothorium (^{228}Th)

$\downarrow \alpha$

Thorium X (^{224}Ra)

$\downarrow \alpha$

Thorium emanation (^{220}Rn)

$\downarrow \alpha$

Thorium A (^{216}Po)

$\downarrow \alpha$

Thorium B (^{212}Pb)

$\downarrow \beta$

Thorium C (^{212}Bi)

α β

Thorium C" (^{208}Tl)   Thorium C' (^{212}Po)

β α

Thorium D (^{208}Pb)

inally designated radium G (206), actinium D (207), and thorium D (208). Chemically, all three were identified as identical to lead (207.19). Ironically, the natural process of transmutation stops with an element the alchemists often started with.

When it comes to classifying substances, it's chemistry that counts. Physicists may poke around in atoms, but elements are the stuff that chemistry is made of. If a number of basic substances have exactly the same chemistry, it is reasonable to conclude that they are all species of the same element. And, since they all fall in the "same place" on the periodic table, Soddy's term, *isotopes,* is an apt one.

To be sure, there were problems with this interpretation. The argument admits that a given element can exist in more than one atomic species, and that these species have different weights. If thorium atoms can exhibit at least six different masses and still act like thorium, *the chemical uniqueness of an element cannot be a consequence of its atomic weight.* Sometimes, in fact, isotopes of different elements have the same atomic weight. For example, lead and bismuth both exist in forms with atomic weights of about 212.

This research was greatly aided by the development of the mass spectrometer, which made possible the "weighing" of individual atoms. By this means, a sample of a pure element previously thought to consist of identical atoms could be shown to contain several isotopes. This was the explanation of the earlier observation of several different q/m values for anode ray particles of a single element. As additional data were collected it became more and more obvious that *most elements naturally exist in a mixture of isotopes.* Some of these isotopes are radioactive, but by no means all of them.

Soddy's work on isotopes, while simplifying things greatly, had raised a most perplexing problem. What structural characteristics, presumably at an atomic level, do all atoms of a single element (regardless of mass) share? In other words, what makes uranium uranium and sodium sodium and hydrogen hydrogen? Certainly something physical must be responsible for the uniquely identifying chemical properties of the elements. And whatever that something is, it is not simply associated with the atomic weight of an element. Besides, the discovery of isotopes proved that an atomic weight measured by conventional methods (which involve billions and billions of atoms) is really an average of the atomic weights of several isotopes. Table 6.1 illustrates how the individual isotopic weights must be weighted by the usual relative distribution of the isotopes in bulk samples of a naturally occurring element. The result is the familiar atomic weight commonly tabulated in textbooks. Understanding of the concept of atomic weight was thus further refined, but the answer to the riddle of elementary identity had to lie elsewhere inside the atom.

Soddy himself made a step toward answering his own question when, in 1913, he stated two rules of radioactive decay. *When an*

TABLE 6.1.   *Isotopes of Magnesium*

Isotope wt.		Relative fractional abundance		
23.99	×	0.7870	=	18.88
24.99	×	0.1013	=	2.53
25.99	×	0.1117	=	2.90
		Average atomic weight	=	24.31

element loses an α particle it is always shifted two places to the left in the periodic table. For example, uranium-238 (the number refers to the approximate isotopic weight) is transmuted to thorium-234 by the emission of an α particle. Clearly, this reaction is not conventional chemistry, and a special radiochemical equation is required to represent it. The relative masses of the atoms and particles participating in the transformation are indicated by superscripts which appear before the appropriate symbols. Thus, we write

$$^{238}U \rightarrow {}^{234}Th + {}^{4}\alpha.$$

The same sort of process also describes the conversion of radium-226 to radon-222 (or radium emanation, as we called it earlier). This shift in the general direction of lower atomic weight is specifically associated with a decrease of about 4 g in the gram atomic weight of the isotope. The particular value of this change is a consequence of the fact that the weight of an α particle on the conventional atomic weight scale is about 4. The relative weights are indicated in the equation.

The loss of a β particle, on the other hand, *always moves an element one place to the right on the periodic table,* but leaves its mass essentially unaltered. Thus, the thorium-234 formed by the α decay of uranium-238 itself undergoes β emission to form another element, protactinium-234 (Pa),

$$^{234}Th \rightarrow {}^{234}Pa + \beta.$$

Not the least remarkable aspect of these amazing elementary transmutations was the colossal energy change accompanying them. Calculations clearly showed that the energy evolved with α, β, and γ radiation far exceeded (on a per gram of original matter basis) the amount of energy released by common chemical reactions such as the burning of coal or the explosion of gun powder. To Rutherford's insightful eye, these results suggested a vast potential source of unprecedented levels of energy. His comments were prophetic:

*All these considerations point to the conclusion that the energy latent in the atom must be enormous compared to that rendered free in ordinary chemical change. . . . There is no reason to assume that this enormous store of energy is possessed by the radio-elements alone.**

According to William Dampier, a fellow physicist, Rutherford's statement had been a "playful suggestion that, could a proper

* E. Rutherford and F. Soddy, Radioactive Change, *Philosophical Magazine* **5**, 591 (1903).

detonator be found, it was just conceivable that a wave of atomic disintegration might be started through matter which would indeed make this old world vanish in smoke." That was in 1903, two years before Albert Einstein expressed the equivalence of mass and energy in his epoch-making $E = mc^2$. The ball of fire which bathed the New Mexico desert in a light "brighter than a thousand suns" on the morning of July 16, 1945, and illuminated Einstein's equation in an unearthly glare made Rutherford's suggestion seem anything but "playful."

There will be more on the subject of atomic energy in the final chapter of this book. For the moment, the issue is atomic structure and elementary identity. By 1913 it was clear that these two must be so interrelated that a solution of either enigma would ultimately solve the other. Somehow, the explanation would have to accommodate a host of signals sent from inside the atom itself—visible spectra, X rays, and α, β, and γ radiation. Scientists had at last arrived at the point where they were confronting the substance of the universe on its most fundamental level.

For thought and action

6.1   The wavelength of the brightest line in the potassium spectrum is 4047.20 Å. Express this length in (a) centimeters, (b) nanometers, and (c) inches. Also calculate the frequency of the light.

6.2   The mean distance from the earth to the sun is 149,674,000 km or 93,003,000 miles. How much time is required for light to travel this distance?

6.3   The star system nearest the earth, the α-Centauri triplet, is 4.29 light years away, that is, light from that star requires 4.29 years to reach the earth. How far away is α-Centauri in kilometers and miles?

6.4   Suppose you were a contemporary of Kirchhoff's. Speculate on the origin of visible spectra.

6.5   How many electrons weigh 1.0 g?

6.6   A current of 15 amp flowing through an iron for 10 minutes corresponds to the passage of 15 amp $\times$ 10 min $\times$ 60 sec/min = 9000 coulombs. How many faradays and how many electrons does this quantity of electricity represent?

6.7   In the 1940's, it was common for shoe stores to have X-ray machines or fluoroscopes for aid in fitting. Suggest why this practice has been discontinued.

6.8   Why are you asked to remove pencils and other metallic objects from your pockets before having a chest X ray?

6.9   Speculate on why α particles are less penetrating than β particles.

6.10   The half-life of radium-223 is 11.6 days. Starting with an initial sample of 1.00 milligram of this isotope, how much would remain (a) after 11.6 days, (b) after 23.2 days, and (c) after 34.8 days?

6.11   Tritium, an isotope of hydrogen, has a half-life of 12.3 years. How many years would be required for the tritium radiation produced in water by a nuclear power plant to fall to 1/16 of its initial value, presuming no new tritium were produced?

6.12   What elements are formed by the following radioactive emissions? Include the atomic weights of the isotopes.
       (a)   Loss of a β particle by hydrogen-3 (tritium).
       (b)   Loss of an α particle by radium-226.
       (c)   Successive loss of a β particle and an α particle by actinium-228.
       (d)   Successive loss of two β particles by thorium-234.

6.13   Complete the following radiochemical equations:
       (a) $^{40}K \rightarrow$ _____ $+ \beta$
       (b) $^{222}Rn \rightarrow$ _____ $+ {}^4\alpha$
       (c)   _____ $\rightarrow {}^{234}Th + {}^4\alpha$
       (d) $^{138}La \rightarrow {}^{138}Ce +$ _____

6.14   Write balanced equations for the following radioactive transformations:
       (a) β emission by carbon-14
       (b) β emission by strontium-90
       (c) α emission by uranium-238

6.15   Strontium-90, a radioactive isotope, is concentrated in milk. From what you know about the periodic table and the elements normally present in milk, suggest why this should be so.

6.16   What do you suppose daughter elements formed by radioactive decay were originally named after the parent elements?

6.17   Suggest how the phenomenon of radioactive decay might be used to determine the age of rocks.

6.18   This chapter has included some examples of the variety of approaches which exist within the "scientific method." Identify some of these and discuss them.

Suggested readings

Andrade, E. N. da C., "Rutherford and the Nature of the Atom," Anchor Books, Doubleday, Garden City, New York, 1964.

Lagowski, J. J., "The Structure of Atoms," Houghton, Boston, Massachusetts, 1964.

Romer, Alfred, "The Restless Atom," Anchor Books, Doubleday, Garden City, New York, 1960.

The architecture of an atom

Of atoms and α particles

This chapter is devoted to a critical study of the evolution of a metaphor for matter. Like a poetic metaphor, a scientific model serves to further understanding and aesthetic enjoyment. Both seek the end ascribed to acting by Hamlet, "to hold as 'twere the mirror up to nature." In order for the metaphor or model to be true and beautiful, its mirror must not distort. But it should do more than literally and slavishly reflect the world outside it, and here the metaphor of the mirror begins to fail. All we require of a looking-glass is clarity and fidelity. A meaningful model of nature also provides the insight which is born of imagination. Of course, the insight must stand the test of experience. Ultimately, it is on the basis of its profundity and perceptiveness that a metaphor, poetic or scientific, stands or falls.

To the author, a model of nature couched in scientific terms seems easier to evaluate than its poetic counterpart. For one thing, there is consensus on the methods which are to be used in the validation procedure. Experimental results, prevailing physical theory, the symbolic system of mathematics, reason, and logic constitute the materials and the means. And perhaps of greatest significance, the predictive powers of a model provide the final pragmatic test of its usefulness and relative truth. All of these were employed in the "discovery"—perhaps "creation" would be a better word—of the nuclear atom. To pretend that the process was a simple one is to misrepresent the facts. The experiments were often elaborate, the mathematics sophisticated, and the concepts difficult. Nevertheless, following this account of model building can be a challenging and rewarding excursion into the nature of matter and the nature of man's mind.

"I was brought up to look at the atom as a nice hard fellow, red or gray in colour, according to taste." So said Ernest Rutherford, looking back over his own education. But he had helped change all that. The experimental evidence amassed in the first decade of this century had made it manifestly clear that such a billiard ball atom was inadequate as a model of matter. To be sure, the analogy was not useless. It remained then, as it does now, a valuable and valid explanation for the mass laws of chemical combination and many of the properties of gases. Nevertheless, it was insufficient in the face of the new data. The time was ripe for a model which would encompass electrons, X rays, radioactivity, and elementary transmutation.

One of the first attempts at a revised theory of atomic structure was the work of Lord Kelvin and J. J. Thomson. In their model, electrons were embedded in a uniform sphere of positive electricity like raisins in a thick English pudding. Nature would not permit the electrons to be randomly distributed within the atom. Their common charges would cause them to repel each other and form certain stable configurations. For example, two negative particles would orient themselves opposite each other along a diameter of the sphere and separated by a distance equal to the radius of the atom. As the number of electrons increased, a series of concentric, spherical shells of seven or eight electrons each was postulated. All this was compatible with the laws of classical physics. Such a structure would normally be stable, but it could also account for the occasional escape of electrons as β particles or cathode rays. On the other hand, the emission of positively charged α particles was more difficult to explain if the atom were indeed much like a miniscule raisin pudding.

The chief failing of the Kelvin-Thomson model, however, was in its inability to account for the scattering of α particles by thin metal foils. The research was again performed under the direction of Rutherford, but by a young German scientist named Hans Geiger, inventor of the Geiger counter. The foils were positioned in the path of a narrow beam of α particles emitted by a sample of radon. The end of the 6-foot vacuum tube was coated with a thin layer of zinc sulfide. Every impact of an α particle on this screen was signaled by a flash of light or scintillation. These flashes could be observed through a microscope in a dark room and could be counted with considerable accuracy (and eyestrain). Moreover, the trajectory of the particles passing through the foil could be precisely measured.

The pattern formed on the screen indicated that the beam was more diffuse after leaving the foil than before it entered it. In other words, the stream of radiation was spread out by its encounter with the thin sheet of metal. This meant that the positive bullets fired at the foil were being deflected or scattered by the metallic atoms of the target. The experiment further indicated that the number of particles deflected through a given angle decreased as the size of the angle increased. Thus, the paths of most of the α particles were only slightly bent as they passed through the metal.

The fact that the great majority of the positively charged particles penetrated the foil with apparent ease and little change in direction was somewhat difficult to rationalize in terms of the Kelvin-Thomson model. Recall that according to this theory, the atom was a sphere of positive electricity containing neutralizing electrons. A sample of a metal, then, should be an array of these units, presumably in close contact—a sea of positive electricity studded with tiny negative nuggets. Even a very thin sheet would be several hundred atoms thick and hence would present a formidable barrier of diffuse positive electricity. It follows logically that the similarly charged α particles should have been stopped after penetrating only a few atomic radii into the metal. But in fact, they were not. The experimental results were at variance with the predictions of the model.

Even more irreconcilable with the idea of matter as raisin pudding were the results obtained when Rutherford's colleagues looked for large-angle reflections of α particles. Previous experiments had concentrated upon the region immediately behind the foil, and the observed scattering had been confined to a relatively narrow angle. In 1909, Geiger and a 20-year-old undergraduate, Ernest Marsden, moved the scintillation screen around to the front and side of the target. In this position deflections of 90° or more could be determined. Much later, Rutherford admitted that he had expected there would be nothing to measure. After all, the minute missiles emitted by the radon were traveling at 10,000 miles per second. It seemed highly unlikely that enough small encounters could occur between an α particle and a whole series of atoms to cause the particle to be ultimately reflected through a large angle. Nevertheless, the unmistakable flashes of light on the zinc sulfide screen soon proved Rutherford wrong. "It was quite the most incredible event that has ever happened to me in my life," he said,

FIGURE 7.1.   *Rutherford's α-particle scattering experiment.*

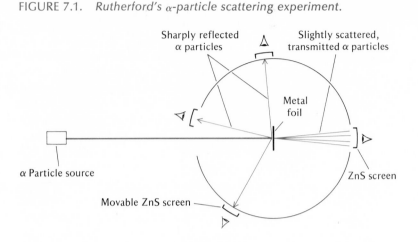

as an old and honored scientist. "It was almost as incredible as if you fired a 15-inch shell at a piece of tissue paper and it came back and hit you."

Faced with the challenge of making the incredible intelligible, Rutherford poured over the data. Geiger had shown that the most probable angle of deflection for a beam of α particles passing through a gold foil 4.0×10^{-5} cm thick was 0.87°. The result was consistent with the distribution calculated by assuming that the observed deflections are the result of multiple small deflections, each averaging about 0.005°. However, this mechanism of multiple scattering could not account for the 1 particle in 20,000 which was found to be reflected from the foil at right angles to the incident beam. Something like 18,000 successive encounters, all in the right direction, would be required to turn the beam 90°. In a foil only several hundred atoms thick, this was so improbable as to be an almost certain impossibility. And there were deflections at even greater angles.

Rutherford realized that the explanation of α particles that bounced back from metal foils like squash balls from the fore-court wall must lie elsewhere. The heart of the matter was literally that—the center of the atom! If, Rutherford argued, the positive charge of an atom were concentrated in a very small region in its center, the vast majority of an atom's total volume would be empty space. Specifically, he estimated the radius of the *nucleus* (as the center soon came to be called) at 10^{-12} cm, compared with a typical atomic radius of 10^{-8} cm. Roughly the same ratio would result if the nucleus were the size of a baseball and the atom were a sphere half a mile in diameter. Clearly, if one were to suspend baseballs half a mile apart and, without aiming, throw other base-balls at them, hits would be extremely infrequent. Similarly, α particles, about the size of atomic nuclei, would encounter little

FIGURE 7.2.   *Interaction of α particles and atoms as predicted by the Kelvin-Thomson model and the Rutherford model.*

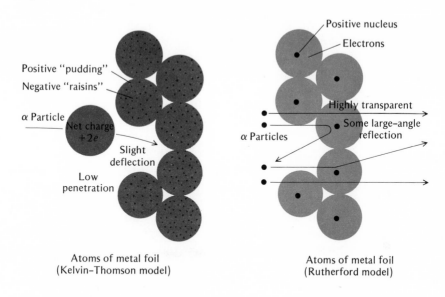

Atoms of metal foil
(Kelvin–Thomson model)

Atoms of metal foil
(Rutherford model)

resistance in their passage through an array of almost-empty atoms. To be sure, similarly charged particles, unlike baseballs, do repel each other at a distance. Direct contact is not necessary. Nevertheless, according to calculations based on the theory, an α particle could come as close as three diameters of the nucleus and suffer deflection of less than 10°. In this manner, the relative transparency of metal foils to positive particles could be explained by Rutherford's model.

But experiment required more than this; a satisfactory theory of atomic structure must also explain large-angle scattering. Again, the nuclear atom was equal to the task. Occasionally, Rutherford continued, the encounter between an α particle and an atomic nucleus was much more direct and more violent than a mere passing nod. In an event which almost approached the old paradox of the confrontation between an irresistible force and an immovable object, the two minute particles would meet head on—or very nearly so. A fantastic repulsive force would build up between the two highly concentrated positive charges—enough force to radically change the direction of the missile and send it ricocheting back toward its source. It was in terms of such direct hits and near misses that Rutherford succeeded in explaining the rarity and the severity of α-particle reflection. This was something that the Kelvin-Thomson model, with its much more diffuse positive charge, could not begin to do.

As a further test of his nuclear atom, Rutherford predicted that the extent of forward scattering (within certain limits) should be directly proportional to the thickness of the target foil. In other words, the number of reflected particles should increase with increasing metal thickness. The basis for this prediction was the single-encounter reflection mechanism and common sense. After all, doubling the thickness of a sheet of metal doubles the number of atoms and atomic nuclei present in a uniform, cross-sectional area. If the number of potential targets is thus increased, the number of reflections should increase proportionately. When the experimental scattering results were found to show just such a dependence, it afforded further proof that large-angle deflection could not be the sum of many small turnings. The latter mechanism predicted that the scattering should increase as the square root of the thickness of the foil, not the first power. It should also be noted that the observed dependence of scattering upon thickness also indicates that the phenomenon is not merely a surface effect. In scattering experiments, α particles actually penetrate into the atoms of the metal.

Rutherford concluded that the nuclei, which he posited at the centers of all atoms, must be different for each element. For one thing, he suggested that most of the mass of an atom was concentrated in its nucleus. Certainly, there was ample supporting evidence that positive particles, for example, α particles and anode ray particles, were much heavier than electrons. Thus, it was not unreasonable that the nuclear weight of an element would be very close to its atomic weight. However, reason reels at the density calculated for nuclear material on the basis of Rutherford's assumptions. For example, the mass of an individual atom of gold is about 3.3×10^{-22} g. For the sake of our argument, assume it is all concentrated in the nucleus. This center, according to Rutherford's estimate, is a sphere with a radius, r, of 10^{-12} cm. Consequently, this value of r can be substituted into the equation for the volume of a sphere, $V = \frac{4}{3}\pi r^3$. The calculation yields 4×10^{-36} cm^3 as the approximate volume of the nucleus. Combining these mass and volume values, the predicted density of the nucleus becomes 3.3×10^{-22} g$/4 \times 10^{-36}$ cm^3, or about 8×10^{13} g/cm^3. The number takes on added meaning when one realizes that a pocket-sized matchbox full of matter this dense would weigh over 2½ billion tons!* In fact, a matchbox full of metallic gold (density 19.3 g/cm^3) really weighs only slightly more than a pound. But then, metallic gold, although one of the densest of elements, is mostly empty space.

All this was fascinating, but as far as the scattering experiment was concerned, the charge on the nucleus was more important

---

* Incredible as it seems, astrophysicists have suggested that the matter in the interior of a collapsed star may reach approximately this density before it explodes as a supernova.

than its mass. Nevertheless, according to Ernest Rutherford's logic, the two must be related. His guess was that the positive charge was proportional to the atomic weight of the element. For example, the charge at the center of an atom of gold (relative weight 197) should be about 7.3 times the nuclear charge of aluminum (atomic weight 27), because $197 = 7.3 \times 27$. Since this charge was a chief factor influencing the deflection of an α particle, a gold atom should present a larger target than an aluminum atom. Rutherford could calculate just how much larger, because the extent of forward scattering is proportional to the square of the charge on the target. This means that reflection from gold foil should be $(7.3)^2$ or about 53 times more intense than that from aluminum. Again, "the Prof's" prediction was confirmed by experiment, at least reasonably well. His guess had been roughly right.

This agreement prompted Rutherford to suggest that the nuclear charge of an atom, expressed in fundamental units of the charge on an electron, e, must be somewhere near one-half the atomic weight of the element in question. Thus, he postulated a value of 100e for the charge at the center of a gold atom. Of course, unlike electrons, the charge was presumably positive. The "presumably" appears in the last sentence because exactly the same scattering behavior predicted by Rutherford's calculations could arise from either a positive or a negative nucleus. A highly concentrated negative charge would pull an α particle back around itself and send it whizzing into space along the same trajectory dictated by an equally strong positive charge. But fairly early in this research, Rutherford began assuming that the nucleus was indeed positive and that the electrons required to electrically neutralize its charge were distributed in the atomic space surrounding it.

A new model had thus evolved, and according to its assumptions, the atom was far from a "nice hard fellow." Instead, it was mostly empty space. At its heart was most of its mass, positively charged and packed into an incredibly small volume. Around this nucleus, arranged in some yet unknown manner and held by some incompletely understood force, were the electrons. The conceptualization seemed to fit many facts admirably well. And, like any truly valuable theory, it was capable of further refinements which would increase its comprehensiveness. That evolution is the subject of the next section.

Of nuclei and numerology

Two years after the appearance, in 1911, of Rutherford's first monumental paper on "The Structure of the Atom," evidence which was used to extend the nuclear model emerged, again from the laboratory at Manchester. The research was that of H. G. J.

Moseley, a young physicist fresh from his bachelor's degree at Oxford. Moseley set out to investigate the spectra of the X rays evolved when a stream of cathode rays falls upon a metal anode in a gas discharge tube. The penetrating rays which Röntgen first observed had been formed by the interaction of electrons with the glass wall of the tube. However, subsequent study had shown that metals also emit X rays when similarly bombarded. Since this form of radiation is, like visible light, electromagnetic in nature, it is possible to determine the frequency and wavelength characteristic of X rays.

The principle employed by Moseley and his collaborator, Charles Darwin's grandson and namesake, was fundamentally that of the Kirchhoff-Bunsen spectroscope, but with some significant experimental modifications. Because the wavelength of X rays proved to be much shorter than that of visible light, a glass prism was not effective in splitting the spectrum into its various individual components. However, such diffraction could be achieved when a crystal of potassium ferrocyanide, $K_4Fe(CN)_6 \cdot 3H_2O$, was used in place of the prism. By this means, it became experimentally possible to analyze X-ray spectra, much as visible spectra had previously been studied. And, as in the case of visible spectroscopy, the pattern of the Röntgen rays emitted by a particular element was found to be uniquely characteristic. In fact, X-ray spectra proved to be simpler than visible spectra; there were fewer lines to interpret.

The first elements Moseley chose to investigate were ten metals, increasing in atomic weight from calcium through zinc (scandium was not initially included). Each of these elements exhibited one strong line and several weaker ones. In every case, the strong line proved to be the radiation specifically associated with the target metal. The weaker lines were soon shown to arise from the presence of small amounts of other, contaminating elements. Indeed, the extreme sensitivity of the X-ray spectra to minute quantities of elements at once led Moseley to recognize the implications of his work for analytical chemistry.

The prevalence of lines due to impurities suggests that this may prove a powerful method of chemical analysis. Its advantage over ordinary spectroscopic methods lies in the simplicity of the spectra and the impossibility of one substance masking the radiation from another. It may even lead to the discovery of missing elements, as it will be possible to predict the position of their characteristic lines. *

* H. G. J. Moseley, The High Frequency Spectra of the Elements, *Philosophical Magazine* **26**, 1024 (1913).

Powerful as the method was for the detection of elements, it had an even greater impact in elucidating the basis of elementary periodicity. The reason lay in the regularity relating X-ray spectra and atomic sequence. It was first noted that the frequency of the observed characteristic line generally increased with the increasing atomic weight of the element which emitted it. Understandably, Moseley was interested in the origin of the X rays, and so he adopted an approach which is really quite fundamental to physical science. The idea is to take two (or more) properties which change simultaneously, for example, frequency of X-ray spectral lines and atomic weight, and determine the mathematical relationship between these variables. If a relatively simple dependence does exist, it is an indication that there may also be a causal relationship. Of course, a mathematical correlation is by no means an unambiguous proof of an underlying physical phenomenon—the connection could be coincidental. But such mathematical maneuvers often do provide important evidence, particularly since the form of the dependence may suggest the mechanism by which the properties are related. For example, you will recall from your reading of the previous section that Rutherford made good use of simple mathematical relationships between measured variables to help establish the model of the nuclear atom.

A convenient way to visualize mathematical and physical dependencies is by drawing graphs. Perhaps Moseley began his analysis of the X-ray data by simply plotting frequency against atomic weight. If so, he soon saw that the resulting graph was a curve, and not a very informative one at that. Straight lines are much nicer. They mean the quantities graphed are simply related to each other, either increasing or decreasing in a regular, constant fashion. And so, Moseley tried to get a straight line ("linearize the data" in the jargon of science) by plotting the square root of the frequency versus atomic weight. Figure 7.3, which is based on his actual data, indicates that this approach came quite close to yielding the desired result. But it was not close enough to permit easy interpretation. Therefore, Moseley graphed the square root of the frequency against the order number of the elements in the periodic table instead of the atomic weight. This *atomic number*, as the young physicist called it, was simply symbolized as N. Hydrogen was given the value 1, helium was 2, lithium was 3, and so on, much as Newlands had suggested to his skeptical colleagues almost 50 years earlier. You will note, from the periodic table of Chapter 5 or the endpapers, that in most cases the sequence of elements thus obtained corresponds to increasing atomic weight. There are, however, several exceptions. For example, cobalt (atomic weight 58.93) comes before nickel (atomic weight 58.71). This apparent inversion makes eminently good sense on the basis

FIGURE 7.3.   *Plot of the square root of the frequency of the major X-ray spectral line vs. the atomic weight and atomic number of the emitting element. Note that the latter points fall on a straight line, but the former do not. (Data from Moseley's original paper.)*

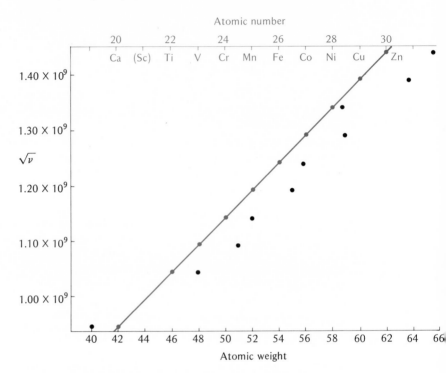

of the properties of the two elements and similarly, it is fully compatible with the X-ray results.

It is obvious, again from Figure 7.3, that plotted in this manner, the data give a beautiful straight line. Moreover, when Moseley applied this treatment to all the elements from aluminum ($N = 13$) through gold ($N = 79$), the same dependence was observed over the entire range. To be sure, X-ray spectra had not been measured for all 67 elements in this series. As a matter of fact, not all of them had been discovered. But, like Mendeleev before him, Moseley left gaps for these. Again, this foresight was justified.

If this linear dependence of $\sqrt{\nu}$ upon N were nothing more than another method of elementary bookkeeping, Moseley's name would probably not have long survived 1915 and his death in World War I. But, two years earlier, at the age of 26, he had recognized that something physical must underlie this disarmingly simple regularity.

*We have here a proof that there is in the atom a
fundamental quantity, which increases by regular steps as
we pass from one element to the next. This quantity can
only be the charge on the central positive nucleus, of the
existence of which we already have definite proof.**

The magnitude of this charge is $+Ne$, where e is the charge on an
electron. Thus, all elements have positive nuclear charges which
are integral multiples of a fundamental unit. *The atomic number
is simply the number of these charge units.*

The concept of atomic numbers immediately illuminated the
issue of elementary identity and simultaneously shed light upon
the related subject of atomic structure. Moseley's generalization
made it quite clear that *nuclear charge and not atomic weight
determines which element a particular atom represents.* Consequently, in a very real chemical sense, atomic number is a more
fundamental property than atomic weight. To begin with the first
and lightest element, all atoms of hydrogen have in common a
nuclear charge of $+e$, that is, $+1.60 \times 10^{-19}$ coulombs. At some
distance from the nucleus (a miniscule distance in absolute terms,
but immense on the scale of the atom) is an electron which balances the positive charge and thus renders the entire hydrogen
atom electrically neutral. Each helium atom contains two electrons
to offset the nuclear charge of $+2e$. Similarly, the common structural characteristic of all lithium atoms is the presence of three
electrons and an opposite nuclear charge of three units. In this
manner, the model-building and element-building continues, the
number of electrons and the magnitude of the nuclear charge each
increasing by one unit at every step of the way from hydrogen
($N = 1$) to hahnium ($N = 105$).

In contrast, atomic weights do not increase with such integral
regularity. Rutherford's estimate that nuclear charge is proportional to atomic weight was, therefore, not exactly correct. A quick
consultation of the periodic table indicates that atomic weight increases more rapidly than atomic number. As a result, the nuclear
charge on gold is $+79e$ and not $+100e$, as Rutherford guessed.
In its basic features, however, the nuclear model remains valid.
Specifically, Rutherford's conclusion that all of the positive charge
and almost all of the mass of an atom is concentrated in its nucleus
is consistent with X-ray spectra. While the charge is fixed for a
specific element (indeed, the charge specifies the elementary identity of the atom), the mass is not. Soddy's isotopes were examples
of just such heterogeneity with respect to atomic weight. Now,
Moseley's data helped explain the contradiction of Daltonian

---

* H. G. J. Moseley, The High Frequency Spectra of the Elements,
Philosophical Magazine **26**, 1031 (1913).

Atomic Theory implicit in the observation that a given element can consist of atoms of varying weights. To use an example we encountered earlier, individual atoms of thorium can exhibit relative masses of 227, 228, 230, 231, and 232. In other words, their nuclei differ in mass, yet the atoms all possess the chemical properties of thorium. What all these isotopic atoms do have in common is a nucleus with a charge of $+90e$, 90 extranuclear electrons, and an atomic number of 90, which signifies all this. Similar arguments can be applied to any element.

Protons, neutrons, and other peculiar particles

Given the substantial experimental evidence that the positive charge on the nucleus is a fundamental and uniquely characteristic property of an atom, the next problem involves assigning these charges to physical entities. After all, the negative charges in an atom are borne by identical particles called electrons. Thus, it seems likely from the integral increase in atomic number that discrete, positively charged particles should also exist. Moreover, the hydrogen ion, H^+, suggests itself as an example of such a particle. Stripped of its lone electron, the nucleus of a hydrogen atom carries a charge of $+e$ and almost all the weight of the atom, 1837/1838 of it to be exact. Since the nuclear weight is thus essentially identical to the atomic weight, we can closely approximate its value as unity on the standard scale. The temptation, then, is to postulate that the nuclei of all atoms of all elements are built up of such particles, that is, of hydrogen nuclei. To be sure, some experimental proof is desirable, and in 1919 just such corroboration was obtained. Rutherford had tried his old trick of bombarding elements with α particles. This time, nitrogen gas and other light elements were the targets. The result of the interactions were new particles, knocked out of the target atoms by the bombarding α particles. Measurements of charge-to-mass ratio quickly substantiated the fact that these products were indeed hydrogen nuclei, or *protons* as Rutherford christened them.

But it was obvious that the elements could not be simply built up from equal numbers of electrons and protons. To be sure, this did work for hydrogen—but for no other element. The two electrons assigned to helium could be neutralized with two nuclear protons, but the resulting atom would have a relative mass of 2 instead of the observed value of 4. The extra mass had to come from somewhere. To account for it, Rutherford assumed that the helium nucleus contained not two but four of the heavy, positive particles. This, of course, would cause a net charge imbalance of $+2e$ on the entire atom. So, in order to return the helium atom to its known state of electrical neutrality, Rutherford postulated the presence of two additional electrons in the nucleus. These electrons would contribute insignificantly to the total mass of the atom while solving the problem of charge.

The idea seemed to fit the structure of any atom. By combining the proper number of electrons and protons, Rutherford was able to account for the atomic numbers and the atomic weights of all elements and all isotopes. The atomic number continued to represent the number of extranuclear electrons and the charge on the nucleus. For example, lithium is characterized by a value of 3 for these two properties. But lithium exists in isotopes of atomic weight 6 and 7. The theory attributed this mass to six and seven protons, respectively. In the former case, the difference between the isotopic mass and the atomic number, $6 - 3$, implied the presence of three electrons in the nucleus. In the latter case, four electrons were required to neutralize the four excess protons and bring the net nuclear charge to $+3e$, the value dictated by the atomic number.

The concept of nuclear electrons was beautiful in its simplicity. According to the model, all the matter in the universe was composed of only two fundamental particles, protons and electrons. From these basic building blocks arose the great heterogeneity of a hundred elements and the even greater diversity of thousands of compounds. But the scientist must consider conformity to experimental fact as well as aesthetic appeal in evaluating his theories. And a very important fact emerged from the Nobel Prize-winning work of James Chadwick.

Around 1930, Chadwick commenced a study of the radiation resulting when beryllium or boron is bombarded with α particles. The induced radiation, which had been known for some time, was originally though to be a form of γ emission. And the assumption was a logical one, since the rays were highly penetrating and uninfluenced by electrical charges. However, Chadwick suggested and proved that the radiation is not part of the electromagnetic spectrum, but rather consists of neutrally charged particles. Each of these *neutrons* was found to have a mass very nearly identical to that of a proton, that is, about 1 on the conventional scale.

The discovery of the neutron removed the necessity of nuclear electrons as a means of balancing the charge of the presumed excess protons. According to the newer theory, the number of nuclear protons is identical to the number of extranuclear electrons. The difference between the atomic or isotopic mass and the atomic number is attributed to the presence of neutrons in the nucleus. Essentially all of the mass of the atom is due to the combined masses of the protons and neutrons. Indeed, the total number of protons and neutrons in an atom is known as its *mass number.* It follows that different isotopes of a single element consist of atoms containing a constant number of protons, but a variable number of neutrons. The isotopic weights and mass numbers vary accordingly.

To return to an elementary example we considered earlier, each atom of lithium has a nucleus which includes three protons. The nucleus of the isotope with atomic weight 6 also contains three neutrons. On the other hand, there are four neutrons in each nucleus of the lithium isotope which has a mass number of 7. Of course, the atoms of both isotopes contain three electrons, well removed from the nucleus. Similarly, hydrogen exists in three isotopic forms. By far the most common species has an atomic weight of very nearly 1. However, one atom in 6500 exhibits a relative mass of about 2. In addition to this deuterium or "heavy hydrogen," there is an even heavier isotope. Tritium, as it is called, has an isotopic mass of approximately 3. The natural abundance of tritium is very low, only one atom out of ten million. But because this species is radioactive, its presence in discharge from nuclear power plants is a cause of considerable concern. The fundamental particles making up the three isotopes of hydrogen are summarized in Table 7.1.

A further indication of the adequacy of this model is the way in which it explains elementary transmutation by radioactive decay. An α particle, the nucleus of a helium atom, consists of two protons and two neutrons. This is fully consistent with its observed mass of 4 (on the atomic weight scale) and its charge of $+2e$. The loss of an α particle by a radioactive isotope should thus reduce the mass of the emitting atom by 4 units and the nuclear charge by 2. For example, uranium-238, an α-emitter, has 92 protons and 146 neutrons in its unstable nucleus. The loss of two protons and two neutrons, that is, an α particle, reduces these values to 90 and 144, respectively. But a nucleus with 90 protons is characteristic of element number 90, thorium. In conformity

TABLE 7.1.   *The Fundamental Particles in the Three Isotopes of Hydrogen*

	Isotopic mass	Outside nucleus	Inside nucleus	
		Electrons	Protons	Neutrons
Ordinary hydrogen	1.0078	1	1	0
Deuterium	2.0141	1	1	1
Tritium	3.0	1	1	2

with Soddy's generalization, the escape of an α particle has transmuted the emitting element into one belonging two places to the left in the periodic table. Moreover, the isotope of thorium formed will have an atomic mass number of 90 + 144 or 234.

Changes of this sort can be conveniently represented in radio-chemical equations similar to those used in the previous chapter. The approximate isotopic mass (the mass number of the atom) is written as a superscript to the elementary symbol, and the nuclear charge (number of protons or atomic number) appears as a subscript. Thus, the equation for α emission by uranium-238 is written*

$$^{238}_{92}U \rightarrow \ ^{234}_{90}Th + {}^{4}_{2}\alpha.$$

Note that the charge is conserved, that is, the subscripts add up to the same value on both sides of the arrow. Moreover, mass appears to be constant, as in a balanced chemical equation. But in point of fact, nuclear reactions are often attended by the conversion of a detectable quantity of matter into energy—a transformation which has been made manifest in mushroom clouds.

The radioactive properties of the product formed in the α emission described above reintroduces the subject of β decay. Thorium-234 spontaneously emits β particles in accordance with the equation†

$$^{234}_{90}Th \rightarrow \ ^{234}_{91}Pa + {}^{0}_{-1}\beta.$$

Since the mass of the β particle or the electron is very small compared to that of the atoms involved, we simply write 0 as its superscript. Consistent with this, the isotopic masses remain essentially unaltered at 234. On the other hand, the emitted β particle does carry a charge of $-1e$, indicated by the subscript. The loss of such a particle thus increases the nuclear charge by $+1e$. And what was formerly thorium (atomic number 90) is now protactinium (atomic number 91). The shift on the periodic table is one place to the right, in keeping with Soddy's second rule. But the explanation of this change is a little awkward. One thing is clear, however, the emitted electron could not have come from outside the nucleus. Nuclear charge gives an element its identity, and this is a nuclear transformation.

At this point it is tempting to resurrect Rutherford's old idea that nuclei consist of only protons and electrons. According to this model, the loss of an electron or β particle would simply mean one more proton with an uncompensated charge, and thus an increase of 1 in the net nuclear charge and the atomic number. But the existence of neutrons necessitates a somewhat different argu-

---

* Because α particles are helium nuclei, they are sometimes symbolized ${}^{4}_{2}He$.
† Because β particles are electrons, they are sometimes symbolized ${}^{0}_{-1}e$.

ment. The most obvious one is to suggest that a neutron can be regarded as a combination of an electron and a proton. Certainly the charge of such a binary particle would be zero and its mass would be 1. Moreover, the escape of an electron from the nucleus would leave behind an extra proton in place of the neutron which was previously there. The atomic number would correspondingly increase by 1. We can, in fact, write an equation for this change. If $_0^1n$, and $_1^1p$, respectively, represent a neutron and a proton, the transformation is

$$_0^1n \rightarrow {_1^1}p + {_{-1}^0}\beta.$$

However, the physical equivalence implied by this equation should not be taken literally. It may satisfy a chemist's needs, but a physicist would regard it as rather naïve.

Modern nuclear physics has demonstrated that the center of the atom is far more complicated than we have here suggested. Under the impetus of atomic research, engendered in part by interest in nuclear energy, some 30 fundamental subatomic particles have been identified or postulated. They include positrons, neutrinos, mesons, leptons, hyperons, and their antimatter counterparts. These particles have decidedly nonchemical properties, even including something called "strangeness." Quite clearly, to a modern physicist, a model nucleus built exclusively of neutrons and protons is as oversimplified as is a billiard ball atom to a contemporary chemist. In fact, today physicists' ideas of the atomic nucleus are about as badly confused and as much in need of clarification as were chemists' concepts of gross atomic structure at the turn of the century. Nevertheless, it is precisely because this proliferation of particles is so "strange" and so nonchemical, that it is beyond the scope of not only this book, but the branch of science it treats. It is in keeping with the economy which characterizes nature and should characterize its investigators, that the model or theory used should fit the problem at hand. For a great majority of chemical problems, electrons, protons, and neutrons are perfectly sufficient to describe matter and its interactions. To include more detail would unnecessarily complicate matters without adding any significant new insight. Indeed, as we had suggested before, for some chemical purposes, even indivisible, integral atoms are adequate.

In the spirit of the last sentence, we might well ask how it was that the Daltonian atom managed to live such a long and useful life. After all, the theory was mistaken on a number of important counts. Contrary to what Dalton thought, all the atoms of a given element need not have identical masses. Moreover, atoms of different elements occasionally exhibit very nearly equal masses. Providing further potential complication is the fact that atoms can change their structure or constitution by radioactive decay and thus be transmuted into other elements. And of course, the nineteenth century theory was silent on nuclear charge or atomic number.

In principle, any of these non-Daltonian aspects of atomic structure could be manifested in such ways as would make simple atomic relationships almost impossible to discern. For example, if isotopic composition varied widely from one sample of an element to another, the mass Laws of Constant Composition and Multiple Proportions would not hold. Thus, the idea of the combination of integral numbers of atoms to yield molecules of compounds could not have been inferred without great difficulty. Even the concept of atomic weight would have lacked authority. This, in turn, would have undoubtedly inhibited the evolution of periodic classification. And, if radioactive decay were more common than it is, elementary transformations would have further obfuscated the issue.

Fortunately for the beginnings of chemistry, to a good first approximation, matter is considerably simpler than it is in its fine structure. In the first place, the distribution of isotopes of any element is generally quite constant, no matter what the source of the substance. Thus, the average atomic weight of a mixture of isotopes is essentially invariant within the limits of conventional mass measurement. This statistical regularity is, of course, manifested in the mass laws of chemical composition. A second example of the convenient cooperativeness of nature is the fact that atomic weight almost always parallels atomic number. The fundamental importance of the nuclear charge was not known until 1913; and Dalton, Mendeleev, and many others assigned to the atomic weight primary responsibility for the determination of elementary identity and chemical and physical properties. Nevertheless, thanks to the relationship between nuclear charge and nuclear mass, the incorrect interpretation made little practical difference for such important developments as the periodic table. Finally, natural elementary transmutations did not present a major impediment to the usefulness of the Daltonian atom. Radioactivity at such high levels that it causes transmutation of large quantities of matter is so rare that it does not significantly influence most analytical results.

But in spite of its durability, the Daltonian atom could not survive the onslaught of experimental data that ushered in this century. By 1913, only 20 years after X rays first signaled the complexity of the atom, the same rays provided a key clue to the structure of the most minute complete entity of any element.

The case of the curious quantum

Perhaps about now, the reader is wondering whatever happened to chemistry—test tubes full of smelly stuff and all that sort of thing. Admittedly, α-particle bombardment and X-ray spectroscopy seem much more the methods of physics. And even the scientific heroes we've been praising lately have called themselves physicists, not chemists. All this is true enough. But it is also true that some of these physicists won Nobel Prizes in Chemistry. The selection committee for those most prestigious of awards realized that fundamental research in physical phenomena was leading to an understanding of the atom which would ultimately illuminate chemistry. Chemists can hardly begrudge giving their prize to an outsider like Rutherford, not when his contributions to chemistry are considered. To paraphrase Newton's famous remark, if chemists have seen farther than others, it has been by standing on the shoulders of giants who were physicists.

In order to see farther into modern chemistry, you still need a bit more of that elevated overview. And therefore, this section is devoted exclusively to physics and the curious *concept of the quantum—the idea that energy, like matter, comes in little bundles.* We, or rather Niels Bohr, will make brilliant use of this theory in the final part of this chapter. There, a mathematically exact model of matter will at last emerge.

The areas in which evidence for the discreteness of energy was first amassed are far removed from conventional chemistry. Indeed, the genesis of these atoms of energy had little to do with more material atoms. There were two specific test cases: the distribution of energy radiated by an incandescent body and the release of electrons which takes place when light falls on certain metals. In both instances, difficulty arose from the inability of the prevailing paradigms to explain empirical evidence. Only when energy was granulated into tiny corpuscles and a new theory thus began to emerge could the experimental results be rationalized.

Initially, Max Planck's quantum hypothesis was introduced to account for the energy given off by radiating bodies. Our discussion of this phenomenon will be necessarily simplified—not because you are intellectually incapable of understanding the subject, but because a substantial background in physics and mathematics is required. You can, however, readily comprehend Planck's conclusions and appreciate their importance. And you have often observed the experimental evidence which prompted them. The heating element of a cold electric burner is dark gray; it absorbs most of the light that falls upon it. When the control is turned to low, the burner begins to emit radiation, at first only infrared rays which can be felt as heat but not seen. As the temperature of the heating element is increased, it starts to glow: at about 500°C it is a dull red, then a cherry color, and finally, at the highest setting of the thermostat, a bright orange.

It turns out that the light which characterizes any particular temperature is not of a single wavelength, nor is it a set of specific lines, as in the atomic spectrum of sodium. Instead it is a continuum of colors, a smooth distribution of light of varying frequency. Thus, it resembles the radiation emitted by the sun or the glowing filament of an incandescent light bulb. Now in none of these instances is the intensity of the light uniformly strong across the entire spectrum. The very fact that the burner glows red at a certain temperature means that more red rays must be emitted than yellow or green or blue. Put in slightly more graphic terms, there is a maximum in the light distribution curve, a most common wavelength corresponding to the observed color. The experimental data incorporated in such a curve are obtained by measuring the energy associated with light of a specific wavelength, and repeating such measurements at a number of other wavelengths. Plotting corresponding values of energy and wavelength yields a curve of the sort reproduced in Figure 7.4. The hump clearly indicates the wavelength associated with the maximum energy.

You will see that the same figure includes curves for three different temperatures. As the temperature increases, the position of the energy maximum shifts to shorter wavelengths. This, of course, is just another way of recording what we noted earlier—the fact that the color of an electric heating element depends upon its temperature. The hotter the burner, the shorter the wavelength associated with the radiation of greatest intensity. Moreover, the trend continues to even higher temperatures. Thus, the "white hot" filament of an ordinary electric light bulb has a temperature of about 2000°C. The white light indicates that most of the visible spectrum is emitted with roughly equal intensity. At still higher

FIGURE 7.4.   *Distribution of wavelengths from a radiating body.*

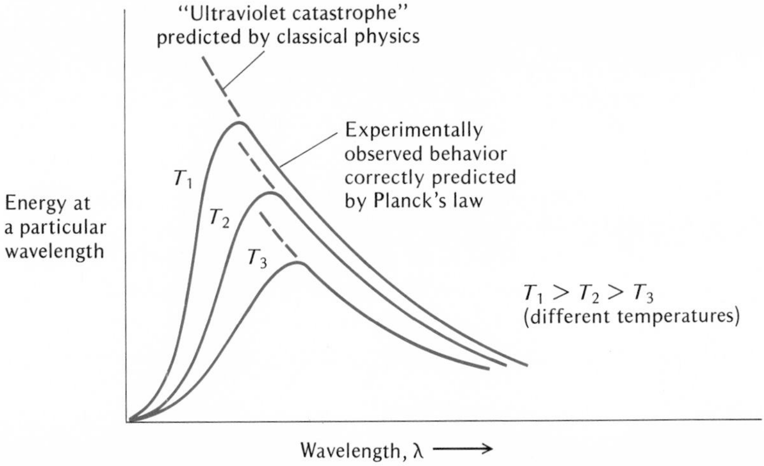

"Ultraviolet catastrophe" predicted by classical physics

Experimentally observed behavior correctly predicted by Planck's law

Energy at a particular wavelength

T_1

T_2

T_3

$T_1 > T_2 > T_3$
(different temperatures)

Wavelength, $\lambda \longrightarrow$

temperatures, such as those of a welder's electric arc, the predominant visible radiation is at the blue-violet end of the spectrum. Other radiation emitted by the arc lies in the ultraviolet region, beyond our limits of visual perception. In spite of the fact that this radiation is unseen, it can be damaging to the eyes, and therefore, welders' goggles contain special glass which filters out high-frequency rays.

It is a great strength of physical science that it can reduce commonplace experiences to mathematical formalization and general causal relationships. We are headed in that direction and will get a good deal more exact about these and other familiar phenomena before this book is over. For the moment, let us continue our consideration of a radiating body. We have suggested that the experimental results summarized in Figure 7.4 are consistent with experience and compatible with common sense. Nevertheless, and this is of extreme importance, such behavior violates classical (in this case, pre-1900) physics. The paradigm which had prevailed for several centuries before that watershed year held that electromagnetic radiation is made up of waves of varying lengths. A wide variety of phenomena were subjected to laboratory investigation and almost all fit neatly into the accepted pattern. But radiation from glowing bodies was an exception. Here experiment and theory were clearly at variance.

On paper, the problem was treated by coupling the traditional wave concept of light with a statistical approach. The derivation made use of the widely held assumption that the energy associated with radiation of a given wavelength could have any value whatsoever. Indeed, to scientists steeped in the established tradition, the idea that energy is continuous was more than an assumption. It approached an article of faith. But here faith failed. Its application simply could not reproduce Figure 7.4. Specifically, there was no way to account for the temperature-dependent energy maximum; according to theory it had no business being there. And yet it was!

Physicists were well aware of this discrepancy and referred to it as the "ultraviolet catastrophe." Of course, it was a catastrophe only insofar as theory and theoreticians were concerned. Nature was working quite well, though no one knew how. After a number of attempts to reconcile classical concepts with experimental fact had failed, it became apparent that this behavior was inexplicable if the game of physics were played according to the generally accepted rules. Only one alternative remained—change the rules. And that was just what Max Planck did in his great quantum jump of 1900.

In retrospect, Planck's hypothesis seems deceptively obvious and not the leap of scientific faith it really was. What the Professor of Physics at Berlin suggested was that the energy of electro-magnetic radiation comes in little packages or *quanta*. These packages can have only certain energy values, values given by equation

$$E = h\nu.$$

It is apparent from this relationship that the size of the energy package, E, varies directly with the frequency of radiation, ν. The constant of proportionality, h, which relates these two variables has a fixed value for the entire spectrum. Under the usual system of units, the energy is expressed in ergs, the frequency in sec^{-1}, and Planck's constant has a value of 6.625×10^{-27} erg sec. Needless to say, this is an extremely small number and it reflects how tiny these bursts of energy are in the visible spectrum.

Consider, for example, the yellow D_1 line of sodium. Its frequency is 5.085×10^{14} sec^{-1}. Consequently, the energy associated with light of this specific color comes in bundles of 6.625×10^{-27} erg sec $\times$ 5.085×10^{14} sec^{-1} or 3.369×10^{-12} ergs each.* According to Planck's hypothesis, the energy possessed by a ray of light of this frequency, no matter how intense, must be some multiple of this characteristic quantum quantity. A greater intensity of radiation merely means more quanta are being emitted per unit time. But always in whole numbers, never in fractional parts. There is no such thing as half a quantum of sodium D_1 light, any more than there can be half an atom of sodium. To be sure, atoms smaller than the sodium atom do exist, but they are those of other elements. Similarly, quanta smaller than 3.369×10^{-12} erg are possible, but they no longer correspond to light of $\nu = 5.085 \times 10^{14}$ sec^{-1}. Light of longer wavelength and lower frequency is proportionately less energetic, as we qualitatively observed earlier. Radiation in the ultraviolet, X-ray, and γ-ray regions of the spectrum comes in considerably larger, more energetic bundles. Again, this is fully consistent with experience. No one ever got a sunburn by sitting in front of a radio (unless the radio was in the sun!) But γ rays can kill.

The size of these packages is less astounding than the fact that they exist—or at least have sufficient correspondence to reality to enable physicists to avoid the ultraviolet catastrophe. For when Planck's identity is introduced into the equation for energy distribution from a radiating body, the experimental results of Figure 7.4 are identically reproduced. There can be no argument with such spectacular success. Nonetheless, the idea that the energy of a ray of light can have only certain values, rather than a con-

---

* An erg is approximately the amount of energy required to raise a straight pin 1/100 of an inch against the earth's gravitational field.

tinuum, as predicted by classical wave theory, was undoubtedly aesthetically distasteful to a physicist steeped in pre-Planckian thought. In fact, Planck himself had misgivings about atomizing energy. But these cautions, bred of classical conviction, were of no avail. After 1900, physics, chemistry, and their description of the universe would never again be the same.

For five years after its appearance, Planck's hypothesis generated, in scientific circles, considerably more heat than light. But in 1905, Albert Einstein successfully applied the idea of quantized energy in his illuminating study of the *photoelectric effect*. This effect is the release of electrons from a metal plate under the influence of a light beam. In contemporary life, it is familiar as the phenomenon underlying the "electric eye." To nineteenth century physicists it had the dubious distinction of being another worrisome ripple in the generally calm and unbroken surface of orthodox mechanical and electromagnetic theory.

The difficulties can be easily summarized. In the first place, not all light is effective in causing the emission of electrons from a metal. Each element has its own threshold frequency. Radiation of a lower frequency, no matter how bright the beam or how long the exposure, cannot liberate electrons. Once this threshold is reached, however, electrons appear immediately, with no time lag between the moment of light-beam impact and electron evolution. Moreover, the number of electrons emitted increases in direct proportion to the intensity of the light. So long as the wavelength of the exciting radiation remains constant, the energy of the escaping electrons is unaltered. But if the frequency is increased above the threshold value, the energy of the electrons also increases. To be a bit more exact, this energy is directly proportional to the

FIGURE 7.5.   *The photoelectric effect. The influence of light intensity and frequency upon photoelectron emission.*

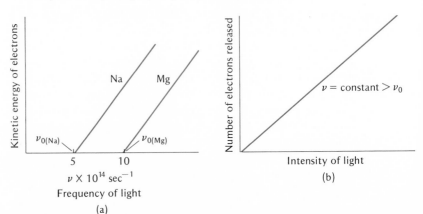

difference between the frequency of the incident light and the threshold frequency of the metal. Both of these behaviors are represented in Figure 7.5.

Let us for a moment consider the properties which the photo-electric effect should have shown if nature were docilely obeying the rules laid down by the classical theorists. The starting premise is one which must hold for any theoretical treatment of the phenomenon: energy is required to pry electrons out of a sheet of metal, and this energy must come from the light beam impinging on it. Classical concepts hold that the energy of a beam of radiation depends upon its intensity, that is, the amplitude or "swing" of its wave motion. A well-behaved metal target should continuously absorb this energy and store it, distributing it among all the electrons in the plate. If the incident light is dim, it might take a long exposure time, but sooner or later, an electron in any metal should acquire enough energy to escape. Other electrons would soon follow. Increasing the intensity of the incident beam would, of course speed up the process. Furthermore, greater brightness should also increase the energy of the emerging electrons. On the other hand, there is no reason to expect a linear dependence of the electron energy upon the frequency of the incident light, or to anticipate a threshold frequency.

Remember that the above paragraph was an exercise in what *ought* to be if classical physics holds for the photoelectric effect. Of course, it isn't that way at all. Each metal has its own threshold frequency, below which no electrons escape. The electron energy is independent of light intensity but a linear function of exciting frequency. And there is no delay time between light exposure and electron emission. Obviously, theoretical physics had flunked the test set by experimental physics. Some new answers were badly needed. They came from a man once classified as "uneducable" by his teachers.

Albert Einstein viewed the light-induced escape of an electron as the consequence of a single energetic event—not a gradual accumulation of energy. Taking his text from Planck, Einstein argued that radiant energy is packaged in quanta or *photons,* each identical in size for any specific wavelength or frequency. Symbolically, we again have $E = h\nu$. Moreover, he suggested that a photon cannot be divided or shared among electrons. The photoelectric effect is thus democratic but capitalistic: all electrons have an equal chance of acquiring a quantum of energy, but once an electron interacts with one of these photons, it does not redistribute the energy gain to its fellows. Rather, if the acquired energy is high enough to overcome the energy with which the electron is bound inside the metal, the electron escapes. This binding energy, often represented with a W, can be simply related to

v_0, the threshold frequency for photoelectric emission, by the equation

$$W = hv_0.$$

Clearly, if the frequency of the radiation is less than v_0, the energy from an individual encounter will never equal W and no electrons will be liberated. If, however, v exceeds v_0, the difference between the energy of a single quantum ($E = hv$) and the binding energy ($W = hv_0$) will be manifested in the kinetic energy (KE) of the escaping electron:

7.1 $$hv - hv_0 = E - W = KE = \tfrac{1}{2}mv^2.$$

The above equation also includes the definition of the kinetic energy of a particle as one-half of the product of the mass of the particle and the square of its velocity. It is just this energy which is plotted in Figure 7.5a. Equation 7.1 is a mathematical representation of the two straight lines and evaluation of their common slope yields Planck's constant. The fact that the two metals exhibit different values of v_0 of course reflects differences in electron binding energies. The electrons of magnesium are more tightly held than those of sodium.

Einstein's theory explains Figure 7.5b equally well. An increase in the intensity of light merely means that a greater number of photons fall on the target per unit time. As a result, there are more opportunities for electrons to gain sufficient energy to escape and the rate of emission increases. But so long as wavelength is fixed, the size of the energy bundle is also invariant. If this bundle is less than hv_0, no electrons will appear, no matter how bright the light or long the exposure. On the other hand, the electrons which will be liberated at any fixed frequency above the threshold limit will all have identical energy, independent of the intensity of radiation. The uncomfortably unfamiliar idea of grainy energy had again graphically demonstrated its scientific usefulness.

Like it or not (and many scientists didn't), Planck's concept as extended by Einstein was solving problems classical physics just could not handle. Ever since Newton devised his elegant scheme for an orderly universe, energy had been considered to be continuous. As far as any experiment could show, an unbroken infinity of values of energy were open to a planet or an apple falling on that planet. There was no reason to expect any part of the universe to behave otherwise. To be sure, matter did act as if it were atomic. But for centuries there had been no need to suggest that energy is also discrete. Now fact had forced just that conclusion on a reluctant Max Planck and a more accepting Albert Einstein. The success of the quantum hypothesis in explaining radiation and the photoelectric effect could not be denied. But unfortunately (perhaps fortunately), the curious concept raised more questions than it answered. Even after 1900, light still acted as if it were a wave: it could be bent, diffracted, and reflected; it had a wavelength and a frequency; and it continued to obey the classical laws of optics. But sometimes radiation seemed to be a string of little energetic corpuscles, arrows of light, if you like. Perfidious behavior for a clockwork universe! No doubt about it, the mechanism was showing definite signs of coming apart. It would have to be further disassembled before it could be reconstructed in a new form. And the twentieth century world would never be as neat as the happily humming machinery of the Enlightenment watchmakers.

The equation of an atom

By 1913 the model of the atom stood ready for final assembly. Although all the parts had not yet been discovered—the identification of protons and neutrons came later—there was general agreement on the basic design. Ernest Rutherford had started the construction by placing a minute but massive nucleus in the center of the atom and endowing it with a positive charge. H. G. J. Moseley had estimated the magnitude of that charge and found that it was the one atomic property which dictated elementary identity. It is true that the actual composition of the nucleus was not exactly known then, nor is it today. But in spite of that ignorance, one could, in 1913, proceed with the next step in model building —adding the electrons. Max Planck and Albert Einstein had provided, in their concept of quantized energy, an ingredient essential to that task. And now the architect of the atom entered the scene to supervise the construction.

He was a tall, young (27) Dane with a massive head and in it, a phenomenal brain. His name was Niels Bohr. Bohr realized full well that although Rutherford's nuclear atom was grounded on the firm experimental foundation of α-particle scattering, it was also a theoretical impossibility. The difficulty was associated with the extranuclear electrons postulated for the outer reaches of each atom. According to the laws of classical physics, they had no business being there. If the electrons were assumed to be fixed in space, electrostatic attraction would pull these negative particles into the positive nucleus. Even the expedient of setting the electrons in motion in circular orbits about the nucleus could not resolve the dilemma. To be sure, this motion would tend to counteract the mutual attraction between the oppositely charged bodies, much as the orbital path of the earth keeps our planet from being pulled into the sun. But this analogy, in spite of its familiarity, soon fails. An atom is definitely *not* a miniature solar system; the resemblance is at best superficial and at worst misleading.

One essential difference is obvious: planets are not electrically charged, electrons are. The consequences of this difference were well known in 1913. According to electrodynamic theory, developed in the previous century, a charged particle moving in a circle should constantly emit electromagnetic radiation. If an electron acted in this accepted manner, it would, as a result of this radiation, lose energy. This, in turn, would mean that the orbit of the radiating, revolving electron would gradually get smaller and smaller. "Gradually" is hardly the right word. In about one 100 millionth of a second the electrons of any atom would all spiral into its nucleus! But electrons are ignorant of the laws of physics. They continue to exist outside the nucleus, even as the earth continues to revolve about the sun in spite of Galileo's recantation. In neither case could the facts be altered or long ignored; the dogma ultimately had to yield.

Niels Bohr's great contribution was to supply an alternative theoretical system which accounted for the existence of nuclear atoms of the Rutherford type. Moreover, and this is of particular significance, Bohr's model beautifully and precisely predicted the visible spectra emitted by thermally excited atoms. You will recall that spectroscopy came into use about the time of the American Civil War. By the first decade of this century, literally thousands of spectroscopic lines had been measured, recorded, and organized. Hydrogen, the lightest and presumably the simplest of elements was subjected to particular scrutiny. Part of the hydrogen spectrum is reproduced at the top of Figure 7.6. Obviously, the spectrum is not continuous. The spacing of the lines is seen to decrease in the direction of increasing frequency or decreasing wavelength. Although it may not be evident to the untrained eye, there is regularity in this pattern, and the regularity can be translated into mathematical terms. The spectral lines can be fit to the equation,

$$v = R\left(\frac{1}{1} - \frac{1}{n^2}\right).$$

Frequency is given its usual symbol, v; R is a constant equal to 3.289×10^{15} sec^{-1}; and n has integral values of 2, 3, 4, 5, ...

From this equation, the frequency of any line in the so-called Lyman series of hydrogen can be calculated. For example, for $n = 2$, the frequency is $3.289 \times 10^{15} \left(\frac{1}{1} - \frac{1}{2^2}\right)$ or $3.289 \times 10^{15} \left(\frac{1}{1} - \frac{1}{4}\right)$. Evaluation leads at once to the observed frequency value for the first line in the series, 2.467×10^{14} sec^{-1}. Simple substitution of integrally increasing values of n likewise yields the higher frequencies of the other Lyman lines.

Observations of other regions of the hydrogen spectrum disclosed three similar series. They too could be summarized by analogous equations:

$$v = R\left(\frac{1}{4} - \frac{1}{n^2}\right) \qquad n = 3, 4, 5 \ldots \qquad \text{Balmer Series}$$

$$v = R\left(\frac{1}{9} - \frac{1}{n^2}\right) \qquad n = 4, 5, 6 \ldots \qquad \text{Paschen Series}$$

$$v = R\left(\frac{1}{16} - \frac{1}{n^2}\right) \qquad n = 5, 6, 7 \ldots \qquad \text{Brackett Series.}$$

An inspection of these four equations indicates that they can all be written in the general form

7.2 $$v = R\left(\frac{1}{n'^2} - \frac{1}{n^2}\right),$$

where $n' = 1, 2, 3$, and 4, respectively, for the Lyman, Balmer, Paschen, and Brackett series. In each case, the corresponding n values are integral numbers equal to $n' + 1, n' + 2, \ldots$ And in every instance, R has the same numerical value.

Equations of this sort are important aids in classifying and organizing data. But it should be emphasized that these are "empirical" equations. They are based solely on converting experimental data into mathematical form. When the equations were first proposed they had no theoretical basis. Nevertheless, when such regularity is encountered, it is seldom coincidental. Science is predicated on the implicit assumption that nature is orderly. And such experimental and mathematical regularity is, most likely, the consequence of physical order. The illucidation of the causative physical phenomenon was the ultimate proof and validation of Bohr's theory of the atom.

What Bohr did was to combine the equations of classical physics and the quantum concept to derive the energy of the hydrogen atom. What emerged was a picture of concentric circular orbits to which the electron is restricted. According to the theory, the space between the orbits is a no-man's land where electrons are not allowed. But an electron can leap these interorbital gulfs as it acquires or loses energy. Each of the orbits represents, in Bohr's terminology, a *stationary state,* and has associated with it a unique energy value. In other words, *the energy of the electron is quantized—only certain orbits and certain energies are allowed. Therefore, a specific change in orbit means a specific change in energy level. And it is this change in energy which is observed in the emission or absorption spectrum of the element.* The ultimate test of the model, then, was whether or not it could predict the experimentally measured spectral lines by accurately reproducing Equation 7.2.

The mathematics involved in the calculation is really quite simple, and you will probably be able to follow it easily. Nevertheless, the derivation does make use of a number of the formulas of physics which may be foreign to you. If you feel that accepting these concepts requires too much credulity for comfort, I suggest that you skip the next subsection and go directly to the discussion of Bohr's results. But the derivation provides an excellent example of the mathematical method of creating a scientific model. Thus, it could richly reward the effort you might expend in studying it.

THE DERIVATION

Bohr began his calculations in a traditional manner, using equations familiar to any beginning student of classical physics. He assumed that an electron moving in a circular orbit would have a centripetal force equal to mv^2/r. Here m is the mass of the electron, v its velocity, and r the radius of the orbit. This centripetal force would tend to pull the electron away from the nucleus, like a rock in a slingshot swung about the head of a giant-killer. But in an intact hydrogen atom the electron does not escape. It is balanced, Bohr argued, by the electrostatic force established between the negatively charged electron and the positively charged nucleus. If the former charge has a value of $-e$ and the latter an equal and opposite value of $+e$, the electrostatic force is $(-e)(e)/r^2$. When the electron is in a stationary state, that is, following a fixed orbit, there will be no net force acting on it. This implies that the sum of the centripetal and electrostatic forces must be zero. Or, in symbols,

7.3 $$\frac{mv^2}{r} - \frac{e^2}{r^2} = 0.$$

Bohr's next step was to calculate the energy of the orbiting electron. The Danish physicist was convinced that this property must be related to the observed spectral lines. After all, light is a form of energy; and since in nature energy is neither created nor destroyed, it must come from somewhere. He reasoned that the radiant energy of the emitting atom must signal a corresponding change in the energy of the electron. Again, his methods were the accepted procedures of orthodox physics. It was well established that a body can possess energy by virtue of its motion (kinetic energy) and its position (potential energy). For a body of mass m moving at velocity v, the translational kinetic energy is simply $\frac{1}{2}mv^2$. On the other hand, the potential energy of a charged particle in an electric field reflects its position relative to other charges. Specifically, the potential energy for the electron in Bohr's model is $(-e)(e)/r$. The total energy then becomes the sum of these two terms,

7.4 $$E = \frac{1}{2}mv^2 - \frac{e^2}{r}.$$

The above equation can be somewhat simplified using a previous result. Equation 7.3 can be rearranged so that $e^2/r = mv^2$. Substituting this identity for mv^2 in Equation 7.4 yields this new expression for the total energy of an electron in a hydrogen atom:

7.5 $$E = \frac{1}{2}\left(\frac{e^2}{r}\right) - \frac{e^2}{r} = \frac{-e^2}{2r}.$$

Thus far there is nothing extraordinary about this derivation. Equation 7.5 suggests that the energy of an electron orbiting about a proton is proportional to the square of the electronic charge and the reciprocal of the radius of the orbit. Since, according to classical physics, r can have any value, E can also have any corresponding value. Thus, the energy of the electron should continually become less negative as the size of the orbit continually increases. Almost any undergraduate student of physics or chemistry in 1913 could have done a similar calculation and deduced similar conclusions. The logic was impeccable and the theoretical axioms on which it was predicated were indisputable. But the result was unable to adequately explain either atomic structure or atomic spectra.

Confronted with this impasse, a lesser mind might have packed up his pencils and tried another field. Instead, Bohr introduced the essential idea—the quantum concept. He made the unprecedented assumption that *the angular momentum of the electron can have only certain values*. Angular momentum, as defined by Newton's laws of motion, is equal to *mvr*. Since v and r can classically assume any values, there is nothing about this equation which restricts the angular momentum to discrete values. But just such a quantum condition was imposed when Bohr supplied the right-hand side of the following equation:

7.6 $$mvr = nh/2\pi \qquad \text{where } n = 1, 2, 3, \ldots$$

This seemingly simple expression is the essence of Bohr's famous paper "On the Constitution of Atoms and Molecules." Translated into words, it makes the assertion that the angular momentum of an electron revolving about a nucleus can only have values that are integral multiples of Planck's constant divided by 2π. Moreover, it follows immediately that if the angular momentum of an electron is discrete and noncontinuous, values of both the radius and the energy must also be quantized. Algebraic manipulation of the above equations is all that is required.

A convenient first step in deriving an expression for r is to square both sides of Equation 7.6.

7.7 $$m^2v^2r^2 = \frac{n^2h^2}{4\pi^2}$$

Next, Equation 7.3 is rearranged to yield

7.8 $mv^2r = e^2.$

Dividing 7.7 by 7.8 eliminates velocity and gives the equation

$$mr = \frac{n^2h^2}{4\pi^2e^2}.$$

Therefore,

7.9 $r = \frac{n^2h^2}{4\pi^2e^2m}.$

The presence of the *quantum number, n,* in Equation 7.9 clearly indicates that *only orbits of certain size are "allowed"* in an atom. The other terms in the expression are all constants and can be readily evaluated numerically. Thus,*

$$r = \frac{n^2 \times (6.63 \times 10^{-27} \text{ erg sec})^2}{4 \times (3.14)^2 \times (4.80 \times 10^{-10} \text{ esu})^2 \times 9.11 \times 10^{-28} \text{ g}}$$
$$= 0.529 \, n^2 \times 10^{-8} \text{ cm.}$$

The smallest orbit corresponds to $n = 1$ and has a radius of 0.529×10^{-8} cm or 0.529 Å. The next permissible radius ($n = 2$) is 4×0.529 or 2.116 Å, the third ($n = 3$) is 9×0.529 or 4.761 Å, and so on.

Each one of these stationary states, with its unique orbit of constant, quantized radius, has associated with it a characteristic amount of energy. To obtain a mathematical representation of that energy, we need only substitute our expression for r (Equation 7.9) into Equation 7.5.

7.10 $E = -\frac{1}{n^2}\left[\frac{2\pi^2e^4m}{h^2}\right]$

The quantity in brackets is a constant, but n varies in integral "quantum jumps." Consequently, E does also. Bohr had atomized energy in order to energize the atom.

THE RESULTS

To recapitulate, the significant final equations of Bohr's mathematical treatment of the hydrogen atom are expressions for the radii (r) of the allowed orbits and the quantized energy (E) associated with these orbits,

$$r = \frac{n^2h^2}{4^2\pi^2e^2m} \quad \text{and} \quad E = -\frac{1}{n^2}\left[\frac{2\pi^2e^4m}{h^2}\right].$$

---

* In carrying out the above calculation, it is necessary to express the electronic charge in electrostatic units (esu) rather than the more familiar coulombs. 1 coulomb $= 3.00 \times 10^9$ esu.

Most of the symbols in these equations represent fixed constants of nature: e is the charge on the electron, m is its mass, and h is Planck's constant. But both r and E also depend upon n. This quantum number can only assume integral, whole-number values, 1, 2, 3, ... It thus follows that the orbital radius and its energy are also restricted to certain values.

The equation for the energy provides the primary point of tangency between Bohr's theory and experimental evidence. "Any emission or absorption of energy," he postulated, "will correspond to the transition between two stationary states." In other words, when an electron "jumps" between two allowed orbits, a photon of energy, $h\nu$ is absorbed or emitted. This energy, Bohr argued, must be identically equal to the energy difference between the two stationary states or *quantum levels* involved. For a transition between an initial state characterized by a quantum number n' and a final state of quantum number n, the energy difference is

7.11 $$h\nu = E_n - E_{n'}.$$

The discreteness of atomic spectral lines is thus a direct consequence of the discreteness of atomic energy levels.

Thanks to Equation 7.10, it is possible to carry out a numerical evaluation of Equation 7.11. Substituting, we have

$$h\nu = -\frac{1}{n^2}\left[\frac{2\pi^2 e^4 m}{h^2}\right] - \left\{-\frac{1}{n'^2}\left[\frac{2\pi^2 e^4 m}{h^2}\right]\right\}$$
$$= \left[\frac{2\pi^2 e^4 m}{h^2}\right]\left[\frac{1}{n'^2} - \frac{1}{n^2}\right].$$

Since it is frequency and not energy which is measured spectroscopically, we finally make the minor modification of dividing both sides of the equation by Planck's constant,

$$\nu = \left[\frac{2\pi^2 e^4 m}{h^3}\right]\left[\frac{1}{n'^2} - \frac{1}{n^2}\right].$$

The above equation seems familiar. It appears to be a theoretical counterpart of the purely experimental equation, 7.2. Just how good a counterpart it is can be ascertained by substituting the appropriate values for the constant term in the first set of brackets.

$$R = \frac{2(3.14)^2 \, (4.80 \times 10^{-10} \text{ esu})^4 \, (9.11 \times 10^{-28} \text{ g})}{(6.63 \times 10^{-27} \text{ erg sec})^3}$$
$$= 3.29 \times 10^{15} \text{ sec}^{-1}$$

Obviously, the presumption inherent in assigning it the symbol R has been fully justified. The result is identical to the empirically determined value. The constant and its equation permit the calculation of all the lines of the Lyman, Balmer, Paschen, and Brackett spectral series for hydrogen. Thus it was that Bohr's theoretical approach reproduced, on paper, the spectroscopic results of the laboratory.

The essential features of the Bohr atom are summarized in Figure 7.6. Here the allowed orbits of the stationary quantum states

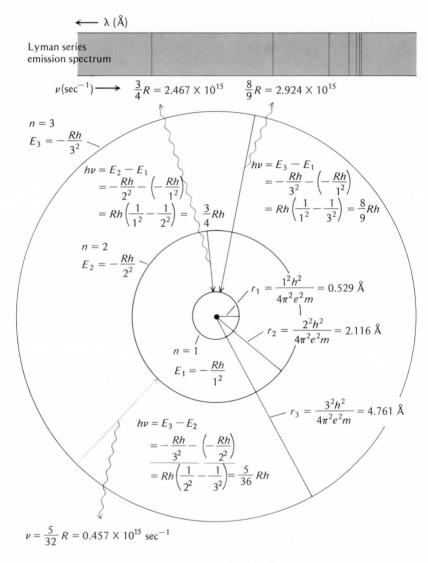

FIGURE 7.6.   *Diagram of the Bohr model for the hydrogen atom, showing the three lowest energy levels (n = 1, 2, 3). Expressions for the radius (r) and energy (E) of each of these orbits are given. The electronic transitions between n = 2 and n = 1 and between n = 3 and n = 1 correspond to the first two lines of the Lyman spectral series reproduced at the top of the figure. The transition between n = 3 and n = 2, represented in the lower left, gives rise to the first line of the Balmer series.*

are roughly drawn to scale and the transitions giving rise to the observed spectral lines are indicated. The smallest orbit represents the *ground state*, that of lowest energy. When the atom is excited, for example, by heat, it absorbs radiation of a characteristic frequency which promotes the electron into one of the more energetic outer orbits. Conversely, when an excited electron falls into an energetically lower level, it emits light. If the lower level involved in any transition is the ground state ($n = 1$), the corresponding spectral line occurs in the Lyman series. Transitions between the $n = 2$ quantum level and higher stationary states are seen in the Balmer series, and so on. In this manner, spectra provide the observable evidence of electronic transitions within the atom. But when the electron is in any given stationary state, it neither absorbs nor emits energy. The classical prediction of an electron spiraling into the nucleus while emitting a continuum of radiation just does not hold. The atom exists in defiance of classical physics.

It was because Bohr, as well as his atom, defied classical physics, that the theory was at first severely questioned. In spite of the unquestioned predictive powers of the model, critics found the admixture of classical and quantum concepts employed in the derivation to be aesthetically objectionable. Bohr himself acknowledged the problem in one of his basic postulates: "The dynamical equilibrium of the system in the stationary states is governed by the ordinary laws of mechanics, while these laws do not hold for the transition of one state to another." This was strange behavior, but it was becoming apparent that the atom was a strange place. Fortunately, some scientists quickly recognized that Bohr's unorthodoxy was a valid and visionary approach to such complexity. The words of one of these men, H. G. J. Moseley, provide an apt conclusion to this chapter:

I should be glad to do something towards knocking on the head the very prevalent view that Bohr's work is all juggling with numbers until they can be got to fit. I myself feel convinced that what I have called the "h" hypothesis is true, that is to say one will be able to build atoms out of e, m, and h and nothing else besides.*

As a matter of fact, the atom was soon to become even more abstract than either Moseley or Bohr had supposed. The refinement of spectroscopic techniques to higher resolution disclosed faint new lines which began to severely tax Bohr's model and require some rather awkward elaborations. Before the third decade of the century was over, another significant symbol, ψ (psi) would

---

* H. G. J. Moseley to Ernest Rutherford, early 1914. Quoted in "Rutherford and the Nature of the Atom" by E. N. da C. Andrade, p. 143, Doubleday, Garden City, New York, 1964.

be added to the alphabet of the atom. Moreover, it would be used in a mathematically complex new language. This method, called quantum or wave mechanics, forms the subject of Chapter 8.

For thought and action

7.1   After this extended exposure to model building, reread the first page of this chapter. What similarities and differences between poetic metaphors and scientific models do *you* perceive?

7.2   Match each of the experimental observations in the left column with the explanation on the right which most directly accounts for the phenomenon. (Which column is more REAL?)

Evidence	Explanation
(a) X-ray spectra	(1) the nuclear atom
(b) α-particle scattering	(2) quantized electron orbits
(c) emission spectra	(3) the quantization of energy
(d) the photoelectric effect	(4) the charge on the nucleus

7.3   Exhibit your ingenuity by inventing a model for the atom other than Rutherford's nuclear version and the Kelvin-Thomson "raisin pudding." (It would be nice if the model were reasonably compatible with experimental fact.)

7.4   Why was Moseley's use of atomic numbers far more convincing than Newlands'?

7.5   Specify, insofar as possible, the number of electrons, protons, and neutrons in an atom of each of the following:
    (a)   carbon   (b)   carbon-12   (c)   sodium-23
    (d) ^9F (e) $_{207}Pb$ (f) $_{226}^{88}Ra$

7.6   The three major isotopes of oxygen have atomic numbers of 16, 17, and 18. State the number of electrons, protons, and neutrons in an atom of each of these isotopes.

7.7   As you know, hydrogen exists in three isotopes. For the sake of clarity, use an H to represent an atom of the common isotope of mass 1. Similarly, symbolize a deuterium atom with a D and a tritium atom with a T. Write the formulas of all the possible diatomic molecules of hydrogen and indicate the relative weight of each molecular species.

7.8   Complete the following radiochemical equations:
    (a) $_1^3H$ $\rightarrow$ _____ $+ \ _{-1}^0\beta$
    (b) $_{89}^{223}Ac$ $\rightarrow$ _____ $+ \ _2^4\alpha$
    (c) $_{27}^{60}Co$ $\rightarrow _{28}^{60}Ni \ +$ _____
    (d)   _____ $\rightarrow _{82}^{206}Pb \ + \ _2^4\alpha$

7.9   Suggest how astronomers can estimate the temperature of stars.

7.10 Rank stars of the following colors in order of increasing temperature: white, red, blue, and yellow. Explain your answer.

7.11 The first spectral line associated with the Balmer series for the hydrogen atom has a frequency of 4.57×10^{14} sec^{-1}.
(a) What is the energy of a single quantum or photon of this frequency?
(b) What is the color of the light?

7.12 At noon on a bright, clear day, each square centimeter of surface exposed to the sun's rays receives about 10^6 ergs of solar energy every second. Of course, this energy arrives in a broad spectrum of frequencies. But for the sake of the calculation, assume it all to be of frequency 5×10^{14} sec^{-1} (a nice yellow color). Now determine the number of photons striking that square centimeter area each second.

7.13 The idea that energy is quantized may not have much of an impact on you. Why do you suppose it shook up many physicists early in this century?

7.14 Explain the origin of the light in a neon sign.

7.15 Calculate a numerical value for the radius of the fourth orbit of the Bohr atom.

7.16 Compute the frequencies of the first three spectral lines of the Balmer series for hydrogen, using the Balmer or Bohr equation.

7.17 Explain to your roommate, an art major who has not had the benefit of this course, the essential features of the Bohr model of the atom. Be accurate, interesting, and entertaining.

7.18 Some textbooks no longer present the Bohr model of the atom because it has been superseded by more comprehensive theories. Should the author of this text have followed that example? Defend your position.

7.19 From your reading of this chapter, distinguish between "juggling with numbers until they can be got to fit" and finding a mathematical form for a physical phenomenon.

7.20 On page 164 you read, "It is in keeping with the economy which characterizes nature and should characterize its investigators, that the model or theory used should fit the problems at hand." What does the related discussion of various versions of the atom imply to you about the nature of scientific "truth"?

7.21 Note the ages of some of the chief players in the drama of discovery described in this chapter: Marsden, Moseley, and Bohr were all on the trustworthy side of 30 when they made their first major contributions. What characteristics make certain disciplines particularly amenable to significant achievement by relatively young men and women?

7.22 On the basis of what you have thus far observed about scientific theorizing, identify the important criteria which a scientific model must meet. How are these standards similar to and different from those operative in your own major field of interest?

Suggested readings

Andrade, E. N. da C., "Rutherford and the Nature of the Atom," Anchor Books, Doubleday, Garden City, New York, 1964.

Gamow, George, "Thirty Years that Shook Physics," Anchor Books, Doubleday, Garden City, New York, 1966.

Jaffee, Bernard, "Moseley and the Numbering of the Elements," Anchor Books, Doubleday, Garden City, New York, 1971.

Lagowski, J. J., "The Structure of Atoms," Houghton, Boston, Massachusetts, 1964.

Romer, Alfred, "The Restless Atom," Anchor Books, Doubleday, Garden City, New York, 1960.

The mathematics of matter

When is a particle not a particle?

It was the idea that energy is discrete, or at least that it can be regarded as composed of minute packets, which provided the crucial key to the development of the atom-according-to-Bohr. The quantum concept was in fact a new way of looking at electromagnetic radiation—not as waves vibrating in emptiness or aether, but as a string of massless "particles." The juxtaposition of the last two words is at once troubling. If radiowaves, light, and X rays are indeed experimentally massless, then "particles" hardly seems an appropriate appelation. Moreover, the apparent wave/photon duality creates conceptual as well as semantic difficulties. But this is only an indication of complexities to come.

Even more startling was the idea which paved the way for the mathematical model of matter which followed Bohr's work. The result was a powerful new method of calculating atomic and molecular structure.

But the intellectual and psychic price paid for this power was im-mense—matter became wavelike, the atom became abstract and unpicturable, and the universe became probabilistic and indeter-minate. Of course, the universe in fact remained unaltered, but not so man's conception of it. And that, is all we know.

The pivotal concept came from a student of medieval history turned physicist. Perhaps the notion was too absurd to be ad-vanced by someone more steeped in and saturated with the laws of physics. At any rate, in 1924, Prince Louis Victor de Broglie, scion of an ancient and noble French family, presented a most original doctoral dissertation. In it he imagined that each allow-able orbit of the Bohr atom has associated with it a wave. The

FIGURE 8.1.   *Schematic representation of the first three Bohr orbits for the hydrogen atom and the corresponding de Broglie waves.*

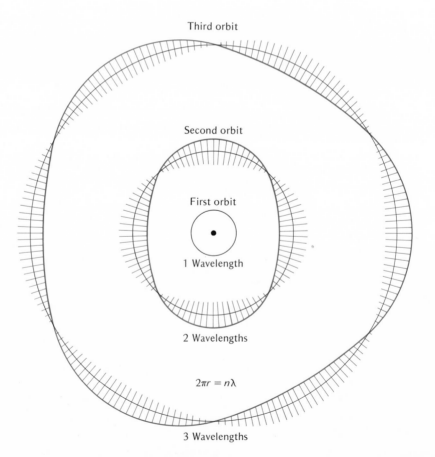

Third orbit

Second orbit

First orbit

1 Wavelength

2 Wavelengths

$2\pi r = n\lambda$

3 Wavelengths

lowest orbit is just a single wave, that is, its circumference is equal to one wavelength. The second quantized orbit is two wavelengths in circumference; the third is made up of three waves, and so on. As usual, what we have been saying in words can be more simply summarized in symbols. The circumference of a circle is $2\pi r$. De Broglie's guess was that each circumference is equal to an integral number of wavelengths, $n\lambda$, where $n = 1, 2, 3, \ldots$ *Ergo,*

8.1 $\qquad 2\pi r = n\lambda.$

The next logical step was to compare this result with the orbital circumference predicted from the Bohr theory. Recall that at the heart of the Dane's derivation was the assumption that the angular momentum of the orbiting electron is quantized according to the equation

$$mvr = \frac{nh}{2\pi}.$$

Rearranging, we obtain at once,

8.2 $\qquad 2\pi r = \frac{nh}{mv}.$

It is trivially simple to set the right-hand sides of Equations 8.1 and 8.2 equal to each other and to obtain the relationship

8.3 $\qquad \lambda = \frac{h}{mv}.$

But the ease of the operation and the simplicity of the solution should not blind you to the fact that Equation 8.3 states one of the most monumental identities in the history of science. The author is aware that such extravagant outbursts run the risk of engendering more skepticism than admiration. Nevertheless, consider the physical meaning of the symbols. Mass, m, and velocity, v, are properties of a particle. Multiplied together they yield momentum. A 3000 pound (1.36×10^6 g) particle called a car, traveling at a speed of 70 miles/hour (3.13×10^3 cm/sec) has a lot of it—4.26×10^9 g cm/sec to be exact. There is no question of its massy tangibility. On the other hand, wavelength, λ, is a property of surf and sound and sunshine. Nevertheless, these apparently unrelated properties of such seemingly disparate aspects of the universe are linked together by Planck's constant. The same constant of nature which earlier helped describe the division of energy was now involved in representing *matter as a continuous wave!* No wonder physicists viewed the Prince's thesis with incredulity. Everyone agreed it looked crazy—even Einstein said so. "But," the master added, "it is really sound!"

De Broglie's equation asserts the apparent contradiction that *a particle can have wave properties and a wave can have particulate properties.* Or, perhaps better, a single physical entity can be both wavelike and particlelike. What to call this schizophrenic entity creates a formidable verbal difficulty. Perhaps "wavicle" or "partave" would do. Our linguistic categories and classifications are just no longer adequate to describe the microcosm of matter. The words and analogies we have invented and evolved for our everyday experience simply do not apply on the minute scale of the atom and the electron. Indeed, according to the physicist D. H. Wilkinson, "Those who ask what an electron is are guilty of a sort of rank anthropomorphism."

In part, the difficulty arises because electrons behave in chameleon-like conformity to the experiment performed. Thus, if an experiment asks if an electron is a particle, it invariably answers "yes." So it did in Rutherford's determination of e/m and Millikan's measurement of e—both of them good, discrete, particulate properties. Of crucial importance to de Broglie's hypothesis was whether or not an electron would also reply affirmatively to an experiment designed to test its wave nature. After all, the thesis merely advanced, in the words of its author, "a formal scheme whose physical content is not yet determined."

Fortunately, the experimental verification was soon forthcoming. Some of the work had actually preceded de Broglie's great guess. Two Americans, Clinton Davisson and L. H. Germer, were investigating the scattering of electrons by the atoms of crystals. Similar experiments with X rays were well known. Basically, the idea was to bounce the radiation off the atoms. In the case of X rays, the father and son team of W. H. and W. L. Bragg had demonstrated that such reflections occur only at certain angles relative to the plane of atoms in the crystal being studied. As a result, the reflected beam forms a series of alternating light and dark spots called a diffraction or interference pattern. Such behavior is fully consistent with wave motion. You can see for yourself the next time you're in the bathtub. When two waves meet "in phase," with their crests together, the crests add up to make extra high peaks. The waves reinforce each other or, to use two apparently contradictory terms, they interfere constructively. This situation corresponds to a bright spot in the pattern made by the X rays reflected from a crystal. But if the waves are "out of phase," the crests of one will coincide with the troughs of the other and they will cancel out into a region of calmness. Similarly, when X ray waves are out of phase and destructively interfere with each other, no spot is seen on the film which records the reflection.

None of this is very surprising. Massless X radiation is part of the electromagnetic spectrum and is supposed to act like a wave—at least most of the time. True, the scattering of X rays has been widely used in determining the arrangement of atoms in crystals and molecules. But what does it have to do with electrons? "A lot!" answered Davisson and Germer. For what they found was the fantastic fact that in a scattering experiment, electrons behave as X rays do. The "particles" are reflected as waves, showing interference patterns with bright spots only at certain angles. Moreover, the electron waves conform to the same simple scattering equation which applies to the more conventional X ray waves. Davisson and Germer were able to use this mathematical relationship to calculate the wavelength of an electron of known velocity. The result was within experimental error of the wavelength calculated from de Broglie's extraordinary equation!

To consider matter as wavelike was a patent absurdity. And yet, here was experiment verifying just this preposterous notion. There was on choice but to award de Broglie's "crazy" thesis a Nobel Prize. Nature had again disclosed herself as easily equal to man's wildest fantasies. In the words of the late British biologist, J. B. S. Haldane, "The universe is not only queerer than we imagine, it is queerer than we can imagine."

The abstract atom

Unorthodox as de Broglie's ideas were, scientists lost no time exploiting them. Erwin Schrödinger, an Austrian theoretical physicist, started with the assumption that the wavelike character of electrons could be incorporated into a model of the atom. But German was not adequate to describe electrons and atoms in this manner. Nor was any other conventional language. The medium had to be that of calculus and the methodology soon came to be known as *quantum mechanics* or *wave mechanics*. It was in this symbolic and syntactical system that Schrödinger formulated his famous wave equation:

8.4 $$\frac{\partial^2 \psi}{\partial x^2} + \frac{\partial^2 \psi}{\partial y^2} + \frac{\partial^2 \psi}{\partial z^2} + \frac{8\pi^2 m}{h^2}(E - V)\psi = 0.$$

Any unfamiliar language looks formidable, doubly so if the letters are foreign. Therefore, Equation 8.4 may engender in the reader the same bafflement that a sentence in Chinese characters evokes in the author. True, both are translatable. But I trust I will be pardoned if I do not devote the rest of this book to providing the translation of Equation 8.4. After all, a typical undergraduate physics or chemistry major does not encounter a detailed duplication of the steps involved in solving the Schrödinger equation until his junior or senior year. And frankly, I suspect that some of you share the sentiments expressed by the French philosopher Auguste Comte in 1830:

Every attempt to employ mathematical methods in the study of chemical questions must be considered profoundly irrational and contrary to the spirit of chemistry . . . . If mathematical analysis should ever hold a prominent place in chemistry—an aberration which is happily almost impossible—it would occasion a rapid and widespread degeneration of that science.*

A lengthy excursion into calculus probably would not change such an opinion, but a summary of the results of quantum mechanics just might.

First a few words about the equation itself. You might decide to memorize it for future use as erudite graffiti, and you should know what the symbols mean. To begin with, it is what those in the numbers racket call a *differential equation*. That's what all the deltas or ∂'s are about. *E* stands for the total energy of the system under investigation (a hydrogen atom in our example) and *V* is its potential energy. Again restricting our arguments to the simplest atom, *m* designates the mass of the electron. Planck's constant, *h*, also appears in Equation 8.4, and π and 8 happily have their usual significance. The "unknown" in the equation is ψ. But the solution is not just a number, as it has been for most of the equations you have heretofore wrestled with. Instead, ψ is a mathematical expression which depends upon *x*, *y*, and *z*. These are merely the spatial coordinates of a point; that is, they specify the location of any point in three dimensional space.

What we are seeking in the solution of the wave equation is a mathematical form which represents the way in which ψ varies with *x*, *y*, and *z*. In other words, we want to be able to calculate a value for ψ at any point. Physically, ψ corresponds to the amplitude of the matter-wave. But more important, you will soon see that for a hydrogen atom, ψ is a measure of where in space the electron is most likely to be found. Hence, it is an important indicator of atomic structure. And deriving a mathematical expression for ψ, the *wave function* or *eigenfunction*, is thus at the heart of the problem.

As you probably guessed, arriving at the correct answer demands a good deal of competence in calculus. But this is more than just an exercise in symbol juggling. The wave function must make physical as well as mathematical sense, a requirement which places certain restrictions upon the solution. When these are invoked, a very significant feature emerges—*the wave functions, ψ, and the energy, E, turn out to be quantized.* Unlike Bohr's ap-

---

* Auguste Comte, "Philosophie Positive," 1830.

proach, the solution of the Schrödinger wave equation makes no
a priori assumptions about the existence of discrete values for
energy or the location of the electron. But they arise nonetheless.

The quantization appears in the form of discrete, integral
quantum numbers present in the wave functions and the corresponding energy terms. Moreover, the wave-mechanical treatment
yields not one, but three quantum numbers. The *principal quantum
number, n,* corresponds to Bohr's single number and can assume
the values 1, 2, 3, ... The letter *l* is used to designate the
azimuthal quantum number. And m_l represents the *magnetic
quantum number*.

All three of the quantum numbers are mathematically related
to one another, and the form of that relationship is specified by
the derivation. The permissible values of *l* depend upon *n*. Specifically, a given value of *n* can be associated with *n* different
values of *l*. They range, integrally, from 0 through $n - 1$. For
$n = 1$, the lowest possible value for the principal quantum number, $n - 1 = 1 - 1 = 0$. Hence, there is only one value for *l*,
$l = 0$. When $n = 2$, the highest acceptable value for *l* becomes
$2 - 1$ or 1. Thus, two values are allowed: $l = 0$ and $l = 1$. Similarly, three values of l—0, 1, and 2—are compatible with $n = 3$.

The permissible values of m_l, on the other hand, are related to
l. They can be both positive and negative and also include 0.
Written in general form, $m_l = -l, -(l - 1), \ldots, -1, 0, 1, \ldots,$
$l - 1, l$. That is, m_l ranges integrally from $-l$ through *l* for a total
of $2l + 1$ different values. The case for $l = 0$ is trivial; there can
be only a single m_l value and it is clearly 0. When $l = 1$, the general relationship leads to three values for m_l: -1, 0, and 1. By
similar reasoning, five different m_l values, $-2, -1, 0, 1, 2,$ are
compatible with $l = 2$. These relationships are summarized in a
more compact form in Table 8.1.

In spite of what it might seem to be, this is not mere numerology. Quantum numbers do have physical significance. In fact, experimental spectroscopic evidence has led to the introduction of
the fourth quantum number.* It is called the *spin* quantum number and is commonly symbolized m_s. The name comes from visualizing the electron as a spinning ball of negative electricity. The
spin can be either clockwise or counterclockwise, and these two
different states are given different m_s designations. The actual
values assigned $m_s = +\frac{1}{2}$ and $m_s = -\frac{1}{2}$. Sometimes the two
spin states are specified as "up" (↑) and "down" (↓). Whatever the
convention, the point to remember is that the states represented
are physically distinct.

---

* The spin quantum number also emerges from a more complicated
version of the Schrödinger equation.

TABLE 8.1. *Relationships among the Three Quantum Numbers,*
n, l, and m_l.

Principal n	Azimuthal l	Magnetic m_l
1	0	0
2	0	0
	1	−1 0 1
3	0	0
	1	−1 0 1
	2	−2 −1 0 1 2
⋮		
n	0	0
	1	−1 0 1
	2	−2 −1 0 1 2
	⋮	
	$(n-1)$	$-l$ $-(l-1)$ ⋮ −1 0 1 ⋮ $(l-1)$ l

[a] Note that the first three specific examples (for $n = 1, 2,$ and 3) are special cases of the general relationship indicated at the bottom of the table.

There are, then, a total of four quantum numbers, n, l, m_l and m_s. One of each, together, form a set which uniquely specifies the quantum state of an electron in an atom. These four numbers are sort of an identifying address of the electron. Both the location of the electron and its energy are determined by this address.

Perhaps the best way to get a feeling for the physical meaning of these numbers and the states which they represent is to consider in greater detail the example of the hydrogen atom. In the atom's lowest lying energy level or ground state, two different sets of quantum numbers are permissible: $n = 1$, $l = 0$, $m_l = 0$, $m_s = +\frac{1}{2}$ and $n = 1$, $l = 0$, $m_l = 0$, $m_s = -\frac{1}{2}$. The energy calculated for these two states turns out to be the same and identically equal to that derived from the Bohr model (Equation 7.10). This result, you will recall, is fully consistent with spectroscopic data. The wave-mechanical approach has thus passed the first hurdle.

Now we turn to the question of the location of the electron as predicted by the wave equation. For this we need the wave function or, more precisely, the square of the wave function; ψ^2 is proportional to the probability density of the electron. It is a measure of where in space the electron is. Because ψ and consequently ψ^2 are functions of x, y, and z, that is, because they depend upon position, the chance of finding the electron will not be uniformly equal across all of space. Assume, for example, that you could observe the instantaneous location of a particulate electron in a hydrogen atom by some sort of super-microscope and photographic film. Each observation would yield a spot, indicating a position. After thousands of exposures you would obtain a cloud, most intense in those regions where the electron was most often located, lighter where it was less commonly found. The density of the exposure would thus indicate the probability of the electron distribution—the property represented by ψ^2. Or, if you prefer to imagine the electron as a smeared out matter-wave, ψ^2 is greatest where the wave is most thickly smeared.

Although the above idea of making a multiple exposure of an atom hopefully helps explain the distribution of an electron, it cannot in fact be achieved. And the reason is not merely the practical limitation imposed by apparatus. It is, rather, a limit which nature appears to place upon our knowledge. As you will learn in the last section of this chapter, it has been argued with convincing conviction that it is impossible to simultaneously exactly specify both the position and the momentum an electron. We cannot be infinitely certain about its instantaneous properties. Therefore, there is always some fuzziness in electron clouds, always some ignorance in atoms, and ψ^2 represents the *chance* of locating an electron, *not* a certainty.

For the hydrogen atom in its lowest energy state, the mathematical expression for ψ^2 describes a sphere with the nucleus at its center. The ball-shaped electron cloud is equally dense in all directions. However, it does have a maximum density at a particular distance from the center of the atom. This distance can be calculated, and it turns out to be equal to the Bohr radius, 0.529 Å. The electron will most likely be found on a spherical surface situated this far from the center. As one moves either towards the nucleus or away from it, the probability of finding the electron decreases.

The fact that both the Bohr and Schrödinger models predict essentially the same, experimentally verified values for the energy and location of the electron is encouraging. But the probabilistic aspect of electron distribution makes talk about specific electron orbits rather meaningless. Instead, the word *orbital* has been adopted to designate the charge cloud of an electron. Unfortunately, experience and common sense fail to provide an apt analogy for an electronic orbital. Indeed, the philosopher of science, F. Waismann, has pointed out that an understanding of quantum mechanics requires "uncommon sense." The electron distribution in a hydrogen atom is not like the discrete wall of a basketball. It is rather a ball which has its maximum hardness at a certain radius, but gets softer as one moves in either direction away from this shell. An attempt at picturing this distribution is made in Figure 8.2. The darkness of shading should be considered as representing the electron density.

The second quantum level is characterized by $n = 2$. For this value of the principal quantum number, l can be either 0 or 1. If it is the former, m_l must also be 0. As usual, m_s can be either

FIGURE 8.2.   *Representation of the electron distribution in the lowest energy state of a hydrogen atom (1s orbital). The drawing on the left is an attempt at a three-dimensional exterior view. The drawing at the right is a cross section through the nucleus. The darkness of the color represents the probability of finding the electron.*

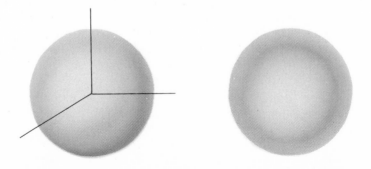

$+\frac{1}{2}$ or $-\frac{1}{2}$. The resulting wave function is similar to that of the ground state of hydrogen. It, too, is a fuzzy sphere, or more exactly, two concentric fuzzy spheres. But the maximum electron density occurs at a greater radius than for $n = 1$. In this respect the model is again much like a blurry Bohr atom—the same sort of effect I get by removing my glasses. And the energy of the second state also agrees favorably with the earlier calculation and with experiment.

Notice that thus far we have been considering states for which $l = 0$. By a convention whose spectroscopic origins are almost forgotten, these are designated as s states. All s orbitals are spherically symmetrical about the nucleus, no matter what the principal quantum number. The value of the latter is indicated by a number preceding the s. Thus, the electronic levels we have just encountered bear the designation 1s and 2s. In the 3s state, that is, $n = 3$, $l = 0$, the greatest concentration of electrons is still farther from the nucleus. The same trend continues for s orbitals of higher energy, the radius of the maximum charge density in the sphere increasing as n gets larger.

For the $n = 1$ level, only 1s electrons are possible because l can only equal 0. But for $n = 2$, $l = 1$ is also an acceptable value. Furthermore, it follows that for $l = 1$, $m_l = -1, 0, 1$. When the spin quantum number is introduced, it becomes apparent that there must be six possible different sets of quantum numbers which involve $n = 2$ and $l = 1$. They are as follows:

n	l	m_l	m_s
2	1	-1	$+\frac{1}{2}$
2	1	-1	$-\frac{1}{2}$
2	1	0	$+\frac{1}{2}$
2	1	0	$-\frac{1}{2}$
2	1	1	$+\frac{1}{2}$
2	1	1	$-\frac{1}{2}$

By convention, the $l = 1$ states are all identified with the letter p, in this case, 2p.

Quantum numbers flow so beautifully from the Schrödinger equation that sometimes they seem to acquire a sort of physical primacy. There is a strong temptation to write, "there are six distinct 2p quantum states available to an electron in a hydrogen atom because there are six different sets of quantum numbers." But a little sober reflection suggests that such a statement is an inversion of art and nature in which metaphor somehow displaces reality. Fortunately, reality, in the form of spectral lines, is around to remind scientists that they did not calculate the universe from a differential equation.

But the fact that the agreement with experiment is excellent can nourish our collective pride. For example, quantum mechanics makes possible the calculation of the energy associated with the 2s and 2p sublevels of the $n = 2$ level. The difference is very small indeed. Nevertheless, when the hydrogen spectrum is carefully examined at high resolution, lines quantitatively corresponding to electronic transitions involving both of these states can be detected. True, these are the same lines the Bohr model was able to account for with the aid of some *ad hoc* hypotheses. But even with the addition of elliptical orbits, the model was unable to explain the lines which arise in a magnetic field.

The wave-mechanical approach rationalizes these as transitions involving sub-sublevels of the electron. Normally, the two s states are energetically identical—there is no detectable energetic difference between a spin of $+\frac{1}{2}$ and a spin of $-\frac{1}{2}$. Similarly, the six p states also have the same energy. However, in a magnetic field, electrons in these various states interact differently with the field. Hence, the states are split; their energies become slightly different. And small as these differences are, they can be detected spectroscopically.

The shape of the electron cloud or orbital for a p electron helps explain its energetic behavior. To find that shape, of course, requires a determination of ψ and ψ^2. There are three distinct wave functions for $n = 1$, $l = 1$; one corresponding to each permissible value of m_l. All three of the ψ's depend upon angles about the nucleus, none of them are spherically symmetrical. This is apparent from Figure 8.3, where ψ^2 is plotted for each of the quantum states. The three p orbitals are dumbbell-shaped. One, designated the p_x orbital, is oriented along the x axis; the p_y orbital lies along the y axis; and the p_z orbital runs along the z axis. Clearly, it seems

FIGURE 8.3.   *Electron distribution in the p orbitals (l = 1).*

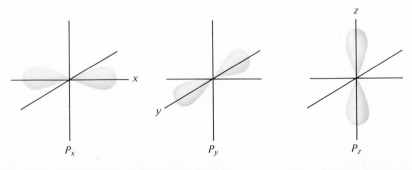

a valid inference that the numerical value of *l* reflects the shape of the electron distribution.

Now suppose a magnetic field is applied across an atom. An electron in a p_x orbital will have a somewhat different orientation with respect to the field than will an electron in a p_y or p_z orbital. Depending upon the orientation, the energies associated with the various orbitals will be raised or lowered by a mechanism which also involves the spin of the electron. In short, the magnetic field induces energy differences between states of identical *n* and *l* but different m_l and m_s. In the absence of such a field, these geo-metrically related energy differences are not detectable—indeed, they do not exist. The states are then said to be *degenerate*.

Similar sorts of degenerate states with normally equal energy are found in the higher electron shells. For $n = 3$, *l* can assume values 0, 1, and 2. The 3s and 3p orbitals are exactly analogous to the 1s and the 2p, respectively. But the $l = 2$ sublevel introduces the *d* orbitals. There are five of these, corresponding to the $2l + 1$ or five values of m_l: ± 2, ± 1, and 0 (see Figure 8.4). The new value of *l* is again an indicator of differences in orbital shape. Four of the *d*-electron density clouds look like two dumbbells at right angles to each other. The fifth looks like a dumbbell with a bagel around its middle. All five orbitals have different spatial orientations, and hence their relative energies are dependent upon their positions in a magnetic field.

For reasons which will soon become apparent, we normally do

FIGURE 8.4. *Electron distribution in the d orbitals (l = 2).*

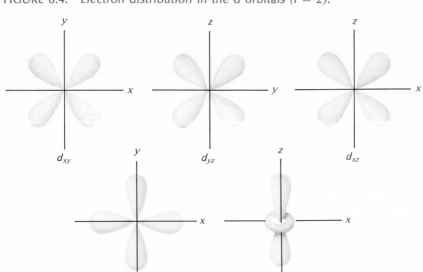

not consider electrons with *l* values greater than 3. These *f* orbitals are first encountered in the *n* = 4 level. The arithmetic predicts seven of them, since for *l* = 3, m_l = ±3, ±2, ±1, and 0. The author has seldom seen an attempt to draw an *f* orbital. We are fast reaching the point where pictures lose their fidelity as representations of nature. Nevertheless, equations and numbers retain theirs. The energy levels calculated by quantum mechanics for the hydrogen atom duplicate, with high accuracy, the results of the spectroscope. This is in itself a considerable achievement, but it might not warrant an entire chapter. Happily, however, wave mechanics has aided our understanding of all atoms.

Building up the elements

In our lengthy discussion of the hydrogen atom we were, of course, considering a system of only one proton and one electron. That electron can occupy only a single orbital at any moment. The others stand like so many ghostly empty rooms, waiting to materialize and enclose the electron when it changes its energy level by absorption or emission. But atoms of other elements contain more electrons, progressively more as the atomic number increases. Unfortunately, the repulsive interaction among these electrons greatly increases the complexity of quantum-mechanical calculations. However, a very convenient fact soon emerges: quantum states similar to those of the hydrogen atom are also accessible to the electrons of other, heavier atoms.

Thus, the process of generating the elements on paper becomes a simple stepwise addition of electrons. One starts at the lowest energy level and works up. It is literally a "building up"—an *Aufbauung,* to acknowledge the German or Austrian origins of much early quantum mechanics and many early quantum mechanicians. Governing the *Aufbau* is an important blueprint prepared by one of those Austrian physicists—Wolfgang Pauli. Pauli's famous *Exclusion Principle* specifies that no two electrons in any single atom can possess identical sets of quantum numbers. All permissible states must differ in at least one of the four numbers. Thus, the electron capacity of each shell or level is limited by the number of different, four-member sets which can be generated from the appropriate quantum numbers.

The lowest lying energy level is the simplest. As we previously noted, here there are only two distinct sets of quantum numbers: *n* = 1, *l* = 0, m_l = 0, m_s = ±½. Therefore, only two electrons, with opposite spin, can be accommodated in the 1s orbital. The one-electron case is obviously hydrogen and the two-electron atom must represent helium. Of course, the helium nucleus must contain two protons; and in addition it contains two neutrons to yield an atomic weight of 4.0. In the shorthand notation of chemists, the *electronic configurations* of atoms and elements are identified by writing the orbitals involved and indicating the population of each sublevel with a superscript. When only a single electron is present in the sublevel, the corresponding 1 is normally not written. Thus hydrogen is symbolized 1s instead of $1s^1$. Helium, on the other hand, has an electronic configuration represented by $1s^2$.

Next in order of increasing energy comes the $n = 2$ shell. Two values of the azimuthal quantum number may characterize this level, $l = 0$ and $l = 1$. The former value is associated with the energetically lower 2s state. Therefore, this orbital is filled before the $l = 1$ or 2p sublevel. Two different sets of quantum numbers can be generated from the values $n = 2$, $l = 0$, $m_l = 0$, and $m_s = \pm\frac{1}{2}$. Thus it follows from the Pauli Principle that the capacity of the 2s state must be two electrons. This means that two elements—one with a total of three electrons and one with four—arise from the stepwise addition of the 2s *electrons*. They are lithium, $1s^2 2s$, and beryllium, $1s^2 2s^2$.

Although beryllium marks the completion of the 2s subshell, the 2p orbitals remain unfilled. You recall that there are three of these, one corresponding to each of the permissible values of m_l: -1, 0, and 1. Since for the 2p sublevel, n is fixed at 2 and $l = 1$, each of the three 2p orbitals represents a different set of *three* quantum numbers. When the fourth, $m_s = \pm\frac{1}{2}$, is added, it becomes apparent that the Exclusion Principle allows six different complete sets of four numbers—those specified on page 195. It should now also be clear why we can stipulate that the maximum capacity of any individual orbital must be two electrons. Moreover, the two electrons must have opposite or "paired" spins in order not to violate the Pauli Principle.

Spin also becomes a factor in the sequence in which the electrons are assigned to the three p orbitals. Obviously there is no ambiguity in the case of boron. A fifth electron is simply added to the beryllium atom to yield a $1s^2 2s^2 2p$ electronic configuration. In the absence of a magnetic field, the p orbitals are energetically identical and there is no spin state distinction. Therefore, it is meaningless to ask whether a single p electron is in the p_x, p_y, or p_z orbital, or whether it has a spin quantum number of $+\frac{1}{2}$ or $-\frac{1}{2}$. However, when a second electron is added to the 2p sublevel to form carbon, the latter question takes on some physical significance. Five possible sets of quantum numbers are available to the electron which converts boron to carbon. One of the five is significantly different from the other four. It is the set which groups the second p electron in the same orbital with the first. If the electron were to enter this orbital, the dumbbell-shaped charge cloud would have its maximum electron capacity of two. This would mean that two particles, each bearing the same negative charge, would occupy roughly the same region of space. To be sure, this situation is possible if the spins of the electrons are opposite. However, this arrangement requires more energy to achieve than one in which electrons with parallel spins occupy different orbitals. In other words, the electron–electron repulsive forces generally tend to keep the particles (or charge clouds) as far apart as possible. And, since it is in the nature of nature to expend as little energy as possible, the two p electrons of carbon do occupy different orbitals and do have parallel spins. If this seems a rather certain statement for a theoretical system, it should be noted that there are experimental methods of counting unpaired electrons. These results fully support the general rule that the order of filling of orbitals of the same l value is such that as many electrons as possible remain unpaired.

The method of designating electron configurations which was introduced a little earlier does not make this spin distinction. The carbon atom is simply written $1s^2 2s^2 2p^2$. However, spin can be represented in *orbital diagrams* such as those of Figure 8.5. Each orbital is symbolized by a square and each electron by an arrow. Arrows pointing in the same direction indicate parallel spins, while two arrows, one pointing up and one down mean two paired electrons of $m_s = +\frac{1}{2}$ and $m_s = -\frac{1}{2}$. The figure depicts the orbital populations of the six elements ranging from boron (atomic number 5) through neon (atomic number 10). All six of these share a

FIGURE 8.5.   *Electronic configurations and orbital diagrams for the elements with atomic numbers 5 through 10.*

	Electronic configuration	1s	2s	2p
B	$1s^2 2s^2 2p$	↑↓	↑↓	↑
C	$1s^2 2s^2 2p^2$	↑↓	↑↓	↑ ↑
N	$1s^2 2s^2 2p^3$	↑↓	↑↓	↑ ↑ ↑
O	$1s^2 2s^2 2p^4$	↑↓	↑↓	↑↓ ↑ ↑
F	$1s^2 2s^2 2p^5$	↑↓	↑↓	↑↓ ↑↓ ↑
Ne	$1s^2 2s^2 2p^6$	↑↓	↑↓	↑↓ ↑↓ ↑↓

common beryllium core, $1s^22s^2$. But the number of p electrons increases from 1 through 6. Note that the three p orbitals are all singly populated in nitrogen. With oxygen, $1s^22s^22p^4$, two electrons must be paired in order to satisfy the Exclusion Principle. Fluorine follows and the period ends when the p sublevel and, hence, the $n = 2$ level become filled to capacity to yield neon.

The next eight elements, sodium (atomic number 11) through argon (atomic number 18), repeat the eight we have just built up. Except, of course, that now electron addition occurs in the $3s$ and $3p$ orbitals. This is clear from the electronic configurations of the two series.

Li	Be	B	C
$1s^2\underline{2s}$	$1s^2\underline{2s^2}$	$1s^2\underline{2s^22p}$	$1s^2\underline{2s^22p^2}$

Na	Mg	Al	Si
[Ne]$\underline{3s}$	[Ne]$\underline{3s^2}$	[Ne]$\underline{3s^23p}$	[Ne]$\underline{3s^23p^2}$

N	O	F	Ne
$1s^2\underline{2s^22p^3}$	$1s^2\underline{2s^22p^4}$	$1s^2\underline{2s^22p^5}$	$1s^2\underline{2s^22p^6}$

P	S	Cl	A
[Ne]$\underline{3s^23p^3}$	[Ne]$\underline{3s^23p^4}$	[Ne]$\underline{3s^23p^5}$	[Ne]$\underline{3s^23p^6}$

Also apparent from this diagram is the similarity between the electronic arrangements characteristic of certain elements. The primary point of resemblance is in the orbital assignments of the outer electrons—those lying farthest from the nucleus. These orbitals are underlined in the table. Thus, lithium and sodium have similar electronic configurations, so do beryllium and magnesium, and so on. In all, there are eight pairs of elements whose atoms seem very much alike. You will recall from our discussion of the periodic table, that just these pairs of elements exhibit similar chemical and physical properties. What we have here, then, is an explanation of elementary periodicity, on the basis of atomic structure. Although Mendeleev was unaware of the fact, he was organizing elements according to their electronic distributions as well as their bulk properties and atomic weights. The fact that the sequence is the same is, of course, hardly coincidental. It is the order of increasing atomic number, the order of stepwise addition of electrons and protons to the atom. This is the underlying principle. After all, the properties of elements must be fundamentally due to the properties of the atoms which constitute them. And, the outer electrons of any atom exert the most profound influence upon its chemical and physical nature. Hence, it is understandable that an organizational system which reflects sequential changes and periodic similarities in atomic structure should correspond to the order among the elements.

This basic and extremely important correlation between structural and chemical periodicity is maintained over the entire table of the elements. To be sure, the situation does get a bit more complicated with element number 19, potassium. Argon, the rare gas which proceeds it, has an electronic configuration of $1s^2 2s^2 2p^6 3s^2 3p^6$. Note that this does not exhaust the capacity of the $n = 3$ shell. We have not yet utilized the $3d$ orbitals—those involving $l = 2$. There are five of these with a total capacity of ten electrons. A logical extension of our atom-*Aufbau* would next place an electron in one of these d orbitals. Potassium would thus be $1s^2 2s^2 2p^6 3s^2 3p^6 \underline{3d}$. But potassium clearly has the properties of sodium and lithium. Since the latter are both characterized by a single outer s electron, $2s$ and $3s$, respectively, there is a temptation to write potassium as $1s^2 2s^2 2p^6 3s^2 3p^6 \underline{4s}$. It turns out that other experimental evidence also supports this intuitive assignment. The $4s$ sublevel has an energy lower than the $3d$. Therefore, the former is filled first. We thus have an atomic rationale for including potassium and calcium in the alkali metal and alkaline earth families, where they clearly belong on the basis of their properties.

On the other hand, the next element, scandium (atomic number 21), does not closely resemble aluminum, which it should if the 21st electron were added to the $4p$ subshell. Instead, scandium introduces a new series—the transition metals. The fact that there are ten of these elements is an important clue. It can be attributed to the ten-electron capacity of the five $3d$ orbitals. The logical conclusion that electronic addition to the $3d$ orbitals precedes addition to the $4p$ is further supported by calculations and measurements which disclose that the energy of the $3d$ sublevel is indeed somewhat lower than that of $4p$ level. Only after the former is fully filled in the form of zinc, $1s^2 2s^2 2p^6 3s^2 3p^6 4s^2 3d^{10}$, do electrons enter the $4p$ orbitals. As one would predict, the six elements which result, gallium through krypton, belong in Groups 3A through 7A plus the rare gases.

The atoms of the second long period, beginning with the alkali metal rubidium, closely approximate those of the first. Electronic addition proceeds in the sequence $5s$, $4d$, and $5p$. And, of course, this addition is paralleled by corresponding increases in nuclear charge. The period ends, as always, with a rare gas, this time xenon—$[Kr]5s^2 4d^{10} 5p^6$. Next in order of increasing energy is the $6s$ orbital, which is filled to form cesium and barium. Lanthanum, the next element, has a single $5d$ electron, analogous to the $4d$ electron of yttrium and the $3d$ electron of scandium. But with element number 58, cerium, the $4f$ sublevel finally begins to be filled. Because of the seven permissible values of m_l associated with $l = 3$, there are a total of seven f orbitals with a maximum capacity of 14 electrons. These are sequentially filled across the lanthanide series. After lutetium, the last of the lanthanides (atomic number 71), electrons again enter the 5*d* orbitals. The period ends with six elements in which the 6*p* orbitals are progressively filled. The final period is much like this previous one, with the electrons entering, in approximate order, the 7*s*, 5*f*, and 6*d* orbitals.

The word "approximate" is inserted into the last sentence because by the time one reaches the bottom of the periodic table, the electronic energy levels lie very close together. Indeed, there are certain overlaps. Moreover, the exact electron configuration is sometimes influenced by such considerations as the particular stability of half-filled and fully filled sublevels. Generally speaking, however, the schematic representations of relative energy levels presented in Figures 8.6 and 8.7 are useful guides to building your own periodic table.

The fact that such a paper construction is possible is perhaps the best validation of the quantum-mechanical atom. After all, the only plans employed in our *Aufbau* were the relationships among quantum numbers, the Pauli Exclusion Principle, and the order of increasing energy levels. These were sufficient to enable us to assemble our starting materials—electrons, protons, and (less specifically) neutrons—into 105 elements. And this, in itself, is quite an achievement. However, more direct experimental evidence also corroborates the "relative rightness" of the model. Energies associated with the transitions of electrons between various levels and sublevels can be accurately determined by spec-

FIGURE 8.6. *Approximate representation of relative energy levels of electronic orbitals.*

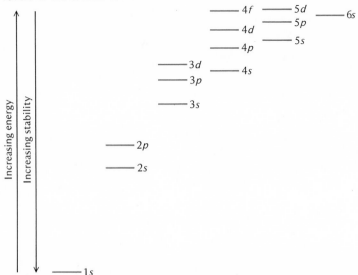

FIGURE 8.7.   *An alternative representation of the relative energy levels of electronic orbitals (see also Figure 8.6). To use this mnemonic device one simply reads down diagonally along the arrows. The orbitals appear in order of increasing energy.*

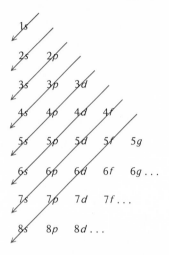

troscopy. Enough experiments of this kind permit energetic assignments of the electrons. Moreover, measurements of the amount of energy required to remove electrons from various atoms also indicate that electrons fall into certain groupings. These correspond to the related orbitals of the model, for example, the two 2s electrons or the six 2p electrons. Finally, the concept of the atom, as conditioned by quantum mechanics, has become a valuable guide in the prediction and explanation of molecular structure. But before we come to that in Chapter 9, there remains a fascinating bit of unfinished business at the electronic level.

Ignorance, uncertainty, and electrons

Thus far this chapter has been devoted chiefly to chemical applications of wave mechanics. That is as it should be. But to terminate the discussion without a brief reference to the philosophical implications of quantum theory would omit one of the most intellectually intriguing aspects of modern science. For the peculiar properties of the particle/wave which we call an electron have raised issues which lie at the very foundations of natural philosophy. These issues transcend linguistic considerations and bear upon the ultimate nature of knowledge, causality, perception, and the universe itself.

At the core of these considerations is the *Heisenberg Uncertainty Principle*. The *Principle of Indeterminacy*, as it is sometimes called, makes the assertion that *it is impossible to simultaneously determine both the position and the momentum of an electron (or anything else) with absolute accuracy*. If one of these variables is measured with complete certainty, the other must of natural necessity be absolutely unknown. More commonly, one would have some degree of uncertainty in both momentum and position. The relationship between the uncertainties in the two variables is represented by a simple equation,

8.5 $\Delta q \times \Delta p \geqslant h.$

Here Δq is the inexactness in position, a measure of length in one direction. The uncertainty in momentum is represented by Δp. Heisenberg's Principle postulates that the product of these two inaccuracies must be larger than or equal to Planck's constant. It can be no smaller, no matter how carefully our measurements are made. The equal sign is as good as we can possibly do. Clearly, if we increase the accuracy with which position is determined, Δq will decrease. There will be less ignorance associated with the result. However, as we gain in knowledge about position, our knowledge of momentum must decrease. In other words, Δp, the indicator of our ignorance and uncertainty of the exact momentum, unavoidably increases.

Admittedly, h is a very small number and consequently Δq and Δp can be small indeed. If you need any convincing of this point, consider this example. Suppose a 30.0 g rifle bullet is traveling at a velocity of 91,500 cm/sec (3000 ft/sec). The momentum of the bullet is the product of the mass and the velocity, or 2.74×10^6 g cm/sec. Let us assume that we know this value to one part in one million, a rather high degree of accuracy. Then our uncertainty in momentum, Δp, would be 2.74 g cm/sec. Now let us calculate the minimum possible uncertainty with which we can measure the position of this moving bullet. Clearly, we are looking for Δq in the equation

Δq (2.74 g cm/sec) = 6.63×10^{-27} erg sec.

A simple division yields $\Delta q = 2.42 \times 10^{-27}$ cm. This is an incredibly short distance—a miniscule fraction of the diameter of an atom. Obviously, the accuracy with which we can measure the position of a moving bullet does not begin to approach this limit. So why all the fuss about a limit of accuracy which we cannot hope to achieve experimentally?

The latter very sensible question fails to recognize the fact that on an electronic scale, where distances are much shorter and momenta are much smaller, the restriction of the Indeterminacy Principle can be detected. In part, this is because all observation is based upon interaction. You see this page because photons of light bounce off it and into your eyes. Compared to the energy of a photon, a book has incredibly high inertia. It just lies on the desk, unmoving (or at least not perceptibly moving) in spite of the bombardment of millions of little arrows of light. The situation remains much the same as the size of the object observed decreases. The photons which strike a crystal on the stage of a light microscope do not alter the crystal's position. However, the dimensions of an object may be so small that they are less than the wavelength of the light used. Then resolution fails and another probe must be used. For example, in an electron microscope, electrons are employed to illuminate and disclose objects invisible in a light microscope.

But when the object to be observed is an electron itself, with what does one observe it? Here I do not refer to instrument design—though none has been developed. Instead I ask, with what sort of energetic probe does one poke an electron? One could use electromagnetic radiation of a suitably short wavelength to resolve the electron, but it would not resolve the problem. For if you were to "see" an electron through such a super-microscope, you could be sure of only one thing—it was no longer where you saw it. The difficulty is that electrons, unlike books, do not passively lie there and let themselves be scrutinized. The interaction between a photon and an electron, an event essential for observation, disturbs the electron and alters its position and its momentum. Moreover, the exact magnitude and direction of these changes cannot be predicted. There is simply no way of observing an electron without producing these unknown perturbations. Now you see it—now you don't!

Other manifestations of indeterminacy arise as a consequence of the dualistic nature of the electron, its particle/wave complementarity (to use Bohr's term). This can be illustrated by considering a beam of electrons passing through a hole and onto a zinc sulfide-coated screen. Assume we are interested in determining the position and momentum of the particle in a direction perpendicular to its path. We will arbitrarily call this direction x. Quite clearly, the hole will permit passage of any electron whose path runs through it. Thus, the position of the particle, at the precise moment of passing through the hole, is determined by the size of the hole. In Figure 8.8 the electron must be somewhere within the gap, Δx, in order to later register on the screen. Just where within Δx we cannot determine, so Δx represents the uncertainty in position. The logical way to reduce this ignorance would be to make the hole smaller. However, when this is done, the spot on the detector gets larger rather than smaller! This is strange behavior for a particle; but, as you have gathered, an electron is a pretty strange particle. In fact, when it starts fanning out after passing through a pinhole, the electron is not behaving like a particle at all. Rather it is acting like a wave, as light would under similar circumstances.

In spite of that, each specific impact on the detector can still be regarded as the arrival of an electron wearing its particulate hat. The distance with which this position is displaced from a line through the center of the hole depends upon the momentum of the electron in the x direction. Therefore, the spreading of the diffraction pattern, as the spot is called, is an indication of increasing uncertainty in the momentum. This increase in Δp_x with decreasing Δx is schematically represented in Figure 8.8. Our efforts at absolute certainty have again been frustrated by the nature of nature.

The distribution pattern on a screen behind a single slit is similar to that produced by a hole—a smear of light decreasing in intensity toward its edges. But when an electron beam is allowed to pass through two slits, close together, side-by-side, the indeterminate behavior of the electrons becomes even more baffling.

FIGURE 8.8. *Illustration of the Heisenberg Uncertainty Principle. Reduction in uncertainty in position (Δx) increases uncertainty in momentum (Δp_x); (Δx)(Δp_x) $\geqslant h$.*

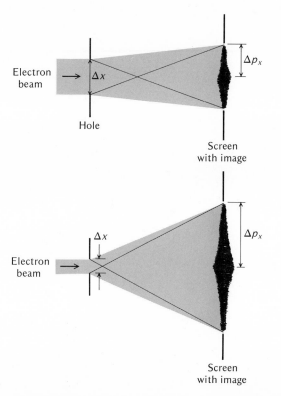

FIGURE 8.9.   *Examples of wave/particle complementarity and the Principle of Indeterminacy in the passage of electrons through slits.*

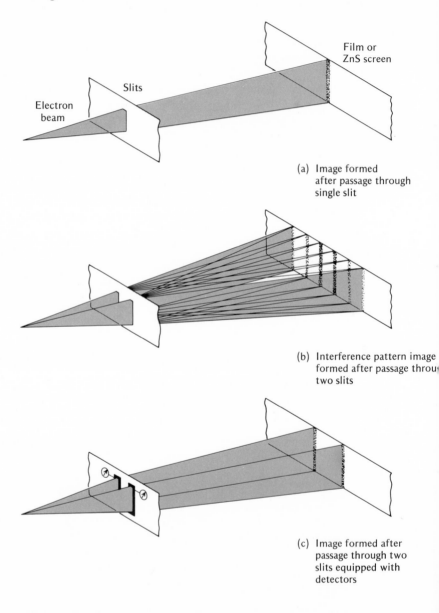

(a) Image formed after passage through single slit

(b) Interference pattern image formed after passage throug two slits

(c) Image formed after passage through two slits equipped with detectors

Our experience with bullets and baseballs convinces us that a particle can pass through either one of these slits, but not both simultaneously. But somehow, an electron passing through a single slit is influenced by the presence of an adjacent slit, the one it did not go through! What we now observe is the alternating series of dark and light bands characteristic of a wave interference pattern (see Figure 8.9). This can only mean that the matter wave has simultaneously and instantaneously passed through both slits. A fundamental particle weighing 9.1091×10^{-28} g and carrying an indivisible charge of 1.6021×10^{-19} coulomb has been in two places at the same time!

Surprisingly enough, we can still continue to pretend that an electron can pass through only one slit at a time. But this subterfuge will not enable us to predict which slit it was from the flashes on the screen. The closer together we make the slits, the more certain we become about the position of the electron. But simultaneously, the diffraction pattern becomes more and more spread out and our ignorance of the momentum grows accordingly. We can even devise detectors to measure the passage of electrons through each of the two slots (see Figure 8.9). Every time one or the other of these detectors (not both) signals, a flash of light is seen on the screen. But there is no way of exactly predicting the position of electron impact. Moreover, once we thus begin treating electrons as if they were exclusively particles, they start acting that way. Although both slits remain open, with the detectors operative the wavelike interference pattern disappears. Instead we see two patterns, each characteristic of a single slit. The very act of noting the passage of electrons through the slits has altered their momenta and wavelengths. Hence, the wave pattern is destroyed.

Some of the best scientific minds of the twentieth century have devised laboratory and thought experiments of this sort in order to test the Heisenberg Uncertainty Principle. But no matter how ingenious the ideas, they all fail to restore man's omniscience. Ignorance seems to unavoidably plague our most fundamental measurements of matter. This inadequacy came as a severe blow to the collective ego of modern science. Until the work of Heisenberg, most physicists and chemists had blithely and optimistically assumed that there was no limit to man's knowledge of the physical world. Never mind the fact that experimental techniques were not infinitely refined. They might never be. But there was no reason to assume that they could not be made absolutely precise. In principle, the position and momentum of any particle could be exactly and unambiguously measured. And with such information and Newton's laws of motion, the past or future state of the particle could be unequivocably determined into infinity. *Ergo*, the entire universe could be extrapolated backwards or forwards into time and space. Everything was determined by natural law.

But according to the Principle of Indeterminacy, chance, not causality, is at the core of the universe. Nature only appears to be determined and causal in its macroscopic behavior because here we observe millions upon millions of particles. The order we see is merely the order of actuarial tables of morbidity rates. It is born of randomness. And the universe is more like a roulette wheel than a clock.

Thus, with the advent of the Heisenberg Principle, man was still further removed from his place with the gods. And, as in the case of the collapse of the geocentric universe, the evidence for evolution and natural selection, and the recognition of the subconscious, man fought for his intellectual supremacy. This time the causal orderliness of nature itself was being defended. That disorder should underly all this beauty was too much for Albert Einstein to accept. "I cannot believe," he said, "that God plays with dice." And yet, the evidence is there.

Some scientists have argued that an undiscovered set of deterministic, causal relationships underly this apparent randomness. Others have attributed our ignorance to the limitations of our experimental methodology and not to the world we investigate and inhabit. But more and more chemists and physicists have come to accept this ambiguity and have learned to live with it. Indeed, some thinkers have found in the freedom of the electron, evidence for the freedom of man's will. If even an electron is independent, they argue, why then man must surely be master of his own fate. Comforting as this thesis might be to some, deriving theology from science is, to the author, as suspect as the widely discredited reverse process. The claims of science should be far more modest. Questions about the absolute truth of quantum numbers and wave functions, about the physical reality of electron clouds and orbitals are the stuff of metachemistry. Whatever the ultimate nature of reality, science, in the final analysis, operates on a different level. The formulas, models, metaphors, and abstractions of physics and chemistry stand or fall on how well they approximate experiment and experience. Their truth is, of necessity, relative. But so, perhaps, is all of the truth accessible to finite human minds.

For thought and action

8.1   Write all the values of l, m_l, and m_s compatible with $n = 4$. How many different sets of four quantum numbers could you make from the allowed values?

8.2   What, if anything, is wrong with the following sets of n, l, m_l, and m_s?

(a) $1, 1, 0, +\frac{1}{2}$ (b) $2, 0, -1, +\frac{1}{2}$ (c) $3, 2, 1, -\frac{1}{2}$

(d) $2, -1, 0, -\frac{1}{2}$ (e) $2, 1, -1, +\frac{1}{2}$ (f) $1, 0, 0, +\frac{3}{2}$

8.3 What, if anything, is wrong with the following electronic orbital diagrams?

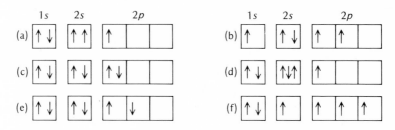

8.4 Write the electronic configurations and orbital diagrams for the elements with atomic numbers 7, 15, and 33. What similarities in atomic structure are evident and how would you expect these to be physically and chemically expressed?

8.5 Provide electronic configurations for atoms of the following elements:
(a) Ca (b) Fe (c) Sn (d) Xe (e) Hg (f) Lu (g) U

8.6 How many unpaired electrons are there in the ground state of the following atoms?
(a) Mg (b) Si (c) Mn (d) Kr (e) I

8.7 Write all the sets of four quantum numbers characterizing the electrons of neon in its ground state. Why is such an assignment more difficult for oxygen?

8.8 Sometimes the correct electronic configuration for lithium is $1s^23s$ or $1s^22p$ instead of $1s^22s$. Explain how this statement can be true.

8.9 There is a good deal of interest in what the properties and electron configuration of element number 104. Suggest why and make a prediction.

8.10 Suppose there were such things as g orbitals corresponding to $l = 4$.
(a) What values of m_l would be consistent with this value?
(b) How many g orbitals would there be?
(c) What would be the total electron capacity of the fully filled g orbitals?
(d) In which principal electronic level, that is, for what value of n, could the g levels first appear?

8.11 Suppose the rules relating n and l were such that l could take on the values $n, n - 1, \ldots, 1$. Assume, however, that the relationship between l and m_l remained unchanged. Use the Pauli Exclusion Principle to write down the acceptable sets of quantum numbers for the levels $n = 1$ and $n = 2$ under the new system.

Also, sketch the periodic table for the first 22 elements as it would appear if these rules held.

8.12   Contrast the concepts of orbit and orbital.

8.13   Explain wave functions, quantum numbers, and their significance to your friend, the art major, who isn't enrolled in the course.

8.14   Suppose in an experiment the position of an electron can be measured with an error of only 0.1 Å. What is the minimum error with which the momentum could be simultaneously determined?

8.15   An electron acts as a wave or a particle, depending on the design of the experiment. Moreover, the act of observing an electron alters its state. These two quantum-linked characteristics illustrate the interaction of the observer and the observed. Comment on the implications of this behavior not only for physical science, but for social science as well.

8.16   The terms "dualism" and "complementarity" have both been used to characterize the wave/particle nature of electrons. Discuss the implications of these two words. What other dualistic or complementary aspects of the universe or of existence can you identify? Do you see any connection between these and the electron?

8.17   In a series of lectures on "Turning Points in Physics," delivered at Oxford University in 1958, F. Waismann stated the following: "The orbits of the electrons within the atom in Bohr's first theory have about as much claim to reality as the hell-circles of Dante." Comment.

8.18   The text suggests three interpretations of the Heisenberg Uncertainty Principle: (1) it is only an apparent failure of causality because of still-undiscovered deterministic order at a lower level; (2) it is an experimental artifact; and (3) it is an accurate representation of the way nature really is. After consulting some of the suggested readings, attempt to evaluate these three hypotheses and their consequences. If experimental science could be used to aid in this evaluation, suggest an appropriate experiment or experiments.

Suggested readings

Barbour, Ian G., "'Issues in Science and Religion," Prentice-Hall, Englewood Cliffs, New Jersey, 1966.

Blin-Stoyle, R. J., and others, "Turning Points in Physics," North-Holland Publ., Amsterdam, 1959.

Bohm, David, and others, "Quanta and Reality," American Research Council, 1962.

Gamow, George, "Thirty Years That Shook Physics," Anchor Books, Doubleday, Garden City, New York, 1966.

Heisenberg, Werner, "The Physical Principles of the Quantum Theory," Dover, New York, 1930.

Heitler, W., "Elementary Wave Mechanics," Oxford Univ. Press, London and New York, 1945.

9

Chemical bonding and molecular structure

Ionic compounds: please pass an electron

At its core, chemistry is concerned with the thousands of combinations and permutations which can be made from the hundred-odd elementary forms of matter. Indeed, we most commonly confront the elements not as pure individual substances, but as constituent parts of compounds. Some of the chemicals essential for life—water, carbon dioxide, sugar, and salt—are simple compounds. Others, which we synthesize within the fantastic factories of our bodies, are far more complex. They include proteins, nucleic acids, fats, and polysaccharides. Compared to these internal creations, the tailormade molecules of modern chemistry—detergents, plastics, miracle fibers, and wonder drugs—sometimes seem almost trivial.

But simple or complex, all compounds are formed of aggregates of atoms combined in some characteristic ratio. When Dalton first formulated this basic idea over a century and a half ago, he was uncertain of the nature of the attractive forces which held the atoms together. He wrote of "atmospheres of caloric" surrounding atoms and molecules, but left the question of the identity of the interatomic "glue" largely unanswered. At first there were no explanations, only quaintly naïve pictures of hooked atoms interconnected with each other. But Dalton and his contemporaries must have recognized the inadequacy of this representation. To be sure, there were advances. That electricity was somehow involved was postulated relatively early. Nevertheless, it is not surprising that a full understanding of the molecular structure of compounds had to await the illucidation of the atomic structure of the elements. It is from this vantage point of contemporary knowledge that we will survey the field of the chemical combination of elements.

THE SALTS OF THE EARTH

Surprisingly, perhaps, we begin with the series of elements which hardly combine at all—the rare gases. The secret of the chemical aloofness of these elements lies with their electrons. Consulting the electron configurations and orbital diagrams of the previous chapter, you will at once observe that all of the almost-inert gases are characterized by fully filled electronic sublevels. In the case of helium, $1s^2$, the first energy level is completely occupied. Similarly, in the ground state of neon, $1s^2 2s^2 2p^6$, there are no vacancies for additional electrons within the $n = 2$ state. True, in the argon atom, $1s^2 2s^2 2p^6 3s^2 3p^6$, the $3d$ orbitals remain empty. And krypton, xenon, and radon also have completely unoccupied energy levels. However, those orbitals which are occupied, are filled to saturation. Moreover, the electrons are held strongly. They are only removed with considerable difficulty. Furthermore, it is energetically difficult to add additional electrons to the neutral rare gas atoms. Quite clearly, here is a case of structural as well as chemical stability. To be anthropomorphic about it, the atoms of the noble gases seem quite content to be as solitary as Greta Garbo or Howard Hughes.

Not so the gregarious families which flank the rare gases—the halogens and the alkali metals. The former group contains some of the most reactive of the nonmetals; the latter is composed of those hyperthyroid metallic elements we encountered earlier. Moreover, the members of these two families have a particular propensity to react with each other. And when they do, it is to form compounds in which the atom ratio of the elements is always one-to-one. Sodium chloride, NaCl, is, of course, the most familiar example. But very likely all 30 possible combinations are known and have been studied. They go by the name of their most ubiquitous member—*salts*.

These alkali metal halides share a number of distinguishing properties. In the first place, they are all crystalline solids with relatively high melting points (600°–1000°C). The fact that a good deal of heat is required to cause melting suggests that the particles which constitute the crystal are held together with considerable force. This force must be overcome in order to disrupt the ordered pattern of the solid and convert it to the more random liquid state. When melting is achieved, the liquefied compound is found to readily conduct electricity. By contrast, the crystalline form does not. This observation has important implications for the structure of these salts.

Electricity travels in units of one electron charge. In the wires of a house circuit, the electrons themselves move. But it is also possible for electrical current to be transported by carriers which are charged or capable of acquiring a charge. In the case of molten alkali metal halides, the nature of these carriers can be determined with relative ease. Let us consider as a specific example sodium fluoride, NaF, the anticavity agent added to fluoridate water systems. When electrodes are inserted into a molten mass of sodium fluoride, both metallic sodium and gaseous fluorine are formed. In other words, the compound is broken down with electricity, just as Davy decomposed the oxides of the alkali metals and the alkaline earths. Sodium always appears at the cathode, the negative electrode at which electrons enter the melt. And, elementary fluorine is always evolved at the anode, the positive electrode from which electrons emerge from the molten mass. The fact that particles of sodium are attracted by the negative electrode and there appear to acquire electrons strongly suggests that in molten sodium fluoride these particles are positively charged. They are not ordinary sodium atoms, but charged atoms or *ions*. Because they migrate toward the cathode they are called *cations*. Similarly, the appearance of fluorine gas at the positive electrode indicates that, in the melt, fluorine must exist as negatively charged ions or *anions*.

The magnitude of these charges can be experimentally ascertained, thanks to Faraday's *Laws of Electrolysis*. Back in Chapter 6, a faraday was defined as one Avogadro number of electrons. It turns out that exactly 1 faraday will displace 23.00 g of sodium and 19.00 g of fluorine from molten sodium fluoride. Significantly, these two numbers are the gram atomic weights of sodium and fluorine. Since each gram atomic weight represents an Avogadro number of atoms, we can at once conclude that in the electron transfer, each sodium ion is acquiring one electron and each fluoride ion is losing an electron. Since we wind up with neutral atoms, it follows that we can represent the ions as Na^+ and F^-. The former has an excess positive charge corresponding to one proton. Or, more exactly, the sodium ion contains one less elec-

FIGURE 9.1.   *Face-centered cubic lattice of sodium fluoride, NaF. The Na⁺ ions are indicated in gray.*

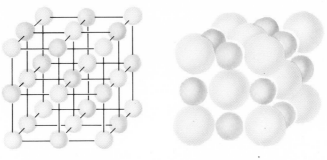

Compound lattice   Packing model

tron than the sodium atom. The fluoride ion, on the other hand, has one electron too many.

These conclusions are supported by additional experimental evidence. Generally speaking, the alkali halides dissolve readily in water. The solvent seems to promote the collapse of the crystal structure and the separation of the particles which make it up. If these particles are really oppositely charged Na^+ and F^- ions, the resulting solution should contain these species. Therefore, it should conduct electricity. And indeed it does. Because of this conductivity property, sodium fluoride and related compounds are called *electrolytes*. Sodium and fluoride ions clearly exist in solutions of NaF as well as in the molten salt. The logical inference would seem to be that the solid, too, must consist of these ions.

Fortunately, we can rely on more than inference. X-Ray diffraction experiments have unambiguously localized the Na^+ and F^- ions. They are located on regularly spaced points in an alternating three-dimensional array called a *crystal lattice*. The "face-centered cubic" pattern of sodium fluoride is represented in Figure 9.1. Although it may not be immediately apparent from the drawing, the lattice consists of equal numbers of sodium and fluoride ions. This, of course, is to be expected from the formula NaF and the electroneutrality of the salt. In the solid the ions are fixed in position because of the strong attractive forces between the electrically opposite charges. But when the crystal melts or dissolves, the ions become free to move about. Note that they do not migrate as "molecules." Indeed, there are no molecular entities of NaF except in the gaseous state. The alkali metal halides are thus considered *ionic compounds*, as distinct from *molecular compounds* like H_2O.*

* This explains why it is more accurate to speak of the "gram formula weight" of an ionic compound rather than its "gram molecular weight." However, both are computed from atomic weights in the same manner.

IONS, OXIDATION, *et cetera*

Having thus established that sodium fluoride is made up of alternating Na^+ and F^- ions, let us consider the electronic arrangements of these species. First recall the orbital configurations of the corresponding atoms. Neon will provide a useful reference.

F	$1s^2 2s^2 2p^5$	9 electrons and	9 protons
Ne	$1s^2 2s^2 2p^6$	10 electrons and	10 protons
Na	$1s^2 2s^2 2p^6 3s$	11 electrons and	11 protons

Now add one electron to the fluorine atom and remove one from sodium.

F^-	$1s^2 2s^2 2p^6$	10 electrons and	9 protons
Ne	$1s^2 2s^2 2p^6$	10 electrons and	10 protons
Na^+	$1s^2 2s^2 2p^6$	10 electrons and	11 protons

You immediately see that the two ions and the neon atom all have identical electronic arrangements. They are said to be *isoelectronic*. To be sure, their nuclear charges are different; but this, of course, is manifest in their total net charges. The fundamental point is that the ionic species, F^- and Na^+, have electron configurations that correspond to the particularly stable fully filled orbitals of neon. The sodium atom readily loses an electron to achieve this arrangement, the fluorine atom readily gains one.*

$$Na \rightarrow Na^+ + e^-$$
$$\tfrac{1}{2} F_2 + e^- \rightarrow F^-$$

Together, the two half-equations add up to represent the chemical combination of the two elements:

$$Na + \tfrac{1}{2} F_2 \rightarrow NaF.$$

In the process, an electron is transferred from each sodium atom to each fluorine atom. And, to again attribute human characteristics to inanimate matter, the elements are far more content as charged upon ions than as neutral atoms.

What we are rationalizing here is the high reactivity of these two elements. Fluorine is "fervid" because it is so intent upon achieving the eight-electron capacity of the $n = 2$ sublevel that it will snatch an electron from almost any other element. Said somewhat more scientifically, fluorine has a "high electron affinity." Sodium, on the other hand, readily donates its single $3s$ electron to any element willing to receive it. Or, to be less animist about it all, sodium has a "low ionization potential." In this process of electron

---

* The second equation takes into account the fact that elementary fluorine exists in diatomic molecules.

transfer, fluorine and sodium are, of course, converted from unstable elements to far more stable compounds.

Exactly the same argument applies to the formation of another familiar compound, magnesium oxide. Only here, two electrons are transferred to yield the neon electronic configuration. This is apparent from the following brief table.

| Mg | $1s^2 2s^2 2p^6 3s^2$ | Mg^{2+} | $1s^2 2s^2 2p^6$ |
| O | $1s^2 2s^2 2p^4$ | O^{2-} | $1s^2 2s^2 2p^6$ |

Again the two ions are isoelectronic with the neon atom. And again, the cation–anion ratio is one-to-one, MgO.

In the direct combination of magnesium and oxygen, the former donates its two $3s$ electrons to oxygen, where they enter and fill the $2p$ orbitals. As before, the overall reaction can be written as the sum of two half-reactions, which of course, occur simultaneously.

$$Mg \rightarrow Mg^{2+} + 2e^-$$
$$\frac{\frac{1}{2} O_2 + 2e^- \rightarrow O^{2-}}{Mg + \frac{1}{2} O_2 \rightarrow MgO}$$

As you are well aware, this is a classical example of oxidation. It occurs spontaneously at room temperature and, at elevated temperatures, it gives off the light and heat of a flashbulb. But whether or not combustion actually occurs, the magnesium is oxidized, and in the process it loses two electrons. This latter atomic/electronic/ionic event has come to be represented by the originally more restricted term, oxidation. Today, *any loss of electrons is called an oxidation reaction.* This is true whether or not oxygen serves as the *electron acceptor or oxidizing agent.* Thus, the reaction of sodium with fluorine is also considered an example of oxidation, because in the process each sodium atom gives up an electron.

Of course, the electron must go somewhere; it must be gained by some other species. *This electron gain is called reduction.* Thus, the oxidation of magnesium is attended by the reduction of oxygen. That is, each oxygen atom acquires two electrons to form an oxygen ion. Similarly, fluorine is reduced to F^- by the action of sodium, a *reducing agent or electron donor.*

Incidentally, the term "reduction" may seem a rather strange choice for a gain in electrons, but there is an historical explanation. The winning of pure metals from their ores, for example, by smelting, has long been called reduction. Thus, Davy first prepared magnesium by electrolytically reducing magnesium oxide.

$$MgO \rightarrow Mg + \frac{1}{2} O_2$$

In this reaction, the Mg^{2+} ions of MgO gain two electrons each to yield atomic magnesium. At the same time, each O^{2-} ion loses two electrons to form an atom (that is, half a molecule) of oxygen. According to our general definition, Mg^{2+} is reduced, O^{2-} is oxidized.*

Granting the risk of inundating the reader with new terms, this is the logical place to introduce several more. The electrical charge on an ion is called its *electrovalence* or *oxidation number (o.n.)*. It represents the "combining power" of the element. For Na^+ its value is $+1$, for O^{2-} it is -2, and so on. Thus, *oxidation may also be defined as an increase in oxidation number*, for example, from 0 to $+1$. Similarly, *reduction is a decrease in oxidation number*, for example, from $+1$ to 0 or from 0 to -2.

All this nomenclature can be summarized by rewriting the equations for the combination of sodium and fluorine and supplying some labels.

Oxidation: Na $\rightarrow$ Na^+ + e^-
             (o.n. = 0)   (o.n. = +1)
Reduction: $\frac{1}{2}F_2$ + e^- $\rightarrow$ F^-
             (o.n. = 0)   (o.n. = -1)
Species oxidized: Na   Species reduced: $\frac{1}{2}F_2$
Oxidizing agent: $\frac{1}{2}F_2$ Reducing agent: Na

Generalizing now from these specific examples, we see that many stable ions exist in rare gas electronic configurations. Because of their similar external or valence electrons, atoms of all alkali metals tend to lose one electron to yield monovalent cations of oxidation number $+1$. Alkaline earth metals form ions of the general type M^{2+}. And in order to become isoelectronic with the noble gases, aluminum and other related elements must give up three electrons. At the nonmetal end of the periodic table, the halogens follow the example of fluorine and readily gain a single electron to fill the p orbitals. Thus their characteristic oxidation state is -1. And the elements of Group 6A often exist as anions with individual charges of -2.

Some elements which form ionic compounds are too far removed from the rare gases to readily achieve the rare gas configuration. For example, members of the transition series would have to lose or gain a large number of electrons to become isoelectronic with the nearly inert elements. But electrovalences greater than $+4$ or lower than -3 are very rare. In fact, most transition metals form cations of charge $+1$, $+2$, or $+3$. This suggests that stable ions can exist which do not duplicate the elec-

---

* A note to file for future reference—the combination of magnesium and oxygen occurs spontaneously while the reverse reaction, the decomposition of magnesium oxide, can only be achieved by the application of work, for example, by electrolysis.

tronic arrangements of the noble gases. Some of these ions are also characterized by fully filled sublevels. These include Zn^{2+}, $[Ar]3s^23p^63d^{10}$; Ag^+, $[Kr]4s^24p^64d^{10}$; and Pb^{2+}, $[Xe]5s^25p^65d^{10}6s^2$. Other ions contain only partially filled orbitals. In these cases, a *priori* predictions of likely ionic states sometimes become difficult.

Nevertheless, the general principles which we have illustrated are of considerable value in predicting stable electronic arrangements in ions, corresponding valence states, and the formulas of compounds. And, at the very least, they permit us to rationalize experimentally determined formulas. The idea is that the compound must be electrically neutral. That is, the total positive charge must equal the total negative charge. This means that the number of cations per formula times the charge on each cation must equal the number of anions times the charge on each anion. The absolute magnitude of the two products will be identical, though the answers will have opposite signs. Or, if it is easier, consider the rule to state that the sum of all products of the number of ions times the corresponding oxidation number must equal zero.

Thus, the formula of barium chloride must be $BaCl_2$. Each formula unit (not strictly a molecule, recall) consists of one Ba^{2+} ion and two Cl^- ions. Multiplying number of ions times ionic charge and adding we obtain: $1 \times (+2) + 2 \times (-1) = 0$. Similarly, potassium oxide must be K_2O, since the charge on each of the two cations is $+1$ and the charge on the single oxide ion is -2. A final example is provided by aluminum oxide. The electrovalence of aluminum is $+3$, that of oxygen is -2. It follows that the simplest way of balancing positive and negative charges is with a ratio of two aluminum ions to three oxygen ions. Then the total positive charge is $2 \times (+3)$ or $+6$ and the total negative charge is $3 \times (-2)$ or -6. Consequently, the formula of aluminum oxide is Al_2O_3. As you would expect, the formulas of ionic compounds are reflected in their crystal structures. Obviously, the relative number of cations and anions in the crystal lattice must correspond to the atomic ratio of the elements involved. This requirement in turn influences the particular ionic arrangement assumed.

Sometimes formulas indicate that a single element can exist in more than one ionic form. Indeed, the phenomenon of multivalency is fairly common; you encountered it in Chapter 3 as the Law of Multiple Proportions. Now we can offer an electronic explanation. A good example is provided by the two chlorides of copper. Cuprous chloride contains 64.19% copper (by weight) and 35.81% chlorine. This corresponds to 63.55 g of copper per 35.45 g of chlorine. Translating to gram atomic weights, the mass ratio is equivalent to one gram atomic weight of copper per gram atomic weight of chlorine. It follows that the formula of cuprous chloride must be CuCl and that the oxidation number of the cuprous ions must be $+1$. On the other hand, in cupric chloride, 31.77 g of copper are found for each 35.45 g of chlorine. That is, one-half gram atomic weight of copper is combined with a single gram atomic weight of chlorine. Since we do not write formulas in terms of fractional atoms, we must express the copper:chlorine atom ratio as 1:2. In formula form, cupric chloride is $CuCl_2$, and Cu^{2+} must correspond to the cupric ion.

Many other examples could be cited, but by now the concept of ionic compounds should be clear. *Ions are formed when atoms lose (oxidation) and gain (reduction) electrons to achieve stable electronic configurations.* The resulting charged species, positive cations and negative anions, exert and experience mutual electrostatic attraction which holds them tightly together in a crystal lattice characteristic of the compound. The dissociation of the lattice by melting or solution releases the individual ions, which can become migrating carriers of electricity. The correlation between bulk properties and electronic phenomena is beautiful. And because it is, exceptions are obvious. There are literally thousands of them—compounds which unquestionably cannot be explained as ionic aggregates. The glue which holds the atoms of these molecular compounds together must be of a somewhat different sort. So we must look further. After we do, we shall also be able to comment on a class of ionic compounds we have thus far ignored—those involving more than two different elements.

Covalent compounds: togetherness through sharing

As important and as ubiquitous as ionic compounds are, there are many more compounds which do not demonstrate the high melting points, high water solubility, and electrical conductivity of the electrolytes. These compounds may be gaseous, liquid, or solid at room temperature, but generally they have relatively low melting and boiling points. Frequently they are almost insoluble in water. And they do not conduct electricity in solution or in the liquid state. There is thus no evidence to indicate that the elements in these compounds exist as ions. Rather, there is considerable experimental proof that such substances are made up of discrete molecules. Hence, they are sometimes called *molecular compounds.*

Much of our empirical evidence concerning molecular structure comes from spectroscopy. In the first place, molecules exhibit a variety of electronic energy levels, just as atoms do. And, as in the case of the emission and absorption of energy by atoms, the frequency of an observed spectroscopic line is an indication of the energetic interval between the two quantum states involved in the transition. Consequently, the electronic spectrum of a molecule can provide important information concerning the orbitals occupied in the ground and excited states.

But in addition to electronic energy levels, a molecule is also capable of other energetic options not open to an atom. These levels arise from the fact that a molecule consists of two or more atoms. The bonds which hold these atoms together, and which will be the chief item of interest in this section, are not absolutely rigid. In fact, they are more like stiff springs. And because they are, the atoms on either end of the bond/spring can vibrate. They can vibrate at a variety of frequencies, but not any old frequency; the energy of molecular vibrations is quantized, too. Therefore, measurements of spectroscopic lines corresponding to transitions between vibrational states provide information about the actual energy levels involved. At the same time, the results can supply a measure of the stiffness of the bonds studied. And finally, additional information concerning the orientation of these bonds and thus the geometrical structure of the molecule can also be inferred from vibrational spectra.

A second and equally informative mode of molecular motion is rotation. The entire molecule can spin in space at a rate of rotation which is also quantized. Differences between these energy levels manifest themselves as lines in the microwave region of the spectrum or, in association with vibrational transitions, in the infrared range. Equations have been derived which relate the frequencies of these lines to the shape of the molecule. Specifically, the data, coupled with the calculations, can provide accurate values for the lengths of chemical bonds. Thus, a chemist can speak with some degree of certainty about the arrangement of invisible atoms in unseen molecules.

ELECTRONIC CEMENT AND FLYSPECK FASTENERS

The nature of the interaction holding these atoms together is reflected in the word *covalent,* the term most commonly applied to such molecular compounds. Its etymological origins suggest shared strength. And indeed, there is sharing in covalent compounds. Electrons are not transferred between atoms to yield oppositely charged ions. Instead, the sharing of a pair of electrons provides the bond which unites the atoms and the elements.

The simplest example of covalent bonding is not even a compound; it is the hydrogen molecule, H_2. As you well know, each hydrogen atom consists of a single $1s$ electron, a single proton, and usually no neutrons. If we represent the lone electron as a flyspeck, an atom of hydrogen is simply $H\cdot$. Two of them, each contributing its single electron to the commonweal, become H∶H. The spins of the two electrons pair and form a *single bond*, which is frequently represented as a straight line, H—H. Unfortunately, such symbols run the risk of implying that both electrons are strictly localized between the two hydrogen nuclei. This is not so. But experiment and quantum-mechanical calculations do indicate that electron density is greatest in this position. The charge clouds overlap as indicated in Figure 9.2.

If you are somewhat skeptical after reading the last few sentences, you may have the makings of a good scientist. After all, why should a pair of electrons act as interatomic cement? The question is perceptive and pertinent. And happily, it has been answered by rather sophisticated calculations. These mathematical machinations have led to results such as those of Figure 9.3. Here the potential energy of interaction between two hydrogen atoms is plotted as a function of the distance between their centers. When this separation is great, the atoms exert a slight attractive force on one another. As the distance decreases, this force increases and, as a consequence, the energy of the system decreases. Finally, at an internuclear separation of 0.74 Å, a minimum in potential energy is achieved. If the atoms are brought still closer together, repulsive forces become increasingly important, and the energy rises sharply. The minimum thus represents a position of maximum stability. Here the electrostatic attraction of the negative electron of one atom for the positive proton of the other far exceeds the repulsive forces generated between the two electrons and the two nuclei. Hence, the system comes to rest in this configuration—the hydrogen molecule. It is rather as if one were to place a marble anywhere along a curved surface shaped like Figure 9.3. Obviously, the marble would ultimately stop at the bottom of the curve —the point corresponding to the position of minimum potential energy. Nature always seeks the easy way out. And it simply is energetically easier to keep a hydrogen molecule together than to keep two hydrogen atoms apart.

FIGURE 9.2.   *The overlapping s orbitals of the hydrogen molecule, H_2.*

FIGURE 9.3. *Potential energy diagram for two hydrogen atoms as a function of their internuclear separation, r. One of the atoms is presumed to be fixed at the origin. Note the stable molecule at the minimum of the curve.*

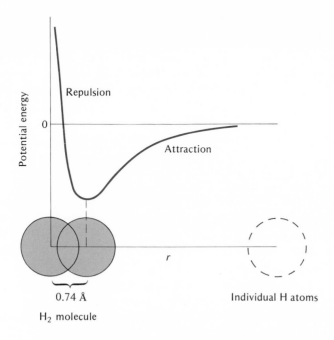

Moreover, one can regard the overlap of atomic orbitals as providing more room for the electrons to move about in a continuous exchange. At one instant, both electrons may be near one nucleus; at the next they may both be on the other side of the molecule. Thus we have an exchange between two transient "ions"

$$H^+ : H^- \longleftrightarrow H :^- H^+$$

Clearly, electrostatic attraction will result. Indeed, the electrons of a hydrogen molecule move about so freely that it becomes pointless to try to identify them with one atom or the other. In a very real sense, they belong to the entire molecule. This view of electrons as common property underlies the quantum-mechanical method of *molecular orbitals*—an approach which has been useful in predicting molecular structure.

We will concentrate upon the rival procedure, the method of *atomic orbitals*. This approach is predicated on somewhat of a fiction, since it speaks of the molecule primarily in terms of the orbitals of the participating atoms. Nevertheless, the method is clear, simple, and generally satisfactory. One of its strengths is in representing the effect of electron sharing upon orbital populations.

Thus, in the case of H_2, each nucleus is essentially associated with two electrons. And two electrons correspond to the stable configuration of helium—an arrangement which also might be expected to confer some stability upon the hydrogen molecule.

The example of hydrogen fluoride, HF, indicates that similar arguments can be easily extended to other molecules. Fluorine, you recall, has as its normal electronic configuration, $1s^2 2s^2 2p^5$. Writing dots for the seven outer elections in the $n = 2$ level, the atom becomes $\cdot \ddot{F} \colon$. Fluorine's affinity and affection for electrons is by now well known to you. But hydrogen holds tightly enough to its single electron that fluorine never manages to steal it completely away. Instead, the atoms compromise and share a pair, $H \colon \ddot{F} \colon$. Both atoms are happy: the hydrogen thinks it is helium and the fluorine imagines itself to be neon. The union may be based upon self-deception, but it works! To be sure, the partners are not equal, as they are in H_2. Fluorine's attraction for electrons, its *electronegativity*, is so great that the bonding electron pair spends more time near that atom than the hydrogen atom. This means that the structure $H^+ \colon \ddot{F} \colon^-$ contributes "partial ionic character" to the covalent bond. The fluorine end of the molecule definitely bears a negative charge compared to the positive charge on the hydrogen. One could push the analogy to the point of a dominant parent drawing children away from his or her spouse. But we are undoubtedly in too deep already!

The presence of eight valence electrons about a single atom is a familiar phenomenon in covalent bonding. It is so common, in fact, that it is enshrined in the *octet rule*. This useful generalization, formulated by the American chemist G. N. Lewis, states that *in many molecules, atoms achieve noble gas structure by acquiring an octet of electrons through sharing*. There are many exceptions; some molecules even include atoms surrounded by an odd number of electrons. But in writing a Lewis dot structure, such as we have been doing, it is a good idea to first try to apply the rule of eight.

The rule works well for the other compounds of hydrogen with the members of the second period. For example, oxygen brings six valence electrons to its union with hydrogen. The orbital assignment of these electrons in the ground-state atom is given by the diagram:

$2s^2$ $2p^4$

| ↑ ↓ | | ↑ ↓ | ↑ | | ↑ |

It is clear that there are two unpaired electrons, each of which is available to pair with a single hydrogen electron. The other two lone electron pairs stick off into space and are not involved in bond formation. The result, in Lewis language, is H:Ö:H. Oxygen enjoys an octet and each hydrogen shares in a pair.

3-D CHEMISTRY

The above molecular structure for water is correct in the two-dimensional world of the printed page. But remember that molecules exist in three dimensions. A literal interpretation of H—O—H suggests a linear molecule, with all three atoms arranged in a straight line. But we are sure that the water molecule is bent. The experimental evidence underlying our certainty is provided by spectroscopy and by the fact that the water molecule has a positive end and a negative end. When placed between two oppositely charged plates, water molecules align themselves. One end points toward the positive plate and the other toward the negative. In the jargon of the trade, water has a permanent *dipole moment*. Now if the water were linear, with the oxygen in the middle, there would be no way of telling one end from the other. Granted, the oxygen, being more electronegative than the hydrogen, would attract the electronic charge clouds toward itself. But the molecule would still be completely symmetrical and hence it could have no permanent dipole moment. To be sure, H—H—O would. But I challenge you to devise even a remotely respectable flyspeck structure for such a molecule. No, the oxygen must be between the hydrogens. And therefore, there can only be one explanation. The molecule must be bent, with the relative charge distribution indicated in the following diagram.

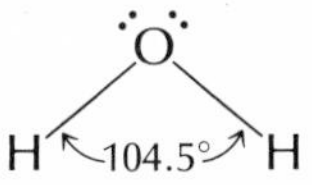

Measurements have told us just how bent. The angle between the two O—H bonds is 104.5°.

The value of this angle suggests an explanation for the bonding. Two unbonded lone pairs of electrons are situated on the oxygen atom. Two more pairs of electrons are involved in the O—H bonds. Since all four pairs bear negative charges, it makes good electrostatic sense to assume that they will repel each other and keep as far apart as possible. For four arms attached at a single pivot point, the furthest possible separation in three dimensions is a tetrahedron. Figure 9.4 is an attempt to depict such a structure. The angle between any two of the four arms is always the same—109.5°. This value is quite close to the observed H—O—H angle of 104.5°. The fact that the two angles are not identical means that there is some distortion from a perfect tetrahedron in

FIGURE 9.4.   *A two-dimensional representation of a three-dimensional structure—four bonds in equidistant tetrahedral orientation. The wedge-shaped line extends out toward you, the dashed line extends back, and the two solid lines are in the plane of the paper. All angles are equal to 109.5°.*

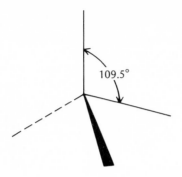

H_2O. The two unbonded pairs of electrons push the bonding pairs a little closer together than the tetrahedral arrangement would predict.

Similar geometrical considerations are evident in ammonia, NH_3. This is the nasty, choking stuff in smelling salts and an extremely important fertilizer and industrial chemical. Its structure comes quite easily. The starting materials are 5 valence electrons from nitrogen (3 of them unpaired) and 1 from each of the three hydrogen atoms. Of these 8 electrons, 6 are used in forming the three electron-pair bonds. We can consider these to be the 3 unpaired electrons donated by the nitrogen and the 3 hydrogen electrons. This leaves 8 − 6 or 2 electrons remaining as a nonbonding lone pair on the nitrogen. Translated into dots, the structure is

$$H:\overset{..}{\underset{..}{N}}:H$$
$$H$$

Experiment indicates that in three dimensions the ammonia molecule looks like a trigonal pyramid. The three hydrogen atoms arranged in an equilateral triangle form the base of the pyramid and the nitrogen atom is centered above them.

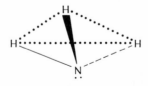

The structure can also be viewed as a tetrahedron with the nitrogen atom at its center and the three hydrogens occupying three of the vertices and the single lone pair of electrons filling the fourth.

The measured bond angles are consistent with this model; each H—N—H angle is 107°. The discrepancy between this value and the tetrahedral angle of 109.5° indicates that the N—H bonds are bent downwards slightly more than normal by the repulsion of the unbonded pair. As would be expected, the ammonia molecule has a permanent dipole moment with its negative pole centered on the nitrogen.

The lone pair of electrons on the nitrogen demonstrates a tendency to bond a proton, that is, to attach a hydrogen nucleus. Thus, an ammonia molecule in aqueous solution abstracts a hydrogen ion, H^+, from a water molecule.

$$NH_3 + H_2O \rightarrow NH_4^+ + OH^-$$

The equation indicates that in the process of proton transfer, two ions are formed. The hydroxyl ion, OH^- exhibits a charge of -1 because it is essentially a water molecule which has lost one proton. A water molecule always includes 10 electrons and an equal number of protons. Removing one of the latter, results in an excess of one electron. The resulting hydroxyl ion obeys the octet rule.

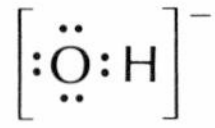

$$\left[:\overset{\cdot\cdot}{\underset{\cdot\cdot}{O}} : H \right]^-$$

The nitrogen of the ammonium ion is also surrounded by eight electrons.

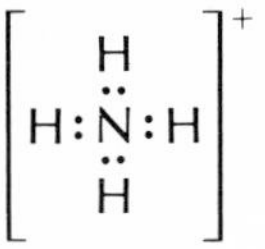

$$\left[\begin{array}{c} H \\ \cdot\cdot \\ H : N : H \\ \cdot\cdot \\ H \end{array} \right]^+$$

Its charge of $+1$ is a consequence of the addition of a proton to the neutral ammonia molecule. Obviously, the proton cannot contribute an electron to the new N—H bond. Instead, both electrons of the bonding pair are donated by the nitrogen atom. To be sure, the hydrogen and nitrogen still share the electron pair, but both of the electrons come from the latter atom. Such instances of inequitable electronic contribution are called *coordinate covalent bonds*. In practice, however, it is impossible to distinguish the one coordinate covalent bond from any of the three normal covalent bonds of the ion. They are all identical in the symmetrical, tetrahedral geometry of the ion.

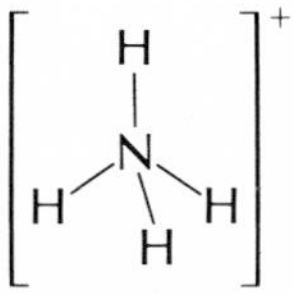

$$\left[\begin{array}{c} H \\ | \\ N \\ H \diagdown \; \diagdown H \\ H \end{array} \right]^+$$

We have been emphasizing the fact that the bonds within this ion and the hydroxyl ion are covalent in nature. But the ions themselves can participate in ionic compounds. For example, the halide salts of the ammonium ion are quite similar to the alkali metal halides. Thus, a crystal of ammonium chloride, NH_4Cl, is a lattice of equal numbers of alternating NH_4^+ and Cl^- ions. When the compound dissolves in water, the cations and anions are dissociated from each other and released. But the covalent bonds of the ammonium ion remain intact and there is no further breakdown to nitrogen and hydrogen atoms or ions. Similarly, the hydroxyl ion is the anion of many common compounds including sodium hydroxide, NaOH, and calcium hydroxide, $Ca(OH)_2$. You may recognize these substances as members of a class of compounds called *bases* or *alkalis*. Their importance is such that we shall encounter them again. The Lewis structures of several other common polyatomic ions are given in Table 9.1. (The concept of resonating double bonds mentioned in the table is explained on pp. 244–245.)

HYBRID ORBITALS: BLENDED BONDS

Methane, CH_4, a major component of natural gas, is isoelectronic with the ammonium ion. That is, it also has ten electrons. Of course, the methane molecule has one less proton and hence it is electrically neutral. But its tetrahedral structure corresponds exactly to that of NH_4^+. The carbon atom occupies the center and the four hydrogen atoms are found at the apexes. Because the molecule is symmetrical, it has no permanent dipole moment. This geometry is again consistent with the idea of repulsion between electron pairs. Moreover, the octet is readily achieved with four covalent bonds and no lone pairs.

At first glance, this all seems very simple and straightforward. But on closer inspection it is really rather curious. The ground-state orbital population of the valence electrons of carbon is given by the following diagram:

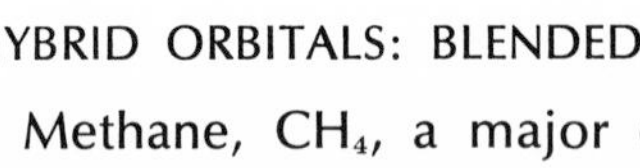

$2s^2$ $2p^2$

| ↑ ↓ | | ↑ | ↑ | |

The assignment implies that only two unpaired electrons are available for sharing with hydrogen atoms. This, in turn, would suggest that the formula of methane should be H:C:H. True, the octet rule would be violated, but the idea of electron pairing to form covalent bonds would be obeyed. On the other hand, the correct formula,

```
      H
      ··
  H:C:H
      ··
      H
```

obeys the former rule but appears to violate the latter.

TABLE 9.1.   *Some Common Polyatomic Ions*

Name	Formula	Lewis structure	Geometry
Sulfate	SO_4^{2-}		Tetrahedron
Nitrate	NO_3^-		Planar (resonating double bond)
Nitrite	NO_2^-		Bent (resonating double bond)
Chlorate	ClO_3^-		Trigonal pyramid
Carbonate	CO_3^{2-}		Planar (resonating double bond)
Phosphate	PO_4^{3-}		Tetrahedron

That it only appears to is explained in terms of the *hybridization of atomic orbitals*. The argument holds that as the electrons of the carbon atom move into their bonding configurations, one of the $2s$ electrons is promoted to the $2p$ state. The resulting orbital population is

Now there are four unpaired electrons, each presumably prepared to join with a hydrogen electron to form four covalent bonds. The fact that the bonds are equivalent—all the same length and all at equal tetrahedral angles to one another—provides further support for the idea of hybridization or the mixing of orbitals. If the individual orbitals retained their characteristic geometry, there should be the three p_x, p_y, and p_z dumbbells and the spherical smear of the 2s electron. Hence one of the four C—H bonds would be different than the other three. But quantum-mechanical calculations indicate that if the s and p wave functions are mathematically mixed, they are modified and made equal. Because this mixture involves one s and three p orbitals, it is called an sp^3 hybrid. The result is the symmetrical four-lobed tetrahedral distribution of Figure 9.5. It is in obvious agreement with experiment.

It would be difficult to overemphasize the structural significance of the tetrahedral carbon atom, chiefly because of the thousands and thousands of compounds which include this element. You will meet some of them soon. But first we should consider a few other molecules with other geometries which can be explained by other hybrids. In some of these compounds the octet rule is not followed. Boron trifluoride, BF_3, is a case in point. The entire molecule includes 24 valence electrons, 7 contributed by each of the fluorine atoms and 3 more from the single boron. Six electrons are required to make the three B—F electron-pair bonds which hold the molecule together. This leaves 18 electrons to distribute among the four atoms as nonbonding pairs. A little effort with paper and pencil should convince you that it is impossible to come up with an octet around each atom in a structure which has three single covalent bonds. The best we can do is this:

Note that the boron atom has six, not eight, electrons around

FIGURE 9.5.   *Electron charge cloud distribution for methane, CH_4, the carbon atom exhibiting a four-lobed sp^3 hybrid orbital.*

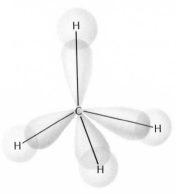

it—only the three bonding pairs. Now suppose, as before, that these three pair want to keep as far apart as possible. How would they do it? The answer is in a plane with each B—F bond separated by 120°. Unlike NH_3, the BF_3 molecule should not be a trigonal pyramid. There are three instead of four electron clouds to repel each other. These three clouds of BF_3 are part of an sp^2 *hybrid* system. The three valence electrons of the boron atom occupy, in the ground state, orbitals represented as

Unpairing one of the 2 electrons and promoting it to the $2p$ level gives rise to three singly occupied orbitals:

Mixing the single s orbital with the two p orbitals yields a trilobed sp^2 hybrid with the geometry of Figure 9.6. Happily, experiment attests to the correctness of this structure.

One more example will be mentioned, but not discussed in detail. Beryllium difluoride, BeF_2, is a linear molecule with the following Lewis structure:

$$:\overset{..}{F}:Be:\overset{..}{F}:$$

The two Be—F bonds, situated at 180° to each other, are expained as the consequence of *sp hybridization*. One of the two electrons in the $2s$ orbital of ground-state beryllium is promoted to one of the normally completely empty $2p$ orbitals. These two electron

FIGURE 9.6.   *Electron distribution for boron trifluoride, BF_3, showing a three-lobed sp^2 hybrid orbital.*

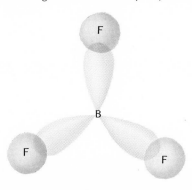

clouds then hybridize to form the oppositely oriented bonding orbitals.

Other hybrid orbitals have been postulated, some with a capacity of as many as six electron pairs. However, sp^3, sp^2, and sp will be enough to enable you to better understand the complexities of covalent bonding in carbon compounds.

Some covalent compounds of carbon

One of the reasons underlying the plethora of carbon-containing compounds is the almost unique ability of carbon atoms to share electrons with each other. As a matter of fact, carbon atoms demonstrate this property even in the absence of other elements. They covalently bond together to form a regular three-dimensional array. Each atom is surrounded by a tetrahedron of sp^3 hybrid orbitals. The orbitals of adjacent atoms overlap and thus create the bonding network which holds the entire structure together. If a molecule is defined as a group of covalently bonded atoms, the result is a giant molecule of carbon. And, generally speaking, the larger this molecule is, the more valuable it is. Because, you see, this three-dimensional array of carbon atoms is better known as diamond. The hardness and high melting point of the substance are at once explicable as consequences of the strength of the electron-pair bonds. Indeed, diamond does not really melt at all. It *sublimes*, that is, it changes directly from the solid to the gaseous state at about 3500°C.

Diamond also undergoes another change at elevated temperatures—a rather uneconomical one. It converts to graphite, the other allotropic form of carbon. Even at room temperature, a humble "lead" pencil is more stable than the Hope Diamond. There is a sort of sweet irony in knowing that, in the fullness of time, even that spectacular rock will be so much soot. In spite of rumor to the contrary, diamonds are not forever. But happily for those whose best friends are diamonds, the conversion is imperceptibly slow under ordinary conditions.

The structure of graphite also correlates well with its properties. It consists of planar sheets of atoms arranged in a hexagonal honeycomb. Covalent bonding occurs only within these planes where each atom is joined to three, not four, others. Thus, each sheet is, in essence, a carbon molecule. Adjacent layers are held together by forces far weaker than those of covalent bonds. Consequently, the graphite crystal can be readily sheared between these planes of atoms. The ease with which the sheets slip over each other underlies the use of graphite as a lubricant and as a means of transferring these words onto paper. Nevertheless,

FIGURE 9.7. *Structures of (a) diamond and (b) graphite.*

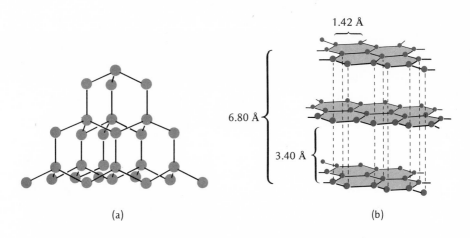

(a) (b)

graphite does exhibit a melting point roughly equal to the subli-
mation temperature of the more valuable form of the element.
Again, this observation reflects the high energies necessary to
break the covalent bonds within the honeycomb sheets.

It should be noted at this juncture that the formation of giant
covalent crystals or super-molecules is not an exclusive property
of carbon. Some compounds also exist in similar arrays of covalently bonded atoms. The most familiar example is the mineral
quartz, a particularly hard and high-melting form of silicon dioxide, SiO_2. Other similar substances find applications as
abrasives.

But in this section we are chiefly concerned with the compounds
of carbon. Some of these contain continuous chains of dozens of
carbon atoms. Such macromolecules manifest themselves in the
cellulose of wood and cotton fiber, in petroleum, and in our skin
and hair and fingernails. Indeed, the biological origins of many of
these substances have provided a name for all carbon-containing
compounds. They are called *organic compounds,* and a vast subdiscipline of chemistry is devoted to their study.

ISOMERISM OR HOW TO MAKE A DEGRADABLE DETERGENT

You have already encountered one of the simplest of organic
compounds—methane. It is the first member of a series of oily
substances called *alkanes* or paraffins. The second, another fuel,
is a two-carbon compound named ethane. It, too, is a hydrocarbon, that is, it consists of only carbon and hydrogen. Its formula
is C_2H_6 and its dot structure is as follows:

```
      H H
      .. ..
  H : C : C : H
      .. ..
      H H
```

In three dimensions, the ethane molecule looks like two tetra-hedra meeting point to point. Figure 9.8 is an attempt to represent this. Each carbon is surrounded by four hybrid sp^3 orbitals. Three of the orbitals on each atom are involved in electron-pair bond formation with three hydrogen atoms. The other two orbitals overlap to form the covalent bond between the carbons.

In a similar fashion, other tetrahedral carbon atoms can be added to form higher alkanes. All of them conform to the general formula C_nH_{2n+2}, where n is the number of carbons in the compound. Propane, familiar as "bottled gas," is C_3H_8. It can exhibit only one reasonable geometrical structure. But with butane, C_4H_{10}, things become more complicated. Two logical but different arrangements of the carbon atoms are possible. They can be placed either in a straight chain or with a branch point. Of course, even the former structure is not really straight because the tetrahedral carbon atoms give rise to a zig-zag backbone.

```
    H  H  H  H                H  H  H
    |  |  |  |                |  |  |
 H—C—C—C—C—H            H—C—C—C—H
    |  |  |  |                |  |  |
    H  H  H  H                H  |  H
                                 |
                              H—C—H
                                 |
                                 H
```

         n-Butane   Isobutane

Both *isomers,* as these forms are called, have the same chemical formula. But because their molecular structures are different, they might be expected to have different physical and chemical properties. And indeed they do. For example, normal or *n*-butane, the more-nearly linear compound, has a boiling point of $-0.5°C$ and a melting point of $-138.3°C$. Isobutane boils at $-11.7°C$ and melts at $-159.6°C$. Notice that both isomers are gases at room temperature, about 25°C.

As the number of carbon atoms in an alkane increases, so do the opportunities for isomerism. A molecule with the formula C_5H_{12}, pentane, can exist in the three possible geometries pictured in Figure 9.9. And by the time the alkane contains eight carbons, there are 18 allowable isomeric arrangements. The specific form of octane used as an "antiknock" standard for gasolines is isooctane.

```
  H₃C              CH₃
     \             /
 H₃C—C—CH₂—CH
     /             \
  H₃C              CH₃
```

         Isooctane

FIGURE 9.8.   *Orbitals indicating the molecular structure of ethane, C_2H_6.*

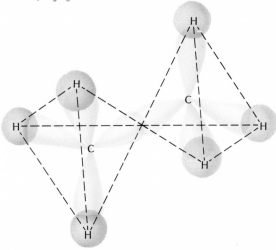

All existing isomers of a compound can be separated from a mixture and can be individually and uniquely characterized. Obviously, isomerism represents an important exception to the Daltonian idea that a specific chemical formula implies a single substance with fixed properties. Properties clearly depend upon molecular shape as well as elementary composition.

FIGURE 9.9.   *Isomers of pentane, C_5H_{12}. Note that the lower two structures are spatially identical.*

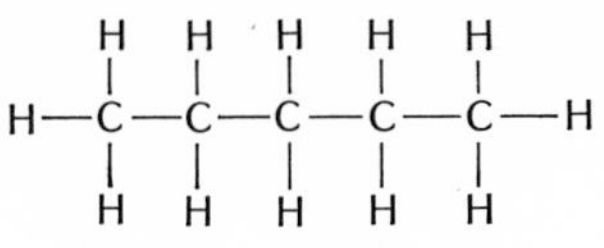

n-pentane

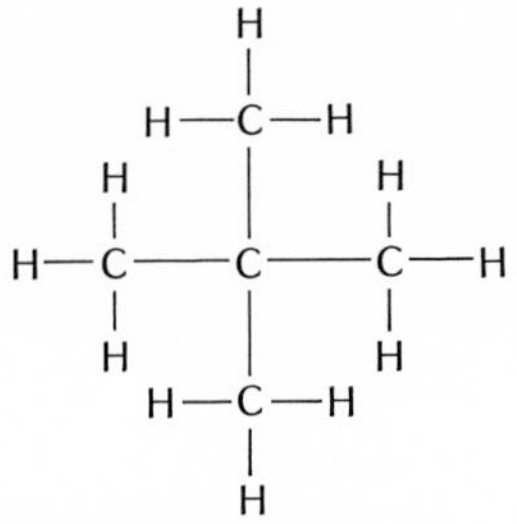

neopentane

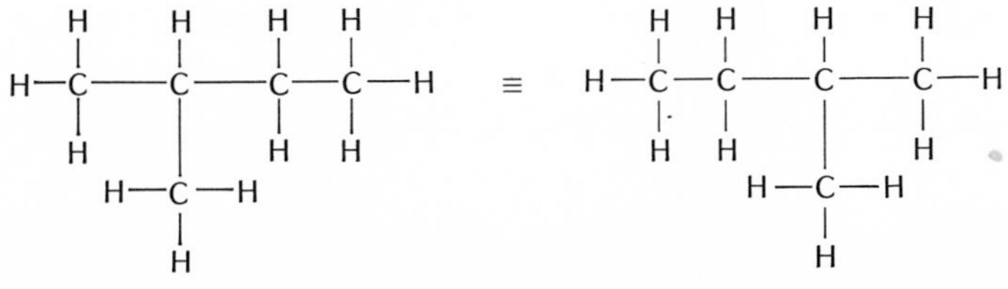

isopentane

You may recall having read about an instance where just such geometrical factors had important environmental implications. After World War II, American soap manufacturers introduced synthetic detergents. These manmade molecules (or more exactly, ions) were decidedly superior to soap in hard water. The calcium and magnesium ions present in many water supplies combine with conventional soap ions to form the insoluble curdy crud of the bathtub ring. By contrast, this precipitation does not occur with synthetic detergents, whose calcium and magnesium compounds are water-soluble. But while making the wash whiter and brighter, these new cleaning compounds were creating problems far underground. After several years of widespread use, synthetic detergents began to appear in unexpected and undesirable places. A glass of water drawn from a faucet in a Long Island kitchen would bear a head of foam more appropriate on a draft beer. Somehow, detergents had found their way into underground water tables and from there into public supplies. For some reason, the detergent ions were not being destroyed in sewage treatment plants, but were passing through with their sudsy potential unimpaired.

Fortunately, chemists were able to discover the cause. It had to do with the geometry of the detergent ions. Ordinary soap is prepared by reacting animal and vegetable fats and oils with sodium hydroxide or potassium hydroxide.* The composition of the biological starting ingredients is understandably reflected in the structure of the large soap ions which are formed. These ions consist mostly of long, straight chains of 16 to 20 carbon atoms with attached hydrogen atoms. At the other end is a group which bears a negative charge. The presence of both a fatty tail and a charged head on the same soap ion provides the essential features for the cleaning function. The oily ends of a number of ions can dissolve in a spot of greasy gravy. The charged ends, however, stick out of the glob and thus cover it with negative groups. These charges promote the solubility of the entire agglomerate, which is carried off in the wash water. Synthetic detergents work exactly the same way, although the charged group is different.

Significantly for the problem of pollution, the hydrocarbon tails of the early synthetic detergents were also different. They were branched rather than linear. This may seem to be a minor distinction but not as far as the digestive systems of bacteria are concerned. For millenia bacteria have been eating straight-chain hydrocarbons with voracious appetites. Consequently, they are capable of breaking down or biodegrading soaps. But these same bacteria find branched chains to be indigestible. Once a branching

---

* An equation for this reaction is given in Chapter 15.

point is reached, the "bug" stops eating, biodegradation ceases, and the detergent escapes mostly intact and still effective.

The solution to the problem was relatively straightforward. In effect, it involved some molecular engineering—designing molecules to have certain desired properties. Detergent companies made their products more appetizing to bacteria by using straight-chain compounds as starting materials in their manufacturing processes. Today commercial synthetic detergents all have linear tails. Their desirable anti-crud properties have been retained and biodegradability has been achieved. Everyone, from housewives to bacteria, should be happy. But as you know, this is not the only problem associated with keeping clean. More of that later.

DOUBLE BONDS IN YOUR DIET

Returning to our survey of carbon compounds, we next encounter another series of hydrocarbons called *alkenes* or olefins. The distinguishing characteristic of these substances is illustrated by ethylene. The formula of this gaseous compound is C_2H_4. Ex-

FIGURE 9.10. *Detergent molecules: their function and fate.*

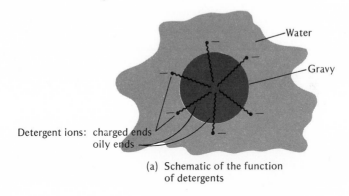

(a) Schematic of the function
of detergents

(b) A bacterium breaking down a biodegradable
detergent molecule

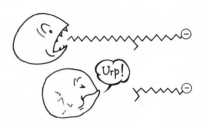

(c) A bacterium suffering from indigestion
because of a nonbiodegradable detergent

periment indicates that the carbon–carbon bond is somewhat shorter and stronger than the corresponding linkage in ethane. Furthermore, two atoms of hydrogen can chemically add to C_2H_4 to yield C_2H_6. Thus, ethylene appears to be *unsaturated;* it can accommodate more hydrogen. In contrast, ethane is *saturated.* This evidence suggests something special about the carbon–carbon bond in the former compound. It appears to involve four instead of two electrons. Thus, ethylene can be represented as

$$\begin{array}{cc} H & H \\ \ddot{} & \cdot \\ :C::C: \\ \cdot\cdot & \cdot\cdot \\ H & H \end{array}$$

Here again is agreement with the rule of eight, but two pairs of electrons are shared between the carbon atoms. These constitute a *double bond,* which is symbolized by two parallel lines.

$$\begin{array}{cc} H & H \\ \diagdown & \diagup \\ C=C \\ \diagup & \diagdown \\ H & H \end{array}$$

The above diagram is a reasonably accurate representation of the geometry of the ethylene molecule. All the atoms lie in the same plane with the H—C—H and the H—C—C bond angles all equal to 120°. This structure suggests that hybrid sp^2 orbitals are utilized in forming the C—H bonds and one of the C—C bonds. Tilting the molecule upwards from the paper, the orbitals have the appearance indicated in Figure 9.11. The end-to-end overlap of the two sp^2 hybrid orbitals form what is known as a sigma (σ) C—C bond. Similarly, the C—H bonds, which are also symmetrical about a line between the nuclei, are given the same Greek designation. Two other electrons are contributed by the carbon atoms,

FIGURE 9.11. *Electron distribution of ethylene, C_2H_4, the two carbon atoms exhibiting sp^2 hybrid orbitals involved in σ bonding. Also note the π bond formed from overlapping p orbitals. (All atoms are in the same plane; the π orbital is above and below it.)*

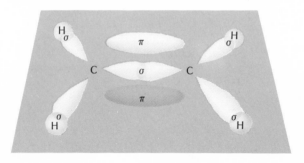

but are not involved in the hybrid. Their *p* orbitals overlap side-by-side above and below the plane of the molecule. Together they form a molecular orbital which looks rather like two sausages placed parallel to the σ bond. Both sausages are part of the same orbital, which has a total capacity of one pair of electrons. This configuration forms what is known as a pi (π) bond, the second of the two carbon–carbon covalent bonds in ethylene.

Another way of viewing the ethylene molecule is as two tetra-hedra meeting side to side. The resulting structure is rigid. There can be no free rotation about a double bond as there is about a single bond. Thus, the two $-CH_2$ (methylene) groups of ethylene are locked in position, while the two $-CH_3$ (methyl) groups of ethane can rotate with relative ease. This concept becomes much clearer when one has an opportunity to use stick-spring-and-ball molecular models. If you have access to such chemical Tinker Toys, you will find that playing with them is well worth any indig-nity you may suffer at such childish diversion.

For one thing, molecular models can aid one's understanding of *cis–trans isomerism*. This form of structural multiplicity is a direct consequence of the above-mentioned rigidity of the double bond. Ethylene itself obviously exhibits only one geometrical form. But suppose two of the hydrogen atoms were replaced with bromine atoms to form dibromoethylene. This can be done quite easily, though not by direct reaction of ethylene with bromine. Note that the substituting bromines could enter this molecule in three dif-ferent arrangements.

```
   Br      Br    Br      H     Br      H
     \    /        \    /        \    /
      C             C             C
      ‖             ‖             ‖
      C             C             C
     /    \        /    \        /    \
    H      H     Br      H     H      Br

    gem            cis          trans
```

In the first place, both substituents could be bound to the same carbon to yield the *gem* isomer. Or, one bromine atom could be found on each carbon. Here there are two possibilities. The sub-stituents can be on the same or on opposite sides of the double bond. As indicated in the drawing, the former isomer is called the *cis* isomer. The latter, cross-molecule configuration is called the *trans* isomer. Because free rotation about the C=C linkage is im-possible, the two forms are not equivalent and cannot interchange unless the double bond actually breaks.

By the way, perhaps you have been wondering if ethylene has anything to do with polyethylene. Indeed it does. The prefix poly-, meaning "many," provides the clue. The ubiquitous plastic which sometimes seems to engulf us is made by combining or polymerizing ethylene molecules to form super-molecules. The double bonds open up and create new C—C single bonds. Thus, the resulting *polymer* or *macromolecule* is, in effect, a long saturated alkane. Moreover, cross-links form which tie adjacent chains together. In the process, the ethylene loses its identity and its gaseous properties to become the familiar stretchy plastic.

Following ethylene in the alkene series is propylene or propene.* Its formula, C_3H_6, corresponds to the general form C_nH_{2n}, which applies to all the monounsaturated hydrocarbons. The mono- means that the compounds contain only one double bond each. In the case of propylene, there is only one place to put it. Admittedly, one can write two structures which seem to be different.

But translated into sticks and balls or atoms and electrons, the two are identical.

Not so with butylene or butene, C_4H_8. Here the double bond can occupy two distinct positions and still be consistent with the formula.

The arrangement on the right really represents two different isomers—the cis and the trans.

And to further complicate things, branching can introduce a fourth isomer.

---

* There is some method to the madness of these names. The prefixes indicate the number of carbon atoms: meth-, 1; eth-, 2; prop-, 3; but-, 4. The suffixes designate the series; -ane, alk*ane;* -ene, alk*ene;* and -yne, alk*yne.* (See Appendix 3 for further information.)

Once the number of sequential carbon atoms in a molecule reaches three or more, it becomes potentially capable of accommodating more than one double bond. The term "polyunsaturated," popularized by television commercials, simply means having a number of double bonds. You may recall that this property is thought to influence the metabolic fate of hydrocarbon chains, particularly those found in fats and oils. Bacteria are not the only creatures living on Long Island that have digestive difficulties with carbon and hydrogen!

SEEING SNAKES

Even more intriguing than the molecules of margarine is a smaller carbon compound with multiple double bonds. Obtained chiefly from coal tar, benzene has become one of the most widely used industrial chemicals. You may have met this liquid as a solvent or spot remover. If you have, you will agree that *aromatic*, the term applied to benzene and its compounds, is obviously a euphemism.

As soon as the formula of benzene was determined to be C_6H_6, the problem of its molecular structure began driving chemists to distraction. How in the world could six carbon atoms and six hydrogen atoms be arranged in a way that made chemical and structural sense? The experimental behavior of benzene suggested the presence of three double bonds. But a straight chain of alternating double and single bonds would certainly not do; its equation would be C_6H_8.

$$H^a_{}\diagdown_{} \quad H^b \quad H^c \quad H^c \quad H^b \quad \diagup H^a$$
$$C{=}C-C{=}C-C{=}C$$
$$H^a\diagup \diagdown H^a$$

Moreover, there was the experimental peculiarity that all the hydrogen atoms of benzene were chemically and thus presumably structurally equivalent. This would obviously not be the case if the structure were linear. In the above molecule, for example, there are three different sorts of hydrogen atoms, marked a, b, and c. Thus, three different isomers could result if a single hydrogen atom were replaced by a bromine atom. But not in benzene! Here only one monosubstituted molecule is formed.

Many ingenious diagrams were drawn, but the correct answer came as a revelation in a reverie. These are the words of the dreamer, August Kekulé.

I was busy writing on my textbook but could make no progress—my mind was on other things. I turned my chair

*to the fire and sank into a doze. Again the atoms were
before my eyes. Little groups kept modestly in the
background. My mind's eye, trained by the observation of
similar forms, could now distinguish more complex
structures of various kinds. Long chains here and there
more firmly joined; all winding and turning with snake-like
motion. Suddenly one of the serpents caught its own tail
and the ring thus formed whirled exasperatingly before my
eyes. I woke as by lightning and spent the rest of the night
working out the logical consequences of the hypothesis. If
we learn to dream we shall perhaps discover truth. But let
us beware of publishing our dreams until they have been
tested by the waking consciousness.**

The ancient alchemical symbol for continuity, the ouroboros, the
serpent biting its own tail, had solved the riddle of benzene. For
Kekulé's dream did survive testing by the "waking consciousness."
Modern laboratory measurements, including the methods of spectroscopy and X-ray crystallography, indicate that the molecule is
a six-membered ring.

There is a temptation to consider the hexagonal benzene molecule as containing alternating double and single carbon–carbon
bonds. Certainly, such a structure would obey the octet rule and
is consistent with some of the properties of the compound. But
this representation cannot be strictly true because the double and
single bonds cannot be localized. In other words, the six carbon–
carbon bonds are indistinguishable. They are equal in length and
strength, with values somewhere between those for single and
double bonds. Because of this fact, any attempt to draw the structure of a benzene molecule runs the risk of being misleading. In
order to illustrate the dynamic nature of the electron distribution,
two sketches are included in Figure 9.12. You will observe that the
double bonds in these *resonance structures* are drawn in different

FIGURE 9.12.   *Resonance between two extreme electronic
arrangements in benzene,* C_6H_6.

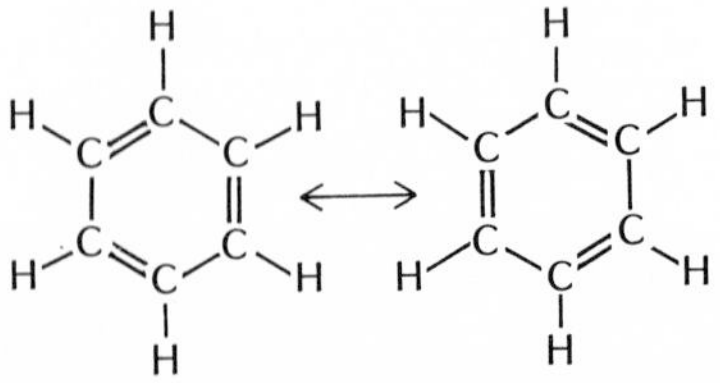

* August Kekulé, quoted in F. J. Moore, "A History of Chemistry,"
3rd ed., p. 291, McGraw-Hill, New York, 1939.

FIGURE 9.13. *Appearance of charge clouds in benzene, C_6H_6.*

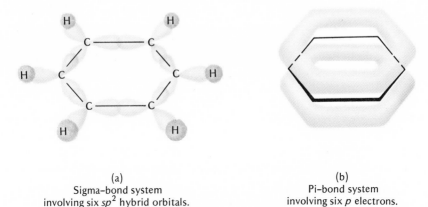

(a)
Sigma–bond system
involving six sp^2 hybrid orbitals.

(b)
Pi–bond system
involving six p electrons.

positions. An earlier interpretation held that the electrons rapidly move or *resonate* between these two extremes. On the average, then, the electrons would be symmetrically distributed around the hexagon. And the stability of the benzene molecule was attributed to this resonance. Indeed, the concept of resonance turns out to apply to a good many covalent compounds (see, for example, Problem 9.17).

But there is danger in taking the resonance forms too literally. They are, in a sense, invented abstractions, and the true structure of benzene is intermediate between them—a real hybrid of two mythical beasts. For these reasons it is preferable to speak of the benzene molecule in other terms. Thus, the results of resonance can be interpreted as the creation, on the average, of a partial second bond between each adjacent pair of carbon atoms. Or, the bonding pattern can be considered in terms of σ and π orbitals. The planar and perfectly regular hexagonal structure suggests that a series of sp^2 hybrids make up the basic C_6H_6 skeleton with σ bonds. A p orbital containing a single electron extends above and below this hexagon from each of the carbon atoms. The orbitals then overlap to form a ring or hexagonally shaped π system on either side of the plane of the atoms. This electron charge cloud must be smeared on the benzene ring like cream cheese spread on both sides of a bagel or, if you prefer, a doubly frosted donut. Figure 9.13 is an attempt to visualize this distribution.

In recognition of the fact that the six π electrons of the benzene molecule are equally shared among its six carbon atoms, some organic chemists have recently been abandoning the long-established practice of writing the double bonds in the structure. In many new texts, the representation on the left (at the top of the next page) has replaced that on the right.

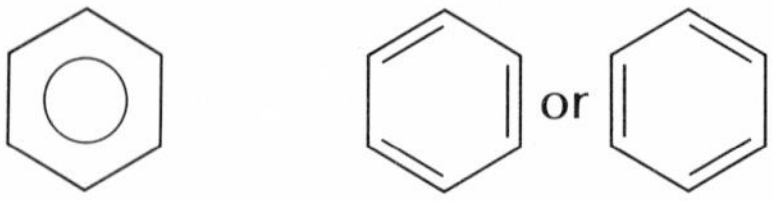

In neither case are the constituent atoms actually written down. It is understood that one carbon and one hydrogen atom occupy each corner of the hexagon. Given the ubiquity of the benzene ring, this shorthand has undoubtedly saved tons of paper and printing ink.

Although this book will adopt the more modern representation of the benzene molecule, it must be admitted that the double-bonded version does have some utility. To be sure, all the carbon–carbon bonds are indeed equivalent. But the fact remains that the molecule can act as though it contained three double bonds. It is as if the reaction singles out and involves one of those instantaneous resonance forms. For example, C_6H_6 can be saturated to C_6H_{12}, cyclohexane, by the addition of six hydrogens to the ring. Many other elements or groups of elements can also be added to the hexagon or substituted for hydrogen. Hence, we again encounter potential for thousands of compounds, many of them with isomers.

There are no isomers of monosubstituted benzene. As we have emphasized, all six of the hydrogen atoms in the molecule are geometrically identical. It follows that one of them can be replaced with, say, a bromine atom and the result will be indistinguishable from any of the other five singly substituted molecules. However, the presence of two substituents leads to three isomeric possibilities. The second group can replace the hydrogen atom (1) adjacent to the first substituent, (2) opposite it, or (3) one removed from it. In presenting these three options, Figure 9.14 should be worth at least a hundred words, if not the legendary thousand. Note, how-

FIGURE 9.14.   *Isomers of dibromobenzene. In writing the benzene molecule, the hydrogen atoms on the ring carbons are usually omitted. In all three of the isomers, two hydrogen atoms have been replaced by bromine atoms. Hence, their common formula is $C_6H_4Br_2$.*

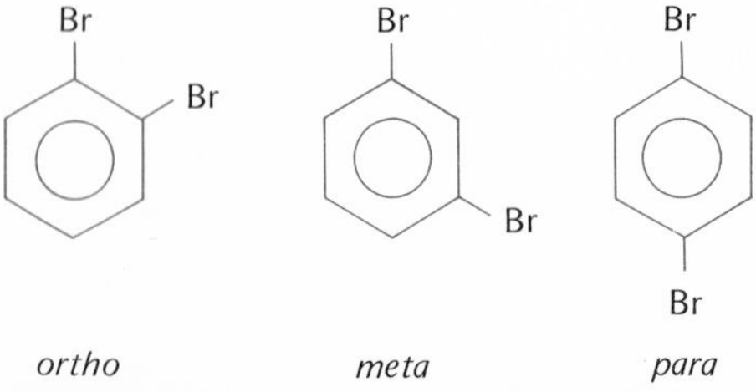

ortho   meta   para

FIGURE 9.15. *Structural formulas of some important molecules containing the benzene ring.*

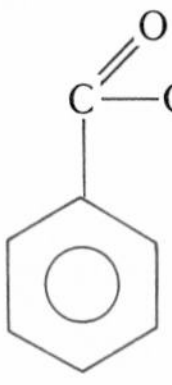

Benzoic acid
(a food preservative)

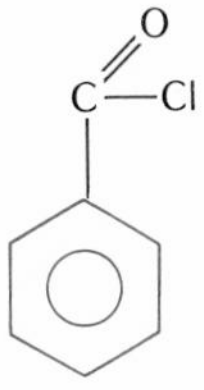

Benzoyl chloride
(a tear gas)

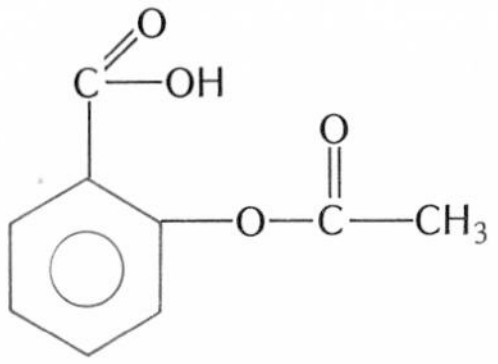

Aspirin

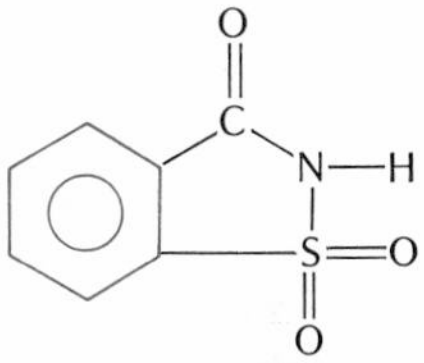

Saccharin

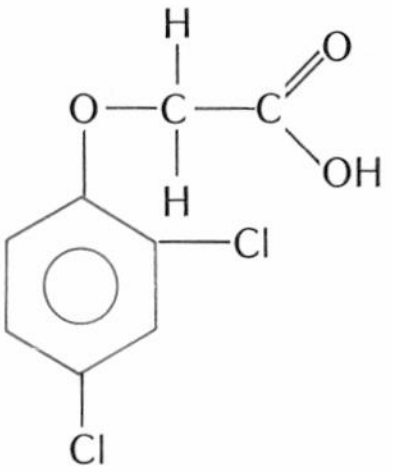

2, 4-D

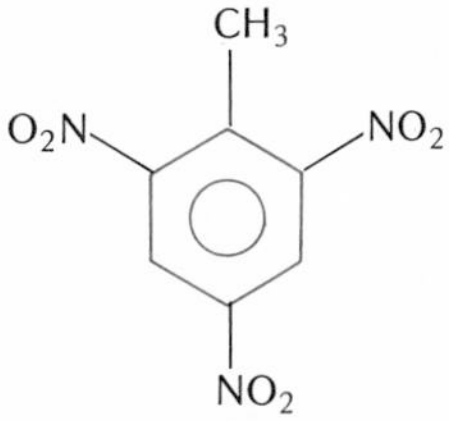

TNT

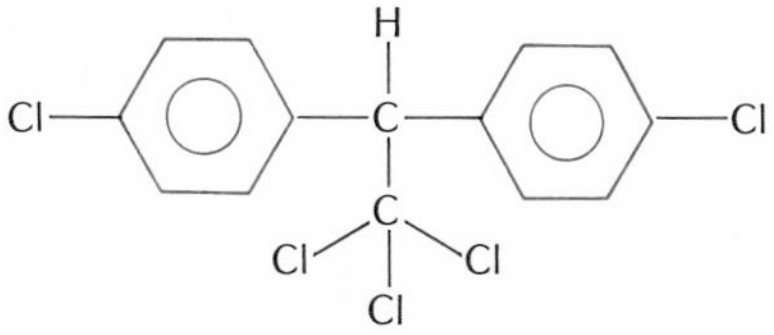

DDT

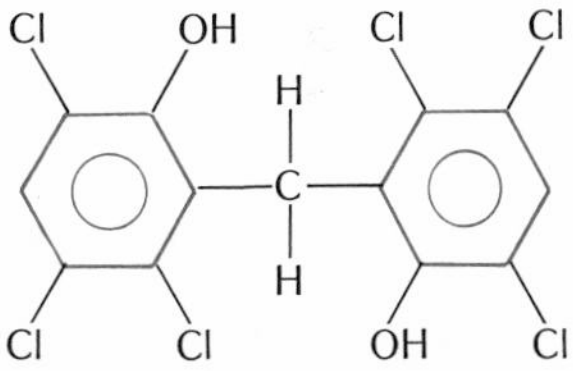

Hexachlorophene

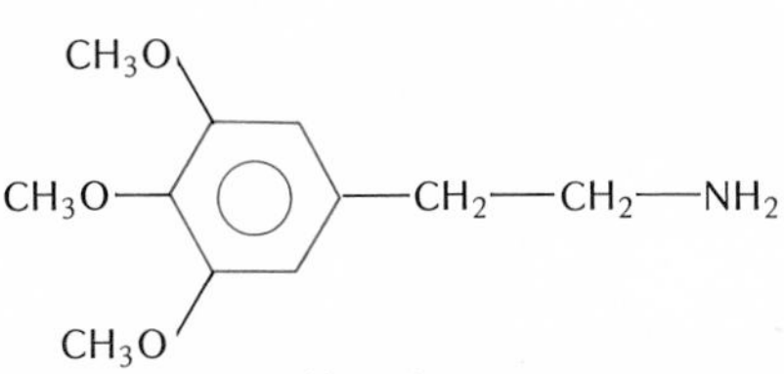

Mescaline

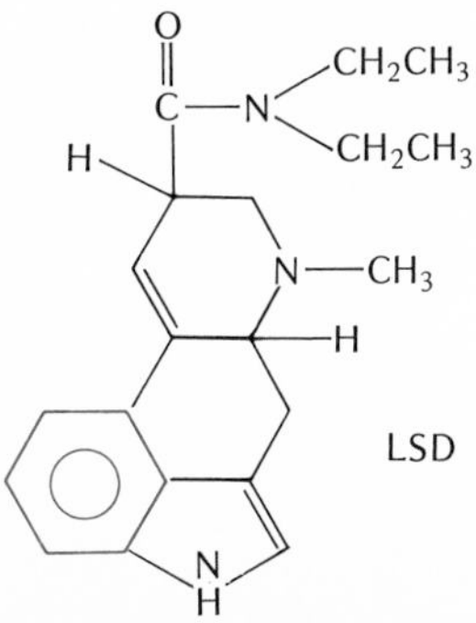

LSD

ever, that there are verbal designations for these isomers: *ortho,
para,* and *meta.* Placing the ortho or meta substituent on the other
side of the initial one does not make a new molecule—additional
proof of the equivalence of the carbon–carbon bonds. But genuine
geometric differences do make a good deal of difference when it
comes to properties.

A great variety of chemical substituents can be associated with
benzene. As a result, literally thousands of compounds are known
which contain this ring system. To attempt to list more than just a
few would turn this text into a table. But the molecular structures
selected for Figure 9.15 should be sufficient to convince you that
substituted benzenes do profoundly influence our lives for better
or for worse.

You have most likely also encountered the first member of the
third and final major series of hydrocarbons, the *alkynes.* The compound in question is acetylene, the gas commonly used in welders'
torches. Its formula is C_2H_2 and its Lewis structure is given below.

$$H : C :: C : H \quad \text{or} \quad H—C≡C—H$$

Here three pair of electrons are shared by the two carbon atoms
to fulfill the octet rule and to yield a *triple bond.* Indeed, the
acetylene molecule can be visualized as the result of two tetra-
hedra meeting face to face with three common points of contact.
Consistent with this model, the molecule is linear; all four atoms
fall in the same line.

Again, hydridization helps explain the structure. Two *sp* hybrid
orbitals are associated with each of the carbon atoms. One of these
orbitals overlaps with the 1*s* electron cloud of a hydrogen. The
other forms a σ bond with an *sp* hybrid of the other carbon. Two
more carbon–carbon bonds arise from the overlap of the two
unhybridized *p* orbitals remaining on each atom. The resulting π
system looks like four electronic sausages grouped around the
central σ bond (see Figure 9.16). Together, the cement of six electrons holds the atoms together even more tightly than the double

FIGURE 9.16.   *Representation of electron distribution in
acetylene, C_2H_2. The drawing indicates a σ C—C bond (in color)
formed from the two sp hybrid orbitals and two π bonds (in gray)
formed from four p electrons.*

bond of ethylene. The triple bond is also somewhat shorter than the four-electron bond.

Higher members of the alkyne series have the general formula C_nH_{n-2} if only a single triple bond is present. More than one would, of course, mean a departure from this ratio. As you no doubt anticipated, structural considerations such as the position of the triple bond (or bonds) and chain branching are again operative and expressed in the existence of isomers. Hopefully, there is no need to cite additional examples to convince you of the fact that the number of organic compounds is far greater than the sum of the compounds of all the other elements put together. And remember that we have thus far restricted ourselves almost exclusively to compounds of hydrogen and carbon, largely neglecting those organic compounds which contain additional elements. They are of such importance that you will encounter many of them as you continue your tour of this planet, this science, and this book.

For thought and action

9.1   Write the electronic configuration of each of these ions. Also indicate the number of electrons and protons in each.
(a) Br^- (b) Al^{3+} (c) Cu^+

9.2   List three ions which are isoelectronic with the Ar atom and note the oxidation number of each.

9.3   What ions would you expect the following elements to form? Specify the charge on each ion and write its electronic configuration.
(a) Be   (b) Rb   (c) I

9.4   From what you know of ionic states, write the formulas of these ionic compounds.
(a)   Potassium iodide (added to "iodized salt")
(b)   Silver bromide (used in photographic emulsions)
(c)   Barium sulfate (swallowed for X-ray diagnosis of ulcers)
(d)   Zinc oxide (a white pigment and constituent of ointments)

9.5   Specify the cations and anions released when the following ionic compounds dissolve in water.
(a)   Potassium bromide, KBr
(b)   Ammonium chloride, NH_4Cl
(c)   Lead nitrate, $Pb(NO_3)_2$
(d)   Calcium hydroxide, $Ca(OH)_2$
(e)   Sodium phosphate, Na_3PO_4

9.6   How many grams of potassium and how many grams of chlorine could be displaced from molten KCl by the passage of 1 faraday of electricity? By 0.1 faraday?

9.7   Aluminum is prepared commercially by a process developed by Charles M. Hall in 1886, while he was a student at Oberlin College. Basically, the idea is to electrolyze molten aluminum oxide, Al_2O_3; hence the large demand for electrical energy. The overall reaction is

$$Al_2O_3 \rightarrow 2Al + \frac{3}{2}O_2.$$

Separate the equation into two parts, one representing a gain in electrons, the other a loss. Indicate the species oxidized, the species reduced, and the changes in oxidation number.

9.8   We have, from time to time, mentioned the smelting of copper. The process can be represented by the equation,

$$2CuO + C \rightarrow 2Cu + CO_2.$$

Identify the oxidizing agent, the reducing agent, the element oxidized, the element reduced, and the changes in oxidation number.

9.9   Tin exists in two oxidation states and forms two chlorides, with the formulas $SnCl_2$ and $SnCl_4$.
   (a)   Specify the valence or oxidation numbers of tin in the two compounds. (The oxidation number of chlorine is -1 in both cases.)
   (b)   Calculate the weight percent of tin and chlorine in the two compounds.
   (c)   Demonstrate numerically how these two chlorides of tin illustrate the Law of Multiple Proportions.

9.10   Recall the two oxides of iron mentioned in Problem 3.6. Oxide A contains 77.8% Fe and 22.2% O, while oxide B contains 70.0% Fe and 30.0% O. Use the method of Chapter 4 to determine the formulas of these two compounds. When you have done so, specify the valence state or oxidation number of iron in A and in B.

9.11   Listed below are several compounds with their melting points. On the basis of this information, attempt to classify the substances as ionic or covalent compounds.
   (a)   Ethyl alcohol ($-117°C$)
   (b)   Magnesium sulfate ($1124°C$)
   (c)   Nitrous oxide ($-102°C$)
   (d)   Sucrose ($186°C$)
   (e)   Cesium chloride ($646°C$)

9.12   Keeping in mind the similarities between elements in the same column of the periodic table, draw the Lewis structure of the nasty, rotten-egg gas, hydrogen sulfide, H_2S. Be sure to describe its shape.

9.13   Carbon tetrachloride, CCl_4, was widely used as a home spot remover and fire extinguisher until the extent of its toxicity was realized. Write a Lewis dot structure for the molecule and draw or otherwise indicate its geometry. What hybrid orbital is involved in the bonding?

9.14 Use orbital diagrams to explain *sp* hybridization in BeF_2 and C_2H_2.

9.15 The octet rule can be used to correctly represent the molecular structure of the gas you are exhaling at the moment, carbon dioxide. Draw a flyspeck molecule of CO_2 and indicate whether or not it has a permanent dipole moment. *Hint:* The carbon atom is in the middle.

9.16 The choking smell of burning sulfur arises from sulfur dioxide, SO_2, a compound which we have already identified as a common air pollutant. Try your hand at drawing its molecular structure and shape. *Hint:* The octet rule is obeyed.

9.17 Sulfur trioxide, the compound which reacts with water to yield sulfuric acid, has a molecular structure which is often written in the following manner.

But experiment shows all the sulfur–oxygen bonds to be the same length. Explain this observation, drawing any additional structures which would help clarify your answer. What implications do your conclusions have for the structure of SO_2? (See Problem 9.16.)

9.18 Explain, on the basis of molecular structure and electronic arrangement, why NH_3 has a permanent dipole moment, but BF_3 does not.

9.19 Which of the following compounds encountered in this chapter have permanent dipole moments and which do not? Apply your knowledge of molecular structure in explaining your answers.
(a) HF
(b) BeF_2
(c) C_2H_6
(d) benzene
(e) *para*-dibromobenzene
(f) *ortho*-dibromobenzene

9.20 See how many different isomers of hexane, C_6H_{14}, you can draw. (There are five in all.)

9.21 Draw and name the isomers of dichloroethylene, $C_2H_2Cl_2$.

9.22 Draw all the possible isomers of tribromobenzene, $C_6H_3Br_3$.

9.23 Differentiate between σ and π bonds, giving examples of each.

9.24 One early suggestion for the molecular structure of benzene was a regular prism with equilateral ends.

One carbon and one hydrogen were assumed to be situated at each corner.

(a)   Show that this model is consistent with the octet rule and the existence of three isomers of $C_6H_4Br_2$.

(b)   What sort of experimental evidence do you suppose led to the conclusion that the benzene molecule was not shaped like a prism?

9.25   Here is a practical problem for peanut butter connoisseurs. According to the labels, conventional supermarket brands, the kind without the oil separation, contain "vegetable oils hardened by hydrogenation." But oil separates from the "pure" peanut butter fancied by health food enthusiasts. Explain the process of "hardening" and the molecular difference between the two kinds of peanut butter.

Suggested readings

Benfey, O. T., "From Vital Forces to Structural Formulas," Houghton, Boston, Massachusetts, 1965.

Freund, Ida, "The Study of Chemical Composition: An Account of Its Method and Historical Development," Dover, New York, 1968.

Lynch, P. F., "Orbitals and Chemical Bonding," Houghton, Boston, Massachusetts, 1966.

Pauling, Linus, and Hayward, Roger, "The Architecture of Molecules," Freeman, San Francisco, California, 1964.

Pimentel, George C., and Spratley, Richard D., "Chemical Bonding Clarified Through Quantum Mechanics," Holden-Day, San Francisco, California, 1969.

10

Force and form: properties of gases, liquids, and solids

Of breezes, brooks, and boulders: a brief beginning

The pages which follow (and you will note that there are quite a few) are chiefly devoted to the macroscopic properties of matter and the forces which dictate the physical state assumed by elements and compounds. Whether a substance is a solid, a liquid, or a gas is a consequence of the interactions between and among the structural units—atoms, ions, or molecules—which make up the bulk matter. These interparticulate forces literally hold things together. But figuratively speaking, they also create a common bond among the three phases of matter. And for this reason, I have chosen not to dissociate this topic into two or three briefer chapters. Individual sections will emphasize the structure and properties of gases, liquids, and solids, but always with reference to the various molecular phenomena which provide a unifying theme.

You have already encountered several instances of interaction within the solid state. In the case of giant crystalline molecules like diamond, the fundamental structural units are atoms. These attract each other by strong covalent bonds to form the hardest, highest-melting, and least-soluble substances known. To melt or sublime a diamond, one must actually break some of these bonds—not an easy task. On the other hand, the electrostatic forces which bind ions together are somewhat weaker than covalent bonds. Less energy is required to separate the anions and cations from their crystal lattice to yield the more disordered liquid state. Consequently, crystals of sodium chloride and most other salts melt at much lower temperatures than grains of sand and other similar substances which are essentially giant molecules. But, as we noted in the previous chapter, the melting point of ionic compounds is still sufficiently high that they are solids at room temperature. Water, however, readily promotes the disruption of the crystal lattice to release individual ions. As a result, electrolytes generally exhibit high aqueous solubility. As you will learn later, metals represent yet another type of bonding.

But most of this chapter will be devoted to the bulk properties of covalent compounds. Clearly, the smallest integral unit of such a substance is a molecule. You have just completed a survey of the structure of some of these molecules and the electron-pair *intra*molecular bonds which hold them together. Before that, you studied matter at a more fundamental level of organization—atomic structure. Now we progress to a third, higher order of organization. Here we are interested in *inter*molecular interactions—the forces which hold the molecular units into forms called solid, liquid, or gas.

Generally speaking, these forces are a good deal weaker than ionic or covalent bonds, and the molecules can be separated with relative ease. As a result, solid covalent compounds (sugar is a familiar example) melt at fairly low temperatures. And many compounds which are aggregates of individual molecules are liquids (e.g., H_2O) or gases (e.g., CO_2) at room temperature. The intermolecular attraction is, of course, greatest in solids, intermediate in liquids, and least of all in gases. It is the latter state of matter which starts this survey of stuff.

Hot air and high pressure

In the gas phase, the forces between molecules are so weak that, to a first approximation, they can be completely ignored. The N_2, O_2, CO_2, and H_2O molecules which make up most of our atmosphere are free to wander about relatively unencumbered and uninfluenced by each other. This is because a room full of air is mostly empty; the matter occupies only a relatively small fraction of the total volume. (We will shortly calculate exactly how small, and at the same time discover that the number of molecules in the same room is, nonetheless, immense.)

A further indication of the weakness of interparticle forces in gases is the ease with which they diffuse. A drop of perfume behind an earlobe sends its molecular messengers to the furthest corner of the room in only a few minutes. And so, unfortunately, do the less agreeable concoctions of chemists. Such migratory behavior reflects the fact that a gas really has no fixed volume of its own, except, of course, the actual volume of the molecules themselves. Instead, a gas expands to uniformly fill any container which surrounds it. Volume is thus imposed upon the gas by the size of the vessel it occupies. Not so with liquids. To be sure, a liquid will assume the shape of its container—a bottle, a glass, or a stomach—but only up to a certain volume. That volume is determined by the forces which hold the molecules together in the liquid state. Solids are even more adamantly independent. A rock retains its size and its shape no matter what the volume or the contours of the container into which it is placed.

Of course, one must exercise care in making any statements about volume. It is more than a little dangerous to talk about the volume of a substance as if it were something as invariant as mass. The volume of all bulk forms of matter depends upon temperature and pressure. For liquids and solids, the dependence is so slight that in some circumstances it is possible to ignore it. But gases are so sensitive to changes in pressure and temperature that the values of these variables must always be specified when reporting volumes.

Fortunately, the equation giving mathematical expression to the relationship of the volume, pressure, and temperature of a gas is simple and readily understood. Underlying it are two familiar laws first formulated in the early days of chemistry. The older one, dating back to 1660, is attributed to the Skeptical Chymist, Robert Boyle. Recall that this was a period when many natural philosophers were investigating and pondering the properties and composition of air. Therefore, Boyle's study of the variation of volume with changing pressure was of particular importance. It is to us as well, but first a few pertinent paragraphs on pressure.

About 20 years before Boyle published "A Defense of the Doctrine Touching the Spring and Weight of the Air," Galileo's secretary invented an instrument which greatly influenced the subsequent study of "pneumatics." This man, Evangelista Torricelli, realized that the blanket of air surrounding the earth has mass and therefore exerts pressure. Moreover, he devised a method of measuring this pressure. A glass tube, about a yard long and sealed at one end, was completely filled with mercury. The open end was then temporarily closed, probably with a finger; and the tube was inverted and the unsealed end was placed in a dish of mercury. When the finger was removed from this end, now under the surface of the liquid metal, mercury began to drain from the tube. But the process soon stopped with a column of mercury about 30 inches high remaining in the tube. The first *barometer* had been built.

In effect, a barometer is a device for balancing a column of air with a column of liquid. Mercury is usually chosen because at room temperature it is the densest of all liquids. Therefore, only a relatively short column is required. Above the mercury level, in the closed end of the tube, is a vacuum. Pressing down on the surface of the mercury in the dish is the atmosphere. The pressure of this air literally supports a column of mercury about 30 inches long. The air pushes the mercury up the evacuated tube and the mercury pushes back. Or, if you prefer, the air keeps the mercury from draining completely out. The draining stops when the pressure exerted by the mercury column is exactly equal to the pressure of the atmosphere. Should the atmospheric pressure decrease, the mercury level falls, often a portent of stormy weather. And, when the pressure of the air increases, the "barometer rises." The height of the column is directly proportional to the atmospheric pressure. And thus, pressure can be expressed in units of millimeters of mercury, mm Hg.*

Perhaps the principle of the barometer can be best understood in terms of weight. After all, pressure is force per unit area which,

FIGURE 10.1.   *The mercury barometer.*

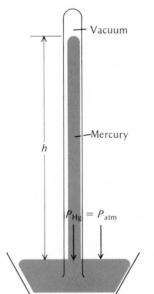

---

* The unit mm Hg is sometimes called "torr."

in this case, is simply weight per unit area. Quite obviously, the height of liquid in a column of uniform cross section must be a direct indicator of the liquid's volume. And, if the density of the liquid is constant throughout the column, its mass is directly proportional to its volume. Consequently, height becomes a measure of mass or weight. And, as we just observed, weight per unit area is nothing less than pressure.

In order to apply this argument to a specific example, let us calculate the pressure equivalent of one standard atmosphere. This is a defined unit corresponding to a column of mercury 760 mm or 29.9 inches high at 0°C. Assume a column with a cross-sectional area of one square centimeter, 1 cm^2, about a quarter the size of a small postage stamp. It follows that 760 mm or 76.0 cm of such a column will have a volume of 76.0 cm × 1.00 cm^2 or 76.0 cm^3 (read "cubic centimeters"). The weight of this much mercury is obtained by multiplying the volume by the density of the element, 13.60 g/cm^3 at the reference temperature of 0°C. Thus, 76.0 cm^3 × 13.60 g/cm^3 = 1034 g or 1.034 kg. This weight of mercury is pressing down on the square centimeter at the bottom of the column. And the same weight of air is pressing down on every square centimeter it touches. In other words, 1 atmosphere (atm) is equivalent to 1034 g/cm^2. Or, if you conceptualize better in our archaic Anglo-American system of units, you can easily convert the answer to 14.7 lb/in^2.

Note that this latter value means that a column of air one inch square and as high as there is air weighs 14.7 lb. So does a column of mercury of the same cross-sectional area, but only 29.9 inches high. And so does a comparable column of water 33.9 feet high.* A longer column of water, that is, a greater volume of it, is required because water, though far denser than air, is considerably less dense than mercury. So much so that water would make an awkward barometric liquid for pressure measurements near 1 atmosphere.

As you know, atmospheric pressure varies not only with weather conditions, but with altitude as well. The higher the elevation of the barometer, the thinner and lighter the blanket of air above it and pressing upon it. This reduced pressure is reflected in lower barometric readings. Moreover, the pressure does not simply decrease in inverse proportion to the altitude above sea level. It decreases more rapidly than that. Related to this phenomena is the fact that our atmosphere is not uniformly dense. Any one who has hiked in the Rockies, the Andes, or the Alps can testify that the air is "thinner" at these heights. But descending to lower altitudes,

---

* Problems at the end of the chapter provide the reader with opportunities to check the correctness of these calculations.

the air quickly becomes considerably denser, that is, it contains more molecules per unit volume. This is because the molecules of air are pulled upon by the force of gravity, and the closer the molecules are to the center of the earth, the stronger the pull. Moreover, air is much more readily compressible than mercury or water. It is squeezable, and more than 20 miles of air is squeezing the bottom layer. The greater the pressure upon it, the more its volume is reduced and its density increased. Which very neatly brings us back to Boyle.

The English investigator's apparatus borrowed freely from Torricelli. It was simply a glass tube bent into a J with the shorter leg sealed at the end. Mercury was poured into the tube through the open end of the longer leg. As a result, air was trapped in the shorter leg. Boyle observed that as the level of mercury in the longer leg was raised, the volume of air above the mercury level in the shorter leg decreased. The entrapped air could not be escaping, so quite clearly it was being compressed by the pressure of the column of mercury.

In order to determine the exact nature of this relationship between volume and pressure, Boyle took some data. First he marked the short leg of his J-tube with equal arbitrary units. Assuming the

FIGURE 10.2. *Boyle's apparatus for measuring the influence of pressure upon the volume of a gas.*

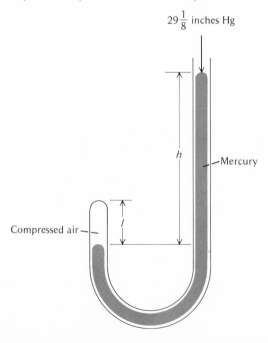

$29\frac{1}{8}$ inches Hg

h

—Mercury

Compressed air —

l

tube to be of uniform cross section, the number of these units between the mercury level and the sealed end of the tube should be directly proportional to the volume of the gas. Boyle recorded this length (*l* in Figure 10.2) as a function of the height (*h*) of the mercury level in the longer leg above the level in the shorter leg. This height represented the pressure exerted by the mercury. Of course, the atmosphere was pressing down upon the open end of the mercury column. On the day of Boyle's famous experiment, it did so with a pressure equivalent to 29⅛ inches of mercury. Therefore, the total pressure on the trapped gas was (29⅛ + *h*) inches of mercury. With the mercury levels in both arms equal, that is, with *h* = 0, the gas in the shorter arm was at the same pressure as the atmosphere. But, as the height of the mercury level in the longer arm increased, the pressure of the trapped gas increased correspondingly. And, simultaneously, its volume decreased.

Figure 10.3 contains two graphs of Boyle's actual data. In Figure 10.3a, *l*, the index of volume, is plotted versus the total pressure, 29⅛ + *h*. The decrease in volume with increasing pressure is obvious. The mathematical form of this dependence is clarified in Figure 10.3b, where *l* is plotted as a function of the reciprocal of the pressure 1/(29⅛ + *h*). The straight line which results indicates that *l, and hence the volume (V) are inversely proportional to pressure (P).* Translated into an equation, the data show that

10.1 $$V = \frac{k}{P}.$$

Here *k* is a constant of proportionality. The fact that it is a constant means that any product of volume and the corresponding pressure must also be invariant. That is,

10.2 $$PV = k.$$

Equations 10.1 and 10.2 express *Boyle's Law.* Although Boyle worked with air, a mixture of gases, the reciprocal relationship between the volume of a gas and its pressure holds for any gas, pure or mixed. But note that the mass of the gas and its temperature must both be held constant. It should be self-evident that the volume of a gas will depend upon the total quantity of the substance studied. And you know from your own experience that heating a gas causes it to expand. This variation of volume with temperature has also been given mathematical form. Two scientists, J. A. C. Charles and Joseph Gay-Lussac, share the honor of having independently carried out the experimental work and generalized their results into a law.

Again, a graph provides an efficient means of visualizing the behavior. This time the plot is of the volume of a fixed mass of

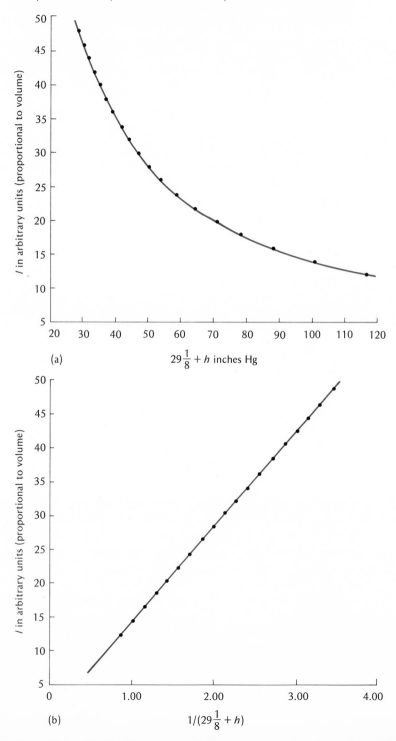

FIGURE 10.3.   *Plots of Boyle's original data.* (a) l vs. 29⅛ + h; temperature is constant. This is equivalent to a plot of volume vs. pressure. (b) l vs. 1/(29⅛ + h); temperature is constant. This is equivalent to a plot of volume vs. 1/pressure.*

(a) $29\frac{1}{8} + h$ inches Hg

(b) $1/(29\frac{1}{8} + h)$

gas (any gas) versus its temperature at constant pressure. Note that the result is another straight line. This line, like any other, can be represented as an equation relating the variables, here V and t. If the temperature is expressed as °C, this equation is

10.3 $V = V_0 + 0.0037V_0t = V_0(1 + 0.0037t)$.

Here V_0 has been introduced to symbolize the volume of the gas at 0°C. Of course, it is possible to make measurements of volumes at lower temperatures. The experimental points still fall on the same line. Therefore, it is reasonable to conclude that the same equation will apply even below the temperatures conveniently attainable in the laboratory. To cover this region the straight line is extended or extrapolated back until it crosses the horizontal axis. As you can see from Figure 10.4a, the temperature corresponding to this point is −273°C. Other arguments indicate that this is the coldest temperature attainable. Therefore, we define −273°C as *absolute zero*, and use it as the lowest point of a new temperature scale which only has positive values. That is, −273°C becomes 0°A or 0°K. The designation °A is read "degrees absolute" and is identical with °K or "degrees Kelvin." The latter name honors the English physicist who established the thermodynamic basis of this scale. To convert °C to °K one merely adds 273° or, to be exact, 273.16°.

°K = °C + 273°

And, to express °C as °K one subtracts.

°C = °K − 273°

We can do this in Equation 10.3. Using a capital T to represent the absolute temperature, we obtain the following:

$$V = V_0[1 + 0.0037(T - 273°)]$$
$$= V_0(1 + 0.037T - 1) = V_0(0.0037)T.$$

Since the mass and pressure of the gas are fixed, its volume at 0°C also has a constant value. Therefore, $0.0037V_0$ is a constant which we will call K. Substituting this symbol yields another mathematical version of the *Law of Charles and Gay-Lussac*.

10.4 $V = KT$.

In words, the law simply states that *the volume of a fixed mass of*

---

* The data are included on p. 53 of Robert Boyle's Experiments in Pneumatics, in "Harvard Case Histories in Experimental Science" (James Bryant Conant, ed.), Vol. I, pp. 3–63, Harvard Univ. Press, Cambridge, Massachusetts, 1966.

FIGURE 10.4.   *Plots of data for Charles' Law experiment. (a)*
Volume of gas vs. temperature in degrees Centigrade; pressure
is constant. (b) Volume of gas vs. temperature in degrees Kelvin;
pressure is constant.

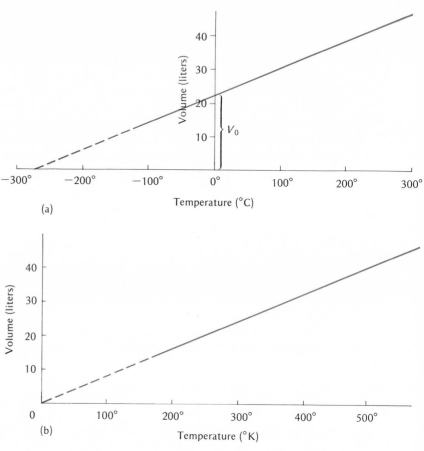

any gas, at constant pressure, is directly proportional to its absolute
temperature. Consequently, the volume of gas divided by its corresponding absolute temperature must be a constant,

10.5 $$\frac{V}{T} = K.$$

The change from the Centigrade to the Kelvin temperature
scales can also be represented graphically, as in Figure 10.4b.
Note that the zero position on the horizontal axis has been shifted
273° to the left. The resulting straight line conforms to Equation
10.4.

A useful utopia: an ideal equation for a perfect gas

The Laws of Boyle and of Charles provide useful generalizations for the behavior of gases. But because the former is restricted to conditions of constant temperature and the latter specifies fixed pressure, their applications are limited. In many practical situations, for example an automobile tire, volume, pressure, and temperature all change simultaneously. Therefore, it is desirable to have an equation which relates all three variables. Such an expression can be derived by the stepwise application of the two laws. Figure 10.5 provides a schematic outline of the procedure. Consider first a process in which a fixed quantity of a gas, let us assume 1 mole, undergoes a pressure change from P_1 to P_2 at a constant temperature, T_1. There will be a corresponding change in volume from V_1 to an intermediate value, V_i. Symbolically summarized, the change is

$$1 \text{ mole}, V_1, P_1, T_1 \rightarrow 1 \text{ mole}, V_i, P_2, T_1$$

Clearly, Boyle's law is operative and

10.6 $P_1V_1 = P_2V_i.$

If the pressure increases, that is, if $P_2 > P_1$, the volume of the gas will decrease, $V_i < V_1$. Conversely, a decrease in pressure will result in expansion.

Now consider a second step in which the pressure is held constant at P_2 while the temperature increases or decreases from

FIGURE 10.5.   *A schematic representation of the two-step application of Boyle's Law and Charles' Law to arrive at the general, Ideal Gas Law.*

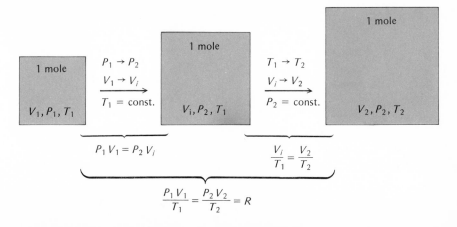

T_1 to T_2. The attendant change in volume is from V_i to the final value, V_2. That is,

1 mole, V_i, P_2, $T_1 \rightarrow$ 1 mole, V_2, P_2, T_2.

Here Charles' Law applies and

10.7
$$\frac{V_i}{T_1} = \frac{V_2}{T_2}.$$

The next operation is to mathematically combine the two steps to yield the overall change,

1 mole, V_1, P_1, $T_1 \rightarrow$ 1 mole, V_2, P_2, T_2.

Since this change does not involve the intermediate volume we eliminate V_i between Equations 10.6 and 10.7. The result is

10.8
$$\frac{P_1 V_1}{T_1} = \frac{P_2 V_2}{T_2}.$$

It follows that PV/T will always be constant, so long as these values of pressure, volume, and temperature simultaneously characterize the state of the gas. If we symbolize this constant as R, the resulting general relationship is

10.9
$$\frac{PV}{T} = R.$$

Recall that because of our initial assumption, this equation applies to 1 mole of a gas. Common sense suggests that at constant pressure and temperature, 2 moles should fill twice the volume occupied by 1 mole. Or, in more general terms, the volume should be directly proportional to the number of moles. If we designate this number by n, Equation 10.9 is simply modified to read

10.10
$$\frac{PV}{T} = nR \qquad \text{or} \qquad PV = nRT.$$

The *Ideal Gas Law*, expressed by this equation, is one of the most important relationships in physical science. It subsumes Boyle's Law and Charles' Law as special cases and it reflects with reasonably high accuracy the behavior of all gases. The numerical value of the gas constant, R, is independent of the gas studied. It does, however, depend upon the units used to express pressure, volume, and temperature. Various commonly used values are summarized in Table 10.1. Obviously, in carrying out a calculation involving the Ideal Gas Law, it is essential that the units selected for R agree with the units chosen for P, V, and T.

Part of the usefulness of Equation 10.10 is that it can be solved for P, V, T, or n, depending upon the problem at hand and the variables specified. As an example, let us carry out the calculation promised on p. 254. Our first task is to determine the number of

TABLE 10.1.   *Values of the Gas Constant,* R, *in Various Units*

$(P$	$\times$ $V)$ $\div$	$(n$	$\times$ $T) =$	R
atm	liters	moles	°K	0.0821 liter atm/(mole °K)
atm	ml	moles	°K	82.1 ml atm/(mole °K)
mm Hg	liters	moles	°K	62.3 liter mm Hg/(mole °K)
—ergs—		moles	°K	8.31×10^7 ergs/(mole °K)
—joules—		moles	°K	8.31 joules/(mole °K)
—cal—		moles	°K	1.99 cal/(mole °K)

moles of air present in a typical small classroom. Assume the room is 10 m wide × 12 m long × 3 m high. (Recall that a meter is about a yard.) Multiplying these dimensions yields a volume of 360 m³. But we would like to express V in liters, so we consult Appendix 2 and learn that 1 m³ = 1000 liters. Therefore, it follows that V = 360 m³ × 1000 liters/m³ = 3.60 × 10⁵ liters. Normal room temperature is about 25°C or 273 + 25 = 298°K. And let us assume a pressure of exactly 1 atm. Knowing V, T, and P, and selecting R = 0.0821 liter atm/mole °K, we can proceed to solve Equation 10.10 for n, the number of moles of gas present in the room:

$$n = \frac{PV}{RT} = \frac{1 \text{ atm} \times 3.60 \times 10^5 \text{ liters}}{0.0821 \text{ liter atm/mole°K} \times 298°K}$$
$$= 1.47 \times 10^4 \text{ moles}$$

The number of molecules is thus 6.023×10^{23} molecules/mole × 1.47×10^4 moles or 8.85×10^{27}.

Note that these molecules could be of any gas or any mixture of gases. Since equal volumes of all gases (at constant pressure and temperature) contain equal numbers of molecules, there is no need to identify the compounds or elements present. However, if we wish to know the mass of the gas in the room, we must specify the molecular weight or the average molecular weight. In our example we must use the latter, because air is a mixture of elements and compounds of various molecular weights. The average value, properly weighted to be consistent with the composition of air, is about 29. Thus, each mole of air weighs 29 g, and 1.47×10^4 moles weigh 29 g/mole × 1.47×10^4 moles or 4.26×10^5 g—about 900 pounds.

Finally, in order to calculate the total volume occupied by the molecules of nitrogen, oxygen, and the other constituents of air, we must estimate an average molecular volume. On the basis of experimental results, a reasonable value is 100 Å³ or 1×10^{-22} cm³. This is equivalent to 1.0×10^{-25} liter, a pretty small volume. But

there are quite a few molecules in a room—we calculated about
8.85×10^{27}. Multiplying these two numbers yields a value of 885
liters for the actual volume of the molecules. On a percentage
basis, this corresponds to (885 liters/3.60×10^5 liters) $\times$ 100 or
about 0.25% of the total volume of the room. Empty space
accounts for the remaining 99.75%.* Thus it is no wonder that
at ordinary room temperature and pressure, molecules generally
manage to keep far enough away from each other to avoid strong
interactions.

At long last we have returned to one of the issues which
prompted this discussion. But before we move on to an exami-
nation of the molecular microcosm of the gaseous state, there is
another important application of the ideal gas law to consider.
And appropriately, it has to do with molecules and their masses.
Because Equation 10.10 includes n, the number of moles, this
relationship can be used as a basis for the experimental determination of molecular weight. One need only to identify n with w/M,
the weight of a sample of gas divided by its molecular weight.
Then Equation 10.10 becomes

10.11 $$PV = \frac{wRT}{M}.$$

Experimentally, one fills a vessel of known volume with the gas
being studied. This is often done by placing some of the substance,
in liquid form, in a bulb with a very narrow neck. The bulb is then
transferred to a bath set at a constant temperature somewhat
above the boiling point of the unknown. Evaporation begins immediately, and the process is allowed to continue until the liquid
phase disappears. Some of the vapors will escape through the
capillary opening of the container, but the diffusion of gas from
the bulb will stop when the internal pressure is equal to the external atmospheric pressure. The bulb is then sealed and weighed.
The result is obviously the weight of the bulb itself plus the gas
it contains. If the weight of the empty, that is, evacuated, bulb is
also known, this value can be subtracted from the weight of the
bulb plus gas. The difference is clearly due to the gas. Thus we
obtain w. P and T correspond to the pressure of the atmosphere
and the temperature of the bath, respectively, and V is the known
volume of the bulb. The only unknown remaining in Equation
10.11 is M, and it is a simple algebraic manipulation to solve for
the molecular weight,

$$M = \frac{wRT}{PV}.$$

* Of course, almost all of the internal volume of a molecule is also
"empty space."

This procedure, called the Dumas method (see Figure 10.6), can be illustrated with an example first cited in Chapter 4. You may recall that way back there we followed the legendary Al Chymist as he determined the formula of a hydrocarbon which turned out to be octane, C_8H_{18}. An important piece of evidence was the molecular weight of the compound. Now let us consider how he might have used the above method to obtain this information. Suppose Al initially introduced several grams of the liquid un-known into a glass bulb with a volume of 235 ml. The bulb was placed in an oil bath maintained at a constant temperature of 120°C. The barometer indicated an atmospheric pressure of 733 mm Hg. At this pressure, the boiling point of octane is about 99°C. Consequently, it vaporized at the higher bath temperature, and the vapor began to diffuse out through the narrow neck. When no further escape of gaseous octane was observed, the chemist sealed the bulb, removed it from its hot bath, and weighed it. Because he also knew the weight of the empty container, he was able to determine that the vapor which filled the vessel weighed 0.803 g. To summarize the data: 0.803 g of gaseous octane was found to occupy 235 ml at 120°C and 733 mm Hg. We can now take over from Al and calculate the molecular weight, using this information and the previous equation. But first we need a

FIGURE 10.6. *Use of the Ideal Gas Law for determining molecular weights—the Dumas method.*

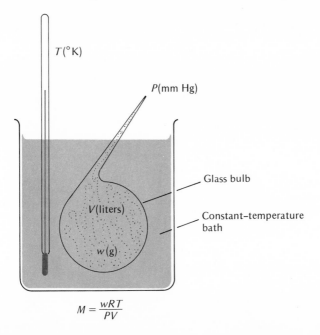

$T(°K)$

$P(mm\ Hg)$

Glass bulb

$V(liters)$

Constant–temperature bath

$w(g)$

$$M = \frac{wRT}{PV}$$

value for the gas constant. For convenience sake we select $R = 62.3$ liter mm Hg/mole °K. In order to bring the volume and temperature into conformity with these units, we convert 235 ml to 0.235 liter, and 120°C to 393°K. The pressure and mass terms are correct as they stand. Substitution of these values yields

$$M = \frac{wRT}{PV} = \frac{0.803 \text{ g} \times 62.3 \text{ liter mm Hg/mole °K} \times 393°K}{733 \text{ mm Hg} \times 0.235 \text{ liter}}$$
$$= 114 \text{ g.}$$

Which is what Al said a lot of pages ago.

Molecules in motion

It may strike you as a bit curious that in spite of the fact that we just calculated a molecular weight, the Laws of Boyle and Gay-Lussac do not postulate the existence of molecules. To be sure, we have argued (in Chapter 3) that the behavior summarized by these generalizations is inferentially consistent with the discreteness of matter. But molecules do not unambiguously and necessarily follow from the gas laws. A more convincing argument is provided by the reverse process. These same general laws of gaseous behavior can be derived from an *a priori* assumption of the molecular nature of gases.

This approach, called the Kinetic Theory of Gases, will not be presented in its full mathematical finery. Rather we will summarize its starting assumptions and its final conclusions. The former are the following:

1. Gases are composed of molecules in continuous, random motion, colliding with each other and the walls of the container.
2. Molecular collisions are elastic. There is no frictional loss of energy as a result of encounters with other molecules or with the wall. But transfers of energy do occur.
3. Bombardment of the walls of the container by the moving molecules results in pressure, i.e., the force of bombardment per unit area.
4. Attractive forces between molecules are assumed to be negligible. Moreover, the actual volume occupied by the particles is also considered to be zero. Thus, molecules are regarded as nonattractive point masses.
5. The average kinetic energy of translational motion exhibited by a molecule is proportional to the absolute temperature only.

These postulates serve as the guidelines for the application of the physical laws of moving bodies. The procedure involves calculating the pressure produced on the walls of a container by molecular bombardment. When this is done, a vaguely familiar equation is obtained:

10.12 $$PV = \frac{2}{3}nNcT.$$

It becomes considerably more recognizable if we associate ⅔ *Nc* with *R*, the gas constant. This turns out to be quite legitimate, since *N* is Avogadro's number, *c* is another constant, and the other symbols have the same significance as in Equation 10.10. Numerical evaluation indicates that the identification is correct. The Ideal Gas Law has emerged from the Kinetic Theory of Gases.

Happily, the Kinetic Theory gives us more than the bulk properties of gases. It provides insight into the microscopic phenomena underlying macroscopic behavior. Thus, the theory advances qualitative explanations for the experimental laws enunciated by Boyle, Charles, and Gay-Lussac. At constant temperature, increasing the pressure will force the molecules of a gas closer together, thus reducing the volume. If pressure is maintained at a fixed level, a rise in temperature causes the molecules to move more energetically and therefore to expand the volume they occupy. Should this volume be kept constant, but the temperature nevertheless increased, the enhanced translational energy of the molecules will result in more frequent and more forceful collisions with the walls of the container. Hence, the pressure will increase.

These arguments are given further credibility by the quantitative applications and extensions of the Kinetic Theory. We can, for example, calculate the average kinetic energy with which a molecule moves at a particular temperature. The fifth of our initial axioms in effect sets the kinetic energy equal to *cT*, where *c* is the same constant which turned up in Equation 10.12. A little fiddling around with these relationships and we discover that

$$E = \frac{3RT}{2N}.$$

Note that there is nothing in this equation which specifies the identity of the gas, for example, the molecular weight does not appear. The absolute temperature of any gas is a direct measure of the average kinetic energy of its molecules. At absolute zero, $T = 0°K$, the molecules have zero translational energy; they do not move. But as the temperature increases, so do the average energy of the individual particles and the total energy of the entire gaseous assembly of particles.

Since kinetic energy is related to velocity, *v*, an increase in the former must also be attended by an increase in the latter. The more energetic the molecules, the faster they move. Again, the Kinetic Theory of Gases enables us to calculate just how fast. But this time, the molecular weight, M, does enter the equation.*

$$v = \sqrt{\frac{8RT}{\pi M}}$$

The presence of M clearly indicates that at any fixed temperature, heavy molecules will, on the average, move more slowly than light ones. And, lighter molecules diffuse more rapidly than heavier ones. Thus molecules are somewhat like men (and vice versa). A 175 lb quarterback probably has an average speed considerably greater than that of the 240 lb guard blocking for him. But it is pretty nearly impossible to generalize the mathematical form of the dependence of velocity upon mass for football heroes. Molecules are much more manageable. Their average velocity is inversely proportional to the square root of the molecular weight.

To illustrate this behavior, consider two specific gases. Methane, CH_4, has a molecular weight of about 16.0, while that of monoatomic helium is 4.0. Applying our equation, we need only write down these numerical values,

$$v_{CH_4} = \sqrt{\frac{8RT}{\pi(16.0)}} = \frac{1}{4}\sqrt{\frac{8RT}{\pi}}$$

$$v_{He} = \sqrt{\frac{8RT}{\pi(4.0)}} = \frac{1}{2}\sqrt{\frac{8RT}{\pi}}.$$

The other terms are constant for both substances, and hence they cancel when we write the two speeds as a ratio:

$$\frac{v_{CH_4}}{v_{He}} = \frac{\frac{1}{4}\sqrt{8RT/\pi}}{\frac{1}{2}\sqrt{8RT/\pi}} = \frac{\frac{1}{4}}{\frac{1}{2}} = \frac{2}{4} = \frac{1}{2}.$$

Since the average velocity of helium is twice that of methane, the former gas will diffuse twice as fast as the latter. This reflects the fact that the molecular weight of helium is only one-fourth that of methane. What we have here is an example of *Graham's Law*, which states that *the rate of diffusion of a gas through an aperture is inversely proportional to the square root of its molecular weight.*

Some of the practical implications of Graham's Law are rather more important than the leaky helium balloon of Problem 5.11. Differences in diffusion rate between compounds of two uranium isotopes provide a basis for separating uranium-235 and uranium-238. The former isotope is fissionable. It will sustain a chain reaction in which the atomic nucleus is split into two smaller fragments and several neutrons. Perhaps you recognize this reaction

---

* In order to calculate velocity in cm/sec, we must use $R = 8.31 \times 10^7$ ergs/mole °K in this equation.

as that of atomic bombs and nuclear power plants, topics which warrant more detailed discussion later in this text. For the moment, the point of interest is the fact that only one uranium atom out of more than a hundred has isotopic weight 235. Since the chemical properties of all uranium isotopes are essentially identical, the concentration and collection of the fissionable isotope presents a formidable challenge. The solution devised by American and British scientists and engineers during World War II was to prepare gaseous uranium hexafluoride, UF_6. Obviously, both isotopes take part in this reaction, forming $^{235}UF_6$ and $^{238}UF_6$. The former species has a molecular weight of 349, that of the latter is 352. The difference in weight is very small, and its effect on diffusion rate is even smaller because the molecular weight enters the equation under the square root sign. Nevertheless, by using a series of thousands of successive diffusion barriers, the separation of the isotopes can be effected. So it was, and still is, in the huge gaseous diffusion plant at Oak Ridge, Tennessee.

It is important to reemphasize that the molecular velocities we have just been discussing and the translational energies we considered a moment ago are *average* values. Not all the molecules in a room or a balloon move with identical speeds—even when only a single gas is present and the temperature is constant. Far from it! There is a broad distribution of molecular energies and velocities. Indeed, although such comparisons are admittedly dangerous, the energetic diversity of molecules is probably as great as the variety among *Homo sapiens*. But two factors save the physical scientist from the vagaries which bedevil his statistically minded colleague in the social sciences. In the first place, the number of molecules in even a small balloon is about 10^{14} times greater than the entire population of the planet. And, in the second place, individual molecules, unlike individual men and women, do not exert disproportionate influence on their fellows' behavior. In short, there are more molecules than men and they behave better—or at least more predictably. As a consequence, the challenge to a chemist or physicist is a good deal simpler than that confronting a sociologist or economist. Statistics about molecules are simply more trustworthy than public opinion polls. The fact that assumptions about average molecular velocities and energies are sufficiently accurate to yield the Ideal Gas Law is a case in point.

This is not to imply that scientists only know average values of molecular velocities. While it is true that we cannot specify the speed of any individual molecule, we do know the form of the entire distribution. We can plot, as in Figure 10.7, the fraction of molecules having a particular velocity versus the velocity. Each point on the line thus represents a specific velocity and the fraction

FIGURE 10.7.   *Distribution of molecular speeds for nitrogen gas at two different temperatures. (Based in part on Gilbert W. Castellan, "Physical Chemistry," p. 68, Fig. 4.9, Addison-Wesley, Reading, Massachusetts, 1971.)*

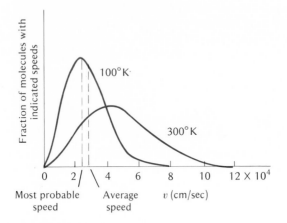

of molecules which exhibit that speed. Do not jump to the conclusion that each gaseous molecule has a fixed velocity at which it moves in perpetuity. Every collision between molecules alters their velocities. Yet, the net gains cancel the net losses, and the distribution, while changed in individual molecular detail, remains constant and characteristic of the gas and the temperature.

Note that there are relatively few molecules at the extremes of the distribution—the very slow ones and the very fast. The largest relative number of molecules have velocities near the average value. The most probable velocity, that is, the velocity exhibited by the greatest fraction of the molecules, occurs at the peak or maximum of the curve. Because the distribution is not exactly symmetrical, this most probable value does not coincide exactly with the average velocity. The former is always slightly lower than the latter.

Both of these indicators of molecular motion are seen to increase with increasing temperature. The curve for 300°K is shifted to include more fast moving molecules and fewer slow ones. The corresponding direct dependence of average energy upon absolute temperature clearly reflects one of the initial assumptions of the Kinetic Theory. The fact that each temperature is characterized by a particular distribution curve carries the interesting and not immediately obvious implication that temperature itself is a statistical phenomenon. It is meaningless to talk about the temperature of a single molecule. Temperature, like pressure, is the result of group effort.

Thus far our group has behaved ideally. It has cooperatively followed the dictates of $PV = nRT$ and the postulates of the Kinetic Theory. Indeed, of the many Platonic abstractions of science, the Ideal Gas Law is one of the most pragmatically useful. Its many applications are proof that it is highly relevant to the "real" world. Moreover, the nature of deviations from this simple regularity help us to understand "real" gases.

These deviations become particularly apparent at high pressure. Molecules are squeezed closer and closer together until their volume comes to represent a significant fraction of the total. Then they can no longer be regarded as point masses; they occupy too much of the space formerly free for movement. Moreover, the proximity of the molecules to one another is expressed in mutual attraction. This pull decreases the push on the walls of the container. Assumption number 4 underlying the Kinetic Theory is thus no longer legitimate and the Ideal Gas Law begins to fail. The beautifully simple equation no longer fits the experimental facts.

Since we cannot change the behavior of gases, our only alternative is to modify our equation to bring it into conformity with experiment. Some of these modifications are based solely upon empirical data. Values of P, V, T, and n are measured, recorded, and tabulated, and a mathematical formulation of the results is calculated. Quite a few of these "equations of state" have been devised. They demonstrate varying degrees of success in representing the observed behavior and predicting new results. But they are not as intrinsically interesting as theoretical calculations which seek to summarize bulk behavior on the basis of molecular events. These latter calculations endeavor to modify the Kinetic Theory so that an equation more widely applicable than $PV = nRT$ will follow.

Models have been invented which better approximate the behavior of gaseous molecules than noninteracting point masses. The simplest change is to give the molecules the volume which is rightfully theirs. In effect, they become miniature billiard balls. This removes an obvious failing of the Ideal Gas Law—the implication that $V = 0$ at $T = 0°K$. It should be self-evident that the volume of the molecules will remain, even at absolute zero where all gases have condensed to liquids and frozen to solids. A somewhat more elaborate model makes the molecules weakly attracting hard spheres. This attraction is represented by a potential energy term which becomes stronger as the particles get closer to each other. Finally, it is possible to further improve the approximation by postulating mutually attracting "soft" spheres. After all, molecules do not really have distinct hard surfaces. Their electron clouds are fuzzy and hence their mutual repulsion at close contact is not the bounce of billiard balls. Rather, in this region the repulsion increases with decreasing distance of separation between the molecular centers. The quantitative result is a potential energy diagram not unlike Figure 9.3 in its general shape.

With increasing sophistication of the model there is, as you might expect, an attendant increase in mathematical complexity. And, if the assumptions and the arithmetic are valid, there is also an increase in the accuracy with which the theory represents actual behavior. Thus, the process of refining our conception of nature proceeds. Part of this refinement is identifying the cause of intermolecular interactions. It is not enough to know how strong they are. The scientist's need to know is such that he also seeks their origin. We will seek it in our study of liquids.

Liquid assets

In many respects the liquid state is more incompletely understood than either of the other two physical forms of matter. Gases are relatively simple because they are completely chaotic. And solids are so well ordered that their internal arrangements of ions, atoms, or molecules can be unambiguously determined. But the structure of a liquid is intermediate between these extremes of order and disorder. In some respects, it is a dense gas; in other ways it behaves as if it were a soft solid. Consequently, it is difficult to invent a theoretical model which accurately mirrors the behavior of a liquid.

For the moment, at least, we will concentrate upon the relationship between liquids and gases and the interface between these two fluid states. As you know from your own experience, this boundary is readily crossed. Water evaporates from wash on a clothesline or a wet swimming suit. It boils away from an uncovered, unwatched pot. And it condenses again in the morning dew. These are all familiar examples of conversions between the liquid and gaseous states of the compound H_2O.

Such behavior suggests that molecules of a liquid have a tendency to escape the physical forces which hold them and to take figurative wing. To be a little less poetic and a good deal more precise, all liquids have *vapor pressure*. This measurable quantity is an indication of that escaping tendency. It is defined as *the pressure exerted in a closed container by the gaseous molecules of an element or compound which are present above the surface of a sample of the liquid form of the same substance*. Note that both phases, gas and liquid, must be present together. Moreover, *they must be in equilibrium*. That is, the rate of evaporation must equal the rate of condensation. We can simply represent this condition by a set of oppositely oriented arrows,

$$H_2O(l) \rightleftarrows H_2O(g).$$

Once equilibrium is achieved there is no net change in the number of liquid or gaseous molecules. Nevertheless, the system is dynamic, not static. Individual molecules continue to leave the liquid phase while others return. But in any given period of time, the number vaporizing will equal the number liquefying. Note that the examples cited earlier are not equilibrium phenomena. Wet clothes dry because the rate of evaporation exceeds the rate of condensation. Dry air continues to carry away molecules of water vapor before they can return to the liquid state. And water "boils away" if the pot is left open. In order to attain equilibrium the vessel must be sealed. Hence, measurements of vapor pressure can be made only in closed containers.

Figure 10.8 indicates that this is done by attaching a pressure gauge. Typically a U-tube of mercury (a manometer) is used. The tube is sealed on one end with a vacuum between the seal and the mercury level. If the vessel attached to the open arm is completely empty and evacuated, the mercury level in both arms will be equal. Now some liquid is added to the container, carefully, to avoid admitting any air. At once molecules will begin to escape into the empty space above the liquid. These gaseous molecules will exert pressure everywhere—the walls of the vessel, the surface of the liquid, and the surface of the mercury. In response to this force, the mercury level in contact with the vapor will drop and the level in the other arm will rise. The pressure of the gas is obviously greater than the vacuum in the closed end of the manometer. How much greater is indicated by the difference in height.

FIGURE 10.8. *Apparatus for the measurement of vapor pressure, P.*

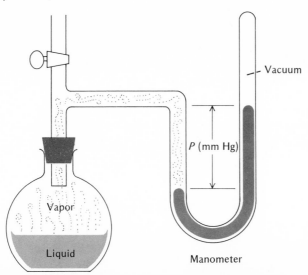

After several minutes at constant temperature, this difference becomes stabilized. Equilibrium has been achieved and the vapor
pressure of the liquid can be read from the manometer.

The condition, "at constant temperature," is an important one.
As temperature increases, so does vapor pressure. This is because
at higher temperatures more molecules acquire sufficient energy
to escape from the liquid state. As with gases, the molecules of a
liquid exhibit a broad distribution of kinetic energies. Only the
most energetic can break away from the liquid surface. Consequently, the liquid loses energy as well as matter—a process which
explains the familiar phenomenon of cooling by evaporation. Heat
is withdrawn from the surroundings to replenish that carried off by
the extraactive escapees. When heat is continuously applied the
temperature increases and the energy distribution shifts. The molecules of the liquid become more and more agitated. As more
escape, the vapor pressure rises.

It may come as somewhat of a surprise that the vapor pressure
of a liquid should be independent of the pressure of other gaseous
substances. But in point of fact, the evaporation of molecules of a
given substance is not inhibited by nonidentical molecules above
the surface of the liquid. Thus, the vapor pressure of water could
be measured in a closed vessel which contained 1 atmosphere of
air. To be sure, the total pressure would now be greater, by 1
atmosphere, than if only H_2O were present. However, the pressure
due to water vapor, its *partial pressure*, would remain unaltered.
This is a manifestation of *Dalton's Law of Partial Pressures*, a generalization which states that *the total pressure of a mixture of
gases is the sum of the pressures of the individual constituent
gases.* It is really not such a foreign phenomenon, since it is obviously exhibited by air.

Air also presses down upon all liquids in open or partially open
containers. As the temperature of a liquid is raised, its vapor pressure more closely approaches atmospheric pressure. Finally it
equals it at a particular temperature, characteristic of the substance. You know it as the *boiling point*. Once it has been reached,
the temperature remains constant, no matter how much heat is
pumped into the pot.* As long as the boiling continues, all the
heat goes into energizing molecules and converting them from the
liquid to the gaseous state.

Thus there is little profit (except for the gas and electric company) in boiling potatoes with ebullient gusto. A gentle bubbling
will be just as hot, if the pot is well mixed. If you want to get water

---

* There is a phenomenon called "superheating" in which the temperature of the liquid rises above the boiling point. This is an example
of "metastable equilibrium."

hotter than 100°C or 212°F you had better increase the atmospheric pressure. Since this is rather difficult to do on a large scale, pressure cookers have been invented. In them, elevated atmospheric pressures are established on a small scale. The liquids they contain can and must be heated to higher temperatures before the vapor pressure will equal the cooker pressure. Consequently, quicker cooking. One of these high-pressure devices might be of considerable culinary usefulness in Denver. There the normal atmospheric pressure is about 630 mm and water will boil at only 95°C or 203°F. The difference is not too great, but it may be enough to make a 3 minute egg taste more like 2½ minutes worth. Certainly the depression of the boiling point is sufficient to cause cake-mix manufacturers to write high-altitude instructions.

Chemistry, which is cooking writ large (or small—depending on your point of view) is also very much concerned with the pressure dependence of the boiling point. Sometimes chemists take good advantage of this behavior to distill under vacuum compounds which decompose readily at higher temperatures. By working under reduced pressure it is possible to lower the boiling point sufficiently to permit distillation at temperatures well below normal. Of course, just what constitutes the "normal" boiling point is a matter of definition and pressure. Often 1 atmosphere, 760 mm Hg, is taken as the standard pressure. But it is advisable to always specify the barometric pressure at which a boiling point was measured. This is of particular importance because boiling points are frequently used as criteria for assessing the purity of compounds or elements and as aids in identifying unknown substances.

It so happens that a surprising amount of information about structure and composition can sometimes be inferred from a boiling point. Especially significant is the fact that the boiling point is an indicator of the strength of the intermolecular interactions within the liquid state. Molecules which strongly attract each other escape the liquid only with considerable difficulty. Hence, the vapor pressures of such liquids are relatively low and their boiling points are high. On the other hand, highly volatile compounds, like ether, evaporate easily. They are characterized by high vapor pressures, low boiling points, and weak intermolecular forces.

Another measure of the intensity of these interactions is provided by the *heat of vaporization*. This experimentally obtainable property is *the quantity of heat necessary to vaporize a fixed amount of liquid (usually a mole) at its boiling point or some other reference temperature.* As I implied earlier, the absorption of the heat of vaporization does not increase the temperature of the liquid. Instead the molecules use it to acquire sufficient energy to escape into the gas phase. The greater the amount of energy required for the getaway, that is, the stronger the mutual molecular interactions in the liquid, the higher the heat of vaporization. For water, it amounts to 540 cal/g or 9700 cal/mole.

More about molecules

Trends in the three indicators of molecular interaction mentioned above—boiling point, vapor pressure, and heat of vaporization—have been related to trends in molecular properties. The result is a better understanding of the phenomena underlying such forces. In the first place, in any series of similar substances, there is a general tendency for boiling points to increase with increasing molecular weights. Although this is an effect of primary importance, we will wait with explaining it until we have considered a secondary influence—that of permanent dipole moments. *When two substances, one with a permanent dipole, the other without, have roughly the same molecular weights, the polar compound generally boils at a higher temperature.* For example, bromine, a liquid made up of symmetrical, nonpolar Br_2 molecules boils at 59°C. But iodine chloride, which exists as ICl molecules, has a boiling point of 97°. Both have molecular weights near 160. The difference in electronegativity between the iodine and chlorine atoms of ICl is sufficiently great that the bulk of the electron cloud is attracted by the chlorine end of the molecule. Hence, it bears a negative charge while the iodine atom becomes the positive pole. These dipolar molecules experience electrostatic interaction, the positive iodine end of one molecule attracting the negative chlorine end of another. Thus, the stabilizing forces are similar to those in ionic compounds, though not as strong because the charges are not as intense.

It turns out that the molecular cohesion arising from electrostatic dipole–dipole interactions in polar compounds is also usually weaker than *dispersion phenomena*. It is this latter type of intermolecular attraction which manifests itself in the previously noted dependence of boiling point upon molecular weight. The fact that such a trend is observed for symmetrical molecules as well as the unsymmetrical indicates that a permanent dipole is not necessary. Nonetheless, dispersion forces are in a sense dipolar in nature. The dipoles, however, are mutually induced and only of transient existence. This behavior is a consequence of the constant motion of electrons in an atom or molecule. At any instant, the distribution of the cloud of negative electricity which is part of the particle can be unsymmetrically arranged. For example, the density of the two-electron smear of the hydrogen molecule can be momentarily greater on one atom than on the other.

As a result, the molecule will become a temporary dipole. This dipole can, in turn, induce a similar dissymmetry of charge in another nearby molecule. The positive end of the first molecule will attract the electrons of the second molecule. By similar successive interactions involving a number of adjacent molecules clusters of temporary dipoles can be created. To be sure, no particular electronic arrangement persists for long. The system is dynamic and continuously changing. But the net result is intermolecular attraction which can be of considerable magnitude.

The strength of these dispersion forces increases with the ease of distortion of the electron cloud. The more readily the electrons be rearranged, the greater the ease of formation of transient dipoles and the stronger the interaction. In small, lightweight molecules and atoms the electrons are tightly held by the nuclei or nucleus. The resultant resistance to deformation or "polarization" leads to only weak attractive forces and correspondingly low boiling points. Hydrogen (b.p. $= -253°C$) and helium (b.p. $= -269°C$) are excellent examples; they are gaseous even to very low temperatures. But the electrons of larger particles are further away from the positive nuclear charge or charges. They are more loosely bound; and because of the floppiness of the electron cloud, temporary dipoles are readily formed. So it is that within a series of related elements or compounds, boiling points increase with increasing atomic or molecular weight. Specific illustrative data for the rare gases, the halogens, and some representative alkanes are included in Table 10.2.

The above argument would clearly predict that the boiling point of H_2O should be well below $-85.4°C$, the boiling point of H_2S. The latter compound, hydrogen sulfide, is one of the most unpleasant of air pollutants. Its unmistakable olfactory properties give rotten eggs their distinctive smell. It is obviously a gas at ordinary temperatures and pressures. Water clearly is not. Hydrogen fluoride, HF, which boils at $20°C$ is just barely a gas at room temperature. Nevertheless, its heavier homologue, hydrogen chloride, HCl, boils at $-85°C$. And ammonia, NH_3, with a boiling point of $-33.4°C$ compared to $-86.2°C$ for phosphine, PH_3, presents a third exception to the rule of increasing molecular weight—increasing boiling point. Without doubt, the molecules of H_2O, HF, and NH_3 are so strongly held that some special attractive forces must be operative.

These interactions have been given the name *hydrogen bonds.* Oxygen, fluorine, and nitrogen are all highly electronegative elements, they exhibit particularly strong affinities for electrons. Hence, their hydrogen compounds are characterized by large permanent dipole moments. The electron clouds are relatively far

TABLE 10.2.   *Boiling Points of Selected Substances,
Illustrating Influence of Molecular Weight*

Substance		GAW or GMW	Boiling point (°C)
Helium	He	4.0	−269
Neon	Ne	20.2	−246
Argon	Ar	39.9	−186
Krypton	Kr	83.8	−153
Xenon	Xe	131.3	−108
Fluorine	F_2	38.0	−188
Chlorine	Cl_2	70.9	−34
Bromine	Br_2	159.8	59
Iodine	I_2	253.8	184
Methane	CH_4	16.0	−162
Ethane	C_2H_6	30.1	−89
Propane	C_3H_8	44.1	−42
n-Butane	C_4H_{10}	58.1	0
n-Pentane	C_5H_{12}	72.2	36
n-Hexane	C_6H_{14}	86.2	69
n-Heptane	C_7H_{16}	100.2	98
n-Octane	C_8H_{18}	114.2	100

removed from the hydrogen atoms. As a consequence, the latter carry strong net positive charges. Because of their charge, the hydrogen atoms are attracted by the nonbonding electron pairs of the oxygen, fluorine, or nitrogen atoms of adjacent molecules. Furthermore, the small size of the hydrogen atom permits fairly close proximity between each pair of mutually attractive molecules. The result is a particularly pronounced dipolar interaction which generates directionally oriented hydrogen bonds of considerable magnitude. To be sure, these bridges are not as strong as the covalent bonds within the molecules themselves. But they are sufficiently strong to account for the anomalously high boiling points of H_2O, HF, and NH_3.

Because each water molecule contains two hydrogen atoms and two lone pairs of electrons, it can potentially participate in up to four hydrogen bonds. Therefore, the effects of such interactions are particularly pronounced in H_2O. Indeed, many of the peculiar properties exhibited by water can be directly attributed to hydrogen bonding. Its heat of vaporation is unusually high because a good deal of energy is required to break these bonds. The same argument explains the high heat capacity of water. And water's relatively great resistance to flow (viscosity) is also a consequence

FIGURE 10.9. *Representation of hydrogen bonding involving four molecules of H$_2$O. Note that the hydrogen bonds (dashed lines) are considerably longer than the covalent bonds (solid lines).*

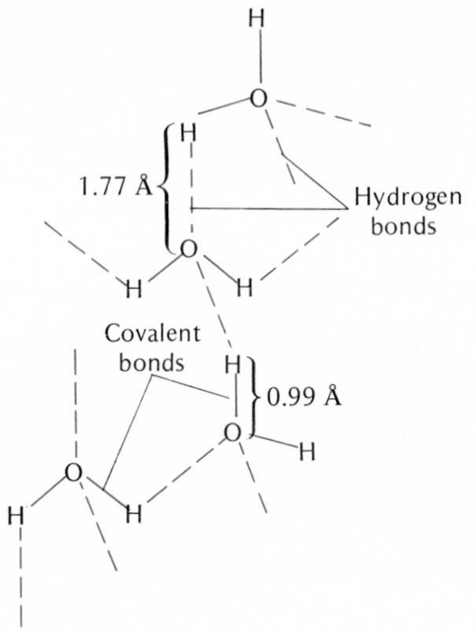

of the strong cohesive forces within the liquid state. Even more interesting is the fact that ice is less dense than liquid H$_2$O at 0°C. For the great majority of substances the solid state is more dense than the liquid. The role of hydrogen bonding in determining this unusual behavior will be considered in detail in a later section. To paraphrase Goethe's statement about blood, *Wasser ist ein ganz besondrer Saft!* ("Water is a most peculiar juice!")

Anomalous water: aqua vera or aqua ficta?

Recent experimental evidence suggests that water may be even more extraordinary than we have heretofore thought. The remarkable new properties were first observed at the Institute of Physical Chemistry in Moscow by B. V. Deryagin and N. N. Fedyakin. In the early 1960's these Soviet scientists carried out a series of disarmingly simple experiments. Their typical procedure was to place extremely narrow quartz or glass capillaries, with internal diameters of 1 to 100 microns (about as fine as a hair), in a closed container filled with water vapor. After a few days the container was opened and the tiny tubes were removed. Usually some of them would be found to contain a liquid, presumably condensed water. Closer investigation suggested that two species were actually present. One, exhibiting the ordinary extraordinary properties of H_2O, could be removed by distillation. But the other remained in the capillary, even though it was heated to 100°C. If this, too, were a form of water, it was indeed "anomalous water." And so it was christened.

Although the Russian physical chemists had made only minute quantities of anomalous water they at once set to work to characterize it. They found that its boiling point was very high, but it could be distilled between 200° and 300°C. When it was, the condensate retained the anomalous properties. Moreover, the wierd water refused to freeze at 0°C like ordinary run-of-the-tap H_2O. Instead, it finally got glassy at −45° to −50°C. And its viscosity was found to be fifteen times greater than normal. To be sure, some of these early experiments were performed in the capillaries. And it is an undisputed fact that surface effects can alter properties in such circumstances, where the surface-to-volume ratio is relatively high. But the water remained anomalous even when it was removed from the tubes. Thus Deryagin and Fedyakin found its density to be 1.4 g/cm³ instead of the oft-measured aqueous value of 1.0 g/cm³.

It was toward the end of the 1960s before reports of this startling discovery began to filter out of the Soviet Union. Generally they were met with a combination of incredulity and fascination. Both responses were in large measure conditioned by the fact that such properties had never before been detected in such a ubiquitous and thoroughly studied compound. Indeed, no other substance has been subjected to such exhaustive investigation. Standards for many of our most commonly accepted and widely used units, including the gram, the liter, the calorie, and the degree, are based upon water. To suppose that a whole new set of properties had been discovered bordered on science fiction. And yet, the competence and credibility of the Russian workers was beyond reproach. If they were right, this was the hottest news from Moscow since the revolution.

A flurry of activity followed in laboratories around the world. In the United States, E. R. Lippincott and his co-workers soon succeeded in duplicating the Russian results. Since then, hundreds of others have apparently also done so. An undergraduate in the author's laboratory prepared several samples with inexpensive equipment.

Unfortunately, the fact that by now the stuff has been made many times over does not prove that it is a newly discovered form of pure H_2O. The samples still remain extremely small and therein lies one of the chief experimental difficulties. Spectra have been measured at a wide variety of wavelengths and molecular weight determinations have been carried out. But all these data suffer from the minuteness of the quantities studied.

Of course, this fact does not deter theoreticians from speculating on the intriguing question of the structure of anomalous water. Many hours have been devoted to complex quantum-mechanical calculations and the true believers have advanced at least a dozen hypotheses differing in structural detail. All of them assume that the peculiar properties result from a network of extremely strong hydrogen bonds. These bonds are postulated to approach or equal the strength of covalent bonds. Thus they are thought to unite the H_2O subunits in a pseudo-polymer or quasi-macromolecule. Indeed, the name "polywater" has often been applied to the stuff. It must be said in defense of these views that many of the properties of anomalous water are consistent with this interpretation. For example, exceptionally strong hydrogen bonding would account for the high boiling point, viscosity, and surface tension.

On the other hand, the nonbelievers insist that polywater is in fact dirty water—a concentrated aqueous solution of something. There is a difference of opinion on just what the something is. Chemical analysis should be able to provide the answer. But here again, efforts at accurate analysis are bedeviled by the small samples with which scientists must work. Deryagin and others have done analyses too, and they maintain that their samples contain only hydrogen and oxygen. Nevertheless, some investigators claim to have found silicates, which could be dissolved from glass or quartz. Others seem to see sodium acetate or a mixture of inorganic compounds. Sweat has even been suggested as the magic ingredient in polywater. But whatever the solution advanced, its advocates see in it an explanation of the anomalies.

One thing is certain, the issue of anomalous water is far from settled. But even as the paucity of definitive experimental evidence has not deterred the theoreticians, neither has it inhibited the speculators. The possible presence of polywater in living things has been discussed at length. Clouds of polywater have been postulated for 50 miles up, where low temperatures and low pressures would cause ordinary water to freeze and/or evaporate. And polywater has even been predicted on the surface of Venus. Obviously, we will have to wait to evaluate that and other hypotheses. Whether or not polywater is a solution of sweat, it will require a good deal more perspiration before the questions raised on these pages can be answered. Perhaps by the time they come from the presses, the controversies will have been resolved. The consensus that anomalous water is something other than a new form of pure H_2O does seem to be growing. But for the present, this fascinating issue remains an eloquent refutation to those who see no ambiguity, no unknowns, and no excitement in science.

Solids of various sorts

As we turn our attention to the solid state, it seems logical to continue our discussion of hydrogen bonding. Although relatively few solids are stabilized by such intermolecular interactions, one of the most common is. And, in spite of the fact that ice is somewhat atypical, it can provide general insights into the solid state and the phenomenon of melting. To begin with, the ice that you have encountered in cubes, on frozen lakes, and on winter streets is only one of at least eight crystalline forms.* All these *polymorphs* (as they are called) differ in the geometrical arrangement of the H_2O molecules which constitute them. With the exception of ordinary ice (ice I), they exist only at elevated pressures. These pressures are so high that they are never encountered outside the laboratory, unless perhaps inside the planet. For example, ice II cannot form at 0°C unless the pressure exceeds 2000 atm. And ice II is a relatively low pressure polymorph. Others are found at pressures as high as 220,000,000 atm, where ice VII melts at 442°C. But such behavior and such pressures are so far removed from common experience that it is pointless to pursue the topic further in an introductory text. It is worth noting, however, that some scientists have suggested that structures identical to those of certain high-pressure ice polymorphs may be present in polywater or, for that matter, in normal liquid water.

At the moment our chief interest is the relationship between liquid water and ice I and the interconversion of these two phases, that is, melting and freezing. The change from the solid to the liquid state can be considered as formally and physically analogous to evaporation. Assume you start with an ice cube from the freezing compartment of a refrigerator. Its temperature is −10°C, well below freezing. You place it in a glass of water which is at 25°C. Predictably (because we live in a universe which obeys the laws of thermodynamics—something you'll soon read about) the temperature of the ice increases and the temperature of the water falls. Some of the heat involved in the kinetic motion of the liquid water molecules is transferred to the molecules of the ice. Hence, the average energy of the molecules in the solid phase increases at the expense of the average energy of the molecules in the liquid state. Some ice molecules thus acquire sufficient energy to escape into the surrounding liquid.

After this process has gone on for some time, preferably with gentle mixing to promote efficient heat exchange, the temperature of the ice is found to have risen to 0°C or 32°F and the temperature of the water has fallen to the same level. You, of course, recognize this as the melting point of ice or the freezing point of

---

* Nine, if you believe Kurt Vonnegut, Jr., and "Cat's Cradle."

water. If the ice and the water mixture were perfectly insulated, it would indefinitely remain at equilibrium at this temperature. The molecular transfer between the solid and liquid states would continue, but the rate of melting would equal the rate of freezing.

$$H_2O(s) \rightleftarrows H_2O(l)$$

In the normal situation, however, the system is not so well insulated. If the room is warmer than 0°C, heat will flow to the ice water, melting more of the ice. As this occurs, no increase in temperature is observed, so long as the liquid and solid are both present and well mixed. All the heat is used to energize the ice molecules so they can escape their hydrogn-bonded lattice. This heat, the *heat of fusion* of ice, amounts to 80 cal/g or 1440 cal/mole. Note that these numbers are considerably lower than the corresponding values for the heat of vaporization of water. The explanation is that the average energy difference between a molecule of liquid water and one in the solid state is decidely less than the energetic difference between a gaseous and a liquid molecule. In other words, a molecule must acquire more energy to escape from the liquid state to the gas phase than is required to escape from the solid to the liquid form. This correctly implies that the intermolecular interactions stabilizing the liquid state are only slightly weaker than those operative in the solid. But attractive forces in gases are a good deal less pronounced than those in either of the condensed phases. Of course, once the ice is all melted the temperature will again start to rise, all the way to the boiling point if enough heat is supplied.

What has thus far been written about the melting of ice could be said for any solid–liquid phase change. Obviously, the melting points differ, and so may the nature of the interparticulate forces, but the basic principles apply equally well to any pure substance. There is, however, an important aqueous anomaly which we noted in the last section—ice floats. The density of the solid phase is less than that of the liquid phase. If one were to take a gram of almost any liquid and measure its volume as a function of temperature, the volume would decrease smoothly with decreasing temperature. It would shrink or contract and become more compact, with its molecules closer together. In other words, the behavior would be similar to the decrease in volume of a gas which, according to Charles' Law, attends falling temperatures. To be sure, the contraction in liquids is much less pronounced than that observed for gases, but it is a consequence of the same phenomenon—decreased molecular motion. The net result of all this is an increase in weight per unit volume, that is, density, which persists all the way to the solid state. As a consequence, the solid forms of most elements or compounds sink in their liquid phases.

If water behaved in the same standard fashion, life would be considerably altered, if there were any life, that is. Lakes would freeze from the bottom up with consequences far more disasterous than no ice skating. Aquatic plants would die or become dormant, fish would die, and the entire ecological balance would be severely perturbed. Pipes and auto radiators, burst by the expansion of freezing water, are an inconvenience and an expense. But it is a modest price to pay.

One of the features related to this nearly unique property of water (a few other substances behave similarly) is the minimum in density which it exhibits at 4°C. Above this temperature, its density decreases with increasing temperature in the normal fashion. But below 4°C the density of water decreases with decreasing temperature. Selected values for several temperatures illustrate the point.

$T(°C)$	0	0	4	25	100
Density (g/cm³)	0.9167	0.9999	1.0000	0.9971	0.9584
	(ice)	(liquid)			

The above data also indicate that ice does not fall on the smooth (though nonlinear) density/temperature curve which characterizes liquid water. At 0°C the density of the solid is significantly lower than that of the liquid. Obviously, a major discontinuity in structure must occur at the melting point. The structure of ice I is well known. Each water molecule is hydrogen-bonded to its four nearest neighbors. Two of these bonds involve the hydrogens of the central molecule, two utilize the lone pairs of electrons on the oxygen. Thus, each oxygen atom is at the center of a tetrahedron of other oxygen atoms. Hydrogen atoms separate them. In its general features, the arrangement is reminiscent of the diamond structure, and ice exhibits the same interlocking hexagons. As is evident from Figure 10.10, these hexagons surround open channels and this empty space accounts for the relatively low density of the solid. It has been suggested that as ice melts, individual molecules enter these channels. This decreases the volume of a given mass of H_2O and hence increases its density.

This density difference also explains the fact that the melting point of ice decreases slightly with increasing pressure. The increased pressure drives the solid–liquid reaction in the direction of the phase which occupies the lower volume. Thus, though the temperature may be below 0°C, if the pressure is enough above 1 atm, the solid state can convert spontaneously to the denser liquid. The skate blades under a 125 lb figure skater bear down on the surface of the rink with a pressure of about 3000 atm. This

FIGURE 10.10.   *The hydrogen-bonded lattice structure of ice I,*
the common form of solid H₂O. Note the open channels which
contribute to the relatively low density of ice. From "The
Architecture of Molecules" by Linus Pauling and Roger
Hayward. W. H. Freeman and Company, copyright c1964.

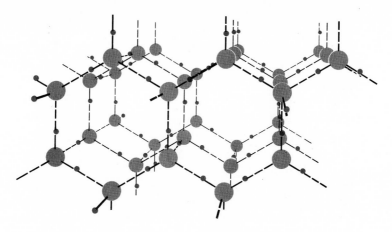

lowers the melting point of the ice to −22.0°C or −8.0°F. The
skater glides smoothly and gracefully on the resulting thin film of
liquid water. She would fall on her face (*et al.*) if she tried skating
on most other solids. Their freezing points increase with increasing pressure.

At temperatures above the freezing point, particularly those
above 4°C, much of the regularity of the ice lattice becomes dis-
rupted by the thermal translational motion of the molecules.
Nevertheless, evidence suggests that "flickering clusters" of ice-
like order do persist. These aggregates are constantly changing
and involving different molecules as old hydrogen bonds break
and new ones form. Thus, they are really not very much like
minute icebergs, although the analogy has been drawn. Whatever
the exact nature of these aggregates, it does seem very probable
that they are responsible for some of the distinctive behavior of
liquid water. It is likely, for example, that quite highly ordered
H₂O is found near enzymes and other biological macromolecules
and accounts for some of the properties of these important
compounds.

Hydrogen bonds are also intimately involved in the internal
structures of proteins and nucleic acids. They frequently hold
different parts of these long, large molecules together. In this manner hydrogen bonds stabilize the molecular shape which is
essential for biological function. The topic is well worth considering in greater depth, but the chapter devoted to biochemistry is
a more appropriate place.

For the moment we turn to molecular solids which do not depend upon hydrogen bonding for their internal cohesiveness. Generally speaking, the same intermolecular interactions which stabilize liquids are operative in the solid state. For any given compound or element, the attractive forces in the solid are greater than those in the liquid. And, as we implied earlier, the particulate building blocks are almost always closer together in the solid state. The same trends observed for boiling points usually pertain to melting points as well. Because dispersion forces increase in intensity with the size and polarizability of molecules, melting points of similar substances rise with increasing molecular weight. And the mutual attraction of permanent dipoles contributes to the elevated melting points of polar compounds. The fact that both dipolar and dispersion forces are weaker than electrostatic interaction explains why covalent compounds are generally lower melting than ionic salts.

The mutual attraction of anions and cations makes bonding in ionic solids readily comprehendable. But far less obvious is the nature of the forces operative in a *metal*. Since in a pure metallic element all the atoms are identical it makes no sense to postulate electron transfer to generate oppositely charged ions. The electron affinity is logically the same for each atom. Nor can the metallic state be attributed to covalent bonding. There simply aren't enough electrons to go around. Indeed, metals are all characterized by relatively few valence electrons—usually one, two, or three. For example, in its crystal lattice each sodium atom has eight equidistant nearest neighbors, but only one electron. Such an atom could contribute an average of only one-eighth of an electron to share with each of its neighbors. This is definitely not enough to form half of a conventional covalent bond. The explanation must lie elsewhere. Moreover, it must be able to account for the properties of metals. They are bright, shiny, and lustrous, and excellent conductors of both electricity and heat. Moreover, they are malleable, easily deformable into spears and swords, pruning hooks and plowshares.

At least two theoretical models have been advanced for the metallic state. Both view the electrons as the common property of all the atoms. The simpler version hypothesizes that a piece of metal is a regular array of positive ions bathed in a *sea of electrons*. Thus, a mole of sodium, only about an ounce, consists of 6.023×10^{23} Na^+ ions and an equal number of electrons. These electrons not only provide the cement which holds the lump together, they are also free to wander about. They readily move to carry electricity or heat energy. They can even be dislodged, as in the photoelectric effect, to leave behind an excess of positive ions. Moreover, the luster of metals can be attributed to the fact that loosely bound electrons would absorb a broad spectrum of light and reemit it, that is, reflect it. Finally, malleability is explained as a consequence of the quasi-fluid nature of the metallic sea. Under the deforming force of a smith's hammer, planes of atoms slide over each other, only to reestablish a structure very like the original.

The *band theory* is really a more sophisticated version of this model. It is based on quantum mechanics and considers the orbitals contributed by each of the atoms. Because each atom is in a slightly different position, relative to all the other atoms of the metal, its energy levels will be ever so slightly different from those of any other atom. When the levels of all the atoms in the sample are added together they form a nearly continuous band of energy. This composite band—a communal, crystalwide orbital—belongs to the entire piece of metal. Because metallic atoms typically contain unfilled orbitals, the band associated with a large number of such atoms is also incompletely filled. For example, each sodium atom has a single 3s electron, though the capacity of the 3s orbital is two. The combination of an Avogadro number of these orbitals yields a super-orbital or band with a capacity of $2 \times 6.023 \times 10^{23}$ electrons. In the electronically unexcited state, only the bottom half of the band is filled. But the differences between the energy levels in the band are so small that excitation to the vacant levels occurs with ease.

Since a broad range of energy levels is accessible to electrons, they can be excited by light of many different wavelengths. And, when the electrons return to their ground state, they reemit this multicolored "white" light. Moreover, a wandering electron entering one end of a wire can be readily accommodated in one of the vacant positions in the band. And, at the other end of the wire, a different electron can be easily withdrawn. Thus, electric current can be conducted without requiring a single electron to migrate the entire length of the wire. High thermal conductivity is also attributed to the closeness of the energy levels and to the fact that only small jumps need be made by the electrons in order to pass on heat.

One of the attractive features of the electron band theory has been its utility in explaining *insulators* and *semiconductors*. Silicon and germanium are good examples of the former class of substances. Atoms of both of these elements contain four valence electrons. In a crystal of pure silicon or germanium, these electrons fully fill the lowest-lying band. To be sure, there are higher, unfilled bands corresponding to more energetic orbitals. But the energy gap separating the lowest of these "conduction bands" from the ground state "valence band" is so great that electrons can make the transition only with a very strong boost. In other words, the electrons are not easily excitable. They behave as relatively localized bonding electrons rather than the rovers of the metallic state. The electrical conductivity of these elements is consequently low.

However, when small amounts of certain impurities are introduced, silicon and germanium become capable of carrying current. The manufacturing process involves "doping" a crystal with atoms containing five or three valence electrons. Arsenic is an example of the former. Each arsenic atom introduces one extra electron which is not involved in covalent bonding. These electrons populate a band near the first conduction band of the parent silicon or germanium. Because of the small energy gap the electrons of arsenic are readily excitable to the empty composite orbital. The

FIGURE 10.11.   *Electron energy band diagrams for conductors, nonconductors, and semiconductors.*

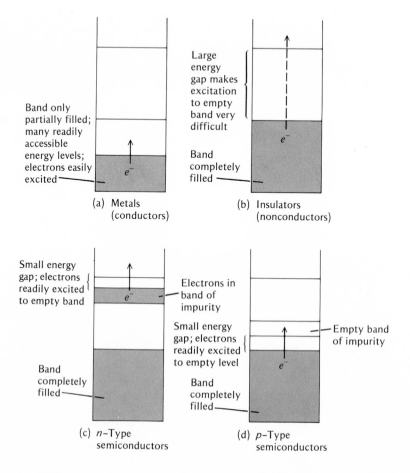

result is known as an *n-type* semiconductor because a negative charge is now free to move. Positive charges migrate in *p-type* semiconductors. Or, if you prefer, "holes" move. The holes are formed when electrons are excited out of the silicon or germanium band into the partially filled band of the additive. Here, the trace element introduced must have fewer than four outer electrons in order to provide unoccupied orbital space. Aluminum and gallium, both with three valence electrons, are commonly used to dope silicon and germanium in the manufacture of *p*-type semiconductors.

The two types of semiconductors have been successfully combined in a variety of sequences to perform specific electrical functions. These include the conversion of alternating current to direct current, the amplification of current, and the transformation of light energy into electrical energy. Known to you and millions of others as "transistors," sandwiched semiconductors have revolutionized electronic technology. Their widespread use as replacements for vacuum tubes has permitted miniaturization undreamed of twenty years ago. Needless to say, the impact of semiconductors and transistors upon the space program, the design of computers, the quality of high-fidelity audio components, the electronics industry, and the economy of Japan has been staggering. And underlying it all is man's increasing understanding of matter and the forces which shape it and determine its form.

For thought and action

10.1   The weather man on the local TV station announces that the atmospheric pressure is "26.8 inches of mercury and falling." Translate this value into millimeters of mercury and atmospheres for the benefit of a European exchange student. (Conversion factors are in Appendix 2.)

10.2   Check the correctness of the calculation on page 257 by converting 1034 g/cm^2 into lb/in^2.

10.3   Demonstrate mathematically that a pressure of 760 mm Hg is equivalent to 10.3 m or 33.9 ft of water. You will find it useful to know that the density of mercury is 13.6 g/cm^3 while that of water is 1.00 g/cm^3.

10.4   Explain to an urban dropout just moved to an abandoned farm why he's having such a hard time pumping water out of his 50 ft well with a pump situated at the surface. Be kind enough to help alleviate his drought with a useful suggestion.

10.5   A quantity of a perfect gas confined to a cylinder occupies a volume of 2.48 liters under a pressure of 1.25 atm. The piston is lowered and the pressure is increased to 2.75 atm, while the temperature is held constant. Calculate the volume of the gas at this new pressure.

10.6   A balloon contains 1.65 liters of air at 20°C. The temperature is raised to 35°C. Assuming the pressure inside the balloon to remain constant, calculate the new volume of gas at this higher temperature. Why is the assumption of constant pressure a little dubious, and what changes in pressure would you expect to see?

10.7   Suppose the pressure in an aerosol spray can is 800 mm Hg at 25°C. The can will burst when the internal pressure reaches 1300 mm. Find the temperature (in °C) at which this sad event will occur, and keep spray cans out of the fire!

10.8   In high school courses you may have made use of the volume of 1 mole of a perfect gas at STP (standard temperature and pressure). These conditions are defined as 0°C and 1 atm. Determine the volume in liters.

10.9   A deep breath of air has a volume of 1.00 liter. Assuming a constant pressure of 750 mm Hg and body temperature, 37°C, find the number of moles of air and the number of molecules in that breath. Incidentally, 20.9% of the molecules in air are oxygen. Approximately how many molecules of oxygen did you just breathe in?

10.10 A 4.24 gal tank of bottled gas contains 20 lb of propane (C_3H_8). The temperature of the tank is 70°F. Convert these units to liters, moles, and °K, and then use the results to determine the pressure of the gas in atmospheres, if it all were gas, that is. In fact, it isn't; some condensation occurs. (1 gal = 3.79 liters.)

10.11 A fully inflated automobile tire has a volume of 27.0 liters and an internal pressure of 32 lb/in². Translate this latter value into atmospheres and calculate the grams of air (average molecular weight 29.0) in the tire at a temperature of 77°F (25°C). (1 atm = 14.7 lb/in².)

10.12 The car rolling on the tire of Problem 10.11 is driven 50 miles over a hot roadway. The pressure in the tire is then measured and found to be 36 lb/in². Assuming that the volume of the tire has not changed appreciably, and that no air has been lost, calculate the temperature of the tire.

10.13 It is found that 0.580 g of cyclohexane vapor completely fills a 225 ml vessel at a pressure of 707 mm Hg and a temperature of 98°C. Determine the molecular weight of the compound.

10.14 Spectroscopic measurements indicate that interstellar space in our region of the galaxy contains one hydrogen atom per cubic centimeter and has a temperature of $-170°C$. Note that the element is present as single atoms, not the familiar H_2 molecules.

(a) Calculate the pressure (in atmospheres) of this extremely dilute gas.

(b) Find the average speed of a hydrogen atom (in cm/sec and miles/hr) at the temperature of interstellar space.

10.15 Much of the sun is made up of isotopes of hydrogen and helium, in atomic or ionic form. Assume that the concentration of these atoms or ions in the interior of the sun is 0.7 mole/cm^3 and that the temperature is $2.3 \times 10^{7}°K$. (Molecules are not stable under these extreme conditions.)

(a) Compute the number of atoms per cubic centimeter of the sun's center, and compare this value with that given for outer space in Problem 10.14.

(b) Calculate the pressure (in atmospheres) of this extremely concentrated gas.

(c) Find the average speed of a hydrogen atom (in cm/sec and miles/hr) at the temperature of the sun's center.

10.16 If you attempted Problems 10.14 and 10.15 you have just been on two astronomical excursions. Not all such trips are equally productive. To which part of the universe, interstellar space or the interior of the sun, are the Ideal Gas Law and the simple Kinetic Theory of Gases more legitimately applicable? In other words, which answers are more likely to be right—and why?

10.17 Determine the *relative* average speeds and average energies of $^{235}UF_6$ and $^{238}UF_6$ molecules at any constant temperature.

10.18 Calculate the number of moles of H_2O vapor in a closed 1 liter bottle which contains 100 ml of liquid H_2O at 25°C. The vapor pressure of water is 23.8 mm Hg at 25.0°C.

10.19 An unwatched pot contains 500 ml of water (about a pint) at its boiling point. How much heat is required to "boil away" all the water? The density of H_2O is 0.958 g/ml at 100°C.

10.20 Suppose you are investigating the properties of three compounds: water, ethyl alcohol (C_2H_5OH), and diethyl ether $(C_2H_5OC_2H_5)$, the "ether" used as an anesthetic. According to published tables of data, the three liquids exert a vapor pressure of 100 mm Hg at these temperatures: H_2O at 51.6°C, C_2H_5OH at 34.9°C, and $C_2H_5OC_2H_5$ at $-11.5°C$. Rank the compounds in order of

(a) increasing vapor pressure at 25°C,

(b) increasing boiling point,

(c) increasing heat of vaporation, and

(d) increasing rate of evaporation.

Now account for these properties on the basis of differences in intermolecular forces, being as specific as possible.

10.21   It sounds a bit odd, but it's true that "ice cools by melting." Explain the statement.

10.22   Indicate the nature of all the interactions (covalent, ionic, dispersion, dipolar, hydrogen bonded, and/or metallic) which must be overcome in order to melt or boil the following substances. Then try ranking them in order of increasing melting points.
   (a)   Argon
   (b)   Diamond
   (c)   Silver
   (d)   Potassium chloride
   (e)   Hydrazine (N_2H_4)
   (f)   Methane (CH_4)
   (g)   Carbon tetrachloride (CCl_4)
   (h)   Chloroform ($CHCl_3$)

10.23   Use your knowledge of the nature of the submicroscopic world of atoms, ions and molecules, and the forces which these particles experience to explain the following familiar properties of bulk matter.
   (a)   Gasoline (C_8H_{18} and similar compounds) is a liquid at room temperature while natural gas (CH_4 and similar compounds) is a gas.
   (b)   Bottled gas (C_3H_8) is really a liquid when it is in the bottle.
   (c)   Salt (NaCl) melts at a higher temperature than sugar ($C_{12}H_{22}O_{11}$).
   (d)   Water pipes burst when they freeze.
   (e)   Burns from steam are more severe than burns from boiling water.
   (f)   Silicon dioxide (SiO_2) is a hard, high-melting solid which makes up most of sand. But as you well know, carbon dioxide (CO_2) is a gas.
   (g)   Pure water is a very poor conductor of electricity. Nevertheless, one's chances of getting a shock while working with electrical wiring are aggravated by standing in water.

10.24   Diamond is an insulator but graphite is a good conductor of electricity. From what you know about the structures of these two allotropes of carbon, advance an explanation of this difference in behavior.

10.25   If possible, describe a controversy in your own major field of interest which is comparable to the anomalous water dispute.

Suggested readings

Conant, James Bryant, ed., Robert Boyle's Experiments in Pneumatics, *in* "Harvard Case Histories in Experimental Science" (James Bryant Conant, general ed.), Vol. 1, pp. 3–63, Harvard Univ. Press, Cambridge, Massachusetts, 1966.

Deryagin, Boris V., Superdense Water, *Scientific American* **223**, 52 (March, 1971). An article in support of anomalous water by one of its discoverers.

Howell, Barbara F., Anomalous Water: Fact or Figment, *Journal of*

Chemical Education **48**, 663 (October, 1971). The author concludes it is probably a figment.

Vonnegut, Kurt, Jr., "Cat's Cradle," Dell Publ., New York, 1963. A novel about scientists, polymorphs of solid H_2O, and the end of the world—by ice, not fire.

11

Solution, pollution, and the pollution solution

Solutions, solvents, and solutes

Thus far, this study of matter has concentrated upon the structure and properties of pure substances—individual elements and compounds. This is as it should be; the basic principles of chemistry are most readily discernible in relatively simple stuff. But as early as Chapter 2 we admitted that much of the matter we meet in ordinary experience exists in mixtures of varying complexity, for example, the ink on these pages and the pages themselves. Now the time has at last come to consider in some detail the nature and behavior of an important class of mixtures—solutions. It will soon become apparent, I think, that what you have learned about elements and compounds can do much to aid this investigation.

A solution can be defined as a homogeneous mixture of two or more substances. The gross aspects of this homogeneity are visually apparent, because under fixed conditions of temperature and pressure, a solution exhibits only a single phase. The phase can be solid, liquid, or gaseous; but

if it is a true solution, there should be no indication of heterogeneity. This uniformity of composition extends to submicroscopic
dimensions. *The particles of the pure substances which make up
the mixture are molecules, atoms, and/or ions*, depending upon
the particular solution. Obviously, a suspension of fine silt in water
does not meet this criterion. Nor does a fog of water droplets in
the air. If the aggregates are visible to the naked eye, they must
consist of thousands upon thousands of individual particles. However, the suspending medium may itself be a solution. This is
clearly true for air we breathe, and it frequently characterizes the
water we drink, cook, and wash in.

Another feature of a true solution is the fact that its constituent
particles are randomly distributed in an arrangement which shows
no significant long-range order. The constant, chaotic motion of
these particles keeps the solution uniformly mixed, and thus there
is no settling out or separation of phases. Generally speaking, all
portions of a solution have the same composition. If, however,
the scale is vast enough, some local differences can arise. You will
recognize that this is so for the oceans and the atmosphere. It is
painfully obvious that the composition of the latter solution depends to some extent upon the gases emitted by automobile
exhausts and factory flues. And garbage scows, raw sewage, and
the effluents of affluence have so altered the chemical and biological composition of certain parts of the sea that life has been
endangered in these areas. But important as these instances of
solution inhomogeneity are to the environment, it would be premature and of questionable value to attempt to discuss them without some prior understanding of the properties of simpler solutions. Idealistic motivation is not enough to clean up the cesspools
of the planet; knowledge is also necessary.

Therefore, we first turn to the basic business of communication
and vocabulary. Commonly, the *solute* is said to dissolve in the
solvent to form the *solution*. Unfortunately, this distinction is
somewhat ambiguous for gas–gas solutions. In the gas phase, the
constituents can mix in any and all proportions. There are no
saturation solubility limits. One can equally well consider air as a
solution of oxygen in nitrogen or, conversely, as a solution of
nitrogen in oxygen. Perhaps the former view is preferable because
the chemical component present in the greater (or greatest) relative quantity is sometimes designated the solvent. Certainly, it
does make some sense to speak of carbon monoxide as a trace
solute in air. And when liquids evaporate or solids sublime into
the atmosphere, their molecules in effect become solutes in a
gaseous solution.

The distinction between solute and solvent becomes clearer in
liquid solutions, particularly if the solute is solid or gaseous. The

solvent is then the pure liquid component. For example, in "salt water" pure H_2O is the solvent and pure NaCl the solute. But the complex composition of sea water testifies to the fact that a solution often contains more than a single solute. Here many other ionic substances are also present. Moreover, in the case of natural waters, dissolved oxygen is a very important solute. Although this gas is only slightly soluble in H_2O, it is normally present in sufficiently high concentration to permit the existence of gill-equipped life.

Some solutions are even more complicated, containing solid, liquid, and gaseous solutes. Beer is an excellent example. The solvent is water; the solid solutes include sugars, salts, and various trace quantities of flavoring substances; ethyl alcohol is the liquid solute; and carbon dioxide is, of course, the dissolved gas which provides the foam and fizz as it escapes from solution.

Finally, although most of the chapter will deal with liquid solutions, you should note that solids can also meet the definitional requirements of a solution. Many alloys are mixtures of this sort. You may have an example in your pocket. A "nickel" is really a solution of nickel (about 25% by weight) in copper. Alas, the days when one could exchange five cents worth of this solid solution for 250 ml of the sudsy solid-liquid-gas solution just cited appear to be gone forever!

Molarity, molality, and more

Qualitatively, we speak of solutions as being *concentrated* or *dilute*. But the distinction is far too vague for scientific purposes. The standards for strength of coffee or sweetness of lemonade are too much a matter of personal preference. Therefore, more precise methods of expressing concentration have been devised. Suppose I introduce some of these with an example. Our starting materials will be 10.0 ml of water and 10.0 ml of ethyl alcohol, C_2H_5OH. The density of the former at room temperature (25°C) is very nearly 1.00 g/ml. Therefore, we can calculate or measure that the weight of the water must be

$$w_{H_2O} = 10.1 \text{ ml} \times 1.00 \text{ g/ml} = 10.0 \text{ g}.$$

The determination of the weight of 10.0 ml of alcohol is almost as easy, but we must use the density of C_2H_5OH, about 0.79 g/ml under similar conditions.

$$w_{C_2H_5OH} = 10.0 \text{ ml} \times 0.79 \text{ g/ml} = 7.9 \text{ g}$$

Our previous experience with matter suggests that we are well justified in anticipating no change in mass upon mixing the two compounds. The resulting solution will weigh 10.0 g + 7.9 g or 17.9 g. From this data we can readily calculate the *weight percent* composition. In other words, we can find the number of grams of alcohol present in 100.0 g of solution.

11.1 $\text{wt \% } C_2H_5OH = \dfrac{7.9 \text{ g } C_2H_5OH}{17.9 \text{ g solution}} \times 100 = 44.2\%$

The remainder is obviously water, so the wt % $H_2O = 100.0\% - 44.2\% = 55.8\%$.

Somewhat surprisingly, the calculation of the composition by volume is not quite as straightforward as you might expect. The complication arises because when 10.0 ml of water and 10.0 ml of ethyl alcohol are mixed, you do not get 20.0 ml of solution. Instead you get 19.3 ml. Unlike the masses, the volumes are not simply additive. Matter is conserved, but here volume appears to disappear. We explain the disappearance by saying that the molecules of H_2O and C_2H_5OH squeeze into a slightly smaller volume than they previously occupied. This in turn is a consequence of the internal structure of these two liquids and the solution formed from them. Not all pairs of liquids behave in this fashion. For some, volume increases are observed on mixing. For others, the solution volume is the sum of the volumes of the two pure liquids. But the water/ethyl alcohol example is an indication that the assumption of volume additivity has sufficient exceptions to make it inaccurate for exact work. Moreover, volume shrinkage or expansion can also characterize the dissolution of solid solutes in liquid solvents. So the best policy in making up solutions is to measure the volume of the solution formed, not the volume of the solvent added.

Concentration can then be based upon the volume of the solution. For example, we could calculate the *volume percent* of alcohol in our solution by determining the volume of C_2H_5OH that would go into making up 100.0 ml of the solution. Since every 19.3 ml of solution contains 10.0 ml of alcohol, we can write

11.2 $\text{vol \% } C_2H_5OH = \dfrac{10.0 \text{ ml } C_2H_5OH}{19.3 \text{ ml solution}} \times 100 = 51.8\%.$

A little arithmetic and a little reflection should convince you that this is not a very satisfactory indication of concentration. You could calculate the volume percent of H_2O in the same manner and come up with exactly the same answer, 51.8%. And 51.8% + 51.8% clearly don't add up to 100.0%. Yet, in spite of this rather obvious failing, volume percent is widely used in expressing concentrations—at least of alcohol/water mixtures. And another familiar unit has a similar basis. For what it's worth, in the United

States "proof" is defined as two times the percent of alcohol by volume.

Happily, scientists have devised concentration units that are somewhat more rational than proof and percent by volume. More-over, like weight percent, these units apply to solid solutes as well as liquid ones. The most widely used is called *molarity*. The molarity of a solution, abbreviated *M*, is defined as *the number of moles of solute per liter of solution*. For our example we first need to calculate the number of moles of ethyl alcohol in the solution sample of 19.3 ml. We know that only 7.9 g of that quantity of solution is ethyl alcohol. And we can readily ascertain from a table of atomic weights that the gram molecular weight of C_2H_5OH (i.e., the weight per mole) is 46.0 g. Thus, we can write

$$\text{Number of moles of } C_2H_5OH = \frac{7.9 \text{ g } C_2H_5OH}{46.0 \text{ g/mole } C_2H_5OH}$$
$$= 0.172 \text{ mole } C_2H_5OH.$$

To reiterate, this result means that there is 0.172 mole of alcohol in 19.3 ml of solution. But we are interested in the number of moles of C_2H_5OH per liter of solution. So we carry out an additional step to find that number. We convert the solution volume from 19.3 ml to 0.0193 liter and divide,

11.3 $\qquad \dfrac{0.172 \text{ mole } C_2H_5OH}{0.0193 \text{ liter of soln}} = \dfrac{8.91 \text{ moles } C_2H_5OH}{1.000 \text{ liter of soln}} = 8.91 \text{ } M.$

The answer is the concentration of ethyl alcohol expressed in moles per liter of solution. We say that it is "8.91 molar."

The procedure is identical for any solution. Simply calculate the number of moles of solute and divide it by the volume of solution (in liters) which contains that quantity of solute. To belabor a point which is sometimes forgotten by students, molarity is expressed relative to 1 liter of solution, not solvent. For example, in making up 1 liter of a 1.00 *M* solution of cane sugar (sucrose, $C_{12}H_{22}O_{11}$), we weigh 1 gram molecular weight of sugar, 342 g, into a 1 liter volumetric flask. Such a flask (one is pictured in Figure 11.1) carries around its long neck a thin etched line which exactly marks the desired volume. We add water or whatever other solvent we are using, swirling the flask to promote the solution of the sugar. Finally, we stop the addition when the level of the liquid reaches this line. Since the liquid is the solution and its volume is exactly 1.00 liter, we need not worry about or correct for any shrinkage or expansion during the process. If the work is done carefully, the result will be the desired 1.00 *M* solution.

Unfortunately, molarity has the same drawback which mars all units and measurements based upon volume. Its value depends upon temperature. This is because the volume of any substance

FIGURE 11.1.   *Procedure for making up a solution of known concentration (e.g., 1.00 M sucrose) using a volumetric flask.*

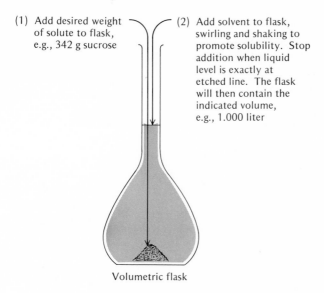

(1) Add desired weight of solute to flask, e.g., 342 g sucrose

(2) Add solvent to flask, swirling and shaking to promote solubility. Stop addition when liquid level is exactly at etched line. The flask will then contain the indicated volume, e.g., 1.000 liter

Volumetric flask

in any physical state varies with temperature. You recall from your study of Boyle's Law that all gases expand with increasing temperature at a fixed pressure. So do most liquids, but to a much lesser extent. Nevertheless, the effect is sufficiently great to detectably reduce the molarity of a solution when the temperature is raised by 10 or 20°. Thus, at body temperature, 37°C, the volume of the specific water–alcohol solution we have been considering will be slightly greater than the 19.3 ml we observed at 25°C. Naturally, the number of moles of alcohol (or water) does not change. In other words, the numerator of Equation 11.3 remains invariant while the denominator increases. As a result, the molarity of the solution must decrease. At 37°C, 1 liter of the solution will contain fewer moles of alcohol than the same volume did at 25°C.

Since most measurements of solution volumes are made at room temperature, this dependence of molarity upon temperature does not present much difficulty to a practicing chemist. However, it is possible to completely eliminate the problem by using another means of designating concentration—one based exclusively on weight. Because weight does not vary with changing temperature, concentration expressed in these terms will also be independent of temperature. The unit bears the name *molality* and a lower case *m* is its symbol. Although the similarity of the word and the symbol are confusingly close to molarity and *M*, there are some significant differences in definition. The molality of a solution is *the number of moles of solute per kilogram of solvent.* Note particularly the last three words; the weight of the solvent, not the volume of the solution, is the reference.

Aqueous alcohol again provides an example. Once more considering the ethyl alcohol to be the solute, we note that in forming the solution, 7.9 g or 0.172 mole of this compound was dissolved in 10.0 g of water. We are seeking the molality of that solution, the number of moles of C_2H_5OH which are associated with 1.000 kg of H_2O. To obtain the answer we need only convert 10.0 g of water to 0.0100 kg and divide this number into 0.172 mole of alcohol.

11.4
$$\frac{0.172 \text{ mole } C_2H_5OH}{0.0100 \text{ kg } H_2O} = \frac{17.2 \text{ moles } C_2H_5OH}{1.000 \text{ kg } H_2O} = 17.2 \, m$$

The solution is 17.2 molal in ethyl alcohol. And, because weight does not vary with changing temperature, this value is also independent of temperature and volume fluctuations.

Like likes like

One of the many uses of molality, molarity, and weight percent is to calculate the concentrations of saturated solutions. You know from your own experience that not all pairs of substances are mutually soluble in all proportions. To be sure, all gases are. Moreover, some pairs of liquids (like water and alcohol) also mix completely and homogeneously, no matter what their composition ratio. But it was observed long ago that "oil and water don't mix"—at least not very well. And certainly you have seen water solutions of sodium chloride or sugar which were saturated. So much solute had been added, perhaps by a heavy-handed cook, that the limit of its solubility in the solvent had been reached. And the excess undissolved solid solute remained in contact with the saturated solution. Undoubtedly you could observe no additional dissolution. But, in fact, an equilibrium had been achieved in which the rate of solution of salt or sugar just equaled the rate at which the crystalline solute separated out of solution.

If we pursue one of the examples just cited we can find in Lange's "Handbook of Chemistry"* that the solubility of sodium chloride at 20°C is 36.0 g NaCl per 100.0 g H_2O. Since the formula weight of NaCl is 58.4 g/mole, 36.0 g of the compound must represent 36.0 g/(58.4 g/mole) or 0.617 mole. This is the maximum quantity of salt which can dissolve in 100.0 g or 0.1000 kg of water

* Published by McGraw-Hill, New York.

at the indicated temperature. It follows that the molality of the saturated solution must be

$$\frac{0.617 \text{ mole NaCl}}{0.1000 \text{ kg H}_2\text{O}} = 6.17 \ m.$$

Of course, if we wished to check the tabulated solubility of sodium chloride we could prepare a saturated solution at the desired temperature, carefully separate it from any undissolved salt, and weigh out a sample of the solution in an evaporating dish or other convenient vessel. We could then evaporate the solution to dryness and weigh the residue of NaCl. The weight percent or molality of the saturated solution could easily be calculated from these data.

Similar solubility determinations indicate that all three of the compounds we have been considering readily dissolve in water. It is logical to conclude that this behavior must be a consequence of the composition of ethyl alcohol, sucrose, and sodium chloride—and the properties of water. At first glance, C_2H_5OH, $C_{12}H_{22}O_{11}$, and NaCl all seem to be quite different from one another. Nevertheless, there must be some unity underlying this apparent diversity, some commonality which these solutes share with water. Moreover, there must be some means to account for the fact that oil, benzene, sand, aluminum, and thousands of other substances are essentially insoluble in water.

The general dictum governing solubility is easily stated: "Like likes like." Substances which are similar mix homogeneously over a wide composition range. On the other hand, some sort of xenophobia keeps dissimilar compounds and elements separated. Since we have defined solutions as intimate mixtures of molecules, atoms, and/or ions, the explanation of solubility behavior must lie at this submicroscopic level. The key concept was introduced in the previous chapter. It is the nature of the forces existing between the ultimate identical particles which make up any pure substance. These interatomic, interionic, or intermolecular interactions must be overcome in both the solute and the solvent in order for a solution to form.

You will recall that these forces are weak in all gaseous substances. Indeed, gases are so much alike in this respect that their physical behavior can be quite adequately described by a single general equation, $PV = nRT$. Thus, the intermolecular forces existing in a mixture of nitrogen and oxygen are not significantly different from those which characterize either of the pure elements. Consequently, the gases are mutually soluble in all proportions. So are all other gases.

Liquid pairs are also completely miscible if they are constitutionally similar. Ethyl alcohol and water provide an excellent example. Both are polar substances exhibiting hydrogen bonding, dipole–dipole attraction, and dispersion forces. The dissolution of alcohol in water necessitates the disruption of some of these inter-actions among identical molecules. For example, hydrogen bonds in both pure alcohol and pure water must be broken. This requires energy. However, new hydrogen bonds can form between the C_2H_5OH and H_2O molecules with the liberation of energy. As a result, the net pattern of intermolecular interaction is not dimin-ished. In fact, the evolution of heat and the decrease in volume which attends this solution process indicates stronger intermolec-ular interaction in the solution than in the two pure component compounds.

On the other hand, water and an oil like *n*-decane, $C_{10}H_{22}$, have little in common. The latter compound consists of long-chain, zig-zag nonpolar molecules which attract each other by dispersion forces. These hydrocarbon molecules find water so distasteful that their interaction is often termed *hydrophobic*. Introduction of an H_2O molecule into such an oily medium would tend to disrupt these forces, and hence such an intrusion is energetically resisted. Similarly, the presence of a $C_{10}H_{22}$ molecule amid water molecules would break up the hydrogen-bonded system. This, too, would require considerable energy, and there would be no compensating energy available from the formation of new attractive interactions between solute and solvent molecules. For this reason, oil spills from tankers and underwater wells spread across harbors, lakes, and seas to leave their sludge and slime deposited on beaches and birds.

Generally speaking, the more oily a compound, the lower is its solubility in water. This is apparent in a series of similar sub-stances like alcohols. The general formula for simple members of this particular class of compounds is $C_nH_{2n+1}OH$. As we have seen, the OH group, which characterizes all alcohols, promotes aqueous solubility because it can form hydrogen bonds with the solvent. But as *n* and the length of the carbon chain increase, the hydro-phobic influence of that chain tends to predominate. The molecule becomes more and more difficult to accommodate within the water structure and its solubility decreases. As a matter of fact, many organic compounds have a good deal of hydrocarbon char-acter, which results in very low aqueous solubility. On the other hand, these same compounds dissolve readily in other water-hating substances. Commercial cleaning fluids are solvents of this sort. Carbon tetrachloride, CCl_4, or dichloroethane, $C_2H_4Cl_2$, can readily solubilize a grease spot or gravy stain that pure water cannot touch.

As you well know, there is a class of compounds which can be added to water to get rid of such oleaginous evidence of good but careless eating. I refer, of course, to soaps and other detergents. The topic has already been treated in Chapter 9, but it is worth reemphasizing that detergents are effective cleaning aids expressly because they combine oil solubility and water solubility in the same molecule or ion. One of their applications has been in dispersing the oil slicks just mentioned.

Predictably, the general rule that organic compounds are insoluble in water has its inevitable exceptions. Cane sugar is one of these. As one would expect, the explanation of its behavior lies in its structure. In a sense sucrose is "like" water. Figure 11.2 indicates that each molecule contains eight OH groups. Consequently, there are many opportunities for hydrogen-bonded interactions with H_2O molecules. The intermolecular hydrogen bonding, dipolar, and dispersion forces within a crystal of sugar can thus be easily overcome and compensated for in the presence of water.

The nature of the forces within a crystal of sodium chloride is, you will recall, somewhat different. The crystal consists of equal numbers of Na^+ and Cl^- ions distributed in a regular lattice. The oppositely charged ions attract each other with electrostatic forces which stabilize the solid structure. For many ionic compounds these forces can be quite readily overcome in an aqueous medium. The polar H_2O molecules surround the ions and tend to keep them separated. But the degree of water solubility does depend, to some extent, upon the magnitude of the interionic forces. Generally speaking, the greater the forces within the crystal, the lower the solubility of the substance. In some instances, silver chloride for example, interionic attractions are so strong that the compounds are only very slightly soluble in water. And crystals characterized by three-dimensional networks of covalent bonds are essentially insoluble. (Silicon dioxide or silica, SiO_2, is an example as widespread as the sand on the seashore.) Thus, solubility becomes another partial indicator of the strength of interparticulate interaction in the solid state.

FIGURE 11.2.   *Molecular structure of sucrose (cane sugar),*
$C_{12}H_{22}O_{11}$. Because of the large number of OH groups, the
molecule interacts readily with H_2O molecules, thus forming
concentrated solutions.

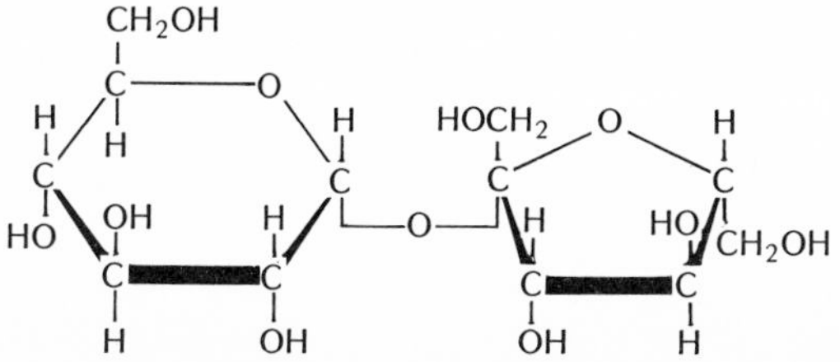

The familiar solution properties of salt and sugar also provide fairly typical illustrations of the dependence of solubility upon temperature. Your experience tells you that in order to dissolve more of either of these compounds in a saturated, aqueous solution you apply heat. In both cases, the solubility increases with increasing temperature, although the temperature dependence is significantly greater for sucrose than for sodium chloride. This correlation of "the higher the temperature, the higher the solubility" is so well known that you may be tempted to apply it to all solutions. Try to avoid the temptation; like many others, it will get you into trouble. To be sure, most solutions of solids in liquids do conform to this rule. During the formation of such solutions heat is absorbed. In some cases you can actually feel the cooling effect as heat flows from your hand to the solution.* Adding more heat merely helps the dissolving process along. But for quite a few pairs of substances, heat is evolved as the solute dissolves in the solvent. Again, this escaping heat can often be detected by touch. Heating such a system will actually lower the saturation solubility. This is just what happens in the case of calcium sulfate. The solubility of $CaSO_4$ in water decreases with increasing temperature. As a consequence, the solid solute may precipitate out of a hot solution. If this separation occurs in the tubes of a boiler or water heater, the resulting deposit of boiler scale can create a major impediment to liquid flow and heat transfer.

The solubilities of many, but not all, gases also exhibit an inverse dependence upon temperature. For example, increasing the temperature decreases the solubility of oxygen in water, an occurrence which can have major ecological impact. If you are not already aware of the fact, you will discover in Chapter 13 that large quantities of "waste" heat are given off by both fuel-burning and nuclear power plants. In most plants, this heat flows to a convenient stream, river, reservoir, or lake. Temperature increases of 5°C or more have been observed in such bodies of water. A change of this magnitude can reduce the solubility of oxygen to the point where the population of fish and other gill-breathing aquatic animals becomes endangered or altered. Moreover, the deleterious effect is aggravated because increasing the temperature of a cold-blooded animal speeds up its internal body chemistry and thus increases its oxygen requirement. Fortunately, not all instances of lowered gaseous solubility are of such consequence. Anyone who ever opened a lukewarm soft drink has experienced ample effervescent evidence that bubbles of carbon dioxide escape more readily from warm soda than from cold.

---

* Commercially available emergency cold packs make use of this phenomenon.

It appears that bubbles tickle the noses and the fancies of people of widely divergent tastes and incomes. Hence, not only soft drinks, but beer and champagne are also "carbonated." "Pop" and most beers are bottled under a pressure of carbon dioxide which is somewhat greater than 1 atm. By contrast, the partial pressure of this gas in ordinary air is less than 0.001 atm. Since the solubility of CO_2 in water is directly proportional to the pressure of the gas above the liquid, considerably more of the gas dissolves in the beer or soda than would dissolve from the air. On the other hand, in the traditional method of making champagne, the increased carbon dioxide concentration arises from the CO_2 generated during a second fermentation of the wine in the sealed bottle. Again, the pressure of the gas in the bottle is significantly higher than the atmospheric value. And, in all instances, the concentration of the dissolved carbon dioxide is greater than it would be under 1 atm of air. Therefore, when the bottle is uncapped or uncorked and its contents exposed to just such a reduced opposing pressure of CO_2, the solubility drops suddenly. The now-insoluble gas collects into bubbles which rise to the surface of the liquid, giving the beverage its "sparkling" quality.

The same phenomenon is used to whip cream of both the edible and the shaving varieties. Once again, a gas, often nitrous oxide or "laughing gas" (N_2O), is dissolved under pressure in the cream or the soap solution. This pressure is maintained by a sealed aerosol can. However, when the valve of the can is opened, the pressure is reduced, the solubility of the gas decreases, and the escaping bubbles whip the cream or the lather and carry it out the nozzle of the can.

While bubbles in the wine may be amusing and bubbles in the whipped cream may be convenient, bubbles in the blood most definitely are neither. Yet, they can form as a consequence of the same general behavior: the fact that the solubility of all gases in all liquids decreases with the decreasing pressure of that gas in contact with the liquid. The painful result of such concentration changes within the body is called the "bends." Of course, the potential problem is created by the increase in solubility which attends elevated pressures. A submerged deep-sea diver is obviously under a pressure substantially greater than 1 atm. Therefore, more air will dissolve in his blood and body fluids than is normal at sea level. This condition may not create any adverse physiological effects in itself, at least not over a reasonable length of time. But when the diver begins to ascend and the pressure on his body and blood starts to decrease, an excess of insoluble gases separate from solution. Bubbles can form and block capillaries, causing impaired circulation and other traumas. The more rapid the ascent, the more severe the bends. Fortunately, the danger to divers can be reduced if they are supplied with a mixture of oxygen and helium instead of ordinary air. Helium is much less soluble in water and blood than is nitrogen. Thus, a smaller quantity of helium dissolves under the high pressure of deep submersion. This in turn means that less gas escapes from solution on ascent, and bends-producing bubbles are less likely to form.

Of antifreeze and osmosis

It is experientially evident that the chemical and physical properties of a solution differ from those of either the pure solvent or the pure solute. Thus, the characteristics of "sugar water" are obviously not identical with those of its constituent components, in spite of certain similarities. Furthermore, the values of many properties, including vapor pressure, freezing point, boiling point, and osmotic pressure also depend upon the concentration of the solution. To be sure, if the chemical composition and concentration of a solution are fixed, its properties will be predictably constant and reproducible. All 1.00 molal solutions of cane sugar in water are identical. But such a solution does not have the same boiling point or freezing point as a $0.50\,m$ or $1.50\,m$ sucrose/water mixture. Indeed, values for these and other properties vary over the entire concentration range accessible to the solute/solvent pair. The quantitative relationships governing dependencies of this sort form the subject of this section.

The general trends can be easily summarized. *Any solution of a nonvolatile solute* (solid or liquid) *in a liquid solvent will boil at a higher temperature than the pure solvent.* Furthermore, *its freezing point will be lower.* And finally, *it will exhibit the phenomenon of osmosis* when separated from a sample of pure solvent by a membrane which is impermeable to the solute. For all solute/solvent combinations, the extent of the elevation of the boiling point and the depression of the freezing point (relative to the values characteristic of the pure solvent) and the magnitude of the osmotic pressure are directly proportional to the concentration of dissolved solute particles in solution. The greater the concentration, the larger the values exhibited by these *colligative* properties.

We begin our survey of colligative properties by comparing the vapor pressure of a solution with that of the pure solvent. You will recall that the magnitude of the vapor pressure is an indication of the tendency of a liquid to vaporize. Highly volatile liquids which evaporate readily have high vapor pressures even at room temperature. Now quite obviously, a pure liquid consists of only one substance and one sort of molecule. It is also trivially apparent that if any solute is dissolved in the liquid, the concentration of the solvent will be less than 100% by weight. For simplicity's sake, let us assume the solute is a nonelectrolyte, a compound which does not dissociate into ions upon dissolution. Moreover, we choose a solute which does not evaporate. Clearly, the molecules of the solute are part of the solution. It follows that the solvent molecules now represent only a fraction of the total molecules present in solution—a fraction which is less than 1. As the concentration of the solute increases, the fraction of the molecules which are solvent must decrease.

You will recall from Chapter 10 that vapor pressure implies a state of equilibrium between liquid and gaseous phases. Therefore, it is not unreasonable to assume that the escaping tendency of solvent molecules from the liquid to the gas phase depends upon the concentration of solvent in the solution. In fact, it is postulated that for ideal solutions, *the vapor pressure of the solvent above the solution is directly proportional to the fraction of solvent molecules in the bulk solution* (see Figure 11.3). Consequently, solutions which contain relatively more solute (and hence relatively less solvent) exhibit lower vapor pressures than do more dilute solutions. Experiment shows that many solutions—particularly at low concentrations—do indeed conform to this ideal behavior.

This depression of the vapor pressure manifests itself most

FIGURE 11.3.   *Schematic representation of the vapor pressure of a volatile solvent above (a) the pure solvent, ○, and (b) a solution of a nonvolatile solute, ●, in the solvent. Note the concentration of solvent molecules in the vapor phase (and hence the vapor pressure) is lower over the solution than over the solvent.*

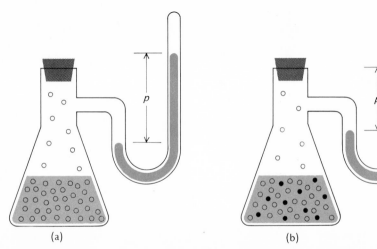

(a)   (b)

clearly in an *elevation of the boiling point.* The boiling point is that temperature at which the vapor pressure of a liquid equals the atmospheric pressure. But at any given temperature, the vapor pressure of a solution is less than that of the pure solvent. Therefore, it is necessary to heat a solution to a higher temperature before its vapor pressure will equal 760 mm Hg or any other fixed value. In other words, the solution boils at a higher temperature than the solvent.

The quantitative dependence of this increase in boiling point upon concentration is simply stated in Equation 11.5:

11.5 $\Delta T_b = k_b m.$

Here ΔT_b is the boiling-point elevation expressed in °C, m is the molality of the solute in the solution, and k_b is the boiling-point constant. The numerical value of this constant of proportionality depends upon the solvent. Therefore, in applying the equation it is important to select the appropriate value for k_b. For water it is 0.52; for most organic solvents it is considerably larger.

On the other hand, Equation 11.5 does not require the identification of the solute. Only its molality appears. The equation is perfectly general for all nonvolatile, nonelectrolyte solutes. Thus, at 1 atm pressure, 1.00 m aqueous solutions of sucrose and glucose should both boil at 100.00 + 0.52°C or 100.52°C. The fact that the experimentally determined value is not exactly the calculated one is a consequence of deviations from ideal behavior which need not concern us further.* What is of interest here is the observation that 1.00 m solutions of the two sugars do indeed boil at essentially the same temperature, in spite of the fact that the molecular weight of sucrose is about twice that of glucose, and thus the weight percent concentration of the former solution is twice as great as the latter. The concentration of solute particles is the same, and that is what counts in colligative properties.

Note that I specified solute *particles,* not solute molecules. To be sure, in the case of these two sugars, I could have used the words interchangeably. But not so for an electrolyte, a compound which causes a solution to conduct electricity. A very common substance of this sort is sodium chloride. According to Equation 11.5, we would predict the boiling point of a 1.00 m solution of NaCl in water to be about 100.5°. But experiment shows that the true boiling point is closer to 101°C. This inconsistency between theory and experiment casts considerable doubt upon the former, until we remember why salt water conducts electricity. The NaCl

---

* Equation 11.5 applies most accurately to very dilute solutions.

dissociates into two sorts of ions, Na^+ and Cl^-. The number of solute particles in a 1.00 m solution of sodium chloride is twice that in a 1.00 m solution of sugar. Therefore, we simply modify Equation 11.5 by introducing a 2.

$$\Delta T_b = 2k_b m$$

Or, to generalize the relationship for any electrolyte which dissociates into n ions,

11.6 $$\Delta T_b = n k_b m.$$

Thus, for $CaCl_2$ which yields 1 mole of Ca^{2+} ions and 2 moles of Cl^- ions for each mole of the parent compound, n equals 3. Similarly, $n = 4$ for $AlCl_3$, and so on. Again, the modified equation is most exactly applicable to dilute solutions.

The other restriction initially placed upon the solute was that it be nonvolatile. A solute which does not readily evaporate cannot contribute positively to the total vapor pressure above a solution; it only serves to reduce its value below that of the pure solvent. Consequently, the boiling point is elevated as indicated in Figure 11.4a. If, however, the solute is itself volatile, molecules of that component will also tend to escape into the gaseous phase. The

FIGURE 11.4.   *The dependence of the boiling point of a solution upon the concentration of the solute. (a) A solution containing a nonvolatile solute, e.g., sugar in water. (b) A solution containing a solute more volatile than the solvent, e.g., ethyl alcohol in water.*

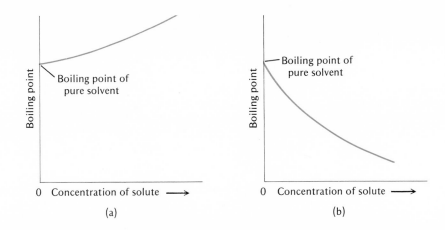

vapor above such a solution thus contains both solute and solvent molecules. And the total pressure of the vapor is the sum of the partial pressures of the two components. This total value can, in fact, exceed the vapor pressure of the pure solvent. As a result, the boiling point of the solution is depressed in a manner illustrated in Figure 11.4b.

Such behavior characterizes water/alcohol solutions. Ethyl alcohol, C_2H_5OH, with a boiling point of 78.3°C has a higher vapor pressure than water at any reference temperature. Over most of the concentration range, solutions of alcohol and water exhibit vapor pressures which are greater than that of pure water, but less than that of pure alcohol. The boiling points of these solutions decrease with increasing concentration of ethyl alcohol. And because the alcohol is the more volatile component, it escapes more readily into the vapor phase than does the water. This makes possible the distillation of alcohol from relatively dilute alcohol/water solutions. The solution is boiled and the vapor, which is richer in alcohol than is the liquid in equilibrium with it, is condensed and caught. The concentration of ethyl alcohol in this condensate or distillate can, in fact, be considerably greater than that in the residual solution remaining in the still pot. By repeating such a

FIGURE 11.5.   *A device for the separation of two volatile liquids present in the same solution, by means of distillation. The more volatile component is concentrated in the vapor and the distillate.*

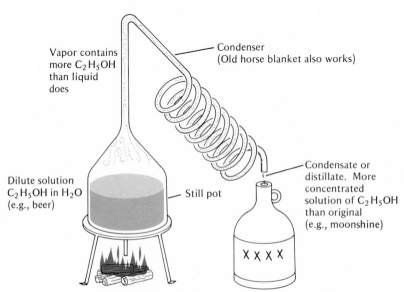

Vapor contains more C_2H_5OH than liquid does

Condenser (Old horse blanket also works)

Dilute solution C_2H_5OH in H_2O (e.g., beer)

Still pot

Condensate or distillate. More concentrated solution of C_2H_5OH than original (e.g., moonshine)

X X X X

process or using specially designed stills, one can almost completely separate the alcohol and water of the original solution.*

Volatility of the solute creates considerably fewer complications for the *depression of the freezing point*. Here the equilibrium exists between pure solid solvent and the liquid solvent in solution. But the presence of the solute in the liquid solution does lower the freezing point below the value characteristic of the pure solvent. The mathematical form of this relationship is directly analogous to Equation 11.5:

11.7 $\Delta T_f = k_f m.$

The only differences lie in the fact that ΔT_f represents the freezing-point depression and k_f is a constant, again characteristic of the solvent, but different in magnitude from k_b. Without exception, the freezing-point constant is larger than the corresponding boiling-point constant. Water, for example, has a k_f of 1.86.

The practical implications and applications of freezing-point depression are many and well known. Most familiar, perhaps, is the use of solutions of ethylene glycol and water as automobile antifreeze. Ethylene glycol, the chief component of essentially all commercial antifreeze liquids, has the formula $HOCH_2CH_2OH$ and a molecular weight of 62.1. Its properties are similar enough to those of H_2O that the two compounds mix completely in all proportions. One can readily calculate that an aqueous solution containing 50.0% of ethylene glycol, by weight, must have a concentration of 16.1 m. Substituting this value into Equation 11.7, we at once determine that theoretically, the freezing point of such a solution should be depressed below 0°C by

$$\Delta T_f = 1.86 \times 16.1\ m = 29.9°C.$$

This translates into a freezing point of $-29.9°C$ or $-21.9°F$. At such high concentrations, the equation loses its accuracy and the experimentally observed value is $-33.8°C$. Nevertheless, the effect is evident and welcome to anyone who has ever endured a Minnesota winter. Because the density of the ethylene glycol/water mixture increases with the concentration of the former compound, the degree of antifreeze protection provided by a given solution can be estimated by measuring its density with a hydrometer.

Winter also provides another opportunity for utilizing the phe-

---

* The "almost" appears in this sentence because water and ethyl alcohol form an *azeotrope*, a solution for which the vapor and liquid phase are identical in concentration. This identity means that it is impossible to further fractionate an azeotrope by additional distillation. For the alcohol–water system, the azeotropic solution contains 95% alcohol by volume, that is, it is 190 proof.

nomenon of freezing-point lowering. As you know, sodium chloride, calcium chloride, and other compounds are often spread upon icy streets and roads. The salts depress the freezing point of water, causing melting. The effectiveness of these compounds is particularly high because they are electrolytes which conform to a modified version of 11.7,

11.8 $\Delta T_f = nk_f m.$

The integer n has its former significance, designating the number of ions formed per formula of the solute compound. A more pleasant and less corrosive use of salt to melt ice occurs in an ice cream freezer. As the ice melts, it absorbs heat which goes to increase the energy of its molecules. Most of this heat comes from the ice cream mixture in the inner can of the freezer. Consequently, the temperature of this mixture drops. Finally, it freezes, though it, too, is a solution with a freezing point well below zero. The freezing point of salt water, however, is still lower. For example, an aqueous solution containing 23% NaCl freezes at $-21°C$.

This same colligative property has recently made possible the successful freezing of whole blood, sperm, and other biological fluids. The expansion of freezing H_2O results in the destructive rupture of cells, as well as engine blocks and radiators. But when blood is mixed with a liquid solute, such as ethylene glycol, ice crystals never form. Instead, the solution merely gets thick and glassy. The fluid can thus be frozen, stored, and subsequently melted without significant cellular disruption. Of course, it is necessary to remove the "antifreeze" before the melted blood can be transfused, but the necessary techniques have been mastered.

But automobile owners, ice cream aficionados, and recipients of banked blood are not the only individuals who have benefited from the fact that a solution freezes at a temperature lower than the pure solvent. In the area of scientific research, freezing-point depression has proved its utility for the determination of molecular weights. Equation 11.7 obviously permits the calculation of the molality of a solution providing that ΔT_f can be measured and k_f is known. If one selects a solvent with a high value for this constant, the freezing point depression can be large enough to permit its determination with a high degree of accuracy, even if the solution is quite dilute. For example, k_f for cyclohexane, C_6H_{12}, has a value of 20.2. Suppose a chemist dissolves 2.40 g of a newly synthesized organic compound in 100.0 g of cyclohexane. He then measures the freezing point of this solution and finds it to be 0.72°C. Since the freezing point of pure cyclohexane is 6.50°C, the freezing point depression must be 6.50° − 0.72° or 5.78°. From this value for ΔT_f and the freezing point constant, it is easy to determine the molality of the solution.

$$m = \frac{\Delta T_f}{k_f} = \frac{5.78°}{20.2} = 0.286$$

In other words, 0.286 mole of the new compound are associated, in solution, with 1 kg of cyclohexane. The chemist also knows the concentration of the solution by weight. After all, he made it up to contain 2.40 g of solute and 100.0 g of solvent. It follows that in a solution of this composition, 24.0 g of the newly prepared substance are associated with each 1000 g or each kilogram of cyclohexane. The ratio is the same. Note that the chemist now has two different expressions for the quantity of solute, relative to 1.000 kg of solvent: 24.0 g and 0.286 mole. Quite clearly, these two figures represent the same amount of stuff; 0.286 mole weighs 24.0 g and 24.0 g correspond to 0.286 mole. To determine the molecular weight, the chemist needs only to set up a ratio,

$$\frac{24.0 \text{ g}}{0.286 \text{ mole}} = \frac{x \text{ g}}{1 \text{ mole}}.$$

Solving for x, he obtains the weight of a single mole.

$$x = \frac{24.0 \text{ g}}{0.286 \text{ mole}} = 84.0 \text{ g/mole}$$

In a somewhat similar fashion, measurements of osmotic pressure can also yield molecular weights. But before we consider that specific application, it might be well to look at osmosis from a more personal perspective. After all, the phenomenon is one of fundamental internal importance. The diffusion of various solution components through cellular membranes is an essential part of the transport of nutrients and wastes in all living organisms. This migration is often the result of osmosis. It can be readily demonstrated by placing red blood cells (erythrocytes) in water. Under a microscope, the erythrocytes are seen to swell until they burst. Quite clearly, water has entered the cells. This inward flow is attributed to the fact that an erythrocyte contains a relatively high concentration of dissolved compounds, particularly the red oxygen-carrying protein, hemoglobin. In contrast to this, only pure water is present on the outside. Obviously, the concentration of the various solutes is higher inside the cell than in its surroundings. And the water concentration is considerably greater *outside* the erythrocyte.

It turns out that there is a natural tendency for adjacent portions of a solution to equalize their concentrations. Thus, if some concentrated sugar water is placed in direct contact with a more dilute solution of the same solute and solvent, the two solutions will tend to mix spontaneously. Water will diffuse from the region of its higher concentration to where its concentration is lower. And sugar will move in the opposite direction, but again from higher concentration to lower. Without stirring, this process may take some time because molecular diffusion can be quite slow. Eventually, however, the composition should become everywhere uniform at a concentration somewhere between the two original values.

As we have noted, in the case of a red cell immersed in distilled water, large concentration differences do exist. And the tendency to equalize these differences is again operative. But because of the presence of the cell membrane, thorough mixing is impossible. The membrane permits water to pass into the cell, but prevents the solute molecules from diffusing out into the surrounding liquid. In a word, it is *semipermeable*. The net result is *osmosis*, the movement of the solvent, here water, through such a membrane from a region of higher solvent concentration to one of lower concentration. This influx of water into the red cell usually continues until the cell wall is stretched to its maximum and bursts.

Osmosis does not always lead to such violent rupture. The process will stop of itself when the internal pressure on the solution side of the membrane is sufficiently great to equal the driving force per unit area with which the solvent migrates through the membrane. This opposing pressure is called the *osmotic pressure*, and it is schematically represented in Figure 11.6. Its quantitative dependency assumes a form already familiar in the ideal gas law,

11.9 $$\pi = \frac{nRT}{V}.$$

Here π designates the osmotic pressure, n is the number of moles of nondiffusing solute in V liters of solution, R is the gas constant (0.0821 liter atm/mole °K if π is to be expressed in atmospheres), and T is the absolute temperature in °K. You recognize n/V to be nothing other than the molarity of the solution, so we can rewrite 11.9 as

11.10 $$\pi = MRT.$$

Now we can use this equation to calculate the osmotic pressure arising from the dissolved hemoglobin in a red blood cell. The internal cell fluid contains about 4.7×10^{-3} mole of this protein per liter of solution. We couple this value of M with the gas constant and 310°K, the absolute equivalent of the normal body temperature of 37°C.

$$\pi = 4.7 \times 10^{-3} \frac{\text{mole}}{\text{liter}} \times 0.0821 \frac{\text{liter atm}}{\text{mole °K}} \times 310°K$$

$$= 0.12 \text{ atm}$$

FIGURE 11.6. *Osmosis. (a) Nonequilibrium condition. Higher concentration of solvent on left side of membrane causes it to diffuse into solution on right. (b) Equilibrium conditions. Osmosis halted by osmotic pressure* π.

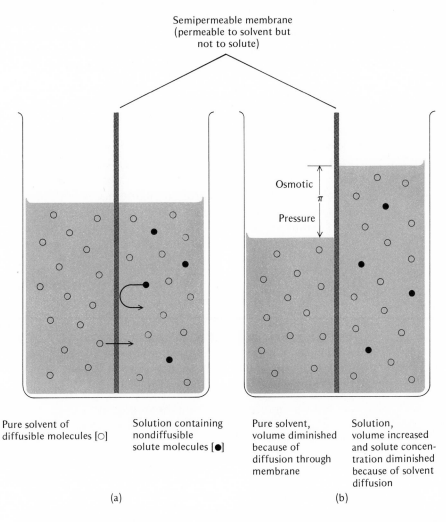

Semipermeable membrane
(permeable to solvent but
not to solute)

Osmotic
π
Pressure

Pure solvent of
diffusible molecules [○]

Solution containing
nondiffusible
solute molecules [●]

Pure solvent,
volume diminished
because of
diffusion through
membrane

Solution,
volume increased
and solute concen-
tration diminished
because of solvent
diffusion

(a)   (b)

An excess internal pressure of 0.12 atm would be required to pre-vent the diffusion of water into an erythrocyte placed in pure H_2O. Or, if you prefer, distilled water enters such a cell with a pressure of 0.12 atm. Since the inside and outside of a cell are both exposed to the pressure of the atmosphere, π is really an additional pressure.

In blood, of course, the suspending liquid is not pure water. It is plasma, a solution of many substances—including electrolytes, proteins, and sugars. Many of these cannot pass through the barrier of the cell membrane, although certain exchanges of material do occur. The net result is a balance between the inward and outward osmotic forces. It should now be evident why the concentration of solutions used for intravenous feeding and injections should be isotonic with plasma, that is, having the some osmotic pressure. If the concentration of nondiffusible solute molecules is lower in the solution than in the cells, an inward aqueous immigration will occur and the cells will swell. On the other hand, if the extra-erythrocyte concentration is greater than the internal level, the outward exodus of water will cause the cells to shrink. It is far preferable to have blood cells which do not resemble either balloons or prunes.

Returning briefly to the matter of molecular weights, we can readily show how Equation 11.10 provides a method for such determinations. Suppose we prepare a water solution of hemoglobin containing 12.5 g of the protein pigment per liter of solution and measure its osmotic pressure against pure water. We find a value of 3.50 mm Hg at 27°C (300°K). Substituting these numbers into the rearranged equation, we calculate the molarity of the solution.*

$$M = \frac{\pi}{RT} = \frac{3.50 \text{ mm Hg}}{62.3 \text{ liter mm Hg/mole } °K \times 300°K}$$
$$= 1.87 \times 10^{-4} \text{ mole hemoglobin/liter solution}$$

This number of moles of hemoglobin must weigh 12.5 g, since the solution was made up to contain that weight of the solute per liter of solution. It follows that the molecular weight must be x in this proportion:

$$\frac{12.5 \text{ g hemoglobin}}{1.87 \times 10^{-4} \text{ mole}} = \frac{x \text{ g hemoglobin}}{1 \text{ mole}}$$
$$x = \frac{12.5 \text{ g} \times 1 \text{ mole}}{1.87 \times 10^{-4} \text{ mole}} = 6.6 \times 10^{4} \text{ g/mole}.$$

A molecular weight of 66,000 is prodigiously large, but not at all unusual for polymeric proteins. The complexity of the molecules which make up our bodies is astounding. And it should be a source of justifiable pride that our understanding of colligative properties and other chemical phenomena has led to detailed knowledge of the structure and function of such significant substances.

---

* Note that because π is expressed in mm Hg, we must use a value of R which involves the same pressure unit.

An optional aside on acidity

One of the most useful concepts in solution chemistry is that of pH. These two letters, shorthand for the French *puissance d'hydrogen* ("strength or degree of hydrogen") are an indication of acidic strength. Generally speaking, an acid is defined as a substance which liberates protons or hydrogen ions, H^+. Thus, in aqueous solution, hydrogen chloride, HCl, dissociates into H^+ and Cl^- ions. The result is hydrochloric acid. The hydrogen ions impart to the solution the properties identified with acids, for example, the ability to turn litmus (a vegetable dye) from blue to red and the characteristic sour taste of lemon juice and vinegar. But because acids also attack skin and many other surfaces, tasting is not generally recommended. They also react with many metals to generate hydrogen gas.

$$Zn + 2HCl \rightarrow ZnCl_2 + H_2$$

And with carbonates, acids cause the effervescent evolution of carbon dioxide.

$$CaCO_3 + 2HCl \rightarrow CaCl_2 + H_2O + CO_2$$

In such reactions, the common significant ionic species is H^+ or, more precisely, the hydronium ion, H_3O^+, formed when a proton bonds to a water molecule. Therefore, the concentration of hydrogen or hydronium ions in solution must be a direct measure of the solution's acidic strength.

In plain distilled water, this concentration is very low, $1.0 \times 10^{-7} M$ at room temperature. Moreover, there is an identical concentration of hydroxyl ions, OH^-. This latter species is responsible for imparting basic or alkaline properties to solutions. Because of the presence of the hydroxyl ion, bases taste bitter, feel soapy, irritate tissues, and turn litmus blue. Furthermore, bases react with acids to form water:

$$H^+ + OH^- \rightarrow H_2O.$$

It is therefore not surprising that the concentration of hydrogen ions in an aqueous solution bears a relationship to the concentration of hydroxyl ions. At room temperature, the product of the molarity of H^+ and the molarity of OH^- equals 1.0×10^{-14},

$$M_{H^+} \times M_{OH^-} = 1.0 \times 10^{-14}.$$

Any solution with an H^+ concentration greater than $1.0 \times 10^{-7} M$ will have an OH^- concentration which is correspondingly smaller, and hence will be acidic. On the other hand, if the molarity of H^+ is less than 1.0×10^{-7} it will be exceeded by that of OH^- and the solution will be basic.

TABLE 11.1.   *The pH Scale*

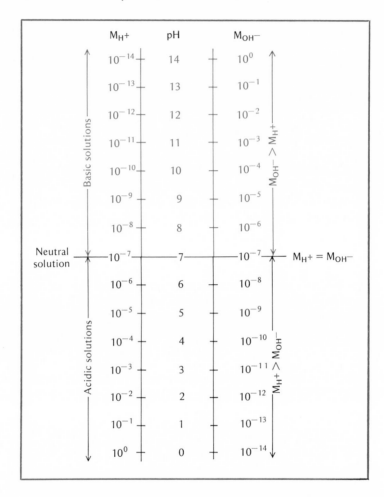

The idea of pH was introduced early in this century by a Danish biochemist, S. P. L. Sørensen, while working on problems associated with the brewing of beer. Sørensen's defining equation indicates that pH is, in fact, another way of expressing the hydrogen ion concentration.

$$pH = - \log M_{H^+}$$

For example, since the H⁺ concentration of pure water is 1.0×10^{-7} M, its pH is easily calculated as follows:

$$pH = - \log(1.0 \times 10^{-7}) = - (-7.0) = 7.0.$$

Such a solution is neither acidic nor basic, but perfectly neutral. An acidic solution, however, will have an H⁺ concentration greater than 1.0×10^{-7} M, and consequently its pH will be less than

7.0. Thus, for 0.01 M HCl, which completely dissociates into ions, $M_{H^+} = 0.010$ or 1.0×10^{-2}. Therefore,

$$pH = - \log(1.0 \times 10^{-2}) = - (-2.0) = 2.0.$$

The more acidic the solution, the lower will be its pH. Conversely, a basic solution will have an H^+ concentration lower than 1.0×10^{-7} M and a pH which is greater than 7.0. The higher the pH, the more strongly basic the solution. In aqueous solutions, the concept of pH is applied over such a concentration range that the values it assumes typically range from 0 to 14. Of course, non-whole-number values for pH arise whenever the hydrogen ion concentration is not simply a whole-number power of 10. Table 11.1 schematically represents the pH scale.

Clean air smells funny

Having acquired some familiarity with the properties of solutions and the principles governing them, we are now ready to attempt the application of this knowledge to problems of greater complexity. There is no need to waste space arguing for the relevance of solution chemistry to the task of cleaning up the environment—examples are far more eloquent. And they are legion. Indeed, there are few aspects of the physical pollution of our planet which do not directly involve solutions. The chief reason, of course, is that we live literally surrounded by two vast solutions—the atmosphere and the oceans.

You are already well aware that the chief components in air are nitrogen and oxygen. At sea level, the latter element constitutes 20.94% of the atmosphere. (Other components are listed in Table 11.2.) But at higher altitudes, air contains relatively less oxygen. Of course, the absolute concentration of all gases is less on a mountain top than at the seashore. We have noted earlier that the pull of gravity gives rise to this dependence upon altitude. But not all gases are equally affected. Because the molecular weight of oxygen, 32, is greater than that of nitrogen, 28, the former element is more strongly attracted toward the surface of the earth. Consequently, the concentration of oxygen decreases more rapidly with increasing elevation than does the concentration of nitrogen.

Differences in gas density also influence the distribution of atmospheric contaminants. Normally, the tendency for gases to mix and interdiffuse is aided by winds and the natural thermal gradient. The usual situation is for the temperature of the atmosphere to decrease with increasing altitude. The lower layers of air, warmed by the surface of the earth, are less dense than the cooler, upper layers. Therefore, the former rise and the latter

TABLE 11.2.   Composition of Clean, Dry Air near Sea Level[a]

Substance	Percent
Nitrogen	78.09
Oxygen	20.94
Argon	0.93
Carbon dioxide	0.032
Other noble gases	0.0024
Methane	15×10^{-5}
Hydrogen	5×10^{-5}
Nitrogen oxides	3×10^{-5}
Carbon monoxide	1×10^{-5}
Ozone	2×10^{-6}
Sulfur dioxide	2×10^{-8}

[a] Source of data: "Cleaning Our Environment: The Chemical Basis for Action," p. 24. Copyright 1969 by the American Chemical Society, Washington, D. C. Reprinted with permission.

sink. As the temperature of the new surface air increases, it expands and begins to rise. Meanwhile the newly elevated air is cooling, contracting, and becoming more dense. Thus, the cyclic mixing process is repeated over and over again. However, in some parts of the world—notably Southern California—meteorological and geographical conditions are such that *temperature inversions* frequently occur. In such cases, the temperature of the upper air is higher than that at the earth's surface. Since the less dense air is already at the top and the more dense air at the bottom, there is little tendency for mixing to take place. Noxious gases, generated at or near ground level are trapped there and a blanket of smog develops.

Among the nastiest compounds found in smog are the two oxides of sulfur, sulfur dioxide, SO_2, and sulfur trioxide, SO_3. To be sure, these substances have been released to the atmosphere for eons by volcanos. But the chief offender is a consequence of man's need for power and the fossil fuels from which he extracts his energy. Many deposits of coal and petroleum contain significantly high quantities of sulfur; 1% by weight is regarded as a low concentration. During the burning of the fuel, the sulfur also undergoes combustion, first oxidizing to form sulfur dioxide.

$$S + O_2 \rightarrow SO_2$$

This gaseous compound can, in turn, react with atmospheric oxygen to form the higher oxide.

$$SO_2 + \tfrac{1}{2} O_2 \rightarrow SO_3$$

When the SO_3 combines with water vapor it yields tiny droplets of sulfuric acid, H_2SO_4.

$$SO_3 + H_2O \rightarrow H_2SO_4$$

Although sulfuric acid is one of the most important industrial chemicals, it is extremely unpleasant stuff, particularly when suspended in the air. It attacks iron, steel, stone, concrete, skin, lungs, and a lot of other exposed surfaces.

An obvious solution to the problem, banning all sulfur-containing fuels, is easier to propose than to put into operation. It is true that prohibiting the long-established practice of burning soft coal in fireplaces did much to clean up London's notorious pea soup fogs and slow down the gradual dissolving of St. Paul's Cathedral. But it is probably impossible to restrict our total fuel consumption to coal and oil with naturally low sulfur contents. Fortunately, methods have been devised for removing the element from petroleum. Of course, such treatment adds to the cost of the refining process, the product, and ultimately the energy derived from it.

An alternative approach is to trap the SO_2 and SO_3 before they escape into the atmosphere. The effluent gas emerging from the fire box is passed through a "scrubber" where an alkaline solution reacts with the acidic oxides to form nonvolatile compounds. These products may even be industrially valuable, a liability thus having been converted into an asset. After all, there is no reason why the profit motive cannot be enlisted on the side of cleaning up the environment instead of assuming its historically more common opposite alignment.

Part of the extreme difficulty in solving problems of air pollution is created by the complex chemical interrelationships of the various contaminants. The system we have just studied is a case in point. The oxidation of SO_2 to the more corrosive SO_3 is promoted by oxides of nitrogen, specifically, nitrogen dioxide, NO_2.

$$SO_2 + NO_2 \rightarrow SO_3 + NO$$

And this is by no means the only instance of the involvement of NO_2 in aggravating atmospheric pollution. The compound is also a chief culprit in initiating the formation of *photochemical smog*. The word "photochemical" reflects the role of sunlight, in the ultraviolet range, which triggers the dissociation of NO_2,

$$NO_2 \rightarrow NO + O.$$

Note the O among the products. It indicates that the oxygen formed in this decomposition exists as individual atoms, not the more familiar O_2 molecules. Because atoms of oxygen are far more reactive than molecules, they readily become involved in dozens of different reactions. Certain of these involve hydrocarbon pollutants from the incomplete combustion of gasoline and other fuels. Under the influence of nitrogen dioxide and atomic oxygen, these compounds are often converted into substances which are particularly irritating to eyes and mucous membrane. And when a thermal inversion traps the air above a city and transforms it into a gas-phase reactor, the choking consequences grow as the chemistry continues.

Of course nature is neutral (although most scientists are not). The challenge is to channel chemistry to contribute toward correcting the problem rather than compounding it. One way to do this is to sever the smog-producing chain by breaking one of its links. Perhaps the strongest link involves the production of nitrogen oxides. Nitric oxide, NO, is formed by the direct combination of elementary N_2 and O_2.

$$N_2 + O_2 \rightarrow 2NO$$

This reaction requires energy which, in nature, is often supplied by lightning. The generation of atmospheric oxides of nitrogen is thus not an exclusive consequence of modern technology. Nor do they always exert a negative influence upon the environment. As a matter of fact, these compounds play an essential role in the nitrogen balance of the earth. Nitrogen dioxide, which we earlier identified as a smog-triggering villain, is again involved. It is formed, in the air, by the further oxidation of NO.

$$NO + \tfrac{1}{2}O_2 \rightarrow NO_2$$

Additional reactions with oxygen and water lead to the production of nitric acid, HNO_3, which is carried to the earth's surface in the rain. There, the acid enters the soil to form compounds called nitrates, essential constituents for field fertility. Of course, too much atmospheric nitric acid can create many of the corrosive problems attributed earlier to sulfuric acid. Whether or not a particular chemical is a pollutant or an essential trace ingredient in the ecology is thus, to a large extent, a question of concentration.

The natural, necessary level of many compounds has been altered by the inventions of man. In the case of NO and NO_2, a major source of the concentration increase has been the internal combustion engine. Nitrogen and oxygen are always present in air, and the energy which promotes their combination to form NO can be supplied in the motor of a car. Ironically, efforts to improve the efficiency of gasoline combustion aggrevate the problem. Thus, the introduction of more air into an engine does attack one link of the smog chain by promoting more complete conversion of the fuel to CO_2 and H_2O. Fewer unburned hydrocarbon fragments escape in the exhaust. But because higher concentrations of nitrogen and oxygen are present in the cylinder, more NO is formed. True, the extent of this reaction could be reduced by decreasing the amount of energy available, that is, by lowering the temperature of the engine's operation. Unfortunately, this expedient also reduces the efficiency with which the gasoline is burned and thus increases the level of hydrocarbon pollutants.

Quite clearly, since we live in a solution of nitrogen and oxygen, we cannot prevent the production of compounds of these two elements. Nor, as we have seen, would we wish to completely eliminate such reactions. One answer to the problem of internal combustion engines (aside from banning them) would be to devise a method of promoting the reverse of reaction 11.11. This might be accomplished by placing certain *catalysts* (agents which influence the rate of chemical reactions) in the exhaust stream of a car. At the lowered temperature of the exhaust, the decomposition of NO into its constituent elements is possible, and the catalyst would speed up the reaction. But even here we must be wary. Gasoline additives like "ethyl" (tetraethyl lead, itself a potentially dangerous pollutant) could react with the catalyst and inhibit its effectiveness.

Hydrocarbons are another essential ingredient in smog formation. If the concentration of these compounds in automobile exhausts could be reduced, the oxides of nitrogen would present a less serious problem. This goal can be achieved by converting cars from gasoline to fuels of smaller molecular weight. The former fuel is a mixture of compounds like octane, C_8H_{18}. Propane, C_3H_8, and butane, C_4H_{10}, burn almost completely to CO_2 and H_2O. Unfortunately, these compounds are gases at room temperature. Although they can be easily liquefied, it is difficult to design an automobile that could carry enough "bottled gas" to travel much over 50 miles without refueling. For this reason, the use of these low molecular weight hydrocarbons has generally been restricted to fleets of urban-based delivery and service vehicles, and not extended to passenger cars.

Another approach which is compatible with present liquid fuels is the introduction of afterburners or catalytic converters in the exhaust line. Both methods complete the oxidation begun in the engine. It is thus possible to convert unburned hydrocarbons to CO_2 and H_2O. Moreover, carbon monoxide, CO, can also be eliminated. This deadly poisonous gas is another consequence of incomplete combustion. But under the influence of the right catalyst the oxidation of CO to CO_2 can be facilitated.

$$CO + \tfrac{1}{2}O_2 \rightarrow CO_2$$

Unfortunately, nothing is simple in our struggle to prevent further contamination of the atmospheric environment. In spite of the fact that CO_2 is a natural product of respiration and combustion, one utilized by plants, it too can become a dangerous pollutant. Carbon dioxide is chiefly responsible for trapping and holding the heat which the earth receives from the sun. The short-wavelength ultraviolet and visible radiation coming from the sun passes through the CO_2 of the atmosphere without diminution of energy. But the same molecules absorb the less-energetic infrared rays radiated by the earth as it cools at night or under cloud cover. Some of this energy is then reemitted back to the surface of the planet. The carbon dioxide thus acts like the roof of a greenhouse. When the atmospheric concentration of the gas increases, this "greenhouse effect" is accentuated. Indeed, if the increase becomes great enough, the average temperature of the earth's surface could rise sufficiently to melt the polar ice caps, flood harbors, and cause climatic changes sufficiently drastic to significantly alter the ecology. For the same reason, even water vapor is a potential pollutant. It, too, is partially responsible for the greenhouse effect. And on a more local scale, the water vapor emitted by power plants can cause dangerous fogs.

It is not particularly reassuring, but the thermal effect of the increasing CO_2 and H_2O concentrations which result from our fires may be compensated by particulate pollutants from the same source. Soot and dust, suspended in the air, can lead to the cooling of the earth. The sun's radiation bounces off these particles and less of it reaches the surface. Of course, this compensation is not sufficient justification for us to tolerate air filled with specks of dirt. Various methods have been devised to trap such particles; for example, they can be collected on electrostatically charged plates. As a consequence of modern technology, the black belch of a smoke stack is fast becoming a rarity. And, thanks to new legislation, so is the smell of burning leaves. At times, one may nostalgically long for that sign of fall. But our children's lungs will be healthier for their ignorance of that odor.

"Water, water, every where, Nor any drop to drink"

You realize, of course, that we have been discussing ways to halt the continued pollution of the air—not ways to clean it up. To actually remove the contaminants already in so vast and dilute a solution is an impossible task. Rather we must try to restrict our gaseous effluents and allow nature to reestablish a new steady-state composition for the atmosphere. We may have a better chance of restoring purity to our streams, rivers, lakes, and seas. Again, we have nature's help. The filtering properties of sand and the absorbance of clays contribute to the regeneration of pure water. And benign bacteria consume organic waste and break it down into simpler and generally nontoxic substances.

So long as human populations were small and widely distributed, these natural processes were generally sufficient to remove the relatively minor quantities of waste released to the waters of the earth. But as the population grew and civilizations advanced, human beings began to exert a disproportionate influence upon the environment—an influence which has irrevocably altered the primitive ecology. The urbanization of man has multiplied many times over the problems of disposing of his natural wastes; and the industrialization of his society has manufactured materials never before encountered by bacteria. In an alarmingly short time, the concentration of contaminants in lakes and rivers in every highly populated region of the world has reached the point where natural processes are no longer effective in purifying the water. Obviously, man must aid nature in this clean-up campaign. And chemistry is conspicuously well equipped to contribute knowledge, skills, and materials to the effort. Perhaps in this manner, chemical technology can make amends for some of its other, not so desirable, contributions to the water problem.

Chemistry must be involved at two vital points in the water system—after the water intake and before the sewage outlet. At the former juncture, steps are taken to purify the water and make it potable. Here, chlorine is often added to kill harmful bacteria. But such treatment does not remove substances which may, at worst, be as dangerous as radioactive wastes, or, at best, as inconvenient as hardness. To completely purify water so that it is H_2O and only H_2O is a relatively costly process which is probably impractical on a large scale. Nor is it generally necessary. For many laboratory purposes, however, pure water is essential. It is most commonly produced by distillation. The water boils, escapes as vapor, and is condensed into liquid. All nonvolatile solutes remain behind in the boiling pot of the still. Indeed, if the operation is done with reasonable care, only volatile substances emerge as possible contaminants of the distilled water. It is worth noting that distillation has been used as a means of purifying sea water and brackish inland waters containing high salt concentrations. And if the sun's radiant energy can be effectively and efficiently harnessed as the source of heat, solar stills could probably provide water for arid countries, particularly those of the Middle East.

Another water purification method which has been suggested as a means of meeting the fresh water crisis, though it is commonly used on a much smaller scale, is *ion exchange*. The ability to exchange ions is common to many natural clays, and synthetic substitutes have been manufactured. Natural or manmade, ion exchangers are ionic substances of a very special sort. Charges of one sign are part of a solid matrix; the opposite charges, which are necessary to keep the entire exchanger neutral, are provided by small ions which are free to migrate. For example, a cation exchanger is a three-dimensional molecular network which has negative charges as part of its structure. A gram of material may include as many as 10^{21} of these charges. Associated with each one is a positive charge carried by a cation. The whole point of a cation exchanger is that these cations can be exchanged for the ions of a solution. Figure 11.7a represents such an exchanger when all the cations are H⁺, in other words, when it is in the hydrogen form. When a solution containing sodium chloride is passed through a column packed with such an exchanger, the Na⁺ and H⁺ ions change places. That is, the latter ions leave the exchanger and enter the solution. Meanwhile, the Na⁺ ions originally in solution

FIGURE 11.7.   (a) *Column packed with cation exchanger in H⁺ form. (b) Column (a) being used to exchange Na⁺ ions in solution for H⁺ ions. (c) Column packed with anion exchanger in OH⁻ form. (d) Column (c) being used to exchange Cl⁻ ions in solution for OH⁻ ions. By combining columns (a) and (c) it is possible to completely deionize a solution, that is, to exchange all of its ions for H⁺ and OH⁻.*

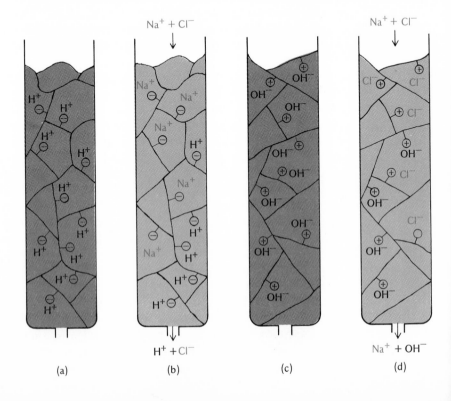

(a)   (b)   (c)   (d)

assume the sites on the exchanger previously occupied by the H^+ ions. As far as the solution is concerned, the sodium ions have been replaced by hydrogen ions. Similarly, other cations could be traded for H^+. And, if the exchanger were in the sodium form, the Na^+ too could be replaced by other ions.

Anion exchangers are similar, except that positive charges form part of the solid and negative anions constitute the exchangeable portion. For example, it is possible for all the negative charges of an anion exchanger to be balanced by OH^- ions. Now suppose this exchanger, in its hydroxyl form, is placed in a column. Consider what happens when a solution of NaCl is allowed to flow through this column. The OH^- ions are displaced from the exchanger by the Cl^-. Thus, the latter are removed from the solution while the former enter it.

It should be apparent that if a sodium chloride solution is successively passed through both a cation exchanger in its hydrogen form and an anion exchanger in its hydroxyl form, essentially all the ions can be removed from solution. Each Na^+ ion is replaced by an H^+, and every Cl^- is replaced by an OH^-. Moreover, the H^+ and OH^- ions emerging from the column react to form water.

$$H^+ + OH^- \rightarrow H_2O$$

No other ions, aside from the low normal concentration of H^+ and OH^-, remain in the column effluent. It should be pure water. Indeed, columns of just this sort are commercially available for the deionization of water. Of course, their capacity is limited. As usage continues, more and more of the H^+ and OH^- ions of the exchanger become replaced by the cations and anions present in the impure aqueous input. When the concentration of hydrogen and hydroxyl ions in the exchanger drops to a certain point, they are no longer efficiently exchanged and contaminating ions begin to pass through the column unarrested. At this point it is necessary to replace the exchanger with fresh material. Alternatively, one can separate the exhausted exchangers and regenerate them with an acid and a base. The former converts the cation exchanger back to the H^+ form and the latter resubstitutes OH^- ions on the anion exchanger.

For home use it is not necessary to remove all ions from water; some are more troublesome than others. The chief offenders are calcium, Ca^{2+}, and magnesium, Mg^{2+}. These two alkaline earth elements are primarily responsible for the hardness of water. For one thing, their compounds can deposit within pipes and boilers. More familiarly, Ca^{2+} and Mg^{2+} form insoluble scum with the large anions of ordinary soap. Even when synthetic detergents are used, thus avoiding this precipitation of bathtub ring, cleaning is inhibited by hardness. One obvious answer is to remove the interfering ions from solution, and here ion exchange is often employed. The ordinary household water-softening system is a tank packed with a cation exchanger. When it is fully charged, all cations associated with the exchanger are Na^+. As the hard water passes through the softener, the Ca^{2+} and Mg^{2+} ions are trapped and held quite tightly to the negatively charged matrix of the exchanger. For each +2 ion so captured, two Na^+ ions are released to the water. Thus, the positive and negative charges in the exchanger remain equally balanced and overall electrical neutrality is maintained. Because the tank contains only a cation exchanger, anions pass through undisturbed. The net result, then, is a two for one substitution of Na^+ for Ca^{2+} or Mg^{2+}. When the anion originally associated with the Ca^{2+} or Mg^{2+} is Cl^- (as it sometimes is), passing the water through the tank in effect increases its NaCl concentration. This explains the taste of salt, which can be detected in very hard water which has been softened with a cation exchanger.

Incidentally, salt is used to regenerate such exchangers. A concentrated solution of sodium chloride is back-pumped through the tank. Now the Na^+ ions force off the Ca^{2+}, Mg^{2+}, and other ions which have been electrostatically bound to the negative matrix. These are washed down the drain, probably to turn up in a washing machine somewhere downstream. And the softener is once again converted to its sodium form, ready for action.

In large-scale water treatment plants, magnesium and calcium ions are sometimes removed by precipitating the insoluble carbonates of these elements, $MgCO_3$ and $CaCO_3$. One of the chemicals used in this process is sodium carbonate, Na_2CO_3. The fact that this compound is commonly called "washing soda" indicates that it has been used for hundreds of years for softening water. Nevertheless, sodium carbonate does have serious drawbacks. It is quite harsh and somewhat toxic. Therefore, the introduction of phosphates was widely acclaimed. These compounds tie up Ca^{2+} and Mg^{2+} ions, keeping them in solution but preventing them from interfering with detergency. Other beneficial properties of phosphates further enhance the effectiveness of laundering.

Unfortunately, while the family wash was getting whiter and brighter, lakes and ponds were getting greener and duller. Studies of these bodies of water, "blooming" with algae, indicate that they contain too much of a good thing. The profuse growth of plant life must be due to an excess of nutrients. For a while, everything is fine, though the scum may make swimming a little unpleasant. The algae even give off oxygen as long as they keep on growing. But once the plant life begins to die, its decomposition starts consuming large quantities of dissolved oxygen. In a word, *eutrophication* occurs. And aquatic animal life ultimately suffocates in a stinking soup.

Identifying the controlling factor in eutrophication is a complex problem, and to attribute it all to phosphates from detergents, or even phosphates in general, is overly simplistic. To be sure, phosphates are essential to the existence of all living things—from simple one-celled green plants to college students. But so are nitrates and so is carbon dioxide. In some specific instances, phosphates are most likely implicated. But these compounds enter lakes from fertilizer runoff, decaying organic matter, and industrial effluents, as well as from municipal sewage treatment plants. And the latter three sources also introduce carbon dioxide and nitrates.

Thus, it seems clear that there is no single cause or cure for eutrophication. We must, however, attempt to do what we can to combat the problem. Restricting the use of fertilizers should help, but there is always the important consideration of crop yields. Reducing the phosphate content of detergents might well be a step in the right direction, but we cannot ignore the need for cleanliness. And we must also take care not to create an even greater hazard by introducing substitutes for phosphates which turn out to be toxic. Obviously, the crisis in the environment requires us to act quickly, but we cannot afford to act irresponsibly, even with the best of intentions. These good intentions must be bolstered with good information.

In part, the problem extends beyond the boundaries of science and technology. The most perplexing questions are those of priorities. In many instances, we already have the technical competence to clean up our effluent before we discharge it. It is a matter of applying chemistry at that other critical point, the sewage treatment plant. For example, we know that phosphate can be removed from waste water by precipitating it as calcium phosphate. But such additional treatment requires money—just as it requires money to install catalytic oxidizers in car exhausts and precipitators in smoke stacks. Ultimately, the consumer must bear the cost in increased charges for water, sewage disposal, automobiles, and electric power. But the alternative, to spend nothing and do nothing, may be much more expensive in the long run—fatally expensive.

This survey of polluted solutions would not be complete without a brief mention of pesticides, compounds which are, by design, toxic to plants or animals. The concentration of these substances in air or water can be very miniscule indeed. And, as with lead, mercury, or any of the other pollutants we have been discussing, the quantitative detection of pesticides demands analytical techniques of great sensitivity and reliability. It turns out, however, that the sensitivity required often decreases as we move up a food chain. Each organism in the chain concentrates compounds which resist biodegradation. The late Rachel Carson painted a bleak picture of a "Silent Spring" in which leaves and compost and earthworms unknowingly compounded pesticides until there were no robins left to sing. And so too, the danger from DDT grows as cows feed on grass, and as humans eat beef and drink milk, and finally as human mothers breast-feed their babies.

Once again, there are no easy answers. It is difficult to over-estimate the benefits of herbicides such as 2,4-D and insecticides like DDT to starving, malaria-infested peoples. The problem of pollution just cannot be approached in isolation, any more than the problem of pest control. All life on this planet is far too inti-mately interwoven for such narrowness of focus. Unquestionably, the threat to our earth and our existence calls for the best chemical information we can obtain and the best chemists we can train. But beyond that, it demands the intelligence and wisdom and compassion of all men, everywhere.

For thought and action

11.1   A biochemist prepares a solution by dissolving 127 g of urea, NH_2CONH_2, in 250 g of water. The solution is found to occupy a volume of 346 ml. Calculate the following:
   (a)   the weight percent of urea in the solution,
   (b)   the molarity of urea in the solution, and
   (c)   the molality of urea in the solution.

11.2   At 30°C a saturated aqueous solution of lead chloride contains 1.20 g of $PbCl_2$ per 100 ml of solution. Calculate the molarity of the solution. Heat is absorbed when $PbCl_2$ dissolves in water. Would you expect its solubility to be higher or lower at 100°C than it is at 30°?

11.3   The concentration of sodium fluoride, NaF, in fluoridated water supplies is about 2 mg per liter of solution. Express this concen-tration in moles per liter of solution.

11.4   In this chapter, instructions are given for preparing a 1.00 M solution of sucrose in water. How would you go about making up a 1.00 m solution of the same compound? You have at your disposal a bottle $C_{12}H_{22}O_{11}$, distilled water, an accurate balance, and any glassware you need. In which solution is the sugar concentration higher, the 1.00 M or the 1.00 m?

11.5   Bud Weiser and Ann Heuser, analytical chemists for a St. Louis firm, report the following data for a 12 oz bottle of the com-pany's product:

wt water	337.0 g
wt ethyl alcohol (C_2H_5OH)	21.5 g
wt sucrose ($C_{12}H_{22}O_{11}$)	8.5 g
wt carbon dioxide (CO_2)	6.5×10^{-3} g

   Assuming that the above ingredients are the only ones present, calculate (a) the molality of the ethyl alcohol and (b) the molarity of the sucrose. (1 fluid oz $=$ 29.6 ml.)

11.6   Predict whether the indicated compounds will be more soluble in water or in benzene, and justify your answers.
(a)   Calcium chloride ($CaCl_2$)
(b)   Ammonia (NH_3)

(c)   Cyclohexane

(d)   Aspirin

11.7   Account for the following observations:
(a) CH_4 (methane) is less soluble in water than in CH_3OH (methyl alcohol).
(b) $CH_3(CH_2)_{10}COO^-Na^+$ (a soap) is more soluble in water than is $CH_3(CH_2)_{10}CH_3$ (an oil).
(c)   Diamond does not dissolve in either water or benzene.
(d)   Aqueous solutions of salt conduct electricity, aqueous solutions of sugar do not.

11.8   Schwartz's self-cooling beer can consists of two concentric metal cans. The inner can contains the beer; the outer ring contains a compound which absorbs heat when it dissolves in water. To use, first puncture the outer can and fill with water. The beer will cool before your very eyes! Why?

11.9   You have no doubt noticed that ice cubes frequently contain trapped bubbles of air. And you may have heard that such bubbles can be avoided by freezing boiled water.
(a)   Explain the origin of the bubbles.
(b)   Assess the reasonableness of the suggestion for avoiding bubbles, indicating any scientific basis it might have.

11.10   A friend of the author (a fellow chemist) once inadvertently "carbonated" eggs and milk by storing them in a closed cooler chest with a piece of dry ice—solid CO_2. What happened?

11.11   The alcohol/water solution used as an example in the body of the text was found to have a concentration of 17.2 moles of C_2H_5OH per kilogram of H_2O. Calculate the freezing point of this solution, and explain why you can't calculate its boiling point.

11.12   In preparation for his first Minnesota winter, a college student from Miami decides to fill his radiator with a solution of ethylene glycol antifreeze, which will remain liquid until the temperature drops to $-40°$ (C or F—both scales are the same at that one point!).

(a) Calculate the molality of the ethylene glycol, $C_2H_6O_2$, in this solution.

(b) Determine the weight percent of ethylene glycol in the solution.

11.13   Methyl alcohol was once widely used as an automobile antifreeze. It mixes well with water and works fine in depressing the freezing point. But it has a serious drawback which "permanent" ethylene glycol antifreeze does not have. Suggest what that drawback is. *Hint:* The boiling point of pure methyl alcohol is $65°C$; that of ethylene glycol is $198°C$.

11.14   A 100 g sample of sea water is evaporated to dryness. The solid residue weighs 3.0 g. Of course, this is really a mixture of many compounds, but it is mostly sodium chloride. For the sake of the calculation, assume it is all NaCl and calculate

(a) the molality of NaCl in the sea water,

(b) the expected boiling point of the solution, and

(c) its predicted freezing point.

11.15   Rank the following aqueous solutions in order of decreasing freezing points.

(a) $0.03\ m$ $C_{12}H_{22}O_{11}$

(b) $0.02\ m$ $MgCl_2$

(c) $0.01\ m$ KNO_3

11.16   A solution is prepared by adding sufficient water to a 0.125 g sample of the enzyme, chymotrypsin, to make the final volume of the solution equal to 10.0 ml. The osmotic pressure of this solution is measured at $27°C$ and found to be 9.35 mm Hg.

(a) Calculate the molarity of the solution.

(b) Compute the molecular weight of chymotrypsin.

11.17   The large drop in blood volume which occurs on heavy bleeding can result in severe shock. A frequently employed procedure for treating this condition is with transfusions of whole blood. If that is not possible, blood plasma (the liquid portion of the blood without the cells) can be administered. During World War II, a group of scientists at Harvard and the Massachusetts Institute of Technology worked on a third method of restoring blood volume. The idea was to isolate soluble blood proteins from donor plasma, where their normal concentration is about 7–8%. These compounds were then redissolved to give aqueous solutions containing 20–25% protein. In combat areas, badly wounded men would be injected with a hundred milliliters or so of this concentrated solution. The resulting restoration in blood volume, however, was considerably greater than the actual volume of solution injected. Explain what was happening within the body of the patient.

11.18 Suppose you have a laboratory ion exchange column which contains both anionic and cationic exchangers. Through extensive use, the column has become saturated with Na⁺ and Cl⁻ ions. You wish to return the cation exchanger to the H⁺ form and the anion exchanger to the OH⁻ form. You propose to do this by treatment with concentrated solutions of hydrochloric acid (HCl) and sodium hydroxide (NaOH). Explain why the anion and cation exchangers must be physically separated in order for this treatment to be effective.

11.19 Statistics gathered in the United States between 1953 and 1964 indicate that the concentration of DDT in human food averaged between 0.005 mg/kg and 0.05 mg/kg. Assuming a daily food intake of 3.78 kg per person, estimate the weight of DDT ingested by an average American in a day and a year during that period. (Keep in mind that knowing the amount of DDT consumed is not enough to indicate whether or not it is likely to be harmful. Tests of the metabolic fate of DDT and the results of long-time high dosage consumption are also necessary.)

11.20 Sometimes concentrations are expressed in parts per million (ppm). For example, a carbon monoxide concentration of 17.1 ppm (as cited in the first chapter) means that out of 1,000,000 molecules of air, 17.1 are CO. Use this concentration and the answer to Problem 10.9 to calculate the number of CO molecules in a deep breath.

11.21 The effluent from a typical municipal water treatment plant contains 25 mg of phosphate ion, PO_4^{3-}, per liter. Recent statistics indicate that the average American served by a sewage system uses about 120 gallons (454 liters) of sewage water daily.
   (a) From these data calculate the number of grams of phosphate in your daily share of water.
   (b) Also compute the weight of calcium oxide (CaO) required to completely precipitate this much phosphate, that is, to remove it from solution. You may assume that the calcium phosphate formed by the reaction represented below is essentially insoluble in water.

$$3CaO + 2PO_4^{3-} \rightarrow Ca_3(PO_4)_2 + 3O^{2-}$$

11.22 Listed below are the pH values exhibited by several familiar fluids. Try your hand at calculating the H⁺ concentration in these solutions. Indicate whether the solutions are acidic or basic.
   (a) Gastric juice   pH 2.0   (c) Blood   pH 7.4
   (b) Sea water   pH 5.5   (d) Cola   pH 2.8

11.23 Find the pH's of solutions with the following H concentrations.
   (a) Vinegar $M_{H^+} = 1.0 \times 10^{-2}$
   (b) Dilute ammonium hydroxide (NH₄OH) $M_{H^+} = 1.0 \times 10^{-11}$
   (c) Pancreatic juice $M_{H^+} = 1.6 \times 10^{-8}$
   (d) Cow's milk $M_{H^+} = 2.5 \times 10^{-7}$

Suggested readings

American Chemical Society, "Cleaning Our Environment: The Chemical Basis for Action," Washington, D. C., 1969. (A valuable source of recommendations and information used by the author in preparing this chapter.)

Carson, Rachel, "The Silent Spring," Houghton, Boston, Massachusetts, 1962. (Also available in paperback from Fawcett Publications, Greenwich, Connecticut.)

Goldman, Marshall I., ed., "Controlling Pollution: The Economics of a Cleaner America," Prentice-Hall, Englewood Cliffs, New Jersey, 1967.

Leinwand, Gerald, "Air and Water Pollution," Washington Square Press, New York, 1964.

Marx, Wesley, "Man and His Environment: Waste," Harper, New York, 1971. One of a series.

12

Energy, enthalpy, and the first law of thermodynamics

The hard law of hard work

One cannot say much of any significance about the science of chemistry without soon encountering *energy*. Had I consciously and conscientiously attempted to avoid introducing this concept until the twelfth chapter, the first eleven would have been woefully incomplete and inadequate. Instead, energy was employed where it seemed appropriate, a generally accepted way of getting things done! No doubt your experience and knowledge are such that this approach did not create any major difficulties. After all, the idea of energy is quite familiar. The "something" described by that term is as essentially intrinsic to the universe as matter itself. And the word, as well as what it stands for, is a part of everyday life.

Precisely for these reasons, it is important that we devote a chapter to a more formal study of energy and its many manifestations. Too often familiar concepts are fraught with misconceptions. Thus, the aim of these pages is to provide a better understanding of energy and an appreciation of its intimate involvement in chemistry and all of existence. Fortunately, there is a vehicle for achieving these ends. It is *thermodynamics*, the specialized science which treats the energetic workings of the world.

Perhaps no other branch of learning exhibits so well the dual aspects of intellectual aesthetics and practical power. But such a statement is best supported by specific examples of the interaction between abstractions and applications. And so we begin with energy itself.

Physics books often call energy "the capacity to do work," a definition which sends the serious student in search of the meaning of *work*. This, he learns is something which is done when "a force causes motion in the direction of its line of action against resistance or opposition." But this definition introduces another new term, *force*, which in turn is related to *mass* and *acceleration*. What soon emerges is a series of connected definitions, each involving some additional concept or concepts. Clearly, the chain must be broken somewhere and dragged into everyday experience.

I choose to make work the point of contact. Although everyone has a sort of gut feeling for all of the terms in the previous paragraph, work probably has the most meaning. Even the most ergophobic of us has done a good deal of it. The idea of energy, on the other hand, is slightly more removed from experience, slightly more abstract. It is the potential which is realized in work. Or, if you prefer, it is the source of work. It is something internal which reflects the state or condition of the part of the universe under investigation. But, because energy is often expressed in work, the two concepts can best be discussed together.

Unfortunately, the term "work" is often imprecisely applied. Some things which seem to be work, holding up an end of a piano, for example, are really not work in the exact physical sense of the word. And other activities which the stern taskmaster of the Protestant Ethic calls play most definitely are work. After all, a baseball pitcher is obviously exerting a force which "causes motion in the direction of its line of action against resistance or opposition." The resistance comes primarily from gravity, partially from the air, and only incidentally from the opposing team. But exhausting as piano-holding might be, it does not conform to this definition. It is a static exercise (to combine two contradictory terms). The piano mover, as long as he just holds up his end, is not moving the piano at all. Paradoxically, he is expending energy keeping his muscles contracted. And the longer he holds the piano, the more the capacity of those muscles to do work temporarily declines. Strictly speaking, however, he is not working, not until he actually lifts the piano higher against the forceful pull of the earth.

The struggle with gravity, which entered into both our examples, is a common and convenient representation of work. Indeed, we can symbolize all work as lifting weights. The activity is unambiguous and experientially familiar to everyone. Furthermore, it can be readily measured or calculated. The quantity of work required to lift an object of mass m a height h against the force of gravity is equal to the product mgh. Here g is the acceleration due to gravity, 981 cm/sec^2. Consider the piano. Suppose it weighs 440 lb, an even 200 kg or 2.00×10^5 g. Lifting it 1 meter (100 cm) uses 2.00×10^5 g $\times$ 981 cm/sec^2 $\times$ 100 cm or 1.96×10^{10} ergs worth of work. Which means that 1.96×10^{10} ergs of energy are required to do the job. This is a good deal of energy and it would be a considerable shame if, after all that effort, it were lost forever. As a matter of fact, it isn't. It is all stored in the piano. The 1.96×10^{10} ergs expended in raising the piano increases its potential energy by just that amount. Because of its new, elevated position the piano is in a more energetic state than when it rested on the floor. It is now capable of doing more work.

All that is required to realize this potential energy is to lower the piano. This process converts the potential energy (here a consequence of position) to translational kinetic energy, a manifestation of motion. If the piano is dropped, the conversion is rapid and may have painful and noisy consequences. But if it is attached to a pulley, its slow descent can be coupled with the ascent of another object. The energy of the descending piano will be transferred to the ascending object. Work will be done. And the potential energy of the second object will increase at the expense of the decreasing potential energy of the piano. But no energy will be lost or found in the process.

This latter fact is of utmost importance because it is a fundamental characteristic of our universe. Indeed, the repeated observation that *energy is neither created nor destroyed* underlies one of the most basic postulates of physical science, the *First Law of Thermodynamics*. The version verbalized by Rudolph Clausius introduces this chapter, *"The energy of the universe is constant."* Paraphrased into the popular patois, the law simply states the hard fact, "You can't get something for nothing." There is no free lunch!

This means that it is impossible, at least in this universe, to devise a perpetual motion machine which will create energy. If this were not so, the public utilities companies would be in a bad, bad way. For example, an electric motor and an electric generator could be coupled together to form a self-contained unit. The idea is to start the motor going and then pull out the plug. The motor will drive the generator which, in turn, will drive the motor. Any left-over energy can be used to do work—to raise weights or something equivalent. It sounds fine, but there's a catch. There just isn't any excess energy. Even under absolutely frictionless conditions all the energy produced by the motor would be required to run the generator. If not, energy would be created. But the way the world works you can't break even. Because of friction, some of the work energy would be converted to heat energy and our ingenious machine would soon grind to a halt.

The canon of caloric meets the cannon of the count

Recognition of the fact that heat and work are both manifestations of energy did not come easily. Of course, the experience and idea of heat must be old as fire. But its understanding is relatively recent. Even the distinction between temperature and heat was not well established until Joseph Black's work in the latter half of the eighteenth century. He was aware that the thermometer registers the "degree of hotness," but it does not indicate the amount of heat in a body. Although Black measured many temperatures, for very good reasons (which you will soon read) he never succeeded in determining the total "heat content" of anything. He did, however, devise a method for measuring the amount of heat attending certain processes. We commonly detect such changes with burned fingers or cold feet. Black realized that the thermometer provided a more reliable instrument for measuring the temperatures induced by heat flow. He also recognized the fact that the direction of this spontaneous flow is, without exception, from the region of higher temperature to one of lower temperature. As a result, the temperature of the warmer body always decreases while that of the colder body increases.

Black made use of this behavior in his experimental design. As his "heat sink" he used a known mass of water. Its temperature was carefully measured at the start of the experiment and it was then placed in contact with the hotter object. The temperature rise observed in the water was taken to be an indication of the amount of heat transferred. As a result of his investigations, Black was able to determine relative values for the heat capacity or specific heat of water and mercury. That is, he found that the two substances differ in the quantity of heat required to raise the temperature of an identical mass by the same number of degrees.* The total amount of heat absorbed by a body thus depends upon the product of its temperature increase, its mass, and its specific heat. Using the symbols Q, m, ΔT, and c_p to denote, respectively, heat, mass, temperature change, and specific heat, we can write

$$Q \text{ cal} = m \text{ g} \times \Delta T {}^\circ C \times c_p \frac{\text{cal}}{\text{g} {}^\circ C}.$$

Recall, for example, that the specific heat of water is 1.00 cal/

* Specific heat was introduced in Chapter 3 in connection with the Law of Dulong and Petit.

g °C. That is, 1.00 cal is required to raise the temperature of exactly 1 g of H_2O by precisely 1°C. Thus, the amount of heat necessary to cause a 5° temperature increase in 10 g of water must be

$$Q = 10 \text{ g} \times 5°C \times 1.00 \text{ cal/g} °C = 50 \text{ cal.}$$

Similarly, 50 cal of heat could raise the temperature of 5 g of liquid H_2O by 10°C. Or, this same quantity of heat could produce a temperature rise of 2.5° in 20 g of water. And, if 50 g of water were to cool by 1°, 50 cal of heat would be liberated in the process.

We will soon be using this same reasoning in other applications. Since it permits the calculation of heat flows, Black's experimental logic was a most valuable contribution. As you know, it has implications for atomic weights in the form of the Law of Dulong and Petit. Moreover, Black succeeded in measuring the amounts of heat absorbed by ice as it melts and water as it boils—the heats of fusion and of vaporization. But in spite of the fact that his numerical results were quite good, even by modern standards, they did not answer a more basic question—what is heat?

Terms like "heat content" and "heat capacity," and the phenomena they symbolize, certainly suggest that heat behaves very much like a fluid. The fact that heat always flows from hotter to colder objects is also consistent with this interpretation. After Black's work there seemed no doubt that as a substance increases in temperature, the amount of heat it contains also rises. Thus, Lavoisier had ample experimental support for including caloric in his table of elements. Many other eminent scholars tended to agree with him, and the notion of heat as a material fluid was widely adopted as a working hypothesis.

Now any hypothesis worth making, whether more-or-less right or more-or-less wrong, must be testable. Granted, theories can never be proved unambiguously and absolutely true, no matter how many positive results are obtained. The next experiment might always yield negative results. But if it is impossible to devise an experiment which could disprove a hypothesis, the idea contributes nothing to human knowledge. It is as indisputable and as devoid of scientific meaning as attributing thunder to Thor's hammer. In the case of caloric, there was general (though by no means unanimous) agreement that a crucial test was weighing. Thus, the issue was not unlike that of phlogiston. Supposing heat to be a material fluid, it was logical, but not absolutely necessary, to expect it to have mass. Consequently many attempts were made to determine how the weight of an object varies with its temperature. The first measurements were inconclusive. Some data suggested an increase in mass with increasing caloric content. Others indicated the opposite, a result more difficult to reconcile with the behavior of ordinary matter. But the experimental difficulties were formidable, and they required a formidable experimentalist.

Such a man was Benjamin Thompson, Count Rumford. In all the history of science there is probably no character more complex or colorful. His fascinating life spanned four countries and easily as many careers. It began in Woburn, Massachusetts in 1753. As a young man Thompson was apprenticed to a storekeeper and later became a schoolmaster. Along the way he demonstrated considerable talent in natural philosophy. He also exhibited political views which were quite unpopular in prerevolutionary New England. A staunch Tory, Thompson even spied for the British, writing accounts of colonial troop strength in invisible ink. Such activities were hardly likely to endear him to many of his fellow Americans, and when the British troops withdrew from Boston in 1776, Thompson left his wife, a rich widow, and wisely went along. He went all the way to London. There he soon rose to a position of undersecretary of state and was knighted. Moreover, he found time to do enough scientific work on gunpowder and firearms to win election to the Royal Society.

Around 1780, Thompson moved to Austria and then to Bavaria. Before long he was again embroiled in affairs of state. For 11 years he served the Elector of Bavaria in a variety of offices, including minister of war, minister of police, and grand chamberlain. He even was instrumental in organizing the Bavarian army. For his numerous contributions to Bavaria, he was made a Count of the Holy Roman Empire, no doubt the first and only native of Woburn, Massachusetts ever to be so honored. But in spite of intrigues and involvements, he remained a scientific amateur of the highest rank, never abandoning his interest in heat.

About 1795, Thompson, now Count Rumford, returned to England, where, among other activities, he helped establish the Royal Institution and install Humphry Davy as its professor. But the wanderlust struck again, and he left for France. There the Count met Lavoisier's widow. Their courtship lasted four or five years, about twice as long as their stormy marriage. Perhaps the union started under a cloud. A London newspaper, announcing the marriage, observed that it was a "nuptual experiment" by which "he obtains a fortune of 8,000 pounds per annum—the most effective of all the Rumfordizing projects for keeping a house warm." Whatever his pecuniary motives, by this time Rumford was ready to settle down and quietly pursue his science. But Madame Lavoisier de Rumford, as she insisted she be called, was much more interested in the Parisian social whirl. Little time passed before the Count was complaining of the "female dragon" in his house, and subsequently the couple separated. Rumford returned to his laboratory, where he continued to experiment and invent such useful articles as a drip coffee maker. He died near Paris in 1814, but not without remembering the land of his birth by establishing the Rumford chair of chemistry at Harvard University.

Rumford's "Inquiry Concerning the Weight Ascribed to Heat" is a classic of beautifully executed laboratory technique.* The weighing of water—hot and cold, liquid and frozen—sounds like a trivially simple exercise. But Thompson had to take extreme pains to avoid specious results arising from convection currents, condensed moisture, and differences in the thermal expansion of the arms of the balance. The final sentence of his paper sums up his findings. "I think we may safely conclude that all attempts to discover any effect of heat upon the apparent weights of bodies will be fruitless." Although not every investigator accepted Rumford's conclusions, his painstaking precision did much to undermine the idea of heat as a material substance. But if it is not that, what is it?

In point of fact, an alternative hypothesis had been propounded well over a century earlier. Its contention was that heat was somehow associated with motion, and its advocates included such illustrious men of British science as Francis Bacon, Robert Boyle, and Robert Hooke. Now Rumford set out to compare this hypothesis with the caloric concept. His method exhibited his characteristic combination of political, military, and scientific interest. His own words best explain his observations and reasoning:

Being engaged lately in superintending the boring of cannon in the workshops of the military arsenal at Munich, I was struck with the very considerable degree of heat that a brass gun acquires in a short time in being bored, and with the still higher temperature (much higher that that of boiling water, as I found by experiment) of the metallic chips separated from it by the borer.

The more I meditated on these phenomena, the more they appeared to me to be curious and interesting. A thorough investigation of them seemed even to bid fair to give a farther insight into the hidden nature of heat; and to enable us to form some reasonable conjectures respecting the existence, or nonexistence, of an igneous fluid—a subject on which the opinions of philosophers have in all ages been much divided.

After this introduction, Rumford goes on to describe his appa-

---

* It is reprinted, in part, with comment in the "Harvard Case Histories in Experimental Science." The quotations from Rumford's work included in this chapter are drawn from that source. (See the list of Suggested Readings at the end of the chapter.)

FIGURE 12.1.   *Illustration of Count Rumford's cannon-boring equipment from his report to the Royal Society, entitled, "An Inquiry Concerning the Source of the Heat Which is Excited by Friction" and published in the Philosophical Transactions 88, 80 (1798). The lower drawing is a detail of the barrel-blank and borer surrounded by a watertight box. (Reproduced from James Bryant Conant, general ed., "Harvard Case Histories in Experimental Science," Vol. 1, Harvard Univ. Press, Cambridge, Massachusetts, 1966.)*

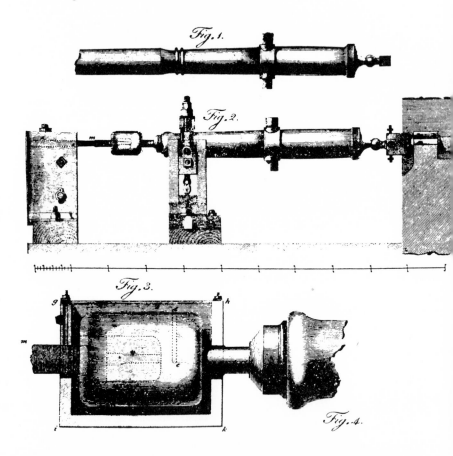

ratus. In short, it was a dull cannon borer and a brass barrel blank, so arranged as to permit the measurement of temperature change and heat flow during the boring.* Such measurements clearly indicated that as the bit chewed through the metal, a tremendous

---

* The Count makes it very clear that he was not guilty of "prodigality in the prosecution of my philosophical researches," since the section he studied would ultimately be discarded anyway, and the barrel itself would be unspoiled.

amount of heat was evolved—far more, Rumford calculated, than could be reasonably squeezed from the brass chips and dust if heat were a fluid. In one experiment he even placed the borer and the blank in a box of cold water and watched with self-admitted glee as the temperature of the water rose. Finally, after 2½ hours of turning, it boiled!

It would be difficult to describe the surprise and astonishment expressed in the countenances of the bystanders on seeing so large a quantity of cold water heated, and actually made to boil, without any fire. Though there was, in fact, nothing that could justly be considered as surprising in this event, yet I acknowledged fairly that it afforded me a degree of childish pleasure, which, were I ambitious of the reputation of a grave philosopher, I ought most certainly rather to hide than to discover.

Again, Rumford was able to calculate the quantity of heat generated by the mechanical friction. It was so large that he concluded that its source must be inexhaustable.

It is hardly necessary to add that anything which any insulated body, or system of bodies, can continue to furnish without limitation, cannot possibly be a material substance; and it appears to me to be extremely difficult, if not quite impossible, to form any distinct idea of anything capable of being excited and communicated in the manner in which heat was excited and communicated in these experiments, except it be MOTION.

I am very far from pretending to know how, or by what means or mechanical contrivance, that particular kind of motion in bodies which has been supposed to constitute heat is excited, continued, and propagated; and I shall not presume to trouble the Society with mere conjectures, particularly on a subject which, during so many thousand years, the most enlightened philosophers have endeavored, but in vain to comprehend.

But, even though the mechanism of heat should, in fact, turn out to be one of those mysteries of nature that are beyond the reach of human intelligence, this ought by no means to discourage us, or even lessen our ardor, in our attempts to investigate the laws of its operations.

That Rumford's words of encouragement were well heeded but hardly necessary is made manifestly clear in the next section.

$$\Delta E = Q - W$$

Rumford's spectacular production of heat from work was soon followed by similar demonstrations by Davy. In one famous experiment he melted two pieces of ice by rubbing them together in an evacuated chamber kept below the freezing point. But the *coup de gras* for the caloric theory was not struck until about 1850. Then, in a series of research papers first received with "entire incredulity," James Joule, a Manchester brewer, proved the fundamental equivalence of work and heat. Joule's procedure and his arguments can be illustrated with modern equipment, data, and units. Figure 12.2 will serve as a schematic guide. Suppose we start our experiment with 100 g of liquid water at 20°C, contained in a vessel open to a constant pressure of 1 atm. We apply heat from a convenient source—a burner or an electric immersion heater will do nicely. Or, we could emulate Black and place the container of water in contact with a larger body of hotter water. Whatever the source of heat, we carefully measure the temperature rise it produces in our sample. When the thermometer reaches 30°C we calculate the quantity of heat (Q) absorbed, again using the equation,

$$Q = m \times \Delta T \times c_p.$$

For our experiment, the mass of water is 100 g, the observed temperature change is 10°C, and the specific heat is 1.00 cal/g °C. Multiplying these three numbers yields the following result:

$$Q = 100 \text{ g} \times 10°C \times 1.00 \text{ cal/g }°C$$
$$= 1000 \text{ cal.}$$

Note that if we use a larger body of hotter water as the heat source, the quantity of heat lost by this surrounding reservoir should exactly equal the heat gained by the 100 g sample. Thus, 1000 g of water at 75°C gives off 1000 cal while cooling 1° to 74°C. Such an experimental arrangement would provide an independent check of our earlier result.

Now we cool the water back to exactly 20°C in preparation for the second phase of our investigation—warming by working. This time we replace the source of heat with a source of work. A paddle wheel or stirrer driven by an electric motor would literally "work well." But suppose, for the sake of calculation, that we choose to drive the stirrer with a system of pulleys and falling weights. As the water is agitated, its temperature is seen to increase. We continue working on the water until it again reaches 30°C. Now we calculate just how much work has been done. We find that in the process, 100 kg of weight have fallen 426 cm. The same gravitational formula we used earlier again applies; work (W) is equal to

FIGURE 12.2. *Sketch of an experiment demonstrating the energetic equivalence of heat and work. In the process pictured above, the temperature of the water is raised 10°C by the application of 1000 cal of heat. Below, the same change in state is produced by doing 4180 joules of work on the system. Note the values for the heat (Q), the work (W), and the energy change (ΔE). The latter is the same in both instances.*

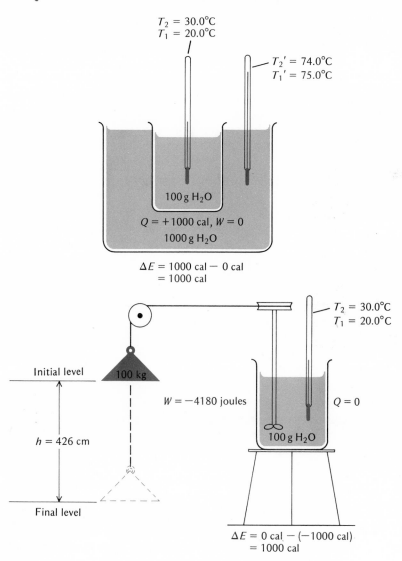

mgh. Converting 100 kg to 1.00×10^5 g and substituting for m, g, and h we obtain

$$W = 1.00 \times 10^5 \text{ g} \times 981 \text{ cm/sec}^2 \times 426 \text{ cm}$$
$$= 4.18 \times 10^{10} \text{ ergs.}$$

This quantity of work was done on the water and was destroyed in the rest of the world.

Observe that the initial conditions of the water are exactly the same in both experiments. And the final conditions are also identical. These conditions specify the *state* of the system. A *change in state* need not necessarily involve a phase change, such as the vaporization of a liquid. In the case of the two experiments just described, the common change involves only a temperature increase:

Initial State   Final State

12.1   100 g $H_2O(l)$ (20°C, 1 atm) → 100 g $H_2O(l)$ (30°C, 1 atm).

There is no way of distinguishing the water heated by the flame from the water heated by the stirrer. A moment's reflection will suggest that the simple phrase "heated by the stirrer" is of profound significance. As a matter of fact, during the second experiment no heat was directly applied—only work was done. Yet the thermometer tells us we are justified in using these intuitive words. The work was indeed the source of the heat.

It follows that the quantity of heat, 1000 cal, and the quantity of work, 4.18×10^{10} ergs, must be identically equivalent:

1000 cal $= 4.18 \times 10^{10}$ ergs
or 1 cal $= 4.18 \times 10^{7}$ ergs.

To honor the man who suggested all this we identify 10^{7} ergs as 1 joule and thus obtain the basic equation expressing the equivalence of heat and work,

1 cal $= 4.18$ joules.

Pretty obviously, a pot of water can be heated in more ways than one. The method used in the first experiment described above utilized heat only; the second exclusively employed work. Therefore, it is reasonable to suppose that the same change in state can also be accomplished by a combination of the two. Thus, we could first do 2090 joules worth of work stirring 100 g of water. As a result, its temperature would rise from 20° to 25°C. Then the container could be transferred to a burner and 500 cal of heat could be applied to increase the temperature of the water to 30°C. Quite clearly, any other combination of work and heat can effect the identical temperature increase, so long as the total energy change corresponds to 1000 cal or its equivalent, 4180 joules.

The last sentence calls for some scrutiny; it says a surprising amount about the way the world works. To begin with, both work and heat are identified with energy. Energy, in turn, is at least implicitly associated with the state of the system. And finally, the Law of Conservation of Energy is tacitly assumed. These three

points are absolutely fundamental to an understanding of thermodynamics and the energetic universe which it describes.

Like any other branch of learning, thermodynamics has its own
jargon. Therefore, it is necessary for us to adopt some definitions
and accept some conventions before we return again to our informative example of heating water. Thermodynamics is of great
pragmatic usefulness to chemistry because it precisely and accurately represents the machinations of the universe. A significant
part of its consistency and coherence is precision in speech and
symbol. Some definitions are almost self-evident. The *system* is
simply that part of the universe under investigation, for example,
the water in our experiment. The system is separated from the
rest of the universe, the *surroundings,* by a *boundary.* This boundary can be real, as in the case of the walls of a vessel, or imaginary.
It can be assumed to be anywhere, but obviously, its location determines what is in it, that is, the contents of the system. Real or
imaginary, the boundary is important because it is at this interface
that we observe the exchange of heat and work between the
system and its surroundings. Measurements of these two param-
eters provide the data from which all other thermodynamic functions are inferred and calculated.

In ordinary speech we say that heat is absorbed or evolved and
that work is done on or by something. Obviously, units and numbers are associated with heat and work. And so are plus and minus
signs. The flow of heat and work is directional. In thermodynamic
changes in state, these manifestations of energy pass back and
forth between the system and the rest of the world. Thus, in order
for calculations to make sense, it is necessary to be able to alge-
braically indicate the direction of transfer. This is done with signs,
the uses of which are dictated by conventions of historical precedent. *Heat bears a positive sign when it is absorbed by the system
from the surroundings.* Its sign is negative if it flows in the opposite direction, from the system to the surroundings. *Work,* on
the other hand, *is given a positive sign if it is produced in the
surroundings,* for example, if weights are lifted there. Thus, positive work "flows" from the system to the surroundings, the exact
opposite of the convention adopted for heat. But if work is destroyed in the surroundings and flows to the system, it carries a
negative sign.

These conventions are schematically depicted in Figure 12.3.
At first sight they may seem contradictory, but the following
example should convince you that they work. Consider first the
experiment in which 100 g of water were heated by absorbing
1000 cal from a heat source. Since the flow of heat was from the
surroundings to the system, $Q = +1000$ cal. There is a strong
temptation to say that the "heat" of the water increased by this

FIGURE 12.3.   *Schematic representation of sign conventions governing work production and heat flow. Because* $\Delta E = Q - W$, *an increase in the internal energy of the system (i.e., a positive* ΔE *is favored by the absorption of heat* $(Q+)$ *and work done on the system* $(W-)$. *Conversely, a decrease in internal energy (i.e., a negative* ΔE) *is favored by the evolution of heat* $(Q-)$ *and work done by the system* $(W+)$.

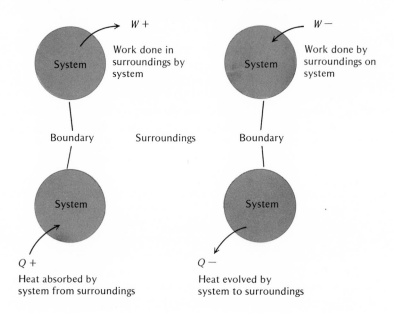

amount. But this is a verbal vestige of the caloric hypothesis. One might just as well say that the "work" of the water increased by an identical quantity, since the same change in state can be effected by doing work on the system. This, of course, is what happened in the second experiment. Here stirring the water was accomplished by lowering weights in the surroundings, that is, by destroying work there. According to our convention, a negative amount of work was done. Thus $W = -4180$ joules. Or, introducing the heat equivalent of work so that both Q and W will involve the same units, $W = -1000$ cal.

In both experiments the system underwent the same change in state—its temperature increased 10°C. But if neither the heat nor the work of the water changed, what did? The answer is, its energy. A rise in temperature signals an increase in energy from an initial value of E_1 to a final value of E_2. The difference between the two, $E_2 - E_1$, represents the change in the energy of the system attending the change in state. We will designate this difference as ΔE. The magnitude of ΔE is related to the amount of heat and/or work involved in the process.

This dependence has been given mathematical form. Since both heat and work are manifestations of energy, Q, W, and ΔE all enter the equation.

12.2 $\Delta E = Q - W$

Translated into words, the equation states that the change in energy of a system is equal to the amount of heat it absorbs from the surroundings minus the amount of work which it does in the surroundings. The signs make sense because heat *absorbed by* the system signifies an *energy increase,* work *done by* the system indicates an *energy loss*. But all the energy is fully accounted for. The $Q - W$ represents the energy change observed as a flow of heat and work across the boundary. Equation 12.2 states that this quantity is exactly equal to the energy change that occurs within the system itself. In summary, the equation reiterates the principle of conservation of energy, the First Law of Thermodynamics.

Now let us return once again to our example. In the experiment in which heat flow was the sole source of energy, $Q = 1000$ cal and $W = 0$ because no work was done. Substituting in Equation 12.2,

$\Delta E = 1000$ cal $- 0 = 1000$ cal.

This means that E_2, the energy of the system in its final state of 30°C, is 1000 cal greater than E_1, its initial energy. No heat *per se* flowed in the stirring experiment, so $Q = 0$; but $W = -1000$ cal. Again we substitute and this time find that

$\Delta E = 0 - (-1000$ cal$) = 1000$ cal.

As before, the answer indicates that as a result of the change in state, the internal energy of the system rose by 1000 cal. In neither instance was the energy expended by heating or stirring lost. Rather it was conserved in the increased energy of the water itself. The third experimental example, the two-step process involving both heat and work, yields exactly the same result. Here $Q = +500$ cal and $W = -2090$ joules or -500 cal. It follows that

$\Delta E = 500 - (-500)$ cal $= 1000$ cal.

The above calculations illustrate an important distinction between energy on the one hand and heat and work on the other. Energy is a *state function*, that is, *it depends only upon the state of the system*. You have already encountered, in this chapter and the last, examples of the thermodynamic meaning of the state of a system. It is a comprehensive term which includes a specification of chemical composition, physical state, temperature, pressure, and so on. Changes in any of these properties will usually alter the internal energy of the system. But, as the example of water clearly indicates, the same change in state can be produced in a variety of ways. Work and heat are *path-dependent functions,* they depend upon the means chosen. But energy is oblivious of the road taken. Thus, for the change in state represented by Equation 12.1, ΔE always equals 1000 cal, no matter what combination of Q and W bring it about.

Of course, to claim that the distinction between state functions and path-dependent functions is an exclusive achievement of thermodynamics would do a great injustice to the keen observation and shrewd analysis of the unknown rustic who first noted "There's more than one way to skin a cat." The skin-on/skin-off conditions which represent the initial and final states of the cat are fixed. But what is easy and effortless for a professional cat-skinner may extract a formidable amount of work from an amateur. And in spite of the fact that the as-the-crow-flies distance from here to Scotland is a constant value, the low road must be shorter than the high. Else why would I get there afore ye?

Enter: enthalpy

Having just established, by example and analogy, that heat and work depend upon the process used in bringing about a change in state, we now advance the thesis that under certain circumstances, Q and W can behave as state functions. First consider a change in which the pressure is constant. This is the most common condition encountered in chemistry, since most reactions are carried out in flasks and beakers exposed to the pressure of the atmosphere. Our specific example will once again be the combination of hydrogen and oxygen to yield water. We begin by writing the change in state. Assume we start with 1 mole of hydrogen and ½ mole of oxygen at 25°C and 1 atm pressure. After the atoms are rearranged in the reaction we will have 1 mole of liquid water. The pressure will still be a constant 1 atm and we will restore the temperature to 25°C. Symbolically, the change is

Initial State
1 mole $H_2(g)$, ½ mole $O_2(g)$ (25°C, 1 atm) $\rightarrow$

Final State
1 mole $H_2O(l)$ (25°C, 1 atm)

It corresponds to the equation,

$H_2(g) + \frac{1}{2}O_2(g) \rightarrow H_2O(l).$

Restoration of the temperature to 25°C requires that heat be absorbed by the surroundings. In a new word, the reaction is *exothermic,* it "gives off" heat. We can ascertain how much by measuring the quantity of heat which flows to the surroundings. For the reaction as written, it amounts to 68.3 kcal. Since this represents heat *evolved* by the system, the number must bear a negative sign in keeping with the convention. The heat effect, Q_p, thus equals -68.3 kcal per mole of $H_2O(l)$ formed. (The subscript designates the condition of constant pressure.)

Other experiments conducted under a variety of conditions indicate first that the numerical value of Q_p is a direct function of the amount of hydrogen and oxygen reacting. Doubling the quantities of both reactants doubles Q_p. Thus, the combustion of 2 moles of hydrogen with 1 mole of oxygen to yield 2 moles of water is attended by the evolution of 2×68.3 kcal of heat; $Q_p = -176.6$ kcal. The result is nothing new; it was noted long ago that two buckets of coal keep a room warm for twice as long as one bucket. Not as obvious is the fact that Q_p also depends slightly upon the temperature and pressure at which the reaction occurs.

Thus, Q_p begins to sound rather like a change in a state function, and so we give it a new symbol, ΔH. H stands for *enthalpy*, sometimes called "heat content." And $\Delta H = H_2 - H_1$, the difference in enthalpy between the final (subscript 2) and the initial (subscript 1) state of the system. Recall that the former involves 1 mole of $H_2O(l)$, the latter 1 mole of $H_2(g)$ and ½ mole of $O_2(g)$. Therefore, the enthalpy of the elements combined to form the compound is less, by 68.3 kcal, than the enthalpy of the uncombined oxygen and hydrogen. The spontaneous reaction has occurred with a decrease in enthalpy, whatever that might be.

It is one of the paradoxes of thermodynamics that Q_p, which is readily measurable and easily understandable as the heat flow at constant pressure, is identified with ΔH, a rather abstract and difficult concept. The very term "heat content" is a little disturbing because it evokes shades of caloric. Moreover, we recall with some unease Rumford's declaration that a body can contain an inexhaustible supply of heat. Fortunately the apparent conflict is avoided if we keep in mind that in measuring Q_p we are measuring a *difference in enthalpy* associated with some particular process. The absolute values of H do not really matter. In fact, we do not know what they are. We merely adopt an arbitrary reference point upon which to base the entire enthalpy scale. In other words, we establish a zero of enthalpy to which all other values can be related.

Our arbitrary assumption is that *the enthalpy of each pure, uncombined element is zero at 25°C and 1 atm pressure, provided the element is in its most stable physical form under these conditions.* Thus, the reference state for both hydrogen and oxygen is the gaseous phase, not the liquid or solid. And, since graphite is more stable than diamond at 25°C and 1 atm, the enthalpy of graphite is set equal to zero at this temperature and pressure. The enthalpy of diamond, relative to this reference state, can be measured as 0.45 kcal/mole. That is, 0.45 kcal is absorbed in the conversion of 1 mole (12.0 g) of graphite into diamond.

Once we have adopted this convention, all enthalpy changes can be referred to it. The most basic of these are *heats of formation*. This term refers to *the enthalpy change which attends the formation of 1 mole of a compound by the direct combination of the elements which constitute it*. You immediately recognize the combination of hydrogen and oxygen as an example of just such a reaction. Since the enthalpies of the elementary reactants are, by universal agreement, zero at 25°C and 1 atm, so is the enthalpy of the initial state, $H_1 = 0$. The value of Q_p or ΔH accompanying the oxidation is -68.3 kcal per mole of $H_2O(l)$ formed. Thus, $\Delta H = H_2 - H_1$ becomes -68.3 kcal $= H_2 - 0$. It follows that $H_2 = -68.3$. The *standard molar enthalpy* of the final state, 1 mole of $H_2O(l)$ at 25°C and 1 atm, must be -68.3 kcal. As Figure 12.4 indicates, this value is numerically identical to the *standard heat of formation* of liquid water.

The word "standard" implies the arbitrarily assumed pressure of 1 atm. In the usual symbol for standard heat of formation, ΔH_f°, the superscript indicates this condition.[*] But since enthalpies and enthalpy changes depend only slightly upon pressure, standard values are commonly also used at pressures other than 1 atm. Enthalpies vary more markedly with temperature. For example, the absorption of heat which occurs when the temperature of water is increased indicates that the molar enthalpy of the water has also increased. Therefore, it is necessary to specify the temperature for which standard heats of formation are determined and tabulated. The most convenient is 25°C, since at this temperature all elements are assumed to have standard enthalpies of zero. This zero value does not hold at other temperatures.

A temperature of 25°C applies to all the standard heats of formation included in Table 12.1. The negative sign which precedes most of these numbers indicates that heat is evolved in the direct formation of most compounds from the elements they contain. This can be attributed to the fact that the chemical bonds in the product compounds are generally stronger and more stable than those in the elementary reactants. Underlying this explanation is the observation that the breaking of chemical bonds always requires energy, while the formation of bonds always liberates energy. Thus, heat must be absorbed in order to free individual atoms of the reactants before they can recombine to form products. For example, the breaking of the H—H and O—O link-

---

[*] The subscript f stands for "formation."

FIGURE 12.4. *Scale representation of molar enthalpies or heats of formation (ΔH_f°) of certain selected substances, expressed in units of kcal/mole under the conditions of 25°C and 1 atm pressure. Two between-state transitions are indicated.*

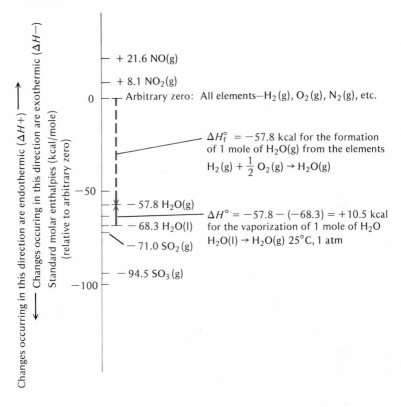

TABLE 12.1. *Standard Heats of Formation (ΔH_f°) (in kcal/mole at 25°C and 1 atm)*

CaO (s)	−151.9	Fe_2O_3 (s)	−196.5	NaBr (s)	−86.0
C (s, diamond)	+0.5	HBr (g)	−8.7	NaCl (s)	−98.2
CH_4 (g)	−17.9	HCl (g)	−22.1	NaF (s)	−136.0
CH_3OH (l)	−57.0	HF (g)	−64.2	NaI (s)	−68.8
CO (g)	−26.4	HI (g)	+6.2	NaOH (s)	−102.0
CO_2 (g)	−94.1	HNO_3 (l)	−41.4	NH_3 (g)	−11.0
C_2H_2 (g)	+54.2	H_2O (g)	−57.8	NH_4Cl (s)	−75.4
C_2H_4 (g)	+12.5	H_2O (l)	−68.3	NO (g)	+21.6
C_2H_6 (g)	−20.2	H_2O_2 (l)	−44.8	NO_2 (g)	+8.1
C_3H_8 (g)	−24.8	H_2SO_4 (l)	−193.9	O_3 (g)	+34.0
$i\text{-}C_8H_{18}$ (l)	−62.0	HgO (s)	−21.7	SO_2 (g)	−71.0
C_2H_5OH (l)	−66.4	MgO (s)	−143.8	SO_3 (g)	−94.5

ages in molecular hydrogen and oxygen is an *endothermic* process with a positive heat flow. But the combination of these individual atoms to form molecules of H_2O liberates heat. In fact, it liberates more heat than is used to separate the atoms in the first place. Therefore, the net heat change is negative. The enthalpy and energy of the system decrease during the reaction, and heat is evolved in accordance with the First Law of Thermodynamics. As a consequence of such behavior, the combination of elements is often a source of heat and energy. If the reaction is uncontrolled, as in a hydrogen/oxygen explosion, it can have disasterous results. But in a charcoal barbeque grill, the combination of carbon and oxygen supplies heat for good purpose.

A great utility of tabulations of heats of formation is that they permit the calculation of the enthalpy changes attending many other reactions. These computations are possible because equations and heats associated with them can be added and subtracted. As long as the initial and final states are what we want, the means employed in making the transition—either in the laboratory or on paper—have no influence on the value of ΔH.

Let us illustrate such an operation with a problem whose answer we already know. The example is really more physical than chemical, it is the evaporation of water,

12.3 $H_2O(l) \rightarrow H_2O(g)$.

A quick consultation of Table 12.1 indicates that ΔH_f° for $H_2O(g)$ is -57.8 kcal/mole. This enthalpy change is associated with the reaction,

12.4 $H_2(g) + \frac{1}{2}O_2(g) \rightarrow H_2O(g)$ $\Delta H_f^\circ = -57.8$ kcal, at 25°C and 1 atm.

We already know the comparable change in heat content for the formation of liquid water from its elementary ingredients,

12.5 $H_2(g) + \frac{1}{2}O_2(g) \rightarrow H_2O(l)$ $\Delta H_f^\circ = -68.3$ kcal.

In order to obtain Equation 12.3, we merely subtract 12.5 from 12.4. And while we are at it, we subtract enthalpy terms as well.

$$\begin{array}{ll} H_2(g) + \frac{1}{2}O_2(g) \rightarrow H_2O(g) & \Delta H_f^\circ = -57.8 \text{ kcal} \\ - [H_2(g) + \frac{1}{2}O_2(g) \rightarrow H_2O(l) & \Delta H_f^\circ = -68.3 \text{ kcal}] \\ \hline \rightarrow H_2O(g) - H_2O(l) & \Delta H^\circ = -57.8 - (-68.3) \text{ kcal} \\ H_2O(l) \rightarrow H_2O(g) & \Delta H^\circ = 10.5 \text{ kcal} \end{array}$$

The elementary forms of H_2 and O_2 very neatly cancel and after a transposition we are left with what we are looking for. The answer, $\Delta H^\circ = 10.5$ kcal, indicates that the change from the liquid to the gaseous phase of H_2O results in an enthalpy increase of 10.5

kcal/mole. The heat necessary to bring about this increase is absorbed from the surroundings in an endothermic process. Thus, $\Delta H° = 10.5$ kcal corresponds to the heat of vaporization of 1 mole of liquid water. (Figure 12.4 should help clarify this process.) The fact that Chapter 10 identified 9.7 kcal/mole as the amount of heat associated with the vaporization of water does not represent an experimental or editorial error. It reflects the slight dependence of $\Delta H°$ upon temperature. The lower value was experimentally observed at 100°C; the higher number was calculated for the standard state of 25°C. It turns out that there are ways of determining $\Delta H°$ at one temperature if its value at another is known, but we will not explore that aspect of thermodynamics. It is more important that you note that the enthalpy change and heat flow associated with the reverse reaction is exactly the same, but with the opposite sign, that is,

12.6 $H_2O(g) \rightarrow H_2O(l)$ $\Delta H° = -10.5$ kcal.

The heat content of liquid water is less than that of the gas. Hence, heat is evolved to the surroundings when water vapor condenses.

You have no doubt noted that the procedure employed in our calculations involved subtracting the heat of formation of the reactant or initial state, $H_2O(l)$, from that of the final or product state, $H_2O(g)$. This explains why reversing the states and the reaction (as in Equation 12.6) simply reverses the sign of the enthalpy change. But more than that, the example also illustrates a general method for calculating enthalpy changes in any reaction. Even if there are a number of different products and reactants, the procedure is the same. First look up the standard heats of formation of all the products, remembering that $\Delta H_f°$ for any pure element in its most stable physical form will be zero at 25°C. Since heats of formation are tabulated for a single mole, multiply each value by the number of moles of that particular product in the balanced equation describing the change in state. Then add up all these heats. The sum will be signified by a capital sigma (Σ), that is, as $\Sigma \Delta H_{f,products}°$. Now do the same for the reactants and thus obtain $\Sigma \Delta H_{f,reactants}°$. The overall enthalpy change for the reaction is a simple difference,

12.7 $\Delta H° = \Sigma \Delta H_{f,products}° - \Sigma \Delta H_{f,reactants}°$.

One more example might be helpful. You will recall that the compound chosen to illustrate the calculations of Chapter 4 was octane, C_8H_{18}. Later you learned that a specific isomer of this compound, isooctane, is used as a standard in rating the "knock" of gasolines. And, of course, it is the combustion of isooctane which is the basis for its use as a fuel. Therefore, it is important to know how much heat can be generated by the reaction summarized in Equation 12.8. (Isooctane is symbolized $i\text{-}C_8H_{18}$.)

12.8 $i\text{-}C_8H_{18}(l) + \dfrac{25}{2}\,O_2(g) \rightarrow 9H_2O(l) + 8CO_2(g)$

According to Equation 12.7, the enthalpy change associated with the process is

12.9 $\Delta H^\circ = \underbrace{[9\Delta H^\circ_{f,H_2O(l)} + 8\Delta H^\circ_{f,CO_2(g)}]}_{\sum \Delta H^\circ_{f,\text{products}}} \; - \; \underbrace{\left[\Delta H^\circ_{f,i-C_8H_{18}(l)} + \dfrac{25}{2}\Delta H^\circ_{f,O_2(g)}\right]}_{\sum \Delta H^\circ_{f,\text{reactants}}.}$

Next we consult Table 12.1 and substitute the appropriate standard heats of formation for all reactants and products. Then we carry out the indicated additions and subtractions:

$$\Delta H^\circ = 8(-94.1) + 9(-68.3) - \left[-62.0 + \dfrac{25}{2}(0.00)\right] \text{kcal}$$

$$= -1305.5 \text{ kcal.}$$

The first thing to note about the answer is the minus sign, an indication that the enthalpy of the system decreases when the isooctane combines with oxygen to form water and carbon dioxide. Obviously, this energy is not destroyed. Instead, it makes itself felt as it flows across the boundary to the surroundings. Stated with the correct common sense of everyday experience—heat is given off when gasoline burns.

Thanks to the calculation, we can be more specific. The complete oxidation of 1 mole of $i\text{-}C_8H_{18}$ to yield 9 moles of $H_2O(l)$ and 8 moles of $CO_2(g)$ is attended by the evolution of 1305.5 kcal of heat. This quantity is equivalent to about one-half of the daily caloric intake of an adult human being. It is enough to produce a rise of 1°C in the temperature of 1,305,500 g of water (well over a ton!). Or, if you prefer, it is sufficient to melt 16,300 g of ice (about 36 lb) at 0°C. No matter how you look at it, 1305.5 kcal is a significant amount of heat energy. And it comes from burning only 1 mole, 114 g, of isooctane. Twice that weight would yield twice the energy, half the weight would yield half. In other words, the quantity of heat evolved, that is, the enthalpy change, is directly proportional to the mass of $i\text{-}C_8H_{18}$ converted to H_2O and CO_2. In a problem at the end of this chapter you are invited to make use of this fact to calculate the heat evolved in burning a gallon of gasoline. Other similar applications are also included. They all help illustrate the utility of enthalpy tables.

In many cases, the values included in such tables have not been determined by measuring the heat flow which accompanies the direct combination of the elements involved. For example, the heat of formation of isooctane cannot be measured directly. The combination of carbon and hydrogen is difficult and yields a large number of hydrocarbon products. Therefore, one must resort to a roundabout method. In practice, it is the heat of combustion of i-C_8H_{18} which is experimentally determined. A known weight of the compound is burned with an excess of oxygen in a device called a calorimeter. The quantity of heat evolved is measured as it flows to the surroundings, and thus one obtains the overall enthalpy change for the reaction: $\Delta H° = -1305.5$ kcal per mole of i-C_8H_{18}. The value of $\Delta H^°_{f,i\text{-}C_8H_{18}}$ is calculated from Equation 12.9. Note, however, that this time, there is a different unknown. Again, an opportunity for a similar computation is provided at the end of the chapter.

It does not require too many such examples to present convincing proof that thermodynamic quantities are of profound importance to people who run power plants and people who eat green plants—in short, to all people. The heat available from coal and cole slaw is a consequence of changes in internal enthalpy which occur when these reactants are chemically converted into products. The ability of science to deal precisely with such thermochemical transformations is an example of its great practical power. But, as you have observed, the study of thermodynamics requires work and heat and energy. Right now, you may feel as though you have expended more than enough of all three. There is, however, an important relationship between energy and enthalpy which might help you understand both a bit better. For that reason, I suggest that you read on.

More energy spent on enthalpy (and vice versa)

Our point of departure for determining the formal relationship between energy and enthalpy is the First Law of Thermodynamics. Let us apply it to another change in state, the expansion of a gas from an initial volume, V_1, to a final volume, V_2. The resulting volume difference, $\Delta V = V_2 - V_1$, can be used to do work. For example, the expanding gas can drive the piston in a cylinder of a steam or internal combustion engine. The amount of work done depends not only upon the volume change, but also upon the resisting pressure against which the gas expands. When the opposing pressure has a constant value of P, the work is given by the equation, $W = P\Delta V$. Since P is always positive, the sign of W depends upon the sign of ΔV. An expansion process, as we have here, means that $V_2 > V_1$ and ΔV is thus positive. So is W, and work is done by the system in the surroundings. If, on the other hand, ΔV is negative, W also bears the same sign. Which only states the familiar fact that work must be done on a gas in order to compress it.

The amount of work which is required to cause compression or is available because of expansion will, most immediately, be expressed in units of pressure × volume. For example, the expansion of a gas from an initial volume of 1 liter to a final value of 2 liters, against an opposing pressure of 1 atm, produces 1 liter atm of work. The unit is an uncommon one. Indeed, it may not be immediately apparent that it has the dimensions of work, mass × length²/time². But it does, and we are thus fully justified in writing its equivalent in more conventional calories,

1 liter atm = 24.2 cal.

Clearly, a conversion of this sort is necessary to get heat and work and energy all in the same units, if we are to use the equation of the First Law. Substituting $P\Delta V$ for W, we obtain

$$\Delta E = Q_p - P\Delta V.$$

Remembering that at constant pressure, $Q_p = \Delta H$, we can once more rewrite the expression as

12.10 $$\Delta E = \Delta H - P\Delta V.$$

Equation 12.10 should help explain the distinction between enthalpy and energy. The only difference between ΔH and ΔE is a little work! Consider once again the vaporization of 1 mole of water (see Figure 12.5). We have calculated that under standard conditions, $\Delta H°$ for the process is 10.5 kcal. Not all this heat goes

FIGURE 12.5.   *Representation of the change in state:*
1 mole $H_2O(l)$ (25°C, 1 atm) → 1 mole $H_2O(g)$ (25°C, 1 atm).
$\Delta H° = 10,500$ cal = increase in enthalpy upon vaporization.
$W = P\Delta V = 24.5$ liter atm or 593 cal = work done by
expanding H_2O. $\Delta E = \Delta H - P\Delta V = 9900$ cal = increase in
energy upon vaporization.

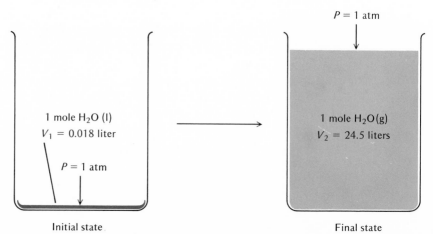

Initial state   Final state

to increase the internal energy of the system, that is, ΔE does not increase by the full amount. Some of the heat is expended as $P\Delta V$, the work of pushing back the atmosphere. To calculate this quantity we first note that the volume change is the difference between V_g, the volume of 1 mole of the gaseous product, and V_1, the volume of the liquid starting state. The former follows from the Ideal Gas Law:

$$V_g = \frac{nRT}{P} = \frac{1 \text{ mole} \times 0.0821 \text{ liter atm/mole } °K \times 298°K}{1 \text{ atm}}$$
$$= 24.5 \text{ liters.}$$

In contrast, the initial volume of 1 mole (18.0 g) of liquid water is only 18 ml or 0.018 liter. Therefore,

$$\Delta V = V_g - V_1 = 24.5 - 0.018 \text{ liters,}$$

or just about 24.5 liters. Thus, $P\Delta V = 24.5$ liter atm or 24.5 liter atm $\times$ 24.2 cal/liter atm $= 593$ cal. It follows at once that

$$\Delta E = 10,500 \text{ cal} - 593 \text{ cal}$$
$$= 9907 \text{ cal, or approximately 9.90 kcal.}$$

This represents the increased energy of the gaseous state relative to the liquid. The small difference between ΔH and ΔE is characteristic of most chemical and physical changes.*

It is also worth noting that for any process performed at constant volume, the $P\Delta V$ term will be zero. Since no expansion or compression work is done, all the heat flow goes to increasing the internal energy of the system. Thus, using a subscript to designate the constant volume condition,

$$\Delta E = Q_v.$$

Again we are concerned with differences in the state function E, not with absolute values. Classical thermodynamics cannot provide us with the means of calculating the amount of energy in a substance or any other system. For such information we must rely upon quantum mechanics and statistical methods. This we did in earlier chapters when we considered the quantized energy levels in atoms or molecules and the energy of translation predicted by the kinetic theory of gases. But measurements of heat absorbed or work done can only give us energy changes within the system. It

---

* The perceptive reader will detect an omission in the example used to illustrate the equivalence of heat and work (pp. 346–348). I did not include the work done when the water expands against atmospheric pressure as it is warmed from 20° to 30°C. But the change in volume which attends this increase in temperature is so slight that the $P\Delta V$ term is negligible. To introduce it would have unnecessarily obscured the issue. The example stands as fundamentally correct.

turns out that such differences are quite enough to explain many chemical phenomena and permit useful predictions.

Envoi on energy

Thus far we have concentrated almost exclusively on energy expressed as work and heat. To end this chapter without alluding to some of its other familiar forms would be a serious oversight. The ultimate energetic source of our universe is nuclear energy, the potential locked up at the very core of matter. In the fiery furnace of the sun and billions of other stars, the nuclei of light atoms like the isotopes of hydrogen combine in *fusion* reactions. In the process, vast quantities of energy are released. One could almost say, in defiance of the First Law of Thermodynamics, energy is created. For indeed, the energy of fusion (and fission) arises from the conversion of matter to energy in accordance with Einstein's famous formula, $E = mc^2$.

This energy is transported through space in α and β particles, but particularly in electromagnetic radiation—cosmic, γ, and X rays; ultraviolet, visible, and infrared light. Fortunately for the inhabitants of this planet, an ozone-rich layer in the lower stratosphere, at an altitude of 10–30 miles, filters out many of the more energetic and hence potentially lethal rays. Thus, only about half of the energy reaches the surface of the earth, chiefly in the near-ultraviolet, visible, and infrared regions of the spectrum. It amounts to a phenomenal 8.6×10^{17} kcal per day!

Most of this energy goes into heating our environment—the atmosphere, the oceans, and the earth's crust. But some of it, a little over 1%, is trapped by green plants. Through the process of photosynthesis these fantastic factories convert light energy into the potential energy of chemical bonds. In so doing they take two simple and plentiful compounds, carbon dioxide and water, and fashion them into far more complex substances, notably the sugar glucose, $C_6H_{12}O_6$. We can write this most significant of all syntheses as

12.11 $6CO_2(g) + 6H_2O(l) \rightarrow C_6H_{12}O_6(s) + 6O_2(g).$

The standard enthalpy change associated with this reaction is 673 kcal. In other words, 673 kcal must be absorbed to produce 1 mole of glucose from 6 moles of CO_2 and 6 moles of H_2O. The source of this energy is the sun. Beyond that, glucose molecules are linked to form chains of cellulose and starch, storehouses of energy in vegetable form.

We and the animals which we feed upon are absolutely dependent upon these stores for our existence. Unable to directly utilize the energy of the sun, we regain it by burning, in our bodies and our power plants, the molecules made by photosynthesis. In effect, we reverse Equation 12.11 to our thermodynamic benefit. To be sure, we have learned to use the rapidly decreasing potential energy of a waterfall to drive a dynamo. We have built nuclear power plants which tap the energy of the atom. We have even succeeded in devising solar cells and batteries which convert light into electrical energy. But by far the largest fraction of the power which we sometimes wantonly consume (95% of that generated in this country) comes from fossil fuels—coal and oil. Thus, we are beneficiaries of the energy husbanded by plants which died and decayed long before our ancestors began to walk erect. It is a finite and irreplaceable store, one which we must jealously and zealously conserve.

Societies of men express and utilize this borrowed energy in a variety of ways, but in no fundamental fashion which is not exhibited by some animal organism. Muscles were used as sources of mechanical energy long before the steam engine was invented. The electric eel generates a current without the benefit of a dynamo. And fireflies produce light far more efficiently than gas lanterns or incandescent bulbs, dissipating much less energy as heat. But particularly in the entrapment of energy into new chemical bonds does the living organism excel. What we can perform internally in an instant may require months and years of work in a laboratory.

All these forms of energy constitute a great continuous chain. Energy is carried through space as photons, through wires as a stream of electrons. It is stored in coiled springs and knotted muscles, chemical bonds and atomic nuclei, gunpowder and gasoline and gum drops. It is made manifest in roaring cataracts, whirling wheels, deafening explosions, and blazing fires. It tills the earth and sends men to the moon. It destroys cities and it builds them. It keeps hearts beating and it silences them. But as energy is endlessly passed from form to form, none is ever lost. It remains constant and conserved. And one of man's great challenges is to use it wisely and well.

For thought and action

12.1   *If* you believe the First Law of Thermodynamics is true (and most people do), *why* do you believe it? If you don't, why don't you?

12.2   The following are examples of statements you're likely to hear any day. Point out why they are inaccurate, and try to rephrase each of them in a way which conveys the probable intended meaning, but is consistent with the laws of thermodynamics.

(a) "Boy am I exhausted, I've used up all kinds of energy today!"

(b) "The proposed power plant will create vast quantities of new energy for our region."

(c) "Since the temperature of the electric burner is higher than the temperature of the potatoes and water, the burner obviously contains more heat."

12.3   Two ice cubes at 0°C, each weighing 35 g, are placed in a glass containing 300 g of liquid water at 30°C.

(a) Calculate the amount of heat given up as the ice melts completely.

(b) Determine the final temperature of the water. *Hint:* The melted ice gets warmer, the original 300 g of water gets cooler, and both end up at the same temperature. Since the glass is an insulated one, you can assume no heat loss to the surroundings. Incidentally, the First Law of Thermodynamics applies.

12.4   Describe, if you can, an experiment that could unequivocally prove that heat is not a massless fluid.

12.5   One of James Joule's predictions was that the temperature of the water at the bottom of a waterfall should be slightly higher than that at the top. Explain why.

12.6   A typical meal provides about 1000 kcal of energy. How high could you lift a 200 kg piano with that much energy—assuming, of course, that you could get all the heat converted into work all at once.

12.7   A beaker containing 100 g of water, initially at 20°C, is placed on a burner. As a result, the temperature of the water increases to 50°C. The beaker is then placed in contact with an ice bath until the temperature of the water has dropped to 30°C. In other words, the water undergoes this two-step change in state:

(1)   100 g $H_2O(l)$ (20°C, 1 atm) → 100 g $H_2O(l)$ (50°C, 1 atm)

(2)   100 g $H_2O(l)$ (50°C, 1 atm) → 100 g $H_2O(l)$ (30°C, 1 atm)

Specify Q and ΔE for step 1, step 2, and the total change, i.e., step 1 + step 2.

12.8   An uninsulated beaker contains 100 g of water initially at 20°C. A quantity of work amounting to 6000 joules is done on the water by stirring it rapidly. As a result, the temperature of the water rises to exactly 30°C.

(a) Indicate numerical values for Q, W, and ΔE for this change in state, taking particular care to get the signs right.

(b) Explain the apparent discrepancy between this problem and the similar example in the text.

12.9   During a particularly strenuous day, an athlete consumes food

with a caloric equivalent of 4000 kcal. He also radiates 3000 kcal of heat into his surroundings and does 2000 kcal of work.
(a) Calculate Q, W, and ΔE for the athlete.
(b) Explain what has happened in order for the energy of the world to be kept constant.
(c) Why would a long series of days like the one described be ill-advised?

12.10 Specify whether the following changes of state are exothermic or endothermic and indicate the sign of the enthalpy change:
(a) The freezing of water
(b) The evaporation of rubbing alcohol
(c) The burning of a log
(d) The decomposition of water into hydrogen and oxygen
(e) The dissolving of salt in water
(f) The rusting of iron

12.11 Arrange the following states in order of increasing enthalpy. Note that on an elementary, atomic level they all represent the same system.
(a) 1 mole $H_2O(l)$ (25°C, 1 atm)
(b) 1 mole $H_2O(s)$ (0°C, 1 atm)
(c) 1 mole $H_2(g)$ + ½ mole $O_2(g)$ (25°C, 1 atm)
(d) 1 mole $H_2O(g)$ (100°C, 1 atm)
(e) 1 mole $H_2O(l)$ (0°C, 1 atm)
(f) 1 mole $H_2O(l)$ (100°C, 1 atm)

12.12 A gallon of isooctane weighs 2.61 kg. Calculate the standard enthalpy change associated with burning this much i-C_8H_{18}. (See the body of the text for the necessary additional data.)

12.13 As you know, propane, C_3H_8, is used to power some cars and trucks. Its complete oxidation corresponds to the equation

$$C_3H_8(g) + 5O_2(g) \rightarrow 4H_2O(l) + 3CO_2(g).$$

(a) Using the table of standard heats of formation, calculate the enthalpy change which characterizes the burning of 1 mole of propane and explain the answer in terms of heat flow.
(b) A gallon of bottled propane weighs 2.14 kg. Determine the quantity of heat which can be obtained from the combustion of this much C_3H_8.

12.14 Isooctane is selling at 32.9¢ per gallon, propane is 47.2¢ per gallon. Consult Problems 12.12 and 12.13, and then calculate the cost, in cents per kilocalorie, of the heat energy provided by the two fuels.

12.15 A good-sized martini weighs 100 g and contains 35.0% ethyl alcohol, C_2H_5OH, by weight. Since this is a very dry martini, you can neglect the vermouth and assume that the nonalcoholic 65.0% is water.
(a) First, find the enthalpy change which occurs when 1 mole

of C_2H_5OH is oxidized at 25°C and 1 atm according to the following equation:

$$C_2H_5OH(l) + 3O_2(g) \rightarrow 3H_2O(l) + 2CO_2(g).$$

(b)   Using your answer to part (a) and the data given above, calculate the heat evolved when said martini is subjected to internal combustion in somebody's insides. The fact that the temperature there is 37°C will not significantly alter the answer. (You will note that this problem does not take into consideration the caloric contribution of the olive. But at 15 kcal, it is definitely not the most fattening thing about a martini.)

12.16   Glucose, an important sugar, is oxidized in a reaction represented by this equation:

$$C_6H_{12}O_6(s) + 6O_2(g) \rightarrow 6H_2O(l) + 6CO_2(g).$$

The enthalpy change which accompanies this process at standard conditions is equal to -673 kcal per mole of glucose. Now suppose that the oxidation of a 10¢ candy bar is equivalent in heat flow and enthalpy change to the oxidation of 25.0 g of glucose.
(a)   Compute the kilocalories of heat absorbed by a chemistry professor who eats such a candy bar.
(b)   If the professor requires 5.75 kcal per minute of lecture, how long can he talk on one candy bar?

12.17   The body of the text reports that about 8.6×10^{15} kcal of energy are trapped each day by green plants. How many kilograms of glucose could be synthesized from $CO_2(g)$ and $H_2O(l)$ if all this energy were used to promote that reaction? (The $\Delta H°$ associated with the formation of 1 mole of glucose by this process is $+673$ kcal.) For the sake of greater familiarity, also convert the answer into tons.

12.18   Calculate and compare the total enthalpy change (in kcal) during (a) a year's worth of eating (at 3000 kcal per day) and (b) a year's worth of driving (15,000 miles at 18 miles per gallon of iso-octane). (See Problem 12.12.)

12.19   One of the components of natural gas is methane, CH_4. The equation which represents its combustion is

$$CH_4(g) + 2O_2(g) \rightarrow 2H_2O(l) + CO_2(g).$$

(a)   Calculate the standard enthalpy change and heat flow which occur when 1 mole of methane is burned.
(b)   Suppose you want to heat 2.00 kg of water (about 2 qt) from room temperature, 25°C, to the boiling point, 100°C. How much heat will this change in state require?
(c)   Finally, how many grams of CH_4 must be burned in order to provide this much heat?

12.20 The reduction of iron oxide has been the subject of problems in several previous chapters. Here it is again. This time you are asked to calculate the standard enthalpy change associated with the process

$$2Fe_2O_3(s) + 3C(s) \rightarrow 4Fe(s) + 3CO_2(g).$$

Also indicate whether heat is absorbed or evolved during the reaction. (Note that the balanced equation specifies that 2 moles of Fe_2O_3 are being reduced. You should also be aware that diamond is not usually used as a reducing agent.)

12.21 The heat of combustion of *n*-butane, the gas used in some cigarette lighters, is 688.1 kcal/mole. In other words, $\Delta H°$ for the following reaction is -688.1 kcal at 25°C and 1 atm:

$$n\text{-}C_4H_{10}(g) + \frac{13}{2}O_2\,(g) \rightarrow 5H_2O\,(l) + 4CO_2\,(g).$$

Couple this result with tabulated standard heats of formation to calculate the standard heat of formation of *n*-butane.

12.22 In Chapter 11 you read that NO is formed by the direct combination of nitrogen and oxygen under the influence of lightning or the high temperatures of an internal combustion engine. Use the tabular value for $\Delta H°_{f,NO(g)}$ to help explain this behavior.

12.23 Distinguish between energy and enthalpy. Why do you suppose chemists spend so much time talking about the latter concept?

12.24 A critic of thermodynamics observes (rightly) that energy is released when coal burns, and cites this fact as evidence of the incorrectness of the First Law of Thermodynamics. Tell him why he's wrong.

12.25 Vegetarians maintain it is more efficient, in terms of energy utilization, to eat wheat germ and rice than hamburgers and hot dogs. Does this mean that energy is somehow destroyed in producing the latter foods? Explain.

12.26 Providing heat for a dwelling is a basic human need in many parts of the world. Suggest a variety of methods of accomplishing this end; and for each one, trace the energy chain back to a primary source, for example, the sun or the atomic nucleus.

Suggested readings

A number of good paperback introductions to thermodynamics have recently appeared. Four are listed below.

Davies, William G., "Introduction to Chemical Thermodynamics: A Non-Calculus Approach," Saunders, Philadelphia, Pennsylvania, 1972.

Goates, J. Rex, and Ott, J. Bevan, "Chemical Thermodynamics: An Introduction," Harcourt, New York, 1971.

Pimentel, George C., and Spratley, Richard D., "Understanding Chemical
Thermodynamics," Holden-Day, San Francisco, California, 1969.

Van Ness, H. C., "Understanding Thermodynamics," McGraw-Hill, New
York, 1969.

A detailed account of Count Rumford's experiments is found in Volume
1 of "Harvard Case Histories in Experimental Science" (James Bryant
Conant, general ed.), Harvard Univ. Press, Cambridge, Massachusetts,
1966. The pertinent section, pp. 117–214, is entitled The Early Development of the Concepts of Temperature and Heat: The Rise and
Decline of the Caloric Theory, and was prepared by Duane Roller.

The entire issue of *Scientific American* for September, 1971 (Vol. **225**)
was devoted to energy and power.

13

Entropy, free energy, and the second law of thermodynamics

Changes: real and reversible

One of the most profound aspects of ourselves and our environment is so much a part of existence that it is seldom marveled at. And yet, it is literally a matter of life, death, and everything between. For we inhabit a universe of inexorable directionality. It is a world in which buds swell and open, green leaves flourish, then turn red or gold or brown, and fall to the ground to crumble and decay. Man, too, follows a similar path. From the fertilized ovum, to "the infant, mewling and puking in the nurse's arms," through "the whining schoolboy," and all the other ages of man until "last scene of all, that ends this strange eventful history, is second childishness and mere oblivion, sans teeth, sans eyes, sans taste, sans everything." This is the panorama of natural change, and the motion picture which records it cannot be run backwards. Dead leaves do not turn green, leap up to bare branches, and shrink into the bud. And the dry bones of Ezekiel's vision do not reassemble themselves in this world. Because the ways of this world are *irreversible*.

An understanding of this irreversibility, an appreciation of its implications, and the predictive application of its governing principles are all achievements of thermodynamics. With the aid of this subscience we can design more efficient power plants, develop new batteries, and predict the direction of untried chemical reactions. Yet, in spite of the pragmatic versatility of thermodynamics, much of its reasoning deals with the realm of the unreal. Once again we encounter a familiar paradox of scientific thought. A leap into the abstract world of the ideal is about to aid our understanding of things as they are.

We start with a system which is comfortably concrete—a fixed quantity of a gas confined in a cylinder by a piston. And the process to be investigated is equally familiar—the production of work by the expansion of a gas. Let us begin our thought experiment (it could be done in a laboratory) by assuming that the piston and the cylinder both have a radius of 10.0 cm, as indicated in Figure 13.1. At the start of the experiment, the piston is held down by two stops, marked s_1. They are placed so that the distance between the bottom of the cylinder and the bottom of the piston is 31.8 cm. These numbers were chosen merely to make the

FIGURE 13.1.   *Representation of the expansion of a perfect gas in a cylinder, the change in state raising a piston from stops s_1 to stops s_2. The initial and final values for volume, pressure, and temperature are specified on the cylinder. See text for details.*

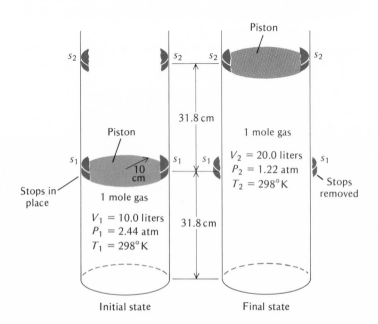

volume under the piston a nice, even 10.0 liters.* This is V_1, the volume of the initial state. We will further assume that it contains 1 mole of a perfect gas. If the temperature is 25°C or 298°K, we can use the ideal gas equation to calculate that the initial pressure, P_1, must be 2.44 atm.

In the loose terminology of normal discourse we would say "the gas is under pressure." If the restraint is released by removing the stops, the gas will expand and the piston will rise until it encounters the second set of stops, marked s_2 in the figure. For simplicity's sake we will position these higher stops so that the piston comes to rest 63.6 cm above the bottom of the cylinder. As a convenient consequence, the volume of gas exactly doubles to a value of $V_2 = 20.0$ liters. During this expansion the temperature is maintained at a constant 25°C, a control which requires the flow of heat from the surroundings to the system. And because T is invariant, the gas behaves in accordance with Boyle's Law. The final pressure, P_2, thus equals 1.22 atm. The total change in state can be summarized as follows:

13.1

$$1 \text{ mole gas } (V_1 = 10.0 \text{ liters}, P_1 = 2.44 \text{ atm}, T_1 = 298°K) \rightarrow$$
$$1 \text{ mole gas } (V_2 = 20.0 \text{ liters}, P_2 = 1.22 \text{ atm}, T_2 = 298°K).$$

We have already commented on the fact that the expansion of a gas can be used to do work. For example, a weight could be placed upon the piston and lifted by the gas as it increases its volume. If the opposing pressure against which the expansion occurs has a constant value of P, this work amounts to $P(V_2 - V_1)$ or $P\Delta V$. Thus, the work done is directly proportional to P. But there is an important restriction which limits the amount of work which can be produced in a single expansion step; P cannot be larger than P_2. If it were, the opposing pressure would halt the expansion short of the second set of stops. Thus the final volume would be less than 20.0 liters and the desired final state would not be attained.

Let us see how this argument applies to our specific example. The final pressure of the gas is 1.22 atm. Therefore, the highest permissible opposing pressure during a one-step expansion must also be 1.22 atm. Using this value of P in the work expression,

$$W_{exp} = (1.22 \text{ atm}) (20.0 - 10.0 \text{ liters}) = 12.2 \text{ liter atm.}$$

Here we have introduced the subscript "exp" to suggest expansion. Converting to calories,

$$W_{exp} = 12.2 \text{ liter atm} \times 24.2 \text{ cal/liter atm} = 295 \text{ cal.}$$

---

* If you wish to check this calculation, recall that the volume of a cylinder is $\pi r^2 h$, where r is the radius and h is its height. Also note that 1000 cm^3 = 1 liter.

This value represents the maximum amount of work which could be done by a single-step transition between the specified initial and final states, a process depicted in Figure 13.2. Some of this work of course goes into pushing back the atmosphere. This quantity, call it W', is again given by $P\Delta V$. Only this time we must use the pressure of the atmosphere as P. If it is exactly 1 atm, we find that

$$W' = (1.00 \text{ atm}) (20.0 - 10.0 \text{ liters}) = 10.0 \text{ liter atm}$$
$$= 10.0 \text{ liter atm} \times 24.2 \text{ cal/liter atm} = 242 \text{ cal}.$$

The difference, $295 - 242 = 53$ cal is available for lifting weights. We can calculate (remembering $W = mgh$) that this much work could raise a weight of 71.0 kg through the distance of 31.8 cm. This, you observe, is just the distance traveled by the piston. In other words, we could place 71.0 kg on the piston (assume the mass of the latter is negligible), pull out the lower stops, and the expanding gas would carry this weight right up to the second set of stops.

In spite of this absolute restriction, we can get more work out

FIGURE 13.2.   *Single-step irreversible expansion of a perfect gas from 10.0 to 20.0 liters against a constant opposing pressure of 1.22 atm. Work done equals 1.22 atm* $\times$ *(20.0 − 10.0 liters) = 12.2 liter atm or 295 cal.*

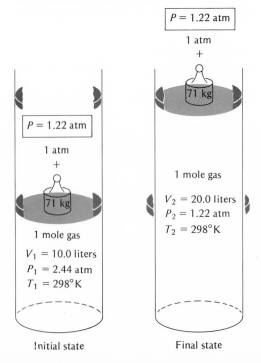

Initial state   Final state

of the same change in state—if we're clever enough. The secret is to do it a little at a time. Suppose we start with a pile of lead shot resting on the piston (see Figure 13.3). There is also a limit on the mass of this shot. The pressure it exerts plus the pressure of the atmosphere can be no greater than the initial internal pressure of the gas, $P_1 = 2.44$ atm. If it were greater, the gas would be compressed to a volume smaller than 10.0 liters, which would alter the initial state. A little arithmetic indicates that this maximum mass is 466 kg or 466,000 g. If each shot weighs 1 g, that amounts to 466,000 little BB's.

Now we again pull out the stops. Nothing happens because the downward pressure of the atmosphere and the shot exactly equals the pressure of the gas. So we remove one shot. The total mass on the piston is now 465,999 g. Consequently, the opposing pressure which it and the air produce is reduced very slightly below 2.44 atm. As a result, the gas expands a little and the piston rises 8.05×10^{-5} cm. Then it again comes to a halt. So we remove another shot, making the mass 465,998 g. Again the gas expands

FIGURE 13.3. *Multiple-step expansion of a perfect gas from 10.0 to 20.0 liters against an opposing pressure which gradually decreases from 2.44 to 1.22 atm. If the expansion is carried out in an infinite number of infinitely small steps, the process is reversible and the amount of work done is 410 cal.*

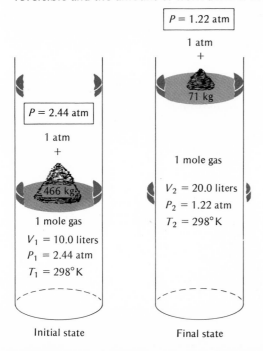

$P = 1.22$ atm

1 atm

+

71 kg

$P = 2.44$ atm

1 atm

+

466 kg

1 mole gas

1 mole gas

$V_1 = 10.0$ liters
$P_1 = 2.44$ atm
$T_1 = 298°$K

$V_2 = 20.0$ liters
$P_2 = 1.22$ atm
$T_2 = 298°$K

Initial state

Final state

and the piston moves up, raising the 465,998 shot another 8.05 × 10^{-5} cm. And so we continue, removing one shot at a time until only 71,000 remain. At each of the 395,000 steps the piston rises 8.05 × 10^{-5} cm. When we finally remove the 395,000th shot, we find that we are at the desired final state. The volume of the gas is exactly 20.0 liters and the bottom of the piston is 63.6 cm above the bottom of the cylinder. The pressure *of* the gas and the pressure *on* the gas are both 1.22 atm. One atmosphere plus 71.0 g of mass (another 0.22 atm) press down on the piston. The change in state is exactly identical to that produced earlier in one fell swoop. But the path followed is quite different. We can now add up all the work done in the 395,000 individual steps, and when we do we find it is considerably greater than 242 cal. We have, in fact, succeeded in getting more work out of the gas by expanding it a little at a time. More mass has been raised in the surroundings.

Such behavior is not beyond common sense and experience. And it is not unreasonable to assume that the smaller the shot, the more work we will be able to extract from the same change in state. But even here there is a limit. If we make the shot infinitely small and take an infinite number of steps to reach the final state, we will do W_{max}, the maximum amount of work possible. The calculation of W_{max} is really quite simple. But you may have to trust my mathematics because the method does require a little calculus. Without going into details, the answer turns out to be 410 cal, significantly greater than the one-step value of 295 cal.

At this point we pause to apply the First Law. We have, in the past, argued that temperature is a measure of internal energy. Therefore, since temperature is constant for this change in state, energy must also be constant. Symbolically, $\Delta E = 0$. It must follow, then, that $\Delta E = 0 = Q - W$ or $Q = W$. Remember that the system had to be heated in order to maintain the temperature at 25°C or 298°K. All of this positive heat was converted into the work done in the surroundings during the expansion. In the first, single-step expansion, $Q = W = 295$ cal. But in our infinite series of steps, $Q_{max} = W_{max} = 410$ cal.

Now we reverse the change in state back to the initial condition of 10.0 liters, 2.44 atm, and 25°C. Again, we can do it in many different ways. The sledge hammer method involves placing 466 kg on the piston and squeezing the gas back to its original state in one step. The pressure due to this mass plus the pressure of the atmosphere equals 2.44 atm. Thus, the work associated with the compression ("comp") is

$$W_{comp} = (2.44 \text{ atm}) (10.0 - 20.0 \text{ liters}) = -24.4 \text{ liter atm}$$
$$= -24.4 \text{ liter atm} \times 24.2 \text{ cal/liter atm} = -590 \text{ cal}.$$

Note that the initial and final states are just the opposite of what they were for the expansion experiment. The new starting volume is 20.0 liters and the stopping volume is 10.0 liters. Thus, follow-ing the rule of final minus initial, we see that $V = 10.0 - 20.0$ liters $= -10.0$ liters. The negative sign is carried over to W_{comp} where it indicates that work is destroyed in the surroundings. Or, if you prefer, work is done *on* (not by) the system. Observe also that -590 cal is a good deal of work.

Chemists, like every part of nature, are lazy. So we chose to do as little work as possible by starting with a weight of only 71.0 g and an opposing pressure of 1.22 atm. This time we add shot, one at a time, gradually increasing the pressure and forcing down the piston. If the shot are infinitely small, the smallest quantity of work need be expended in the process. We calculate it to be $W_{min} = -410$ cal. After this amount of work has been done on the system, it is returned to its initial state. Thus we have converted W_{min}, a negative term, to its equal, Q_{min}. And, since Q_{min} is also negative, heat must flow to the surroundings to keep the temperature con-stant. ΔE is again zero.

The net result of our *cyclic* process of infinitesimal changes is no change whatsoever, either in the system or its surroundings. The gas has returned to its original state. So have the weights and so has the heat.

$$W_{cycle} = W_{max} + W_{min} = Q_{max} + Q_{min} = 410 + (-410) \text{ cal} = 0 \text{ cal}$$

In a word, the entire operation is *reversible*. We are right back where we started from and the universe is no different than it was before. To which the critical reader might well ask, "So what?" If all our infinitesimal efforts produced no net change, why all the fuss and 12 paragraphs of verbiage?

The paradoxical point is simply this: the example of a reversible, cyclic process is important because nature never works that way! *All the spontaneous changes which actually occur in this universe are irreversible.* In order to explore the consequences of this state-ment, let's have another look at an extreme example of irrevers-ibility. We have already calculated the work done or "evolved" by a one-step expansion of a mole of gas between the states in-dicated in Equation 13.1. Recall that it was 295 cal. In marked con-trast, the work required for a one-step compression of the gas from 20.0 liters to its original volume of 10.0 liters was -590 cal. Adding up these two values to obtain the work associated with the entire cycle, we see that we are involved in a losing game:

$$W_{cycle} = W_{exp} + W_{comp} = 295 - 590 \text{ cal} = -295 \text{ cal}.$$

We have put 590 cal worth of work into compressing a gas and have only managed to get 295 cal of work out of its expansion. The result is rather like buying eggs for seven cents apiece and selling them for five cents apiece, a procedure which is profitable only in the mad world of "Catch-22." In this universe you always lose. As far as the First Law of Thermodynamics goes, you can't win, but you should at least break even. And it is true enough that in any change in state, irreversible or reversible, real or imaginary, energy is conserved. It unquestionably was in all our examples, no matter how the process was done or what path was followed. But because of another, still more vexing property of nature, *you can never break even.*

This pessimistic little homily is somewhat more elegantly enshrined as the *Second Law of Thermodynamics.* It is an observation about the workings of the world which explains such apparently diverse phenomena as the direction of heat flow, falling objects, and chemical reactions. We will soon encounter all of these examples, though armed with a few new concepts to make the meeting more meaningful.

Introducing entropy: $\Delta S = Q_{rev}/T$

Our attempt to accurately formulate the directionality of the universe has thus far pointed up a significant characteristic of all real, spontaneous transformations—they are irreversible. This at once suggests that something changes in such a process—something which cannot be fully restored by running it backwards. We start our search for this imprecisely defined entity by reexamining an example of the previous section. Anyone who has ever changed a flat tire can readily testify that the expansion of a gas against a lower opposing pressure certainly fits the requirement of spontaneity. Air simply refuses to flow, by itself, from the atmosphere *into* a punctured tire. But the opposite behavior is common enough. So we look at it, as exemplified by the now-familiar change in state,

1 mole gas (10.0 liters, 2.44 atm, 298°K) →
1 mole gas (20.0 liters, 1.22 atm, 298°K).

We have already calculated that when this process occurs in a single step against an opposing pressure of 1.22 atm, 295 cal of heat are absorbed and converted to work. Of course, Q_{exp} can be smaller, if the opposing pressure is smaller. And it can be larger, up to the limit imposed by the irreversible process, $Q_{max} = 410$ cal. In other words, for this change in state, the heat flow associated with any real and, hence, irreversible process must be less than the heat flow attending an ideal, reversible expansion. Adopting subscripts which identify the irreversible and reversible pathways, we see that

$$Q_{irr} < Q_{rev}.$$

Since we required that the change in state occur at a constant temperature, T, we can also write

$$\frac{Q_{irr}}{T} < \frac{Q_{rev}}{T}.$$

To again belabor the obvious, Q_{irr}/T is not a constant. It depends upon the number and nature of the steps taken in the transformation. But Q_{rev}/T can have only one value for this change in state, 410 cal/298°K or 1.37 cal/°K. No matter how this particular volume increase is achieved, Q_{rev}/T always equals 1.37 cal/°K. In other words, it behaves as if it were the difference between two values of a state function characterizing the system.

This new function appears to be what we've been looking for. So we give it a name, *entropy*, and a symbol, S. We represent the difference between the entropy of the final state and that of the initial state in the usual manner, $\Delta S = S_2 - S_1$. Thus, our defining equation for an entropy change, under the special conditions of constant temperature, becomes.

13.2 $$\Delta S = \frac{Q_{rev}}{T}.$$

We can generalize that for any constant temperature process for which Q_{rev} is positive, ΔS will also bear the same sign. In other words, the entropy of the system will increase. On the other hand, if Q_{rev} is negative, the entropy will decrease. It does so in the compression which returns our system to its original condition.

1 mole gas (20.0 liters, 1.22 atm, 298°K) →
1 mole gas (10.0 liters, 2.44 atm, 298°K)

For this change in state, $Q_{rev} = -410$ cal and $\Delta S = -1.37$ cal/°K or -1.37 eu.*

Note that the transformation from 20.0 to 10.0 liters does not occur spontaneously. In order to compress a gas we have to do work on it. This behavior is in marked contrast to the spontaneous work-producing expansion with its attendant entropy increase. The differences are obvious in the following tabulation:

Expansion	Compression
Work done by system ($W > 0$)	Work done on system ($W < 0$)
Heat absorbed by system ($Q > 0$)	Heat evolved by system ($Q < 0$)
$\Delta S > 0$	$\Delta S < 0$
Spontaneous	Nonspontaneous

* The unit cal/°K is defined as an entropy unit or eu.

Quite clearly, the above observations must bear on the question of the directionality of nature. The evidence suggests that when a system undergoes a naturally occurring spontaneous change, its entropy increases, that is, ΔS is positive. On the other hand, during a nonspontaneous process, the entropy of the system decreases and ΔS bears a negative sign.

To further explore this idea we now look at the combined cyclic process of expansion and compression. For our specific system, $\Delta S_{exp} = +1.37$ eu and $\Delta S_{comp} = -1.37$ eu. The entropy change experienced by the gas during the complete cycle is the sum of these two values.

$$\Delta S_{cycle} = \Delta S_{exp} + \Delta S_{comp} = +1.37 + (-1.37) = 0$$

This result is fully consistent with the fact that the combination of expansion and compression returns the gas to its initial state: $V = 10.0$ liters, $P = 2.44$ atm, and $T = 298°K$. Since there is no net change in the state of the system, ΔS_{cycle} for the system must be zero, Moreover, the same answer is obtained whether the individual expansion and compression processes are reversible or irreversible. ΔS_{exp}, ΔS_{comp}, and ΔS_{cycle} are state functions. Their values are independent of the path taken.

Nevertheless, we have seen that reversible and irreversible transformations between the same states are not identical in their *total* effects. In the reversible cycle studied earlier, 410 cal of work were obtained by expanding the gas and 410 cal were required to compress it. The net result of the entire process was to leave the surroundings, as well as the system, unchanged. But in the case of the irreversible cycle, 295 cal of work were obtained from the expansion step and 590 cal were expended in the compression. Thus, the state of the *surroundings* was, in fact, altered during the irreversible cycle, although the *system* was returned to its initial conditions.

The previous paragraph reflects the fact that all systems, except those which are completely isolated, thermodynamically interact with their surroundings. Consequently, the total entropy change associated with any process potentially includes contributions from both the system under investigation and the rest of the universe. And we must bear this in mind in the continued analysis of our example. As we noted earlier, for the system, ΔS_{cycle} always equals zero, regardless of whether the cycle is reversible or irreversible. In the case of the reversible cycle, the corresponding entropy change in the surroundings will also be zero. The universe is not altered by the process. However, the situation is significantly different for an irreversible process. Under these conditions, a permanent change does occur in the surroundings—*its entropy increases*. Thus, the total cyclic entropy change for the *system plus the surroundings* is positive, $\Delta S > 0$. (Figure 13.4 is an attempt to schematically represent these changes.)

It is important to note that an increase in the entropy of the universe also characterizes the individual irreversible expansion and compression steps. This seems self-evident for the spontaneous expansion. But it might not be readily apparent in the case of compression, where the entropy of the gas actually decreases by 1.37 eu, that is, $\Delta S_{comp} = -1.37$ eu. The negative sign suggests a nonspontaneous situation. And it is certainly true that the gas would not by itself shrink from 20.0 to 10.0 liters. But we work on it until it does. This work is converted to heat which is transferred to the surroundings in order to maintain the gas at a constant temperature of 298°K. And this phenomenon leads to an increase in the entropy of the surroundings—an increase which is greater than the decrease in the entropy of the system. Hence, the net change must be positive. We have succeeded in making a gas perform an "unnatural act." We have caused it to decrease its volume (and its entropy!) when its natural tendency is to do

FIGURE 13.4. *Schematic representation of reversible and irreversible changes in the expansion and compression of a gas at constant temperature.*

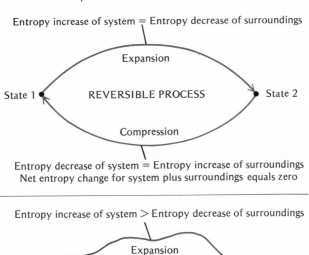

Entropy increase of system $=$ Entropy decrease of surroundings

Expansion

State 1   REVERSIBLE PROCESS   State 2

Compression

Entropy decrease of system $=$ Entropy increase of surroundings
Net entropy change for system plus surroundings equals zero

Entropy increase of system $>$ Entropy decrease of surroundings

Expansion

State 1   IRREVERSIBLE PROCESS   State 2

Compression

Entropy decrease of system $<$ Entropy increase of surroundings
Net entropy change for system plus surroundings is an increase

just the opposite. But, in order to accomplish this, we had to use force and we had to do work. In fact, we had to pile so much weight on the piston that the normally impossible event became the spontaneous one. And in the process, entropy increased.

This behavior can readily be generalized and expressed in mathematical notation. In essence, one applies the definition of entropy, the First Law of Thermodynamics, and logic to any system and its surroundings. The result is a deceptively unassuming inequality,

13.3 $T\Delta S > 0.$

Yet, simple as it is, this relationship summarizes one of the most important aspects of nature. It states, in the words of Clausius, which again introduce the chapter, *"The entropy of the universe strives toward a maximum."* Thus, the net effect of any process is to increase entropy: ΔS for the system plus its surroundings is always positive. To be sure, individual systems can be made to decrease their entropy. This is what we do when we compress a gas at constant temperature, what I do as I write about it, and what you do as you read my words. But such localized decreases in entropy only occur at the expense of larger increases somewhere else in the universe. Only in isolated systems maintained exactly at equilibrium is entropy kept constant. Then, of course, there is no net change in any thermodynamic properties.

Entropy, energy, efficiency, ecology

By now you may have reached the point where you are willing to grant that the idea of entropy seems to work. In all the examples we have investigated, an increase in entropy somewhere in the universe has attended natural processes. Unfortunately, however, this demonstrated usefulness does not speak to the question, "What is entropy?" Equation 13.2 is hardly a satisfying definition. Perhaps your initial response to entropy is not unlike my first meeting with the concept. Entropy seems to be another abstract invention, an intellectual monster created by a bunch of white-coated Frankensteins in order to supposedly explain things. But its introduction appears to be an argument *obscurum per obscurius, ignotum per ignotius*—an explanation of the obscure with the more obscure, the unknown with the more unknown.

In an effort to dispel some of this honest doubt, ignorance, and confusion, let us return briefly to the First Law of Thermodynamics,

$\Delta E = Q - W.$

For the reversible change in state, Q and W will have a pair of uniquely characteristic values, Q_{rev} and W_{rev}. Thus, we can rewrite the First Law statement as,

$$\Delta E = Q_{rev} - W_{rev}.$$

Introducing the definition of entropy, $T\Delta S = Q_{rev}$, leads at once to the following important equation:

13.4 $$\Delta E = T\Delta S - W_{rev}.$$

At this point I should pause to acknowledge that it might seem to you that Equation 13.4 applies only to an unattainable, reversible process. And in a sense, I suppose it does. But you will recall that ΔS characterizes a change in state, *not* the means of producing that change. Its numerical value depends upon differences between the initial and final states. Similarly, ΔE and W_{rev} have specific values which are independent of how the process is actually carried out. Therefore, the following arguments, based on Equation 13.4, have significance for both an ideal, reversible transformation and a real, irreversible change.

An inspection of the equation indicates that if a particular energy change could be completely realized as work, ΔE would equal $-W_{rev}$, and $T\Delta S$ would equal zero. But the presence of $T\Delta S$ in the equation suggests that this term represents *energy which cannot be utilized to do work*. To be sure, work can be produced by a spontaneous change in state. But such a change will always be attended by an increase in the entropy of the universe. In other words, *there will be a decrease in the amount of energy available for doing work*. Not a destruction of energy, mind you! The First Law of Thermodynamics still holds. All energy is conserved and all forms of energy are equivalent, but some forms of energy are more equal than others.

The bit of double-think which concludes the last paragraph cries out for elaboration, but what it says is fundamentally true. Recall that $T\Delta S$ is a heat term; it was defined as Q_{rev}. Therefore, we are tempted to say that energy is "wasted" or entropically dissipated as heat. Heat energy is of a "lower grade" than work energy. There are no restrictions on the complete transformation of work into heat. But unless the temperature is constant, it is impossible to entirely convert heat to work. Note that such a total conversion did occur in the constant temperature expansion of a gas which we have been considering. But even there, an entropy increase attended the natural process of expansion. Moreover, we were unable to obtain any net work, even when the cyclic expansion–compression was carried out reversibly. Under these utopian circumstances we would break even. In any real process, we wind up doing more work compressing the gas than we can get out of its expansion.

FIGURE 13.5.   *Examples of some spontaneous natural changes attended by the increasing entropy of the universe.*

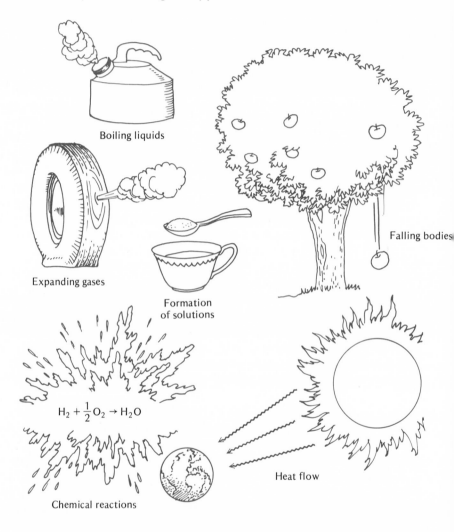

Boiling liquids

Expanding gases

Formation
of solutions

Falling bodies

$$H_2 + \frac{1}{2}O_2 \rightarrow H_2O$$

Heat flow

Chemical reactions

Often spontaneous processes involve changes in temperature. Recall the example of Chapter 12. A 100 kg weight was attached through an arrangement of pulleys to a stirrer placed in 100 g of water. The weight spontaneously fell, turning the stirrer which heated the water. After the weight had dropped 426 cm, the temperature of the water had risen 10°C. In the process, 4180 joules of work were completely converted to 1000 cal of heat. In fact, the frictional heating of the pulleys themselves would also utilize some energy, but that does not alter our argument that the complete conversion of work to heat is an acceptable spontaneous operation. But what about the reverse procedure? Certainly there is nothing in the First Law which denies its possibility. Yet, I submit that we could watch that water from now until doomsday and we would never see it suddenly start to cool, set the paddle wheel turning in the opposite direction, and thus wind the weight back up to its original height. Obvious? Perhaps so, but hardly trivial!

Fortunately, there are more efficient ways of transforming heat into work. But none of them is absolutely efficient. Consider, for example, one of the many power plants which energize the wheels and circuits of our technological age. In spite of our sometimes sloppy speech, such an energy factory does not really "generate" or "produce" energy. A power plant merely converts energy from one form to another, most commonly effecting the change we have already identified. And in the process, it produces entropy. In nuclear reactors the heat comes from the fragmenting hearts of unstable atoms. More conventional power plants utilize burning coal or oil as their energy source. In either case, the process involves the conversion of potential energy, stored in the atomic nucleus or in chemical bonds, into the increased thermal motion of matter. Typically, the heat is used to boil water and superheat the resulting vapor to temperatures well above 100°C. For example, it is not uncommon for the steam to have a temperature as high as 300°C (573°K) and pressures as great as 70 atm. Obviously, in order to attain such conditions, the H_2O which constitutes the system must absorb a good deal of heat.

The actual transformation of *some* of this heat into work occurs in a gas turbine. As the water vapor spontaneously expands, it does work by driving the turbine, which is in turn connected to an electric generator. In the process, little if any heat flows to or from the system. But the temperature, pressure, and internal energy of the water vapor all fall sharply. Indeed, the work appears at the expense of this decreasing energy. The gas which emerges from the turbine is further cooled to below the boiling point, typically to temperatures around 40°C (313°K). This naturally causes the vapors to condense. Cooling and condensation both give up heat which must flow from the system to the surroundings, often a convenient body of water. The condensed, circulating water in the closed power plant pipeline has now been restored to its original conditions. It is therefore pumped back to the boiler and the entire cycle begins again.

Now let us consider the net result of the entire cyclic process. A very important characteristic of a power plant is its efficiency, an indication of how much work we can get out of the heat we put in. Specifically, the efficiency is defined as the amount of work obtained from the system divided by the quantity of heat absorbed from the fire box. Fortunately, it is possible to apply the equations

FIGURE 13.6.   *Schematic diagram of a power plant for the conversion of heat into work.*

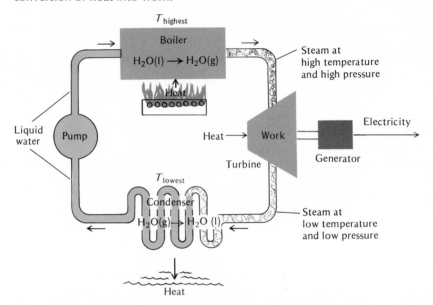

of thermodynamics to an ideal "heat engine" operating in a cycle of reversible steps. The result of such calculations is an equation which relates the efficiency (ϵ) to the highest absolute temperature of the steam ($T_{highest}$) and the lowest absolute temperature of the condensed water (T_{lowest}):

13.5
$$\epsilon = \frac{T_{highest} - T_{lowest}}{T_{highest}} = \frac{\text{work obtained}}{\text{heat absorbed}}.$$

An inspection of the above equation makes it clear that the maximum possible value for ϵ is 1. Moreover, such a perfectly efficient conversion of heat to work can be achieved only when $T_{lowest} = 0°K$. But absolute zero cannot, in fact, be attained. The lowest temperature of the cyclic system must be no lower than the temperature of the surroundings, since heat must flow from the system to the surroundings in the condensation step. Hence, the numerator of the above fraction will always be smaller than the denominator. And the fact that the efficiency is thus always less than unity means that the heat absorbed at the highest temperature of operation will never be totally transformed to work. As a result, the entropy of the universe increases.

Thanks to Equation 13.5, it is easy to find the maximum possible efficiency of a power plant operating between the two extremes of temperature cited in the above example: $T_{highest} = 573°K$ and $T_{lowest} = 313°K$.

$$\epsilon = \frac{573°K - 313°K}{573°K} = \frac{260°K}{573°K}$$
$$= 0.45$$

Even under the ideal conditions of perfect reversibility implied in the equation, less than half of the heat absorbed can be realized as work. In the real, irreversible world, friction and heat loss will further reduce the efficiency so that a value of 0.30 is considered quite high for a modern power plant. Obviously, the conversion of heat to work is a very wasteful process. Only a fraction of the energy released by the combustion of coal, oil, and gas is ultimately expressed as work.

The situation is further aggravated by the fact that these fossil fuels, present only in finite and rapidly diminishing supplies, are far too valuable to burn. For they also form the raw materials for many of the molecular marvels of modern chemistry—plastics, fibers, solvents, detergents, and drugs. One solution is to conserve our natural resources by conversion to nuclear power plants. But here we encounter the problem posed by the need to dispose of hazardous radioactive wastes. In the very legitimate public concern about this particular aspect of potential pollution, many have lost sight of the dangerous environmental contaminants which issue from the smokestacks of power plants fired by fossil fuels. Your reading of Chapter 11 should have convinced you that such atmospheric effluents cannot be ignored.

Ideally, we might couple our power demands with the need to dispose of solid wastes. Today, golfers tee off on manmade hills filled with smashed tin cans, broken beer bottles, old newspapers, torn galoshes, and coffee grounds. Or, such trash is towed out and dumped into the septic tank called the sea. It would be far preferable if solid rubbish could be burned as a source of heat for the generation of power. I am tempted to observe that this would kill two birds with one stone, but the metaphor is all wrong for an ecological subject. Nevertheless, the fact remains that if the potential problems of air pollution can be solved, such hybrid incinerator/power plants could be of great aid in cleaning the environment. Furthermore, not only would the fuel be literally dirt cheap, it might also provide valuable by-products. If the operating temperature of the incinerator is high enough, glass and metals in the trash (chiefly iron, aluminum, tin, and copper) will melt. These could be separated and sold for reuse. Conceivably, the ash might also be of commercial value.

But even if it were possible to design a completely pollution-free power plant, a fundamental thermal effect would still plague us. Every plant, nuclear or conventional, must "dump" the waste heat not converted to work, the heat given off when the expanded steam condenses. We have already noted that this heat is commonly allowed to flow to a river, reservoir, or lake. The resulting increase in temperature of the heat sink has been termed *thermal pollution*. Its influence on lowering the concentration of dissolved oxygen and thus decimating the fish population was mentioned in Chapter 11.

In spite of all these difficulties and dangers, there are few, if any, who would advocate the shutdown of all power plants. To do so would deprive modern society of its essential motive power. Yet, the implications of the entropy which is always produced along with work must be kept clearly in mind. Advances in exhaust emission control and plant design can do much to minimize environmental pollution and improve efficiency of operation. But, as we have seen, definite limits are placed upon the latter by the way in which the universe happens to function. In the final analysis, it is the directionality of natural change, described by the Second Law of Thermodynamics, which drives our power plants. And increasing entropy is the unavoidable, inevitable price.

In the long run, the most promising answers to the great energetic requirements of the twentieth century and beyond lie in tapping four vast natural sources of power—water, wind, sunshine, and underground heat. For here the entropy increases would be minimal. The energy associated with tides, waterfalls, and winds is, of course, mechanical energy. Thus, the inefficient conversion of heat to work is not part of the energy transfer process. Instead, a water turbine or a windmill can be directly harnessed to drive an electrical generator. Sunlight can also be converted into electrical energy through the intervention of photoelectric solar cells. Such devices have already proven to be of great usefulness on space satellites and stations. Moreover, solar radiation has been successfully used to heat buildings, at least in moderately temperate climates. Surely, more efficient methods can be devised to collect and concentrate some of the energy which the sun sends us daily. Only a tiny fraction would be quite sufficient to meet all our forseeable needs. Thus we could drastically reduce our dependence upon the million-year process of energy transfer which culminates in coal and oil. Finally, the hot interior of our own planet is another great reservoir of energy. We have a furnace in our basement, and what we need to know is how to pound on the pipes in order to get the landlord to send the heat up to the first floor. Thus, the limitations are not so much with the supply of energy. For all practical purposes it is boundless. The present limits are imposed by our technological skills. But imagination, intelligence, and thermodynamics will someday change all that.

Far more familiar than the functioning of a power plant is the final example of spontaneous change which we now consider. It is an instance of a profound truth cloaked in a most prosaic event. Without doubt, you personally experienced the event and consciously or unconsciously inferred the generalization at a very early age. For a child once burned quickly learns the Second Law of Thermodynamics. One of its versions simply states the frequently felt fact that *heat always flows, of itself, from hot to cold*. No matter how often you stick your finger into boiling water, it's always the same. Your finger gets hotter and the water gets cooler, never the other way around. Men with ordinary thermal properties find it impossible to make tea by dangling their fingers in a pot of lukewarm water. And mortals such as we cannot make ice cream just by sitting on the freezer. Of course, some men do seem to have peculiar heat-transfer properties. Remember the traveler who knocked at the door of a peasant hut one cold winter night? The old couple hospitably invited him in and soon he sat by the fire, blowing on his hands "to warm them." The peasants were much amazed. But their amazement turned to fear when the stranger blew upon the steaming soup they served him. It was, he said, "to cool it." The peasants, ignorant of the Second Law, turned the traveler out into the night, having no wish to traffic with a demon who could blow both hot and cold!

You may object, with good reason, that in spite of fables, folklore, and formulas, it *is* possible to transfer heat from a colder body to a hotter one. Refrigerators and air conditioners do it all the time. True enough, but they don't do it by themselves. Work is required to pump the heat against the thermal gradient, just as work must be expended in order to pump water uphill. And a refrigerator winds up releasing more calories in the room than it carries away from the beer and salami. Energy is conserved, but some of it is dissipated as heat. The local decrease in entropy which occurs when a cool object is made still cooler is more than equaled by the entropy increase which takes place when the warm surroundings are heated.

If it were not for this fact of universally increasing entropy we could profitably operate a heater/cooler which simultaneously warmed our toes and cooled our brows without the intervention of the electric company. Or, we could build a perpetual motion machine by coupling a heat engine to a refrigerator. The former would extract heat energy from a high-temperature heat source and convert it completely into work. This work, in turn, would be used to drive the refrigerator which would convert all the work back into heat, thus restoring the temperature of the heat reservoir to its original value. Granted, because of the First Law, we wouldn't have any extra energy left over, but we would at least have a free refrigerator. But alas, our marvelous machine runs afoul of the Second Law. The conversion of heat to work can never be complete in this not-quite-best-of-all-possible worlds. We can't even use the virtually limitless supply of heat in the ocean to drive the motive machinery of a ship. The net result of such a process would correspond to simultaneously raising a weight and cooling an object. And this, according to Lord Kelvin's statement of the Second Law, is impossible. In each step, some energy will always be unavailable and the entropy of the universe will increase correspondingly. Entropy always wins!

Order and odds

In the preceding sections of this chapter, two of the words most frequently employed by chemists did not appear so much as a single time. But if you failed to note the complete absence of "atoms" and "molecules," you were merely reflecting the fact that thermodynamics has no need of such miniscule particles. In their inception during the last century, the First and Second Laws were based exclusively on observations of the macroscopic properties of bulk matter. Even today, it is not necessary to introduce the microscopic structure of stuff in order to describe the flow of heat and work to and from a system. The concepts of energy, enthalpy, and entropy are in no way predicated upon the discreteness of matter.

This independence of thermodynamics is epistemologically and aesthetically rather attractive. The entire subscience is a structure of more-or-less-real observables like heat and work and more-or-less-imaginary inventions like enthalpy and entropy, all held together by rigorous mathematical logic. It ranks along with quantum mechanics as one of man's supreme intellectual achievements, worthy of the same respect commanded by Kant's philosophy or Aquinas' theology. Like those two great systems of thought, thermodynamics is an attempt to describe man and his world. It is this relationship which we explore in the present section. For one of the great beauties of thermodynamics is the way in which it helps explain the microscopic details of the universe. And, you will soon find that the ideas of atoms and molecules reciprocate by furthering our understanding of the concepts and principles of thermodynamics.

We begin by considering some specific instances of natural change and seeking a common physical denominator. In other words, we are searching for the microscopic physical phenomena which underly increasing entropy. One of the first examples we encountered was an expanding gas. Here the direction of spontaneous change is one in which volume increases. If a helium-filled balloon is punctured, the atoms of the gas rapidly escape into the surrounding atmosphere. Before the puncture, we can say with certainty that the helium is confined within the balloon. If the gas is pure, every particle in the balloon is a helium atom. But once we burst the balloon, we lose this certainty. Now the helium atoms become mixed up with the molecules of the surrounding air. We no longer have the same sort of assurance that a particular particle is a helium atom. It could be a molecule of oxygen or nitrogen or carbon dioxide or any of several other substances. We must examine it to know its identity. In time, the molecules of air will even diffuse into the deflated balloon, so that its contents becomes indistinguishable from the air around it. Clearly, our ignorance has increased with entropy. The world has become more uniformly mixed up. There is now more randomness, more chaos than there was before the balloon was punctured.

The Second Law of Thermodynamics arises because this randomness cannot reverse itself. The helium atoms cannot all sort themselves out and flow back into the puncture, refilling the balloon to its original volume and pressure. This would be a most unlikely event. And natural changes move from the less probable to the more probable. Granted, we can and do separate the helium atoms from a gaseous mixture such as air by liquefying the gas and distilling it. Helium has a specific boiling point ($-269°C$) and can be selectively isolated in the distillate. But this is hardly a spontaneous process. It requires compressors, stills, condensers, and, above all, work! We can only restore order and certitude at the high cost of an entropy gain elsewhere in the universe.

The same arguments apply when only a single substance is involved. Recall, for example, the expansion of 1 mole of a perfect gas from 10.0 to 20.0 liters at a constant temperature of 25°C. At the start of the experiment the gas molecules were all localized in the lower half of the cylinder. After the expansion they were spread over twice the original volume. Thus, there was more uncertainty about the location of any particular molecule in the final state of the system than there had been under the initial conditions. The entropy increase of 1.37 eu was attended by a decrease in our microscopic knowledge of the gas. Fortunately, we were somewhat compensated for this loss because the process of expansion could be used to do work. But if we are intent upon regaining our previous information we must be willing to work for it. The natural increase in entropic ignorance can only be reversed by compressing the gas to its original state. And this restoration of the initial order requires the expenditure of more work than we gained in the expansion process. It must be bought at a price of increasing disorder in the surroundings. The gas will not condense of itself.

What you and I are doing right now is thermodynamically equivalent to compressing a gas. We are also fighting the spon-

FIGURE 13.7.   *Illustrations of processes characterized by
increases in entropy, randomness, disorder, and ignorance.*

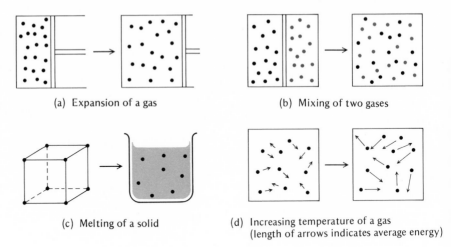

(a) Expansion of a gas   (b) Mixing of two gases

(c) Melting of a solid   (d) Increasing temperature of a gas
                                     (length of arrows indicates average energy)

taneous tendency for disorder and entropy to increase. Judging
from the following passage in "The Two Cultures and the Scientific
Revolution," C. P. Snow has frequently observed this tendency
in a social setting:

*A good many times I have been present at gatherings of
people who, by the standards of the traditional culture,
are thought to be highly educated and who have with
considerable gusto been expressing their incredulity at the
illiteracy of scientists. Once or twice I have been provoked
and have asked the company how many of them could
describe the Second Law of Thermodynamics. The
response was cold: it was also negative. Yet I was asking
something which is about the scientific equivalent of:*
Have you read a work of Shakespeare's?*

Unfortunately intellectual chaos and ignorance of the Second
Law are far more probable than order and knowledge. And to
turn the tide we must be willing to do work. For example, we must
be prepared to think hard about why randomness is more likely
than order. If you have ever participated in the pastimes popular
in Las Vegas and Monte Carlo, your chance of success will be
considerably greater. Consider, for example, a fresh standard deck
of 52 playing cards. Suppose they are arranged by suits in order

* C. P. Snow, "'The Two Cultures: And A Second Look," p. 20, Mentor
Books, New American Library, New York, 1959, 1963. Reprinted by
permission of Cambridge University Press.

of increasing value from deuce through ace, and packed in the sequence clubs, diamonds, hearts, and spades. There is obviously only one such arrangement, and we will take it as our state of highest order and minimum entropy. Interchanging any two cards will introduce into the deck an element of randomness and uncertainty. If we shuffle the cards thoroughly, the entropy will, most likely, increase even more.

There are, in all, 52! different sequences for 52 different objects, that is, $52 \times 51 \times 50 \times \ldots \times 2 \times 1$. This multiplies out to be something like 10^{68}—ample evidence of the appropriateness of the exclamation point for designating factorial! Each of these arrangements is equally probable (or perhaps one should say equally improbable). No *specific* sequence is more likely than any other. But there is only one perfectly ordered state, that which we described above. And there are billions of different more-or-less random arrangements. Thus, if you bet on your ability to duplicate the sequence of a new deck by repeated shuffling, you would be well advised to demand pretty favorable odds or be prepared to shuffle for a long time. The probability of restoring the cards to their initial order is 1 in 10^{68}. Clearly, you are far, far more likely to come up with some random sequence. Indeed, such a game would hardly be a game of chance, it would be more of a certainty.

The chance of winning in craps is much higher. Two dice can yield only 36 possible combinations. The least probable are "snake eyes" (1 and 1 to the uninitiated) and "box cars" (6 and 6). Obviously, there is only one way to throw either of these combinations. We need not see the dice in order to know how a total of 12 or 2 was constituted. The probability in both cases must thus be 1 in 36. In the language of thermodynamics, the high degree of order and the relatively low probability indicate that the entropy of either state is low. By contrast, the most likely sum, 7, can be "made" in three different combinations of numbers: 1 and 6, 2 and 5, and 3 and 4. These really represent six different arrangements, since each of the three numerical combinations can be thrown in two different ways. For example, the first die thrown could show a 1 and the second a 6, or vice versa. Simply being informed that a 7 has been thrown cannot tell us by which of the six possible arrangements it was done. Thus, there is more randomness, more ignorance, and more entropy in a 7 than in a 2 or a 12. There is also a higher likelihood of it happening. The probability of hitting a 7 is 6/36 or 1 in 6. If you can get anyone to bet on a 2 and give you even money on any other combination, put all your cash on 7 and shake. You won't even need good luck. Entropy will be on your side.

Thermodynamic craps: the house always wins

Before you rush to try your newly acquired knowledge of thermodynamics on some unsuspecting innocent, may I suggest that you first extend it a bit by considering some other physical phenomena. All the examples we have thus far encountered involved increasing positional randomness. But it would be a mistake to conclude that only increasing volume can influence entropy. The fact that entropy can also reflect energetic uncertainties is illustrated by the effect of temperature upon a system. As the temperature of a substance increases, so does its energy and its entropy. An easy way to appreciate this relationship is to think of a gas confined at constant volume. As the temperature is increased, the molecules, on the average, acquire more energy. The distribution of speeds and energies becomes more spread out, as illustrated in Figure 10.7. Thus, the number of energetic states accessible to a molecule increases with increasing temperature. And because there are more energetic options available to each molecule, the higher temperature state is more probable than that corresponding to a lower temperature. On a microscopic level, we know progressively less about the specific energetic properties of an individual molecule as the temperature rises. Things get more mixed up, and hence ΔS bears a plus sign. Conversely, cooling an object results in a decrease in the energetic randomness and the attendant entropy.

This argument can be extended to explain why heat spontaneously flows from a hotter object to a cooler one. Clearly, this is an example of an energy transfer. The faster-moving "hotter" molecules and the slower, "cooler" ones collide or otherwise exchange energy. The result is a redistribution of the energetic wealth. On the average, the less energetic particles will acquire energy at the expense of their more generously endowed fellows. And as the energy and temperature of the initially colder body increase, so does its entropy. Simultaneously, the energy and entropy of the object which was originally at the higher temperature decrease.

The First Law requires that the net energy change must be zero, while the Second Law postulates an entropy increase for the entire transformation. It follows that the entropy gain experienced by the warming object must be greater than the entropy loss which attends the cooling process. This difference is rationalized by saying that the same energy change produces a greater entropy change at a lower temperature than at a higher one. Indeed, absolute temperature can be defined as an indication of the amount of energy which must be absorbed in order to cause a fixed increase in entropy. The increase in energetic randomness which occurs when an object is warmed from 300° to 350°K is greater than the decrease which takes place when an identical object is cooled from 400° to 350°K. Thus, if two such bodies, one at 400° and one at 300°K, are placed in contact with each other, the natural tendency will be to increase randomness and entropy by a transfer of energy from the hotter to the colder. While this spontaneous transfer is going on, some, but not all, of the energy can be trapped and converted to work. The heat flow will cease when both bodies attain the same entropy level and the same temperature, 350°K in this instance. This represents a condition of thermal equilibrium.

Although we have seen that all spontaneous changes carry with them the capacity to do some work, we have also argued that a complete conversion of heat to work is impossible. Molecular motion can also help clarify this behavior. To observe it, merely push a large number of molecules, like this book, off the top of your desk. I am quite willing to wager all my royalties on the outcome: the molecules will at once start moving downwards together. After they reach the floor, watch carefully until the process reverses itself. Unless the Second Law of Thermodynamics has somehow ceased to apply to your particular part of the universe, you have a long wait ahead of you. So you might as well use some of your time to think about what happened. To begin with, the book had a certain potential energy by virtue of its elevated position. As it fell, this potential energy was converted to kinetic energy. If you remembered that business with the piano in the last chapter, you could have used this energy to do work. But once the book is down, it's too late. All the kinetic energy becomes crashingly converted into heat energy. An accurate thermometer would inform you that the temperature of the book and the floor both rose very slightly as a consequence of the impact. This, of course, means that the molecules in both have gained in energy. Not only did their motion become more chaotic, but their energies were spread over a wider distribution of values. In short, the entropies of the floor, the book, and the universe increased.

Now let us consider what molecular events would have to occur in order to reward your lengthy vigil. In the first place, all this randomized thermal energy and motion would have to be converted into concerted kinetic energy. All the molecules of the floor would have to get together and agree to give their newly acquired energy back to the book by shoving in the same direction. And the molecules of the book would have to take this energy, plus their own increased thermal motion, and all jump together. Of course, this fantastic increase in organization would destroy a good deal of entropy. And, as the book sprang mysteriously back to the desk, heat would be completely converted to work.

Obviously such an event is a patent absurdity. Experience tells us it is impossible. And practically speaking, it is. But the Second Law is not absolute on the microscopic level. It is a probabalistic generalization based upon the statistical behavior of trillions upon trillions of atoms and molecules. The chances of an individual particle occasionally disregarding it are reasonably high. Thus a molecule of a gas might by chance move from a region of lower pressure to one of higher pressure. Or, a slow moving "cold" molecule may now and then migrate into a part of the system where the average molecular energy (that is, temperature) is considerably higher. A few molecules may even try to jump from floor to table. But at any instant, such maverick nonconformist behavior is exhibited by only a miniscule fraction of the total number of molecules which make up any reasonably sized quantity of matter. When you are dealing with a population of 10^{23}, you can tolerate quite a few radicals without upsetting the natural process. The vast majority of molecules are decent, law-abiding citizens, and so the odds are on the side of the establishment and entropy.

As we have repeatedly emphasized, natural change is always in the direction of increasing randomness, disorder, chaos, ignorance, probability, and entropy. This observation suggests that there must be some state of complete order and certitude (at least within the limits of the Uncertainty Principle), free of all randomness and characterized by zero entropy. This condition is defined by the *Third Law of Thermodynamics*. The last of the three great generalizations about the energetics of the universe postulates that *a perfect crystalline solid has zero entropy at absolute zero*. In such a solid all the atoms or molecules would be exactly localized at perfectly regular lattice sites. Thus there would be no migration, no mixup, and no positional disorder. Moreover, there would be no energetic randomness because at this absolute minimum of temperature all the atoms or molecules occupy the single lowest accessible energy level.

This reference state and experimental data, used with the proper equations, permit the determination of so-called "absolute entropies" of various substances. From tables of these values it is possible to calculate the expected entropy increases or decreases associated with certain chemical or physical changes in state. Moreover, the absolute entropy of a system or a substance under a given set of conditions can be considered as an indicator of the probability of that state relative to all other related states. Thus, entropy provides one of the most important points of contact between the microcosm of atomic and molecular properties and their bulk manifestations. A branch of physical chemistry called statistical mechanics or statistical thermodynamics supplies the conceptual and mathematical apparatus for exploring this relationship. It has proved an invaluable aid in understanding matter, both in its particulate particulars and its gross behavior. Indeed, without introducing the computational complexities of the statistical method, we have here utilized its results in an attempt to make entropy comprehendable and believable.

One thing more needs to be said. The complexity of our bodies, our minds, our institutions, and our machines makes them all highly improbable. Our position on this planet is tenuous, fraught with dangers both to ourselves and our environment. Our evolution as a species, as individuals, and as societies is a continuing struggle against entropy. And our "progress" has occurred only at a tremendous price extracted from our surroundings. While the thermodynamic system "man" has decreased his own entropy, he has filled the universe about him with the entropic disorder and chaos of polluted air and contaminated water, mountains of rusting car bodies, and rotting paper. The laws of thermodynamics are indeed inexorable. We will never succeed in reversing them. Rather we must learn to live within them, to understand their consequences, and, if necessary, to redefine "progress" in terms more compatible with the good of the universal system of which we are only a part.

$\Delta G = \Delta H - T\Delta S$: *it's free!*

In spite of the fundamental importance of entropy, it is not the most useful concept for predicting the direction of many spontaneous changes, including chemical reactions. That honor falls to something called the *free energy*. Our point of departure for developing this idea is Equation 13.4, a mathematical combination of the First and Second Laws of Thermodynamics for changes in state at constant temperature and pressure.

$$\Delta E = T\Delta S - W_{rev}$$

Recall that W_{rev} is the maximum work which can be obtained from an energy change, ΔE, in a reversible process. Part of this work will always involve expansion against the constant opposing pressure, P. We know that this "pressure–volume" work must equal $P\Delta V$. But $P\Delta V$ is not identical with W_{rev}. There is always some work left over after a reversible reaction has occurred and done the required expansion work. This "useful" work can be expressed in a variety of forms—electrical, mechanical, and so on. Because we pretend to be in that efficient and ideal land of reversible processes, the value of this second work component will, like W_{rev}, have a maximum value. If we represent it as U_{rev}, we can write

$W_{rev} = U_{rev} + P\Delta V$. Substituting this in Equation 13.4, we immediately obtain the following:

13.6 $\quad \Delta E = T\Delta S - U_{rev} - P\Delta V$.

We have already implied that the useful work term, U_{rev}, has a constant value for any specific reversible process. To enter the subjunctive case, *if* a particular change in state were carried out reversibly, work equal to U_{rev} would be done in the surroundings. We can very nearly approach this ideal situation in the operation of a voltaic cell or battery, a device for the conversion of chemical energy into electrical energy. The cell is arranged in such a manner that the two halves of the spontaneous reaction occur at different places. For example, in a Daniell cell, a strip of zinc is placed in a solution of zinc sulfate, $ZnSO_4$. A porous barrier designed to retard diffusion separates this solution from a solution of copper sulfate, $CuSO_4$. This latter chamber also contains a strip of copper. The two pieces of metal are the electrodes and they are connected by a wire. In the zinc or anode chamber, the metal is oxidized to Zn^{2+}.

13.7 $\quad Zn \rightarrow Zn^{2+} + 2e^-$

Simultaneously, the Cu^{2+} ion is reduced to metallic copper in the cathode chamber.

13.8 $\quad Cu^{2+} + 2e^- \rightarrow Cu$

Neither of these half-cell reactions can occur alone. Together they add up to yield the total cell reaction,

13.9 $\quad Zn + Cu^{2+} \rightarrow Zn^{2+} + Cu$.

The electrons released by reaction 13.7 flow through the wire from the zinc anode to the copper cathode where they are utilized in reaction 13.8. Along the way, the electrons can be tapped to do work, for example, to turn a small electric motor. Because the total reaction proceeds very slowly it is never far from equilibrium and the process is almost ideally reversible. This means that the useful work generated by the cell is very nearly the theoretical maximum, U_{rev}. Note, however, that the identical change in state could also be carried out irreversibly with the production of less than the maximum permissible amount of useful work. As a matter of fact, the output of useful work is reduced to zero if the cell arrangement is destroyed. Thus, a strip of zinc spontaneously dissolves to form Zn^{2+} when it is placed in a solution of $CuSO_4$. At the same time, Cu^{2+} is reduced to metallic copper which precipitates or plates out of the solution. The reaction still conforms to Equation 13.9. But the transfer of electrons associated with the reaction is

FIGURE 13.8. *One reaction, Zn + Cu²⁺ → Zn²⁺ + Cu, carried out in two ways: (a) in an open beaker with no useful work done and (b) in a Daniell cell with very nearly maximum useful work produced. Note that because the overall change in state is identical in both cases, so are the values of ΔE, ΔH, ΔG, and ΔS.*

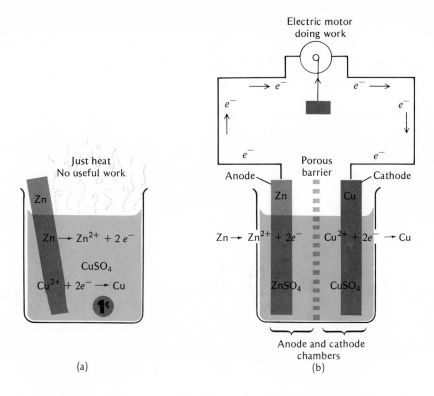

not utilized to do work; only heat is evolved. Yet, in spite of the fact that U for this latter process is zero, the U_{rev} which would accompany the same change in state *if it were done reversibly* remains unaltered. It is a maximum value characteristic of a difference between the initial and final states of the system. In order words, U_{rev} is equal to a change in an internal state function.

It seems logical that this new state function should be the immediate source of this work. Hence, it must be an energy term. Moreover, this particular part of the energy must be available or "free" for complete conversion to reversible, useful work. And so we choose to call it the *free energy*. You will soon find ample evidence that the concept of free energy is one of the most useful ideas of modern chemistry. Its "inventor" was unquestionably one of the most brilliant theoretical scientists ever produced by this country. It would be difficult to exaggerate the significance of his contributions to thermodynamics. In spite of that fact, the name of J. Willard Gibbs is hardly known outside the scientific fraternity. His monument is the letter, G, which identifies the free energy. The change in free energy which attends any change in state is the difference between G_2, the value which characterizes the final state, and G_1, that for the initial conditions. Symbolically, $\Delta G = G_2 - G_1$.

Since the reversible useful work is done at the expense of the free energy, a positive quantity of work done by the system means that the amount of free energy in the system has decreased correspondingly. This means that a negative sign must be involved in the relationship between free energy and work.

13.10 $-\Delta G = U_{\mathrm{rev}}$

In other words, the appearance of U_{rev} in the environment reflects a simultaneous disappearance of free energy internally. Whether or not work is *actually* done depends upon the particular experimental arrangement. But the value of ΔG is independent of the process by which the change of state takes place. Thus, the identical free energy decrease applies to a particular change in state carried out either in a hard-working Daniell cell or an indolent open beaker. If it is thermodynamically possible for a change in the state of a system to do useful work (even though it might not do so in a specific experiment), the free energy of the system must decrease in the process.

It is the *ability* to do useful work which is a characteristic of and criterion for spontaneous changes. Therefore, we conclude that *at constant temperature and pressure, all naturally occurring events exhibit negative values for* ΔG. The direction of spontaneous change is such that the free energy of the system tends towards a minimum. Thus, the amount of energy available to do useful work is decreased. By contrast, *a positive value of* ΔG *implies that the free energy of the system increases during the change in state.* This requires that work must be done on the system in order for the change to occur—clear *evidence of a nonspontaneous process.* Finally, *if* ΔG *is zero, the system is at equilibrium.* Neither the backward nor the forward reaction dominates. Both progress at identical rates by steps which are assumed to be fully reversible. Hence, the conditions of reversibility apply to equilibrium situations.

We will soon consider several examples illustrating the use of the change in free energy as an indicator of natural directionality. But perhaps those examples will be more informative if we first devote a little more time to the concept of free energy. We can introduce it into Equation 13.6 by using the identity of 13.10.

$\Delta E = T\Delta S + \Delta G - P\Delta V$

Some transposing yields

$$\Delta G = \Delta E + P\Delta V - T\Delta S.$$

Back in Chapter 12 you learned that enthalpy and energy changes at constant pressure are related by the expression $\Delta H = \Delta E + P\Delta V$. We now use this definition to obtain the following.

13.11 $\Delta G = \Delta H - T\Delta S.$

Equation 13.11 indicates that changes in two state functions determine the direction of any process at constant temperature and pressure. A negative value for ΔG, a sure sign of spontaneity under these conditions, is favored by a negative value for ΔH and a positive value for ΔS. The former represents an enthalpy decrease. And, since $\Delta H = Q_p$, the sign means that heat is evolved by the system during the change in state. This is true of so many naturally occurring reactions (for example, the burning of carbon or hydrogen in oxygen) that for some years a negative value for ΔH or Q_p was considered the only criterion for spontaneity. But after only a moment's thought, you should be able to give several counterexamples. Thus, when sodium chloride spontaneously dissolves in water it absorbs heat. Or, to cite an even more familiar example, at temperatures above 0°C, ice melts spontaneously in spite of the fact that it must absorb heat in order to do so.

Thanks to Equation 13.11 it is quite obvious that entropy as well as enthalpy must be considered in making predictions about natural changes. An increase in the randomness and disorder of a system means a positive value for ΔS. This will contribute to a negative ΔG. If the magnitude of the entropy increase is large enough, it can overcome an enthalpy increase with the net result of a free energy decrease. In other instances, like the freezing of water, a large negative ΔH can cause a reaction to proceed spontaneously in spite of the fact that the entropy of the system decreases during the change in state. This last statement may, at first hearing, seem to contradict our earlier argument that entropy increases for all natural changes. But recall that there we were speaking of the entire universe. The heat which must be evolved in order for ΔH to exceed $T\Delta S$ contributes to the increased entropy of the surroundings. The outcome is that the overall entropic change for the system plus the surroundings is still positive.

Now let us apply this argument a little more explicitly to the phase change we have already cited—that between solid and liquid H_2O. The melting of a mole of water at a pressure of 1 atm and a temperature T can be written

1 mole $H_2O(s)$ (1 atm, T°K) $\rightarrow$ 1 mole $H_2O(l)$ (1 atm, T°K).

The heat of fusion, ΔH for the process, is 1440 cal/mole at 273°K

(0°C) and it exhibits little temperature dependence. The entropy change at the melting point is +5.27 cal/°K, the positive sign reflecting the increase in disorder which attends disruption of the crystal lattice. ΔS itself does not vary significantly with temperature, but the influence of temperature upon the water–ice system is obviously expressed in the $T\Delta S$ term.

We can demonstrate this dependence by evaluating ΔG at three different temperatures. (Figure 13.9 should help clarify this calculation.) At 273°K, Equation 13.11 takes on the following values:

$$\Delta G = 1440 \text{ cal} - 273°K \times 5.27 \text{ cal/}°K$$
$$= 1440 \text{ cal} - 1440 \text{ cal}$$
$$= 0.$$

The zero free energy change indicates a system at equilibrium, in this case, at its melting/freezing point. Now we try a higher temperature, 10°C. Since ΔH and ΔS are essentially constant we use the same values employed above, but substitute 283°K for T.

$$\Delta G = 1440 \text{ cal} - 283°K \times 5.27 \text{ cal/}°K$$
$$= 1440 \text{ cal} - 1490 \text{ cal}$$
$$= -50 \text{ cal}$$

The negative sign for ΔG implies that at 10°C an ice cube will spontaneously melt, something you've probably suspected for some time. At this temperature $T\Delta S$ is numerically greater than ΔH, and hence the entropy increase wins out over the enthalpy increase which tends to work against spontaneity. The situation is just reversed at temperatures below 0°C. For example, at −10°C or 263°K, we calculate a positive value for ΔG.

$$\Delta G = 1440 \text{ cal} - 263°K \times 5.27 \text{ cal/}°K$$
$$= 1440 \text{ cal} - 1.390 \text{ cal}$$
$$= +50 \text{ cal}$$

The presence of the plus sign indicates that the reaction as written is nonspontaneous. Because T is smaller than in our previous examples, $T\Delta S$ can no longer equal or exceed the unfavorable influence of ΔH, and ΔG is positive. This is a thermodynamic representation of the fact that ice does not melt at −10°C and 1 atm pressure.

You know perfectly well that water freezes below 0°C. In other words, the reverse change in state is a natural one.

1 mole H_2O(l) (1 atm, 263°K) → 1 mole H_2O(s) (1 atm, 263°K)

Since this reaction is the exact opposite of our previous example, all we need do to obtain values for the changes in its thermodynamic state functions is to reverse their signs. The enthalpy change becomes −1440 cal, the minus sign specifying that heat is

FIGURE 13.9.   *Plot of ΔH and TΔS vs. temperature for the
change in state*

   $H_2O\ (s) \rightarrow H_2O\ (l)$

*assuming ΔH to be a constant 1440 cal/mole and ΔS to be a
constant 5.27 cal/°K. The graph indicates values for*

   $\Delta G = \Delta H - T\Delta S$

*at three different temperatures—263°, 273°, and 283°K. The
signs of the calculated free energy change indicate that liquid
water spontaneously freezes below the freezing point of 273°C
and ice spontaneously melts above 273°C.*

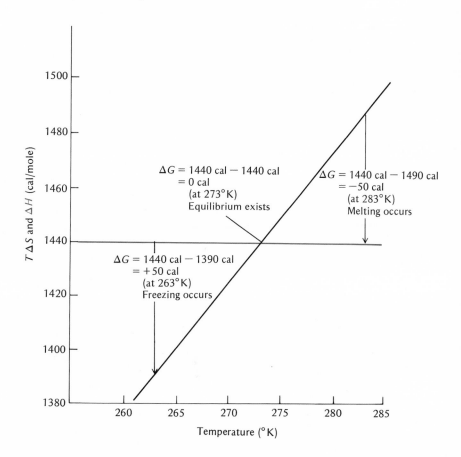

given up when water freezes. This is an influence favorable to the
spontaneity of the change in state from liquid to solid. On the
other hand, entropy tends to work against this process. The new
value of $\Delta S = -5.27$ cal/°K suggests that the solidification of
water is an entropically unnatural process because it reduces randomness and replaces it with order. But the magnitude of the

exothermic enthalpy decrease is sufficient to overcome this entropic opposition and ΔG for freezing at 263°K is −50 cal. All you have to do to freeze water is get it cold enough.

Granted, it was hardly necessary to calculate ΔG in order to reach the above conclusions. But not all systems are as familiar as H_2O. We may not experientially or intuitively know the spontaneous direction of an untried change in state. Fortunately, however, free energy calculations permit such predictions without the need of actually carrying out the experiment. As you might expect, these computational capabilities and the resulting power of prognosis are very valuable in the study of chemical reactions and the direction of their natural occurrence. Indeed, the applications of the concept of free energy to chemical changes is so significant that it will be a major theme in the next chapter. It is to that general consideration of chemical reactions that we now proceed.

For thought and action

13.1   Assure yourself, by a simple calculation, that the volume of a cylinder with a radius of 10.0 cm and a height of 31.8 cm is indeed 10.0 liters.

13.2   Here's another chance to catch the author in an error. One mole of a perfect gas is confined to a 10.0 liter cylinder at 298°K. Calculate the pressure of the gas in atmospheres.

13.3   "We can calculate" (it says somewhere in the pages past) that 53 cal of work could raise a weight of 71.0 kg a height of 31.8 cm. Demonstrate that you really can. By the way, $W = mgh$.

13.4   Suppose the gas of the example used in the first section of this chapter expands from a volume of 10.0 to 20.0 liters against an opposing pressure of 1.00 atm.
   (a)   Calculate the amount of work done in the process. Express your answer in liter atmospheres, calories, and joules.
   (b)   During the above expansion, the temperature of the gas is maintained at a constant 25°C. Calculate Q, ΔE, and ΔS for the system.

13.5   Distinguish between path-dependent and state-dependent thermodynamic functions and give examples of both.

13.6   A steam engine utilizes steam superheated to 150°C and exhausts it at 50°C. Calculate the maximum efficiency of the engine. Incidentally, why do you suppose the steam is called "superheated" and how is it done?

13.7   The radiation-resistant metals which must be used in nuclear power plants cannot withstand steam pressures as high as those encountered in most coal-fired burners. How would this fact influence the relative maximum efficiencies of the two types of power plants?

13.8   Electric heat is often claimed to be "cleaner" than gas, oil, or coal heat. Critically evaluate this claim, keeping in mind efficiency of operation.

13.9   The author recently learned of a college student who is making plans to live simply and self-sufficiently, disturbing his environment as little as possible. In other words, he hopes to keep his entropy production at a minimum. His home will be a rural area in a climate which will require heat during the winter. Outline a plan of action for existence under these circumstances.

13.10 Explain why water doesn't flow uphill. Or, if you prefer, explain the thermodynamics of a waterfall.

13.11 Listed below are a number of alternative statements of the Second Law of Thermodynamics. All but the last two are taken from a convenient compilation in "Chemical Systems: Energetics, Dynamics, Structure," by J. Arthur Campbell, W. H. Freeman and Company. Copyright © 1970. Explain why each of the statements reflects the same general characteristic of the universe. Then try your hand at making up your own version.
   (a)   "The state of maximum entropy is the most stable state for an isolated system." (E. Fermi)
   (b)   "It is impossible in any way to diminish the entropy of a system of bodies without thereby leaving behind changes in other bodies." (M. Planck)
   (c)   "Spontaneous processes (i.e., processes which may occur of their own accord) are those which when carried out under proper conditions, can be made to do work. If carried out reversibly, they will yield a maximum amount of work; in the natural irreversible way the maximum work is never obtained." (J. A. V. Butler)
   (d)   "Every system which is left to itself will, on the average, change toward a condition of maximum probability." (G. N. Lewis)
   (e)   "Heat cannot spontaneously pass from a colder to a warmer body." (R. J. E. Clausius)
   (f)   "Entropy is time's arrow." (A. Eddington)
   (g)   "All the king's horses and all the king's men, Couldn't put Humpty together again." (M. Goose)
   (h)   "Things are getting more screwed up every day." (Anon.)

13.12 The concept of entropy is one of the most ellusive in all of science. Explain it to the art major you've been talking to in previous problems.

13.13 Rank the following poker hands in order of increasing entropy, i.e., probability.
   (a)   Any pair
   (b)   A pair of aces

(c)   A full house (a pair and three-of-a-kind)
(d)   Two pair
(e)   Three-of-a-kind

13.14   Reread the quotation from C. P. Snow found on page 390. Defend or refute the legitimacy of the equivalence which he draws.

13.15   Demonstrate that you are equal to Snow's challenge by describing the Second Law in a manner appropriate for a literary cocktail party.

13.16   Rank the following in order of increasing entropy:
(a)   1 mole $H_2O(l)$ at 25°C and 1 atm
(b)   1 mole $H_2O(s)$ at 0°C and 1 atm
(c)   1 mole $H_2O(g)$ at 100°C and 0.5 atm
(d)   1 mole $H_2O(g)$ at 100°C and 1 atm
(e)   1 mole $H_2O(l)$ at 0°C and 1 atm
(f)   1 mole $H_2O(l)$ at 100°C and 1 atm

13.17   When water boils at 373°K it absorbs 9700 cal of heat per mole. Calculate the entropy change which attends this process and explain its significance. *Hint:* Remember that $\Delta S = Q_{rev}/T$.

13.18   Provide a specific example of the following:
(a)   An endothermic spontaneous change
(b)   An exothermic spontaneous change
(c)   A spontaneous change in which the entropy of the system increases
(d)   A spontaneous change in which the entropy of the system decreases

13.19   A chemist reports that he has found that the entropy of a particular system decreases during a specific spontaneous endothermic change in state. Evaluate the statement.

13.20   If you have already done Problem 13.17 you know that for the reaction,

$$H_2O(l) \rightarrow H_2O(g)$$

$\Delta H = 9700$ cal and $\Delta S = 26.0$ cal/°K at 1 atm pressure. You may assume that both these values are constant over the temperature range which concerns us here.
(a)   Determine ΔG at 110°C and predict the direction of spontaneous change at this temperature.
(b)   Repeat part (a), but for 90°C.

13.21   For the vaporization of benzene, the change in state summarized by the following equation, $\Delta H = 8.2$ kcal/mole and $\Delta S = 23.2$ eu at 1 atm and 25°C, but these same values hold over a reasonable temperature range.

$$C_6H_6(l) \rightarrow C_6H_6(g)$$

(a)   Calculate the boiling point of benzene, that is, the temperature at which ΔG for the above reaction is zero at 1 atm pressure.

(b) Calculate ΔG at 1 atm and 25°C, and comment on the significance of the answer.

13.22 The antientropic aspects of the origin of life and the evolution of the species have been cited by some authors as evidence of divine intervention in the workings of the universe. What is your assessment of this argument?

13.23 Science fiction fans should welcome this suggestion. Outline a story in which the Second Law of Thermodynamics suddenly stops working.

Suggested readings

Angrist, Stanley W., and Helper, Loren G., "Order and Chaos: Laws of Energy and Entropy" Basic Books, New York, 1967.

Bent, Henry A., "The Second Law: An Introduction to Classical and Statistical Thermodynamics," Oxford Univ. Press, London and New York, 1965.

Blum, Harold F., "Time's Arrow and Evolution," Harper Torchbook, New York, 1962.

The four introductions to thermodynamics recommended in Chapter 12 also treat the topics included in this chapter.

14

The constants of chemical change

Which way are you going?

Of all the manifestations of the dynamic, directional universe, undoubtedly one of the most important is chemical change. If you require evidence, just look around you and within you! Chemical reactions produce penicillin, "the pill," and polio vaccine; nitroglycerine, napalm, and nerve gas. They give rise to the vilest pollution and permit the purification of the foulest water. Reactions are responsible for all the food you eat, the liquids you drink, and the medicines you take. And once inside your own personal chemical factory, these ingested agents are subjected to a staggering series of chemical transformations. No wonder, then, that much of the science of chemistry is devoted to studying chemical change.

This involvement should be readily apparent from a brief review of what has preceded this page. Chapters 2 and 3 attempted to show that our most fundamental concepts—elements, compounds, atoms, and molecules—have evolved from observations of chemical change. The shorthand representation of such changes constituted the contents of the subsequent chapter. Chapter 5 revealed that much of the evidence for the classification of the elements comes from their chemical properties. And so on; you can supply many other examples. The recurrence of the subject makes its influence on chemical knowledge manifestly clear.

In short, chemical changes are so intimately woven into the warp and woof of this science that it is effectively impossible to isolate the topic for investigation. Nevertheless, the goal of this chapter is to bring together some very basic questions about chemical reactions: Why do they occur? How far do they go? How fast do they take place? In attempting to supply answers, I will provide examples of various types of chemical changes. But this chapter should be read in the context of what has gone before. And the earlier material should, in turn, be scrutinized under the light which will hopefully be shed here.

A solution to the fundamental riddle, "Why do chemical reactions occur?" has already been advanced, implicitly if not explicitly. The driving force of all chemical reactions is the tendency for the entropy of the universe to increase. It is this same movement toward a more probable, more random state which motivates all natural changes. Viewed from the reference point of the substances which constitute the reacting mixture (the system), chemical reactions will proceed in such a direction that the combined changes in the enthalpy and entropy will result in a minimization of the free energy. This decrease in the free energy of the matter involved means that less energy is available to do work after the reaction than before it. Chemical reactions, like water, run downhill. And ΔG is the indicator of the grade of that hill.

Every potential chemical reaction has associated with it a free energy change. This value of ΔG represents the difference between the free energy of the final state and that of the initial state. Since G depends upon variables such as temperature and pressure, so does ΔG. But so long as T and P are kept constant and the initial and final states are fixed, ΔG will be a constant characteristic of the change in state and independent of the actual reaction process.

Because of the nature of the concept of free energy, ΔG conveys one of the most important pieces of information about a chemical change: whether or not it can take place. If ΔG for the reaction under the specified conditions is negative, the change will occur spontaneously. Given the definition of ΔG, this means that the reaction *could* (in principle) be carried out to yield useful work. But if ΔG is positive, work must be done *on* the system to force it to react. Without such coersion, the *reverse* reaction will occur spontaneously. Finally, if $\Delta G = 0$, the reaction mixture is at equilibrium and no net change will be observed. The rate of the forward reaction will equal that of the reverse reaction.

In an attempt to illustrate and illuminate this rather abstract argument, let us once more turn to a familiar and fruitful example, the reaction of hydrogen and oxygen to form water.

14.1 $H_2(g) + \frac{1}{2}O_2(g) \rightarrow H_2O(l)$

What we are interested in is the free energy change which would attend this chemical transformation. Recalling the fact that free energies vary with pressure and temperature, we restrict ourselves to a system in which they are both fixed. Each substance is assumed to be at a pressure of 1 atm and a temperature of 25°C or 298°K.

If we could somehow measure the amount of free energy in 1 mole of $H_2O(l)$, 1 mole of $H_2(g)$, and $\frac{1}{2}$ mole of $O_2(g)$ under these conditions, we would immediately have an answer:

$$\Delta G^\circ = G^\circ_{H_2O(l)} - \{G^\circ_{H_2(g)} + \frac{1}{2}G^\circ_{O_2(g)}\}.$$

Here each G° represents the free energy of 1 mole of the substance identified by the subscript. The superscript again specifies a pressure of 1 atm.

The similarity of this notation to that used for enthalpy in Chapter 12 is more than coincidental. As in that instance, it turns out to be impossible to make thermodynamic measurements of the absolute free energy of any compound or element. So we adopt the same arbitrary standard state used for enthalpies. *The free energy of all pure elements at 25°C and 1 atm in the physical state most stable under these conditions is assumed to be zero.* Here $G^\circ_{H_2(g)} = G^\circ_{O_2(g)} = 0$. But fortunately, we can experimentally determine *differences* in free energies. In the case of the above reaction, ΔG° turns out to be -56.7 kcal at 25°C. And thus we say that the *standard free energy of formation of liquid water* at 25°C has this numerical value, $\Delta G^\circ_{f,H_2O(l)} = -56.7$ kcal/mole of $H_2O(l)$. Since the standard free energies of the elementary reactants are, by convention, set at zero, the *standard molar free energy* of $H_2O(l)$ also equals -56.7 kcal.

This large negative value for $\Delta G^\circ_{f,H_2O(l)}$ immediately signifies that under standard conditions, the reaction will spontaneously proceed in the direction specified by Equation 14.1. In other words, hydrogen naturally combines with oxygen to form water. You will recall from our earlier discussion that it often does so with an explosive evolution of heat. However, the same reaction can be made to occur slowly and almost reversibly with the production of useful work. To be sure, the upper limit of 56.7 kcal is never quite achieved in our irreversible universe. But the combination of these elements can be controlled in fuel cells which convert the decreasing free energy of the system to electrical energy.

By contrast, the reverse reaction—the decomposition of water into gaseous hydrogen and oxygen—is decidedly nonspontaneous. Its standard free energy change is exactly opposite, $\Delta G^\circ = +56.7$ kcal. The process is as unnatural as if water were to run uphill. To be sure, pumps achieve the latter feat every day; but only with the expenditure of work. Similarly, the application of energy is necessary to reverse reaction 14.1; 56.7 kcal are required for every mole of $H_2O(l)$ decomposed. In an electrolysis experiment, this energy is supplied in the form of electricity.

The operational analogy between enthalpy and free energy can be extended further. The ΔG_f°'s of Table 14.1 can be added or subtracted along with the equations to which they correspond. Thus we can write a faintly familiar equation:

14.2
$$\Delta G^\circ = \Sigma \Delta G_{f,\text{products}}^\circ - \Sigma \Delta G_{f,\text{reactants}}^\circ.$$

Our first application of this relationship will again be to an all-aqueous system, the reaction between liquid and gaseous H_2O. The change in state is simply this:

$$H_2O(l) \text{ (1 atm, } 25^\circ C) \rightarrow H_2O(g) \text{ (1 atm, } 25^\circ C).$$

As written, $H_2O(g)$ is the product and $H_2O(l)$ the reactant. Therefore, substitution of the appropriate free energies of formation in 14.2 yields the following:

$$\begin{aligned}
\Delta G^\circ &= \Delta G_{f,H_2O(g)}^\circ - \Delta G_{f,H_2O(l)}^\circ \\
&= -54.6 \text{ kcal} - (-56.7 \text{ kcal}) \\
&= +1.9 \text{ kcal.}
\end{aligned}$$

The positive value for ΔG° represents the fact that the conversion of liquid water to its gaseous phase does not occur spontaneously at $25^\circ C$. Instead, the reverse, condensation process is the naturally occurring change in state. This transformation is attended by a decrease in free energy, $\Delta G^\circ = -1.9$ kcal.

Although the above example represents a physical change, the same rules apply to chemical reactions. Indeed, it is here that free

TABLE 14.1.   *Standard Free Energies of Formation* (ΔG_f°)
(in kcal/mole at 25°C and 1 atm)

CaO (s)	−144.4	$C_2H_5OH(l)$	−41.8	MgO (s)	−136.1
C (s, diamond)	+0.7	Fe_2O_3 (s)	−177.1	NaCl (s)	−91.8
CH_4 (g)	−12.1	HBr (g)	−12.7	NaF (s)	−129.3
CH_3OH (l)	−38.7	HCl (g)	−22.8	NaOH (s)	−90.6
CO (g)	−32.8	HF (g)	−64.7	NH_3 (g)	−4.0
CO_2 (g)	−94.3	HI (g)	+0.3	NH_4Cl (s)	−48.7
C_2H_2 (g)	+50.0	HNO_3 (l)	−19.1	NO (g)	+20.7
C_2H_4 (g)	+16.3	H_2O (g)	−54.6	NO_2 (g)	+12.4
C_2H_6 (g)	−7.9	H_2O (l)	−56.7	O_3 (g)	+39.1
C_3H_8 (g)	−5.6	H_2O_2 (l)	−28.2	SO_2 (g)	−71.8
$i\text{-}C_8H_{18}$ (l)	1.5	HgO (s)	−14.0	SO_3 (s)	−88.5

energy calculations are most useful, for they permit predictions
of the direction of chemical changes which may never have been
attempted in a laboratory. A reaction which could prove to be of
crucial environmental importance eloquently illustrates the
method.

It has been suggested that the following reaction might ultimately result in the blinding of all animals with unprotected eyes
and the possible death of all land plants.

14.3 $NO(g) + O_3(g) \rightarrow NO_2(g) + O_2(g)$

Nitrogen oxide, NO, is one of the exhaust products of the proposed supersonic transports which would fly at an altitude of
about 12 to 13 miles in the ozone-rich layer of the upper atmosphere. Harold Johnston of the University of California at
Berkeley has argued that as a consequence of the above reaction,
the concentration of ozone, O_3, would be rapidly reduced by a
fleet of SST's.* Since ozone filters out much of the sun's ultraviolet
radiation, such depletion would expose the surface of the earth
to highly energetic rays of blinding intensity.

In order to predict whether or not such a reaction would indeed
occur under standard conditions, we use the procedure applied
earlier.

$$\Delta G^\circ = \Delta G^\circ_{f,NO_2(g)} + \Delta G^\circ_{f,O_2(g)} - \{\Delta G^\circ_{f,NO(g)} + \Delta G^\circ_{f,O_3(g)}\}$$
$$= 12.4 + 0 - (20.7 + 39.1) = 12.4 - 59.8 \text{ kcal}$$
$$= -47.4 \text{ kcal}$$

The large negative value for ΔG° is interpreted to mean that the
reaction is strongly spontaneous at 25°C and 1 atm. Of course,
the temperature and pressure of the lower stratosphere are quite
different, and so is ΔG under these conditions. But because we
know the mathematical form of the dependence of free energy
upon these variables, we can also estimate ΔG for high altitudes.
Again, it bears a negative sign.

Not surprisingly, Professor Johnston's calculations are far more
complicated than has been suggested here. He takes into account
many other possible reactions which might compete for the same
substances. Moreover, he also considers the rates or speeds of
these reactions. The complexity of the chemistry is such that no
definitive answers have yet been found. Walter Sullivan, the dis-
tinguished science editor of the *New York Times* has stated the
problem with characteristic perception: "In any case the argument
is a reminder that some of the factors that make the world habit-

* Harold Johnston, Reduction of Stratospheric Ozone by Nitrogen
Oxide Catalysts from Supersonic Transport Exhaust, *Science* 173, 517
(August 6, 1971).

able for all higher forms of life are fragile. Their care and suste-
nance must be mastered before we endanger their survival—and
our own."* One thing is certain; the science of thermodynamics
and the concept of free energy will be of pivotal importance in
this and in all other instances of man's efforts to live in harmony
with his surroundings.

Why are you going?

I suspect that some of you may object that the question, "Why
do chemical reactions occur?" has not yet been adequately
answered. Granted, the statement, "chemical reactions take place
spontaneously when the free energy of the reacting system decreases" does not appear to explain very much at all. But you
must admit that it works! Besides, I am at somewhat of a loss to
know how to answer the question in any other way. Perhaps looking more closely at the individual thermodynamic contributions
to ΔG will help clarify the issue.

To begin with, remember that changes in both enthalpy and
entropy contribute to the free energy change.

$$\Delta G = \Delta H - T\Delta S$$

Obviously, a negative ΔG is favored by a negative ΔH and a positive ΔS. In other words, the spontaneous tendency of a chemical
change will be enhanced if the reaction is attended by an evolution of heat and/or an increase in disorder. Ultimately, of course,
these changes in enthalpy and entropy must arise from events
occurring on a molecular or atomic level. So we once more enter
the microcosm of matter.

One feature characterizes all chemical reactions: the making
and/or breaking of bonds between atoms or ions. Fortunately,
there are methods for determining the energies associated with
chemical bonds. For covalent compounds they often involve
spectroscopic measurements of the frequency of the light (visible
or invisible) absorbed by the molecules. The energy of the radiation corresponds to the quantized energy with which the atoms
vibrate. Appropriate equations then permit calculation of the
energy of interaction between the atoms, in other words, the
energy required to break the bond. By such means scientists have
found the following bond energies for isolated, gaseous molecules:

* The *New York Times*, Sunday, May 30, 1971, in the "News of the
Week in Review" section. © 1971 by the New York Times Company.
Reprinted by permission.

H—H	104 kcal/mole
O—O	118 kcal/mole
H—O	111 kcal/mole

You will note that the values are expressed in kilocalories per mole. This means, for instance, that 1 mole, that is, one Avogadro number, of O—O bonds has a total energy of 118 kcal associated with it. Thus, to break all of these bonds and convert the 6.023×10^{23} O_2 molecules into $2 \times 6.023 \times 10^{23}$ O atoms would require 118 kcal. Breaking all the bonds in ½ mole of oxygen gas would take half that energy, 59 kcal. Similarly, we see at once that breaking 1 mole of H—H linkages requires 104 kcal.

Such bond-breaking may be regarded as a necessary first step in the reaction of hydrogen and oxygen. Whether or not the reaction in fact occurs by this mechanism really doesn't matter. It's the overall difference that counts in energy changes. Therefore, we must also consider the reuniting of the individual H and O atoms to form the new H—O bonds of the product, H_2O. The process of bond-making releases energy. Thus, 111 kcal are evolved for each mole of H—O bonds formed. And, since each molecule of H_2O contains two of these linkages, 2×111 kcal or 222 kcal of energy will be liberated when 1 mole of water is formed from hydrogen and oxygen atoms.

Now for a little elementary energetic bookkeeping, bearing in mind the sign conventions.

Process	*Energy Change*	
$H_2(g) \rightarrow 2H(g)$	104 kcal absorbed	$\Delta E = \quad 104$ kcal
$½ O_2(g) \rightarrow O(g)$	59 kcal absorbed	$= \quad 59$ kcal
$2H(g) + O(g) \rightarrow H_2O(g)$	222 kcal evolved	$= -222$ kcal
$H_2(g) + ½ O_2(g) \rightarrow H_2O(g)$	59 kcal evolved	$\Delta E = - \quad 59$ kcal

The net result, $\Delta E = -59$ kcal, is the energy change predicted for this reaction from the bond energy calculation. More available energy is "gained" from forming the H—O linkages than is "lost" in breaking the H—H and O—O bonds. You can see that the numerical value agrees quite well with the tabulated $\Delta H°$ of formation of gaseous water from elementary hydrogen and oxygen, -57.8 kcal/mole at 25°C and 1 atm. True, one quantity is an energy change, the other an enthalpy change, but since the $P\Delta V$ work term is very small, the comparison is quite legitimate.

On the basis of this example, one is tempted to generalize that chemical reactions often occur in a direction such that the overall effect is to generate stronger, more stable bonds. And the facts frequently fit the conclusion, since many spontaneous reactions are exothermic, that is, evolve heat. But the use of this rule must be tempered by an understanding of the entropy contribution.

FIGURE 14.1.   *The origin of energy changes in a chemical reaction. The total energy change is the sum of the changes associated with the various individual steps.*

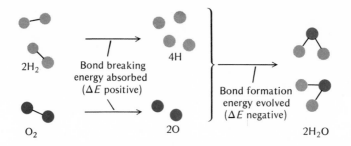

In this particular instance, the entropy effect works against the system; $\Delta S° = -10.7$ cal/°K or -0.0107 kcal/°K at 1 atm and 25°C. Since the temperature is 298°K, $T\Delta S°$ must equal 298°K $\times$ $(-0.0107$ kcal/°K) or -3.2 kcal. Obviously, this decrease in entropy is not nearly enough to offset the energy "lost" as heat. The standard free energy change remains negative and the reaction is spontaneous.

$$\Delta G° = \Delta H° - T\Delta S° = -57.8 \text{ kcal} - (-3.2 \text{ kcal}) = -54.6 \text{ kcal}$$

The entropy decrease attending the formation of $H_2O(g)$ from $H_2(g)$ and $\frac{1}{2}O_2(g)$ is typical of a chemical change in which the number of moles of gaseous products is smaller than the number of moles of gaseous reactants. In general, the greater disorder associated with the gaseous state means that the entropy of gases is high relative to that of solids or liquids. This same effect also explains why the entropy change associated with the direct formation of liquid water from hydrogen and oxygen is more negative than the value which characterizes the formation of gaseous water. For the reaction,

$$H_2(g) + \frac{1}{2}O_2(g) \rightarrow H_2O(l),$$

$\Delta S°$ is -390 cal/°K at 25°C. We have already made use of the fact that at the same temperature, $\Delta S°$ is -10.7 cal/°K when $H_2O(g)$ is the product. The values indicate that a greater increase in order accompanies the formation of liquid water from the elements.

On the other hand, in Equation 14.3, the number of moles to the left of the arrow equals the number of moles on the right. It follows that the $\Delta S°$ associated with this reaction is probably very near zero, and the major factor contributing to $\Delta G°$ must be $\Delta H°$. The accuracy of this prediction can easily be checked by calculating the enthalpy change from the standard heats of formation found in Table 12.1.

Before finally abandoning this attempt to explain why chemical reactions take place spontaneously, it might be informative to investigate one which does not. Consider, for old time's sake, iso-octane. We have physically and figuratively made use of the combustion of this hydrocarbon on numerous other occasions. Now let us explore our chances of reversing the process, that is, of carrying out the reaction,

14.4
$$8CO_2(g) + 9H_2O(l) \rightarrow i\text{-}C_8H_{18}(l) + \frac{25}{2}O_2(g).$$

After all, it would be very profitable if we could produce our own gasoline from carbon dioxide and water. To test the feasibility of this plan, under standard conditions of 25°C and 1 atm, we calculate $\Delta G°$ for the process.

$$\Delta G° = \Delta G°_{f,i\text{-}C_8H_{18}(l)} + \frac{25}{2}\Delta G°_{f,O_2(g)} - \{8\Delta G°_{f,CO_2(g)} + 9G°_{f,H_2O(l)}\}$$

$$= 1.5 + \frac{25}{2}(0.0) - \{8(-94.3) + 9(-56.7)\}\text{ kcal}$$

$$= +1266.2\text{ kcal}$$

The result is not very encouraging! Rechecking the calculation of Chapter 12 indicates that the standard enthalpy change associated with the forward, combustion reaction is -1305.5 kcal. It follows that $\Delta H°$ for the reaction represented by Equation 14.4 must be $+1305.5$ kcal. The necessity of absorbing this much heat in order to make 1 mole of $i\text{-}C_8H_{18}(l)$ presents a huge impediment to our chances of success.

Moreover, the entropy change associated with the process will do little to help us. We can readily compute its numerical value. Since

$$\Delta G° = \Delta H° - T\Delta S°,$$
$$T\Delta S° = \Delta H° - \Delta G°,\text{ and}$$
$$\Delta S° = \frac{\Delta H° - \Delta G°}{T}.$$

Substituting the appropriate values into the latter expression we find that

$$\Delta S° = \frac{1305.5\text{ kcal} - 1266.2\text{ kcal}}{298°K} = \frac{39.3\text{ kcal}}{298°K}$$

$$= +0.132\text{ kcal/°K} \quad \text{or} \quad +132\text{ cal/°K}.$$

The fact that the sign of $\Delta S°$ is positive signifies that an entropy increase would occur along with the reaction—a tendency towards disorder which tends to promote reactivity. But alas, the enthalpy change is far too large, and we have no hope of overcoming it with the increasing entropy of the system. And so we must abandon our hopes of synthesizing gasoline in our own chemical plant. For that we are dependent upon other plants—green ones, which have been dead for millions of years. To be sure, these plants did not make isooctane directly. In the ages which have elapsed since they flourished, additional chemical reactions have occurred which rearranged their carbon and hydrogen atoms into molecules like C_8H_{18}. But the unique property which started it all was the ability of these prehistoric plants to use sunlight to force a nonspontaneous chemical change to actually take place. It is a talent which their modern cousins exhibit every day to their—and our—benefit.

How far are you going?

Chemists have become so skillful at communicating with matter that they can actually ask the above question of a chemical reaction. Naturally, the method of expression is not verbal (nor is it psychokinetic!). Instead, it involves the special language of experiment. And when the question is properly posed, the reaction answers "As far as I can get, but not all the way." Furthermore, we can induce the reaction to tell us just how far that is.

In 1913, a German academic chemist named Fritz Haber asked just these questions of a mixture of nitrogen and hydrogen. The inquiry was innocent enough. But the answer had repercussions which literally shook Europe for five years. For the result of this research was the *Haber process* for producing ammonia. To the German military machine, deeply committed to World War I and blockaded from South American supplies of nitrates, compounds necessary for the manufacture of explosives, the discovery was of incalculable importance. It provided a synthetic source of chemically combined nitrogen which could substitute for the natural minerals. And thus, Haber's breakthrough was undoubtedly instrumental in lengthening the war.* The case is still eminently worth studying for what it teaches about the use and misuse of science. But in order to begin to comprehend the political issues, you should know something about the technical problem. Therefore, we turn to the reaction which started it all.

The direct combination of nitrogen and hydrogen to yield ammonia can be represented by the following equation:

$$N_2(g) + 3H_2(g) \rightarrow 2NH_3(g).$$

---

* Ironically, because of his Jewish parentage, Haber was forced to flee Germany in the 1930's.

We can easily calculate the free energy associated with this reaction at 25°C and 1 atm pressure. It is simply twice the standard molar free energy of formation of $NH_3(g)$.

$$\Delta G° = 2\Delta G°_{f,NH_3(g)} = 2(-4.0) = -8.0 \text{ kcal}$$

The negative value indicates that under these conditions, the reaction will tend to move in the direction of the arrow. And, if we place 1 mole of N_2 and 3 moles of H_2 in a closed container, we do indeed find that NH_3 is produced, particularly if a little iron powder is around.* As the concentration of the ammonia increases, the concentrations of the nitrogen and hydrogen decrease. But (and this is worth underlining) *not all the reactants are converted to products*. The rate of appearance of NH_3 gradually slows until it stops. At this point, some uncombined N_2 and H_2 are still present in the reaction vessel. And they remain unreacted no matter how long we wait. No further chemical change appears to take place. But as you will soon see, the system is, in fact, in dynamic *equilibrium*. The rate of the combination of nitrogen and hydrogen just equals the rate of the decomposition of ammonia.

Fortunately, it is not too difficult to determine the constant concentrations of N_2, H_2, and NH_3 present in an equilibrium mixture after the reaction has "stopped." Whenever we repeat the experiment in exactly the same way, starting with the same predetermined concentrations of nitrogen and hydrogen, we get the same results. The concentrations of NH_3, N_2, and H_2 exhibit uniquely characteristic values when the reaction grinds to a halt. This fact suggests that for a fixed initial composition, the composition of the equilibrium mixture is always the same, at least so long as the temperature is kept constant.

To further test the reaction, we can repeat the experiment using various values for the starting concentrations of the reactants. Moreover, the molar ratio of N_2 to H_2 need not have its stoichiometric value of 1:3, it can vary all over the place. In each case, the reaction again proceeds to an equilibrium point and then stops. Now we discover that the final concentrations of the three participants depend upon the initial concentrations of the reactants. But this variation does not indicate the absence of a unifying principle which applies to all the specific instances. When the results of many such experiments are analyzed, a common characteristic of the equilibrium state emerges.

The trick is to relate the equilibrium concentrations of the three gases in a particular manner. Suppose we represent these specific concentrations by brackets, $[N_2]$, $[H_2]$, and $[NH_3]$. For every equilibrium mixture, we can calculate a numerical value for the following expression:

---

* The important role of this catalyst will be discussed later.

14.5 $$K = \frac{[NH_3]^2}{[N_2][H_2]^3}.$$

It turns out that no matter what the original concentrations of N_2 and H_2, this ratio will have a constant value. And, since a common feature of all the specific systems studied is the existence of a state of equilibrium, we conclude that this constant is representative of that state. Consequently, we call it the *equilibrium constant* for the reaction and give it the symbol traditionally employed for constants by kemists.

We will soon return to the equilibrium constant and the equilibrium condition, but by way of the back door and another experiment. Thus far, all the experiments have started with only N_2 and H_2 in the vessel. This time we introduce only pure NH_3 into the evacuated reaction chamber and wait. What happens might surprise you. Analysis shows that the NH_3 concentration decreases while that of N_2 and H_2 increases. Since matter is neither created nor destroyed, there can be only one explanation: the following chemical change is taking place,

$$2NH_3(g) \rightarrow 2N_2(g) + 3H_2(g).$$

After a while the reaction stops, and the concentrations of the three gases remain constant. So we determine what they are and plug the values into Equation 14.5, the expression for the equilibrium constant. When we finish the calculation we discover that K has exactly the same value determined earlier. It does not matter whether we start with only the reactants (N_2 and H_2) or only the product (NH_3). The reaction always winds up at the same place—the position of equilibrium. In fact, any mixture of nitrogen, hydrogen, and ammonia will always yield the same equilibrium constant as long as the temperature is fixed.

FIGURE 14.2.   *A schematic representation of chemical equilibrium in the nitrogen–hydrogen–ammonia system.*

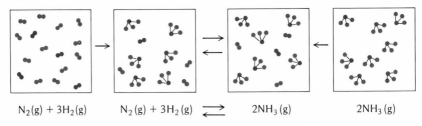

$N_2(g) + 3H_2(g)$ $N_2(g) + 3H_2(g) \rightleftharpoons$ $2NH_3(g)$ $2NH_3(g)$

(a) A nonequilibrium mixture of N_2 and H_2 molecules

(b) An equilibrium mixture of N_2, H_2, and NH_3 molecules

(c) A nonequilibrium collection of NH_3 molecules

Experiments and arguments such as those just described lead to the conclusion that at equilibrium, the rates of the forward and backward reactions are identical. When this particular reaction is at equilibrium, the speed with which N_2 and H_2 combine to form NH_3 is the same as the speed with which NH_3 dissociates into N_2 and H_2. We can symbolize this equality with a pair of opposite arrows,

$$N_2(g) + 3H_2(g) \rightleftharpoons 2NH_3(g).$$

You may recall that there is a precedent for this notation. Back in Chapter 10 I used it to represent the equilibrium between liquid and gaseous water at a constant temperature where the rate of evaporation just equals the rate of condensation. To be sure, that was a case of equilibrium between physical phases; no chemical changes are involved in the vaporization or condensation of H_2O. But the same general principles govern both physical and chemical equilibria. Therefore, the concept and its quantitative implications also apply to solubility phenomena. As you know, many solute–solvent pairs have definite saturation limits. The saturation point is, in fact, a position of equilibrium at which the rate of dissolving is equalled by the rate of separation from solution. And the process can be described by an equilibrium constant. I will resist the temptation to do so. After all, this is a chapter specifically devoted to chemical, not physical changes. That fact will not prevent me from returning to examples of solubility equilibrium in a subsequent discussion of the influence of temperature upon the position of equilibrium. But first we had better spend a bit more time on the constant which represents that condition.

Briefly stated, *the equilibrium constant for a reaction is a quantitative indication of the composition of the system at equilibrium.* It is the answer to the question "How far are you going?" As I have attempted to indicate with the example of the nitrogen–hydrogen–ammonia reaction, K does not reflect specific, individual equilibrium concentrations of the substances participating in the reaction. What it does designate are the *relative* quantities of reactants and products. But K is no simple ratio. In our example, the equilibrium concentration of the product, $[NH_3]$, appears in the numerator; and the reactant concentrations, $[N_2]$ and $[H_2]$, are multiplied together in the denominator. Moreover, the exponents are of particular importance. The equilibrium concentration of NH_3 is squared, its exponent is 2. And $[H_2]$ is cubed, it is raised to the power 3. You will at once recognize that these exponents are the coefficients of NH_3 and H_2 in the balanced equation for the reaction. The corresponding coefficient of N_2 is 1, and hence it is not written in either the equation or the equilibrium constant.

Of course, this is only one instance of thousands upon thousands of chemical reactions. We can represent all of them in this equation:

$$aA + bB \rightleftarrows cC + dD.$$

The equilibrium constant is obtained by a generalization of the specific case. The exponents corresponding to the equilibrium concentrations, [A], [B], [C], and [D] are a, b, c, and d, respectively. When each concentration is raised to the appropriate exponent and the reaction products are placed above the reactants we obtain the following general form for the equilibrium constant:

14.6
$$K = \frac{[C]^c[D]^d}{[A]^a[B]^b}.$$

Our example is merely one particular application of this expression. And you can readily write the equilibrium constant of any other reaction. All you need to know is the correctly balanced equation.

It is not much more difficult to actually evaluate K, providing, of course, you know the values of the equilibrium concentrations. Let me demonstrate with some data. Analysis shows that at 200°C, a 1.0 liter vessel contains 0.10 mole of N_2, 0.30 mole of H_2, and 1.30 moles of NH_3 at equilibrium. It follows that the equilibrium concentrations (molarities) are

$$[N_2] = 0.10 \text{ mole/liter},$$
$$[H_2] = 0.30 \text{ mole/liter},$$
$$[NH_3] = 1.30 \text{ moles/liter}.$$

I now introduce these values into Equation 14.5 and carry out the indicated arithmetic operations.

$$K = \frac{[1.30]^2}{[0.10][0.30]^3} = \frac{1.69}{(0.10)(0.027)} = 625$$

Knowing the value of the equilibrium constant was important to Haber, because it enabled him to determine how far he could get with his reaction and what he could get out of it. The fact that K is larger than 1 indicates that when equilibrium is attained at 200°C, the concentration of ammonia is greater than that of nitrogen or hydrogen. This is a favorable sign for commercial feasibility. Starting with pure N_2 and H_2 in a 1 to 3 molar ratio, the forward reaction will convert most of these elements into NH_3 before equilibrium halts the process. True, K does place a limit on the extent of this chemical conversion, but at this temperature it is a fairly generous one.

On the other hand, it is not nearly as generous as the equilibrium limit imposed upon other reactions. Once more we turn to the combination of hydrogen and oxygen,

$$2H_2(g) + O_2(g) \rightleftharpoons 2H_2O(l).$$

The equilibrium constant for this reaction is immense! At 25°C it has the following value:

$$K = \frac{[H_2O]^2}{[H_2]^2[O_2]} = 10^{83}.$$

The magnitude is so great that hardly any molecules of H_2 and O_2 are left at equilibrium. All but a very, very few have combined to form H_2O. In fact, we can say that the conversion of the elements to the compound is for all practical purposes 100% complete. To be strictly accurate about it, equilibrium does exist, but it strongly favors water.

Other reactions have very low equilibrium constants. The combination of nitrogen and oxygen to form nitric oxide is a case in point. We have already made several references to the reaction

$$N_2(g) + O_2(g) \rightleftharpoons 2NO(g).$$

At 25°C, its equilibrium constant is

$$K = \frac{[NO]^2}{[N_2][O_2]} = 10^{-30}.$$

The miniscule value for K means that the reaction hardly has a chance to start before it is stopped by equilibrium. Only very few N_2 and O_2 molecules combine. Nevertheless, as I have previously implied, enough NO is formed (particularly at higher temperatures) to create serious problems of air pollution. It does not take much nitric oxide to promote the atmospheric production of sulfur trioxide and photochemical smog and perhaps even the destruction of the ozone layer. In fact, one NO molecule out of ten million is enough to be considered a seriously high level.

We will return once more to this highly significant system and the nitrogen–hydrogen–ammonia example in the next section, which looks at the influence of temperature upon equilibrium. But before we move on, it might be well to add a few more words about K. The difficulty is to know which words to choose, and how to restrict them to only a few. For the concept of the equilibrium constant is one of the most important in chemistry. If you decide to study more of this science you will soon learn that K can be usefully applied to a vast range of chemical phenomena. These include the strengths of acids and bases and the solubilities of ionic compounds, as well as all sorts of reactions.

From a knowledge of the equilibrium constant for a particular reaction we can predict the direction of the chemical change which will occur in any known nonequilibrium mixture of reactants and products. We can even calculate the concentrations of all the participants in the reaction when equilibrium is finally achieved. Moreover, we can determine exactly what will happen if we add more of a reactant or a product to a system which is already in equilibrium. I will omit the quantitative details, but the qualitative aspects are easy enough to understand in the following example.

Suppose that a mixture of nitrogen, hydrogen, and ammonia is in equilibrium

$$N_2(g) + 2H_2(g) \rightleftarrows 2NH_3(g).$$

Adding either N_2 or H_2, or both of them, will cause more of the elements to combine, forming additional NH_3. Conversely, if more NH_3 is introduced to an equilibrium system, some of it will dissociate into N_2 and H_2. In both cases, the chemical change will continue until equilibrium is reestablished, again characterized by the same K. The phenomenon is a general one: *adding reactants generates more products; adding products yields more reactants.*

I have already used more than a few words and have not yet mentioned the relationship between thermodynamics and equilibrium, $\Delta G°$ and K. It all fits together and makes beautiful sense, but for that you will have to take my word—or another chemistry course.

Some like it hot—some don't

The thesis of this section can be stated with simple directness: "How far a chemical reaction will go depends upon how hot you get it." Admittedly, the wording lacks a certain elegance, but at least it has the virtue of correctness. If you prefer more formal phraseology, you could say that the equilibrium constant depends upon temperature. Significantly, however, K does not always depend upon T in exactly the same fashion. For some reactions, the constant is larger at higher temperatures. For others, it is smaller. The next few pages will be devoted to exploring the consequences of this behavior.

As a matter of fact, you have already encountered an analog to the temperature dependence of the extent of reaction. Back in Chapter 11 there was a brief discussion of the way in which solubility varies with temperature. At that time you learned that for solute–solvent pairs which *absorb heat* during solution formation, the *solubility increases* with increasing temperature. Conversely, *solubility decreases* with increasing temperature if the dissolution process is accompanied by the *evolution of heat*.

The situation is fundamentally the same for chemical changes.

If a reaction is endothermic, that is, if its ΔH is positive and heat is absorbed, the equilibrium constant will increase as the temperature increases. Consequently, the forward reaction will proceed further at higher temperatures than at lower ones, and relatively more product will be formed. The obvious parallel is the increasing solubility which attends a temperature rise for an endothermic solution process. On the other hand, if ΔH is negative and heat is exothermally evolved during the chemical change, K will fall as the temperature climbs. That is to say, the equilibrium concentration of products decreases with increasing temperature. Here the physical analogy is clearly the decrease in saturation solubility which accompanies increasing temperature if solution formation evolves heat.

Two reacting systems encountered earlier in this chapter provide excellent examples of the two types of temperature dependence. First let us consider the combination of nitrogen and oxygen,

$$N_2(g) + O_2(g) \rightarrow 2NO(g).$$

You will recall that at 25°C the equilibrium constant for this reaction is extremely small, $K = 10^{-30}$. A mixture of the three gases in chemical equilibrium at this temperature contains, relatively speaking, very few molecules of nitric oxide. However, if the temperature of this mixture is raised, the equilibrium is disturbed. In response, the reaction moves in a forward (left to right) direction, more of the elements combining to form the compound. This process continues until another equilibrium state, characteristic of the elevated temperature, is established. In the resulting new equilibrium mixture, the NO concentration will be greater than that which was found at 25°C. And the N_2 and O_2 concentrations will be correspondingly lower. Given the expression for K,

$$K = \frac{[NO]^2}{[N_2][O_2]},$$

the equilibrium constant at the higher temperature will obviously be larger.

This behavior indicates that the forward reaction must be endothermic. We can check it easily enough by consulting Table 12.1. There we learn that the standard heat of formation of NO(g) is +21.6 kcal/mole. It follows that $\Delta H°$ for the formation of 2 moles of NO (as in the equation) must be 2×21.6 or 43.2 kcal. This value will vary slightly with temperature, but it will always be positive. The reaction is decidedly endothermic; heat must be absorbed if the indicated change is to take place. By increasing the temperature and adding more heat, we have driven the reaction to establish a new equilibrium, relatively richer in product than before.

Thanks to the equations of thermodynamics, it is possible to calculate exactly how K varies with T. For one thing, the rate of change depends upon ΔH. But the relationship between K and T is not a simple proportionality. The equilibrium constant increases far more rapidly than that. Doubling the absolute temperature does much more than just double the value of the constant. Thus, K for the N_2–O_2–NO reaction has a value of 10^{-14} at $313°C$ ($596°K$ or $2 \times 298°K$). And at $2500°C$ ($2773°K$), the equilibrium constant is about 10^{-2}.

The latter temperature was selected for a good reason. It is the temperature of the gases in the cylinder of a typical internal combustion engine. Because air is mixed with gasoline and then compressed, the concentration of N_2 and O_2 here is very high. Obviously, much more nitric oxide is produced under these conditions than at $25°C$. This leads to an increase in air pollutants. Because of the way in which K varies with T, a reaction which is of minor concern at normal temperatures becomes seriously troublesome at elevated engine temperatures. And we have no way of preventing it from taking place. Reducing the temperature of engine operation is not a very viable solution. Indeed, the motors which burn fuel most efficiently and completely run at the highest temperatures and use a lot of air.

Of course, equilibrium works both ways. Lowering the temperature reduces the value of the equilibrium constant for this system. Therefore, some nitric oxide should decompose back to nitrogen and oxygen as the exhaust gases cool. This, in turn, should help alleviate the potential pollution problem by reducing the NO level. Unfortunately, however, it takes some time for this reverse reaction to take place. But that is a problem associated with the speed or rate of the reaction, not with equilibrium. The answer is to speed up the rate at which the low-temperature equilibrium composition is attained. Just how that can be done will be discussed in the next section, which concentrates upon reaction rates.

Before we consider that topic, let us look at a reaction whose equilibrium constant decreases with increasing temperature—the combination of nitrogen and hydrogen. This, of course, is the heart of the Haber process,

$$N_2(g) + 3H_2(g) \rightarrow 2NH_3(g).$$

Again consulting the table of standard heats of formation, we learn that $\Delta H° = -22.0$ kcal for this chemical change. It is an exothermic process in which heat is evolved. We calculated earlier that at $200°C$, K has a value of 625. Increasing the temperature of a mixture of N_2, H_2, and NH_3 which is initially at equilibrium at $200°C$ will cause some ammonia to break down into its two con-

FIGURE 14.3.   *A drawing depicting the dependence of the equilibrium constant and extent of reaction upon temperature for an endothermic reaction (above) and an exothermic reaction (below).*

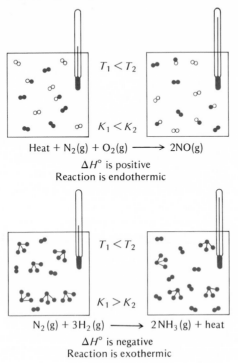

Heat + $N_2(g) + O_2(g) \longrightarrow 2NO(g)$

$\Delta H°$ is positive
Reaction is endothermic

$N_2(g) + 3H_2(g) \longrightarrow 2NH_3(g)$ + heat

$\Delta H°$ is negative
Reaction is exothermic

stituent elements, again indicating that the higher temperature is characterized by a lower equilibrium constant. At 400°C, K has fallen to 0.5, and at 600°C it has a value of 0.014. Qualitatively, these numbers mean that the extent of combination of nitrogen and hydrogen becomes significantly lower as the temperature is raised.

Remember what Haber was after—a way of synthesizing ammonia by the direct combination of nitrogen and hydrogen. Thus the problem is just the opposite of that posed by our earlier example. There the goal is to keep the production of the product, NO, as low as possible. In this case, the aim is to get as much NH_3 as possible. Clearly, the dependence of K upon T for the nitrogen–hydrogen–ammonia reaction suggests that the maximum yield of the desired product will decrease as the temperature of the reaction is increased. The obvious approach would seem to be to work at lower temperatures. But here again, problems associated with reaction rates complicate the issue. At low temperatures, the combination of nitrogen and hydrogen takes place very slowly. Although the amount of ammonia which could potentially be formed under these conditions is quite high, the chemical change occurs so gradually that in practice, the system might never arrive at that advantageous equilibrium condition. To resolve the dilemma, Haber had to know something about reaction rates. And so do we.

How fast are you going?

In this chapter and for most of the previous one, I have made frequent use of the word "spontaneous." In a way, it is an unfortunate choice, but it is the thermodynamically accepted one. To most people, "spontaneous" implies a rapid, self-propelled process. But in point of fact, just because a chemical change is thermodynamically spontaneous does not mean that it will occur easily. Consider once again the combination of hydrogen and oxygen to yield liquid water. The reaction has associated with it a whopping negative standard free energy change of $\Delta G° = -56.7$ kcal/mole of $H_2O(l)$ produced. Consistent with this fact, a huge equilibrium constant indicates that chemistry strongly favors the almost complete conversion of the two separated elements to water. By every thermodynamic criterion, the reaction is spontaneous—very spontaneous!

Yet, in spite of these indications of the great natural tendency for hydrogen and oxygen to react, a mixture of the two elements can stand around for years and no significant quantity of water will be formed. This apparent contradiction between the spontaneity we ascribe to the reaction and its failure to take place at first seems troublesome, but it is easily explained. An analogy helps. Consider a boulder, balanced on a ridge at the rim of a valley. The potential energy of the boulder is much greater than it would be on the valley floor far below. Consequently, there is a strong spontaneous thermodynamic tendency for the rock to roll downhill, converting its potential energy to crashing kinetic energy and disordered thermal motion. But the boulder is prevented from falling by a rise in the earth which blocks its path to the precipice. All it takes to start the boulder on its downward trajectory is a push. Once we get it over the impediment of the mound of earth, gravity will take over. And far more energy will be released by the falling rock than is required to start it moving.

In a similar fashion, it takes a push to start many chemical reactions. For a mixture of hydrogen and oxygen, the impetus can be supplied by a spark. An explosion often follows—ample evidence of the spontaneous evolution of heat and energy. Other spontaneous systems require more extensive priming. Anyone who ever lit a backyard barbeque grill can testify that a good deal of heat must be applied in order to start charcoal burning. But once the combustion process has begun, it proceeds without the need of adding more heat. Indeed, the whole idea of burning charcoal is to get more heat out of it than you put in. Otherwise it hardly pays.

While it may be a nuisance to struggle with charcoal lighter fluid, it is probably for the best. The world would be quite a different place if charcoal immediately exercised its thermodynamic prerogative to combine with oxygen. We couldn't keep it around.

FIGURE 14.4. *An illustration of the energy barrier to chemical reaction (below) and an analogy (above). ΔE^* is the activation energy for the uncatalyzed reaction, ΔE_c^* is the activation energy for the catalyzed reaction, and ΔE is the overall energy change for the reaction, that is, the difference in energy between products and reactants.*

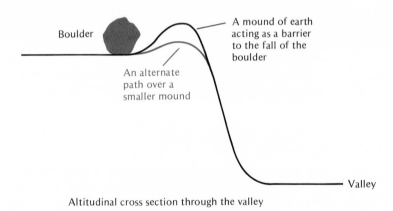

Altitudinal cross section through the valley

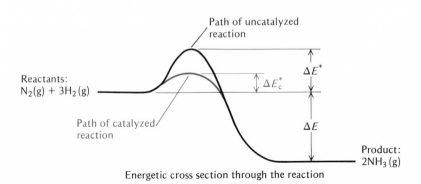

Energetic cross section through the reaction

Moreover, such thermodynamically unstable stuff as paper, wood, petroleum, and coal would also burst into flame upon exposure to air, were it not for the fact that their rates of oxidation are extremely slow at room temperature. These fuels must be heated to their kindling points before they begin to burn and sustain a spontaneous reaction.

Such behavior suggests that an energy mound lies between the reactant ridge and the product plain. In order to start the chemical process moving, the reactants must acquire sufficient *activation energy* to overcome this barrier (see Figure 14.4). If the energetic mound is high, the reaction will proceed only very slowly. But if the barrier is quite low, the reaction will need little initial urging. Generally speaking, then, the greater the activation energy for a reaction, the slower its rate.

I have already indicated the most common way of helping a reaction over this hurdle—applying heat. Chemistry always speeds up as the temperature is increased. For instance, a turkey roasts much faster in a 350°F oven than it does at 250°F. And, while you're in the kitchen, you might also note that increasing the temperature increases the rate at which bread dough rises and wine ferments. Or, look at a cold-blooded animal for additional evidence. The animal's internal chemical reactions are accelerated if the temperature is raised. Its heart beat, respiration rate, and food utilization all increase. Conversely, a cold-blooded animal can be cooled to the point where its biochemical processes are proceeding at such slow rates that it approaches a state of suspended animation. Under these conditions, the animal's demand for oxygen, water, and food is greatly diminished.

Chemists have made many accurate studies on the rates of chemical reactions in a subdiscipline called *kinetics*. The experimentation has led to the discovery of the mathematical form which relates reaction velocity and temperature. These calculations indicate that the speed of most reactions roughly doubles for each 10°C increase in temperature. Moreover, we also know that the rate of virtually all chemical changes depends upon the concentration of the reactants. But here the dependence does not always follow the same pattern. For some systems, doubling the starting concentration doubles the speed of the reaction. For others, the rate may quadruple when the concentration of reactants is doubled. And some systems show still other dependencies. Reactions that are over in a fraction of a second and those that take days to complete have been investigated. Usually the details must be worked out for each individual chemical change. But the methods of kinetics make it possible to obtain quantitative information.

Fortunately, you can appreciate the consequences of kinetics without actually carrying out any of the computations. We have already identified rate-related problems for both of the reactions which have held our attention in the previous two sections. Recall that in the case of the nitrogen–oxygen–nitric oxide system, we would like to be able to speed up the breakdown of the compound. At exhaust and atmospheric temperatures, there is a spontaneous tendency for this dissociation to occur. But precisely because the temperature is low, the reaction proceeds very slowly. Thus, the relatively high concentration of NO formed at engine temperatures of 2500°C is not significantly reduced by cooling. And increasing the temperature of the exhaust gas is obviously not an acceptable solution. This would only tend to again increase the equilibrium concentration of the polluting oxide. Somehow another means must be found to get the reaction over the energy barrier which inhibits its rapid progress.

Just such a means is provided by a *catalyst, a substance which alters (usually accelerates) the rate of a chemical reaction without being permanently changed itself.* In effect, a catalyst provides an alternate route for the reaction. It is as if one were to discover a way around the mound of earth which separates the boulder from the brink. Perhaps there is a much smaller mound of dirt blocking this new path. This should make it considerably easier to push the boulder over the edge of the hill. In essentially the same way, a catalyst lowers the activation energy of a chemical reaction, and hence increases its speed.

In some cases, the catalyst has the same physical state as the reactants. Thus, it may be a gas or dissolved in solution. Frequently a homogeneous catalyst of this sort is actually involved in an important step in the reaction sequence. It undergoes a chemical change which helps promote the overall conversion of products to reactants. However, a subsequent step regenerates the original form of the catalyst and hence there is no net, permanent change in the substance. Many other catalysts are solids, often in powder form which provides a large surface area. Indeed, the surface of these heterogeneous catalysts is the site of their activity. Reactant molecules in the gas or liquid phase are temporarily bound to the surface. As a result, interatomic covalent bonds within these molecules may be weakened and thus made more susceptible to reaction. In other cases, the binding of the reactant molecules to the catalyst may alter the shape of the former in such a way as to promote the desired chemical change. Some *enzymes* (biological catalysts) also appear to work in this latter fashion. After the reaction has taken place, the product molecules are released from the catalyst, which is again ready to bind more reactants.

A suitable catalyst, placed in the exhaust stream of an automobile, could greatly increase the rate at which the high-temperature equilibrium mixture of N_2, O_2, and NO reaches the new equilibrium composition characteristic of the lower temperature. In other words, the catalyst would speed the conversion of nitric oxide back to nitrogen and oxygen. It would not, of course, alter the equilibrium constant. Some progress has recently been made in this direction. But the installation of catalytic converters in all new cars would unavoidably increase their cost. For the sake of the environment and man's place in it, it is probably a price worth paying.

Our other example is one of many instances of catalysis in chemical manufacturing. Fritz Haber's major contribution was to develop a catalyst capable of accelerating the combination of nitrogen and hydrogen at relatively low temperatures. True, increasing the temperature of the reaction would have the same effect. But you will recall that increasing the temperature also lowers the equilibrium constant, that is, it reduces the maximum possible yield of NH_3. Thanks to Haber's catalyst—a mixture of iron with small quantities of metallic oxides—reasonably rapid reaction rates can be attained at temperatures low enough to permit the production of a significant amount of ammonia.

By thus optimizing the equilibrium and kinetic effects, the synthesis of NH_3 has been made industrially and economically feasible. The specific conditions typically employed in the modern Haber process include temperatures of 400° to 600°C and pressures of 200 to 400 atm. The use of elevated pressure helps by increasing the fraction of the nitrogen and hydrogen which combine.

One more point is worth making, particularly if this account has led you to damn chemistry as a science of carnage. Admittedly, by providing Germany with an independent supply of chemically combined nitrogen and the munitions made from it, the Haber process prolonged World War I and its death and destruction. But the same catalyzed chemical change has produced low-cost nitrogen fertilizer in abundant supply. And thus, Haber's discovery has probably saved the lives of millions of men, women, and children who might otherwise have starved to death. The message should be clear; scientific and technological achievements can be utilized for good or evil or both. And the application or misapplication of science is not solely the responsibility of the scientist—it falls on all of society.

Nowhere are these issues more sharply defined than in the vexing dilemma which modern science has created for modern man by releasing the awful and wonderful energy locked in the atomic nucleus. That is the subject of the final chapter of this book. But before we commence that study, we turn to the fantastically complex chemical compounds and reactions which constitute life as we know it and personify it.

For thought and action

14.1   Earlier in the text you read that magnesium burns spontaneously in oxygen or air with the evolution of light and heat, a property which has led to its use in flashbulbs. But Davy had to apply electrical energy to prepare the metal from its oxide, MgO. Use your knowledge of chemical reactions and chemical thermodynamics to account for these observations. Table 14.1 may be useful.

14.2   The standard free energy of formation of $Al_2O_3(s)$ is -337 kcal/mole, that of $Fe_2O_3(s)$ is -177 kcal/mole. Use these values to explain why iron oxide is more easily reduced than aluminum oxide.

14.3   The values of ΔG_f° quoted in Problem 14.2 suggest that aluminum should be more reactive than iron—it should oxidize easier. Yet, in spite of what the free energies imply, steel "tin" cans rust away much more quickly than aluminum ones, and hence the latter pose more of a disposal problem. Suggest a reason (or reasons) for this behavior.

14.4   You meet a smooth-talking salesman who is looking for investors in a sure-fire money-making scheme for synthesizing ethyl alcohol from carbon dioxide and water. He says he plans to use the reaction,

$$2CO_2(g) + 3H_2O(l) \rightarrow C_2H_5OH(l) + 3O_2(g),$$

which will yield oxygen as a valuable by-product. Use your knowledge of chemical reactions to frame your response to this get-rich-quick business opportunity.

14.5   A fuel cell utilizes the controlled oxidation of carbon as its source of energy:

$$C(s) + O_2(g) \rightarrow CO_2(g).$$

(a)   What is the maximum amount of useful work (in kilocalories and joules) which one could hope to get from 1 mole of carbon (graphite) via this chemical change?

(b)   Compute the minimum weight of carbon (in grams) which would have to be oxidized in this cell to yield 1000 kcal worth of useful work.

(c)   Explain why in practice this ideal efficiency would not be achieved.

14.6   I suggested on page 415 that the value of ΔG° for the following reaction, -47.7 kcal, is chiefly due to the contribution of ΔH°. See for yourself by calculating ΔH° from the data of Table 12.1.

$$NO(g) + O_3(g) \rightarrow NO_2(g) + O_2(g)$$

Also calculate the value of ΔS° associated with the reaction at $25°C$.

14.7 Ammonia and hydrogen chloride are both gases at room temperature. They react to form a solid, ammonium chloride, which
frequently forms a film on laboratory bottles.

$$NH_3(g) + HCl(g) \rightarrow NH_4Cl(s)$$

(a) Calculate $\Delta G°$, the standard free energy change associated
with this reaction.
(b) Calculate $\Delta H°$, the standard enthalpy change for the process.
(c) Use your answers to (a) and (b) to compute the entropy
change, $\Delta S°$ at 1 atm and 25°C. Indicate why this answer is
qualitatively consistent with the equation.

14.8 Part of the free energy decrease which attends the metabolic
oxidation of glucose, $C_6H_{12}O_6$, and other foods is stored in the
chemcal bonds of a compound called adenosine triphosphate or
ATP. When ATP breaks down to form adenosine diphosphate and
phosphoric acid, a considerable amount of energy (about 7.3
kcal/mole) is released. Suggest why this process of energy
transfer is particularly useful to you and me and others of our
kind.

14.9 You should now be able to offer thermodynamic and structural
explanations of why graphite is more stable than diamond at
25°C and 1 atm. Please do so.

14.10 A chemical skeptic has overheard your answer to Problem 14.9.
Unimpressed and unconvinced, he demands answers to the following questions. Provide them.
(a) "If graphite is more stable than diamond, how did diamonds
ever form in the first place—a reversal of the Second Law
of Thermodynamics, maybe?"
(b) "If graphite is more stable than diamond at room temperature, how come is it that the Hope diamond isn't so much
soot after all these years?"

14.11 The energies of the H—H, N—H, and N≡N bonds are 104, 93,
and 225 kcal/mole, respectively.
(a) Use the above information to calculate ΔE for the reaction

$$N_2(g) + 3H_2(g) \rightarrow 2NH_3(g).$$

(b) Remembering that $\Delta H = \Delta E + P\Delta V$, predict whether ΔH
will be larger or smaller than the ΔE just calculated.
(c) Now check the accuracy of your prediction by looking up
$\Delta H°_{f,NH_3(g)}$ and calculating $\Delta H°$ for the reaction.
(d) Finally, predict the sign of the entropy change which accompanies this reaction and explain your answer.

14.12 Your young nephew asks why the flashlight cells motivating his
mechanical model of King Kong run down. Use your new-found
knowledge to explain what's happening.

14.13. Write expressions for the equilibrium constants of the following
reactions.

(a)   The burning of methane ("marsh gas"):

$$CH_4(g) + 2O_2(g) \rightleftarrows CO_2(g) + 2H_2O(g)$$

(b)   The combination of hydrogen gas and iodine vapor:

$$H_2(g) + I_2(g) \rightleftarrows 2HI(g)$$

(c)   The breakdown of phosgene (a poison gas used in World War I) to carbon monoxide and chlorine:

$$COCl_2(g) \rightleftarrows CO(g) + Cl_2(g)$$

14.14   One mole of N_2, 1 mole of H_2, and 1 mole of NH_3 are introduced into an evacuated one-liter container at 200°C. Predict what will happen and explain your logic.

14.15   Carbon monoxide and water react to form carbon dioxide and hydrogen in the following reaction:

$$CO(g) + H_2O(g) \rightarrow CO_2(g) + H_2(g).$$

At 420°C, 1 mole of CO and 1 mole of H_2O are introduced into an evacuated 10.0 liter chamber and the system is allowed to come to equilibrium. Analysis of the equilibrium mixture shows the following number of moles of the four participants:

CO	0.25 mole	CO_2	0.75 mole
H_2O	0.25 mole	H_2	0.75 mole

(a)   First write the expression for the equilibrium constant for the reaction.

(b)   Next calculate the equilibrium concentration of each of the products and each of the reactants.

(c)   Finally, combine your answers to (a) and (b) to calculate a numerical value for K.

14.16   The oxidation of sulfur dioxide to sulfur trioxide is an important step in the commercial manufacture of sulfuric acid and a contributor to atmospheric pollution.

$$2SO_2(g) + O_2(g) \rightarrow 2SO_3(g)$$

(a)   Write an expression for the equilibrium constant of this reaction.

(b)   Calculate $\Delta G°$, the standard free energy change for the reaction.

(c)   On the basis of your answer to part (b), predict whether the numerical value of K at 25°C will be greater than 1, less than 1, or about equal to 1.

14.17   Analysis of an equilibrium mixture of SO_3, SO_2, and O_2 yields the following concentrations:

$$[SO_3] = 0.5 \text{ mole/liter}$$
$$[SO_2] = 4.2 \times 10^{-7} \text{ mole/liter}$$
$$[O_2] = 2.1 \times 10^{-7} \text{ mole/liter}$$

Calculate a numerical value for the equilibrium constant. (See Problem 14.16.)

14.18 The maximum percentage yield of SO_3 obtained from the reaction of SO_2 and O_2 is greater at 25°C than it is at 250°C. Predict the sign of $\Delta H°$, and check your prophesy by calculating its value.

14.19 Another chemical change contributing to air pollution is the oxidation of nitric oxide to nitrogen dioxide,

$$2NO(g) + O_2(g) \rightarrow 2NO_2(g).$$

Under standard conditions $\Delta H°$ for this reaction is -26.0 kcal and $\Delta G°$ is -16.6 kcal. Starting with a mixture of NO, O_2, and NO_2 at equilibrium, indicate what would happen in each case if the following changes were made on the system:
(a) Additional NO_2 is added to the equilibrium mixture.
(b) Additional NO is added to the equilibrium mixture.
(c) The temperature of the mixture is raised.

14.20 Given the generalization about the dependence of reaction rate upon temperature which appears in the text, estimate the percent increase in the rate of metabolism of a trout when the temperature of his stream increases from 15°C to 20°C.

14.21 The oxidation of sulfur dioxide to sulfur trioxide is sometimes commercially catalyzed by NO_2. The same reaction, represented by the following equation, also occurs in the atmosphere:

$$SO_2(g) + NO_2(g) \rightarrow SO_3(g) + NO(g).$$

Recall that a catalyst undergoes no permanent change during the process which it accelerates. Therefore, NO_2 must be regenerated by another reaction. Write the equation for the regenerative reaction. Then show that the sum of this equation and the one written above represents the combination of sulfur dioxide and oxygen.

14.22 In the contact process for manufacturing sulfuric acid, the oxidation of $SO_2(g)$ to $SO_3(g)$ is catalyzed by solid platinum metal or vanadium pentoxide. Suggest how this process might work.

14.23 As you know, carbon monoxide is a poisonous gas formed by the incomplete combustion of carbon-containing fuels. Research is being done to develop catalysts to speed up the following reaction:

$$2CO(g) + O_2(g) \rightarrow 2CO_2(g).$$

(a) Calculate $\Delta G°$, the standard free energy change for the reaction.
(b) Use your answer to part (a) to evaluate whether or not it is reasonable to expect this reaction to occur under standard conditions.

(c)   Consult Table 12.1 and calculate $\Delta H°$, the standard enthalpy change associated with the reaction.

(d)   Your answer to part (c) should enable you to predict whether more CO_2 can be formed from a given quantity of CO at cylinder temperature or tail-pipe temperature. Do so, and use your answer to make a practical suggestion about where in the exhaust system the catalyst should be placed.

14.24   One phase of the reaction between hemoglobin (Hb for short) and carbon dioxide can be written

$Hb + CO_2(g) \rightarrow HbCO_2$.

Actually, there are four similar reactions, since each hemoglobin molecule can react with four O_2 molecules or four CO_2 molecules. The equilibrium constants for these reactions indicate that CO_2 can be fairly easily replaced by O_2 in the lungs. Carbon monoxide also binds to hemoglobin, but K for the reaction

$Hb + CO(g) \rightarrow HbCO$

is very large. On the basis of this information, explain the toxicity of carbon monoxide.

14.25   You happen to be a thermodynamically unstable system. What keeps you together?

14.26   How can the executive and legislative branches of the government and the general public obtain the understanding necessary to evaluate studies such as Professor Johnston's work on the nitrogen dioxide–ozone reaction? (The best answer will be forwarded to Washington, D. C.)

14.27   Defend one of these propositions:
(a)   Scientists, such as Fritz Haber, bear particular responsibility for the applications of their discoveries.
(b)   Scientists, such as Fritz Haber, have no more responsibility for the applications of their discoveries than do any other citizens of the society.

Suggested readings

Campbell, J. Arthur, "Why Do Chemical Reactions Occur?" Prentice-Hall, Englewood Cliffs, New Jersey, 1965.

Fischer, Robert B., and Peters, Dennis, G., "Chemical Equilibrium," Saunders, Philadelphia, Pennsylvania, 1970.

King, E., "How Chemical Reactions Occur," Benjamin, New York, 1964.

Lindauer, M. W., The Evolution of the Concept of Chemical Equilibrium from 1775 to 1923, *Journal of Chemical Education* **39**, 384 (1962).

*There is no function in the animal
economy upon which the science of
chemistry cannot throw some light.*
Jean-Antoine Chaptal (ca. 1790)

15

The chemistry of life

The many molecules of organic chemistry

The chief concerns of chemistry are the properties and reactions of inanimate matter. Nevertheless, chemistry is also a science of life. For the great majority of compounds known to chemists are carbon-containing and ultimately derived from living organisms. You have encountered such substances at many stages of this study. For example, most of the molecules used in Chapter 9 to illustrate chemical bonding included atoms of carbon. But these are only a miniscule fraction of the total number of organic compounds. Over a million different carbon compounds have been purified or prepared. And there may be a million times that number which have not been thoroughly studied.

Quite obviously, then, to reserve only a single chapter for the molecules of life is a flagrant underemphasis of a vastly important part of chemical science. Nor can we hope to discuss in any detail more than a mere handful of compounds. But the selection will be made to reflect some of the major types. Thus prepared, we will then consider, particularly in the subsequent sections of this chapter, some of the substances which actually constitute living things and the processes which are part of life itself.

TABLE 15.1.   Classes of Organic Compounds, Their General Formulas, and Some Specific Examples[a]

Alcohol	R—OH	CH_3CH_2—OH	Ethyl alcohol
Ether	R—O—R′	CH_3—O—CH_3	Methyl ether
Aldehyde	$R-\overset{\displaystyle O}{\overset{\|}{C}}-H$	$CH_3-\overset{\displaystyle O}{\overset{\|}{C}}-H$	Acetaldehyde
Ketone	$R-\overset{\displaystyle O}{\overset{\|}{C}}-R'$	$CH_3-\overset{\displaystyle O}{\overset{\|}{C}}-CH_3$	Acetone
Acid	$R-\overset{\displaystyle O}{\overset{\|}{C}}-OH$	$CH_3-\overset{\displaystyle O}{\overset{\|}{C}}-OH$	Acetic acid
Ester	$R-\overset{\displaystyle O}{\overset{\|}{C}}-OR'$	$CH_3-\overset{\displaystyle O}{\overset{\|}{C}}-OCH_3$	Methyl acetate
Amine	R—NH_2	CH_3—NH_2	Methylamine

[a] In most cases, R and R′ represent carbon-containing groups which may be the same or different in a single molecule. However, in formaldehyde and formic acid, R is a hydrogen atom, H.

You will observe in Table 15.1 that a specific *functional group* characterizes each class of organic compounds listed. It is the function of this group which imparts certain common chemical and physical properties to all the members of the family. Thus, as I mentioned in an earlier context, the *hydroxyl group*, —OH, is found in all *alcohols*. The hydrogen and oxygen atoms are bound to each other by a shared pair of electrons, that is, a covalent bond. This leaves five of the seven outer electrons originally contributed by the two atoms. In flyspecks, the electron distribution is $\cdot\overset{\cdot\cdot}{\underset{\cdot\cdot}{O}}\!:\!H$. Note that in addition to the pair of electrons which forms the oxygen–hydrogen bond, there are two more electron pairs on the oxygen atom. But one electron stands alone. This lonesome loner is eager to mate with an oppositely spinning electron, to complete an octet, and to simultaneously form another single bond. And just this opportunity exists because a group called R introduces another eligible electron. The result is an alcohol with the general formula R—OH. R can represent a great variety of molecular fragments, but in an alcohol the oxygen of the hydroxyl group is always covalently linked to a carbon atom.

Not surprisingly, some of the properties of the alcohols depend upon the nature of R. If it is a methyl group, CH_3—, the compound is the simplest member of the series, methyl alcohol, or methanol. The "-ol" ending identifies all alcohols, but this particular one is more commonly known as "wood alcohol," after its original source. You have also met ethanol, C_2H_5OH. Like these two, most alcohols are liquids at room temperature. But as the length and weight of the R group increase, so do the melting and boiling points of the compounds. Furthermore, as R becomes larger it exerts a greater influence upon the properties of the alcohol. The —OH group becomes less important and the entire compound is less like water and hence likes water less. (You will recall that this same point was made during the discussion of solubility in Chapter 11.)

Increasing the number of carbon atoms in an alcohol molecule also means that there are more ways to arrange those atoms. In other words, various isomers are consistent with a single general formula. Consider, for example, the isomeric options open to butyl alcohol, C_4H_9OH. If the carbon atoms are all lined up in a row (really a zigzag chain) the compound is known as *normal* or *n*-butyl alcohol. But there are three other geometrically distinct ways in which the same atoms can be connected. All are pictured below.

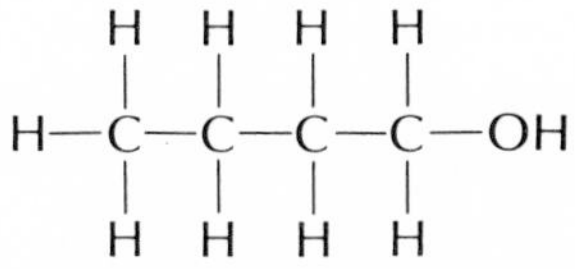

normal-Butyl alcohol

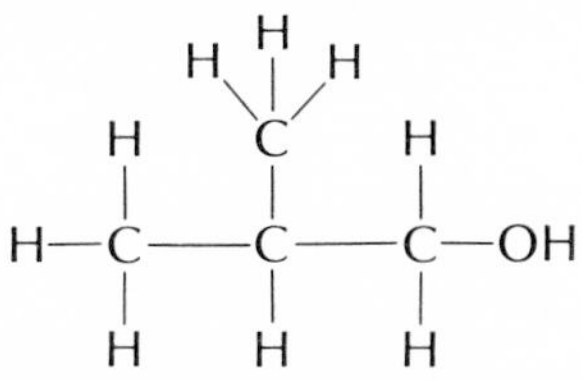

Isobutyl alcohol

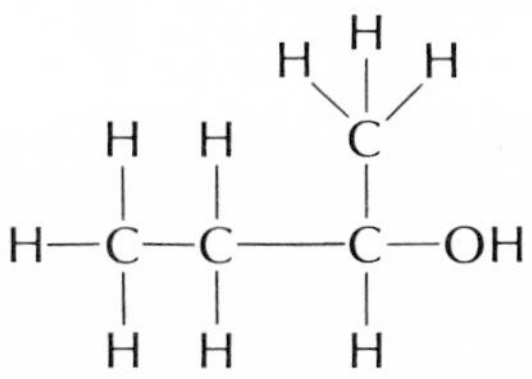

secondary-Butyl alcohol

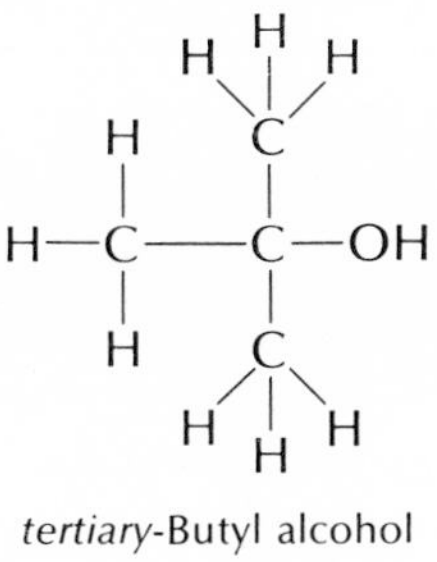

tertiary-Butyl alcohol

Note that in both *normal*-butyl alcohol and isobutyl alcohol, the —OH is attached to a —CH_2 group; the carbon bears two hydrogen atoms. In the nomenclature of organic chemistry, this is an indication of a *primary* alcohol. If the carbon which carries the hydroxyl group is bonded to only one hydrogen, the compound is classified as a *secondary* alcohol. And when there are no hydrogen atoms on this carbon, the alcohol is designated *tertiary*. All of these isomers have been isolated and characterized. They are by no means identical. For example, the boiling points of the four butyl alcohols range from 82.6° to 118.0°C. Moreover, their chemical properties also differ slightly.

To further complicate the issue, there are three compounds which also have the elementary composition indicated by the formula $C_4H_{10}O$, but do not react like alcohols. Hence, they must not contain the hydroxyl group. Experiment clearly indicates that in all of these substances, the oxygen atom is bonded to two carbons, —C—O—C—. This structural feature characterizes a class of compounds known as *ethers*. The general formula is R—O—R'. In this particular instance, the three isomeric ethers have the following atomic arrangements:

$$CH_3CH_2-O-CH_2CH_3, \quad CH_3-O-CH_2CH_2CH_3, \quad \text{and} \quad CH_3-O-\underset{\underset{CH_3}{|}}{\overset{\overset{CH_3}{|}}{CH}}$$

In the compound on the left, the R and R' are identical; both are ethyl groups. Therefore, the isomer is called diethyl ether or, more simply, ethyl ether. In common, inexact usage, the substance is still more simply known as just plain "ether," for this compound is the familiar sweet-smelling anesthetic. The unsymmetrical isomer in the center is named methyl propyl ether after the two groups flanking the oxygen; and the substance on the right is called methyl isopropyl ether. There are, naturally, simpler ethers (as in Table 15.1) and more complicated ones. But most of them exhibit fairly high volatility and low boiling points.*

The oxidation of a primary alcohol yields an *aldehyde*. Compounds of this class contain $\overset{\diagdown}{\underset{\diagup}{C}} = O$, a structure known as a *carbonyl* group. Note the four-electron double bond between the carbon and the oxygen. The general formula, $R-\overset{\overset{O}{\|}}{C}-H$, indicates that a hydrogen atom is always bonded directly to the carbonyl carbon. The other substituent, designated R, can be hydrogen or a heavier carbon-containing group. In the former case, the compound is $H-\overset{\overset{O}{\|}}{C}-H$, a well-known preservative called formaldehyde. Acetaldehyde, the next member of the series and the one

---

* One could argue with good justification that $H_3CH_2C-O-CH_2CH_3$ is a more accurate representation of diethyl ether than the formula given above. After all, the oxygen does lie between the two carbon atoms. Nevertheless, the convention used in the body of the text is the commonly accepted one.

listed in the table, is an important intermediate in chemical synthesis.

The *ketones* form a class of compounds similar to aldehydes in their properties and structure. Again, a carbonyl group is always present. But it is invariably linked to two carbons in the general

form R—$\overset{\displaystyle O}{\overset{\|}{C}}$—$R'$. The two flanking groups, R and R′, can be the same or different. If both are methyl groups, the resulting compound is acetone, a common industrial solvent. Of course aldehydes and ketones can exhibit isomerism much as alcohols and other organic compounds do. But by now the variety of possible permissible structures should need no further comment.

Simple organic *acids* are characterized by the presence of the

carboxylic acid group, $—\overset{\displaystyle O}{\overset{\|}{C}}$—OH. The hydrogen atom of the group comes off as an H^+ ion or proton and gives rise to acidic properties. This leaves behind a large, negatively charged ion of the general

formula, R—$\overset{\displaystyle O}{\overset{\|}{C}}$—$O^-$. The equation for the entire process is

$$R—\overset{\displaystyle O}{\overset{\|}{C}}—OH \longrightarrow R—\overset{\displaystyle O}{\overset{\|}{C}}—O^- + H^+.$$

The more readily this reaction occurs, i.e., the easier it is to break the oxygen–hydrogen bond, the greater the strength of the acid. Compared to inorganic acids such as sulfuric acid, carboxylic acids

are quite weak. Acetic acid, $CH_3\overset{\displaystyle O}{\overset{\|}{C}}$—OH, is the active ingredient in vinegar; and other related compounds, including citric acid and tartaric acid are also found in foods.

The result of the reaction between an organic acid and an alcohol is the formation of an *ester*, a compound having the gen-

eral formula, R—$\overset{\displaystyle O}{\overset{\|}{C}}$—OR′. Many esters are fragrant and flavorful. A simple example is ethyl acetate, the product of the chemical combination of acetic acid and ethyl alcohol.

$$CH_3\overset{\displaystyle O}{\overset{\|}{C}}—OH + HOC_2H_5 \longrightarrow CH_3\overset{\displaystyle O}{\overset{\|}{C}}—OCH_3 + H_2O$$

As the equation indicates, water is eliminated in the esterification process.

An important class of biological compounds, fats and oils, are esters, but somewhat more complicated ones. The alcohol involved, glycerol (or glycerine) contains three —OH groups. Each of these can react with a long-chain fatty acid to yield an ester which is chemically known as a triglyceride. The following equation includes a typical fat, tristearin.

$$
\begin{array}{llll}
\text{CH}_3(\text{CH}_2)_{16}\overset{\displaystyle O}{\overset{\|}{\text{C}}}\text{—OCH}_2 & & & \\[2ex]
\text{CH}_3(\text{CH}_2)_{16}\overset{\displaystyle O}{\overset{\|}{\text{C}}}\text{—OCH} + 3\text{NaOH} \longrightarrow & 3\text{CH}_3(\text{CH}_2)_{16}\overset{\displaystyle O}{\overset{\|}{\text{C}}}\text{—O}^-\text{Na}^+ + & \begin{array}{l}\text{HOCH}_2 \\ \text{HOCH} \\ \text{HOCH}_2\end{array} \\[2ex]
\text{CH}_3(\text{CH}_2)_{16}\overset{\displaystyle O}{\overset{\|}{\text{C}}}\text{—OCH}_2 & & \\
\end{array}
$$

| Tristearin | Sodium hydroxide | Sodium stearate | Glycerol |

This sort of reaction is essentially the reverse of esterification. Here the ester is broken down through the action of a strong alkali, sodium hydroxide. The process is called saponification because the products formed include soaps, the sodium compounds of the fatty acids originally present in the fat. You will observe that the soap, sodium stearate, has the characteristics we identified earlier as necessary for good detergency. The hydrophobic 17-carbon tail readily dissolves in gravy and grease, while the negatively charged carboxylic acid (or carboxylate) group promotes the solubility of the resulting droplet or micelle in the aqueous medium of the dishpan or washing machine.

It should be noted that fats, oils, and other *lipid* materials have biochemical functions besides those made obvious in excess adipose avoirdupois. Certainly, they do provide a chemical means of storing energy in an organism—sometimes rather too much. But fatty compounds are also indispensable structural constituents of cell membranes, nerve sheath, and the brain itself. Sex hormones and the synthetic steroids used as antiovulation contraceptive agents are lipidlike, too, though quite different from triglycerides. And finally, a number of essential vitamins, including vitamins A, D, E and K, are chiefly hydrocarbon and fat-soluble. The structures of some of these molecules are illustrated in Figure 15.1.

Other organic compounds exhibit alkaline or basic properties. Typical of this group are the *amines*, R—NH$_2$ and related substances. The —NH$_2$, or *amino*, group can grab a hydrogen ion away from a water molecule, leaving a hydroxyl ion behind. As illustrated by methylamine, the reaction is

$$\text{CH}_3\text{NH}_2 + \text{H}_2\text{O} \longrightarrow \text{CH}_3\text{NH}_3^+ + \text{OH}^-.$$

FIGURE 15.1. *Some biochemically important lipidlike compounds. Carbon atoms are not individually indicated in the three lower structures, but they are found at each corner of the ring systems.*

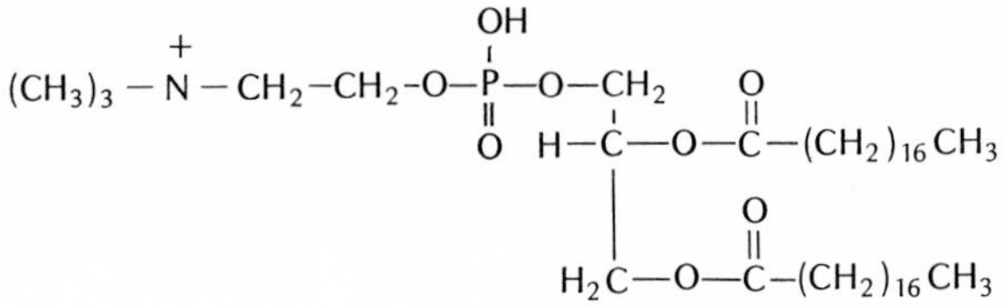

Phosphatidal choline ("lecithin")

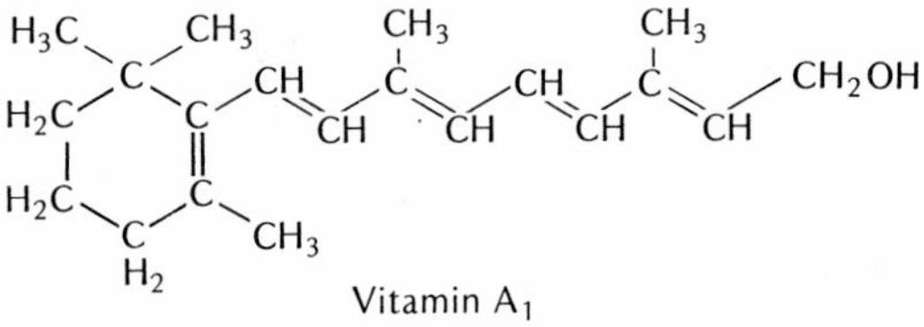

Vitamin A_1

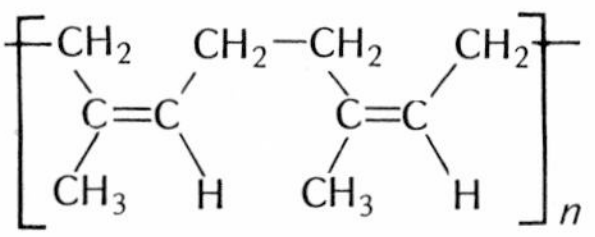

poly *cis* – isoprene (natural rubber)

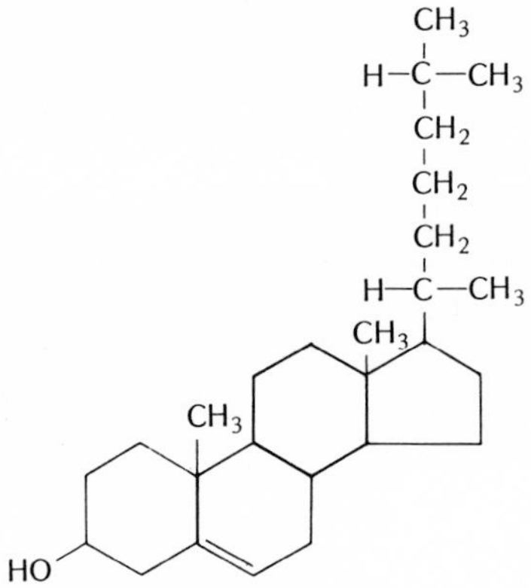

Cholesterol

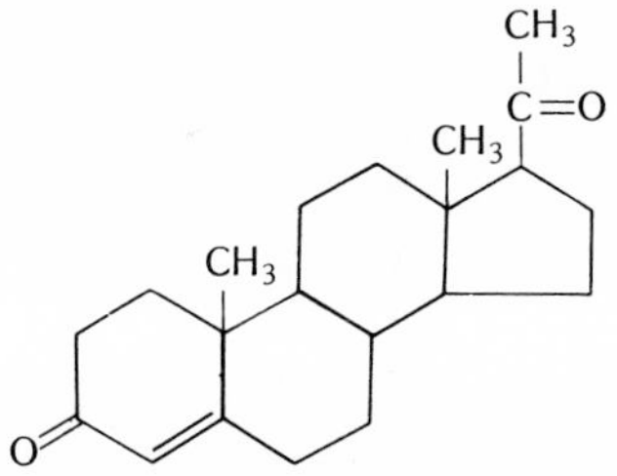

Progesterone
(a female sex hormone)

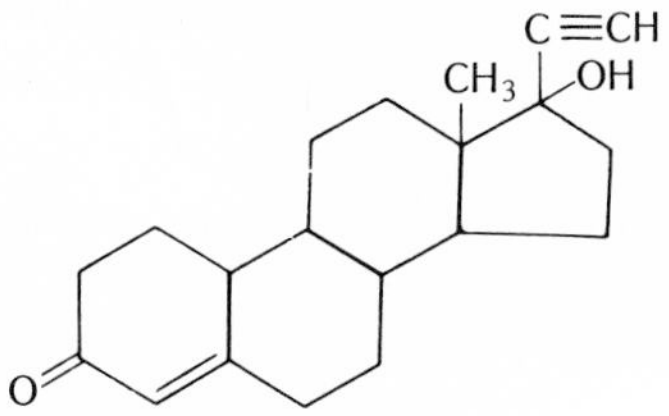

Norethindrone ("Norlutin")
(an oral contraceptive)

You will recall that free OH⁻ ions confer basic properties upon
water solutions. Biological bases are often far more complicated
than the simple amines. Among the most important are the purines
and pyrimidines, which are found in the nucleic acids of genetic
materials. We will touch these bases several times during this
game.

Before you move on, however, please note this cautionary reminder. In summarizing the major classes of organic compounds,
I may have seemed to speak about the unseen world with oracular
omniscience. Let me assure you that my knowledge of molecular
structure is ultimately based upon experiment—not my own, to
be sure, but experiment nonetheless. The limitations of time and
space and the nature of this text have prevented me from presenting the physical and chemical evidence for the correctness of the
structures and formulas I have written. But such evidence does
indeed exist. It is the only way in which we can know the exact
organization of matter on the molecular level.

Amino acids (and Alice)

One of the reasons for the multiplicity of organic compounds
is the fact that a single molecule can include several functional
groups. These groups can all be the same, as in the case of glycerol,
or they can be different. The latter situation is illustrated by three
major classes of biological building blocks: amino acids, sugars,
and purine and pyrimidine bases. We will consider each of them in
turn and the more complicated substances formed from them.

As the name implies, an amino acid is a compound containing
a basic amino group and a carboxylic acid group. In the biologically essential sort of amino acids, both groups are attached
to the same carbon atom. Since this atom is called the α-carbon,
the compounds are known as α-amino acids. They conform to the
general formula

$$
\begin{array}{c}
\quad\ \ \overset{\displaystyle H}{\overset{|}{}}\ \ \overset{\displaystyle O}{\overset{\|}{}} \\
H_2N - C - C - OH \\
\quad\ \ \overset{|}{\underset{\displaystyle R}{}}
\end{array}
$$

Again, R can assume a variety of identities. But in proteins, only
twenty different groups are usually observed.

In the simplest amino acid, glycine, the R is just another hydrogen atom. But in all others, it is something else. This means that
there are four different groups attached to the α-carbon, —H,
—NH₂, —COOH, and —R. Whenever such a situation of *asymmetric substitution* arises, the attendant phenomenon of *optical*

activity is also observed. Simply stated, an optically active compound is one which can rotate a beam of polarized light. Of course, the statement is simple only if you happen to know about polarized light. Normally, the waves of light vibrate in all directions about the line of propagation. But when the beam is passed through certain substances like calcite (calcium carbonate) crystals or Polaroid sheet, only those waves vibrating in a particular plane are able to emerge. The other waves are deflected or absorbed. What gets through is *plane-polarized light*.

All this is far easier to draw or observe than to write about. Therefore, I suggest you consult Figure 15.2. It would be even better if you could get hold of a broken pair of Polaroid sunglasses. Think of each lens as made up of a whole series of narrow, parallel slits. Only the light waves whose vibrations coincide with these slits will be permitted passage. Now place the lenses on top of each other and slowly rotate one of them while looking through both. You will note an orientation where a maximum amount of light is transmitted by the lenses. Obviously, this must mean the "slits" in the Polaroid material are parallel to each other. But in another position, 90° from the former one, practically no light gets through. This suggests that the slits are now at right angles to each other and hence block out essentially all the light.

An optically active substance interacts with a beam of plane-

FIGURE 15.2.   *A representation of polarized light and the rotation of the plane of polarization by an optically active compound. The drawing is a rough schematic of a polarimeter.*

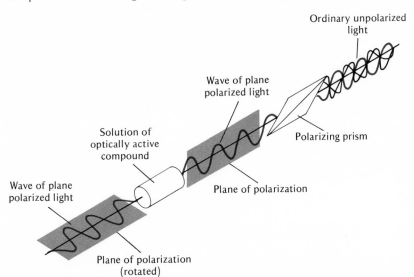

polarized light and twists it. The emergent beam is still vibrating in only one plane, but the plane is rotated from its original orientation. The extent of rotation, and whether it is to the left or to the right can be determined with an instrument called a *polarimeter.* The optically active compound is dissolved in an inactive solvent, such as water, and placed in a tube with glass windows at either end. A beam of polarized light is passed through this tube. The angle through which the plane of vibration is turned during passage is indicated by an optical system involving another polarizing crystal. The size of this angle depends upon the wavelength or frequency of the light used, the length of the light path through the solution, the concentration of solute in the solution, and the structural nature of the solute molecules.

The actual mechanism responsible for the phenomenon of optical activity is complicated enough that many chemists do not understand it. But all of them must know the sort of molecules which rotate polarized light. They are *molecules whose mirror images are not superimposable.* If you'll pardon the pun, you can illustrate the last phrase quite handily. For your hands are asymmetric. The left is a "reflection" of the right, but the two are not identically superimposable. Just hold them palm-to-palm and then palm-to-back if you need proof. As Alice discovered some time ago, the Looking Glass world is different from this one.

It is also relatively easy to see that the same features characterize a tetrahedral atom with four different substituent groups. Molecular models would be ideal here, but Figure 15.3b can help, too. The compound depicted is the amino acid, alanine. It is probably apparent to you that the molecule marked L is the mirror image of that designated D. Now try mentally turning L about its vertical axis in an attempt to make it look exactly like D. Remember that in their original orientations, the —CH₃ group is in the plane of the paper, the —H sticks out toward you, and the —NH₂ group points back. If you turn the L form 180° until the —CH₃ is again in the plane of the paper, but on the left, the location of this group will be the same as that pictured in D. But you will not have duplicated all of D, because now the —NH₂ and —H will be interchanged relative to their positions in the mirror image molecule. In fact, no matter how you twist it and turn it, you cannot orient a molecule of L-alanine so that it will be spatially identical to a molecule of D-alanine. The two are structurally distinct, chemically and physically separable *optical* or *stereoisomers.* By contrast, a molecule of glycine (Figure 15.3c) is superimposable upon its mirror image, since two of the substituents on the α-carbon are identical.

As you might expect from their great structural similarity, the chemical and physical differences between optical isomers are very slight. The chief point of distinction, appropriately enough,

FIGURE 15.3.   *Examples of stereoisomers with nonsuperimposable mirror images (a) and (b), and a compound which does not show optical isomerism (c). The bonds represented by solid lines are in the plane of the paper, those represented by wedges point forward, and those represented by dashed lines point backward.*

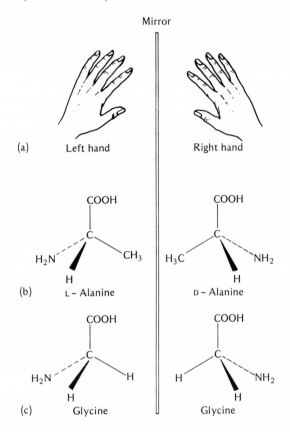

involves optical activity. For example, if one measures the extent of rotation produced by otherwise identical solutions of D-alanine and L-alanine (using the same light source and path length), the two observed angles are found to be equal in magnitude, but opposite in sign. It is also possible to experimentally determine the arrangement of the four substituents on the α-carbon atoms of these stereoisomers. But somewhat surprisingly, the nature of the correlation between the direction and extent of rotation, on the one hand, and the absolute molecular configuration, on the other, is not well understood. In other words, it is risky to predict one from the other.

Yet, in spite of the fact that it is often difficult to distinguish between stereoisomers in a chemist's laboratory, the differentiation is easily made in a chemist's body. In a word, biochemistry is highly stereospecific. All the amino acids that occur in earthly proteins conform to the L-configuration illustrated in Figure 15.3. Proteins in some other part of the universe might well be made of D-amino acids. At any rate, nature obviously knows her left hand from her right.*

Now we can take a look at the 20 common R groups found in ordinary protein. They conveniently fall into the classes indicated in Table 15.2. There is no point in repeating all the categories here. Note, however, the considerable variety of reactive groups. Serine and threonine contain alcoholic —OH's, and cysteine has a somewhat similar sulfhydryl or —SH group. Of course, all amino acids contain at least one —COOH, but aspartic acid and glutamic acid each carry an additional one in their tails. The latter compound is familiar in the form of monosodium glutamate (MSG), a flavor-enhancing substance believed to be responsible for "Chinese restaurant syndrome," the headaches and hot flashes which sometimes follow MSG-rich wonton soup. Extra basic groups are borne by lysine, arginine, and histidine. Each of these extra functional groups is, of course, capable of reacting in ways which reflect its chemical propensities.

All of the common α-amino acids have been prepared in experiments designed to duplicate their original synthesis on the prebiotic earth. It is now believed that the atmosphere of the newly born planet consisted chiefly of nitrogen, hydrogen, carbon monoxide, carbon dioxide, hydrogen sulfide, and water vapor. This mixture of gases was subjected to daily bombardment by visible and ultraviolet light; cosmic, X, and γ rays; α and β particles; and flashes of lightning. Under this energetic influence, the building blocks of biomolecules are thought to have formed. The evidence, though obviously inferential, is quite convincing. Laboratory studies have identified sugars, biologically important acids, and purine and pyrimidine bases, as well as amino acids, among the products of irradiated model primordial atmospheres. These compounds presumably dropped into the warm water which covered most of the earth's surface. Gradually, over millions and millions of years, the concentration of the broth increased. A multitude of reactions occurred. Finally, perhaps 3.2 billion years ago, there emerged a primitive proto-cell with the ability to utilize food, maintain its existence, and replicate itself. One of the most significant steps in that slow, unsteady climb culminating in life was the synthesis of proteins.

---

* It is interesting to speculate whether the great preponderance of L-amino acids means that nature is sinister or just *gauche*.

TABLE 15.2. *R Groups for the 20 α-Amino Acids Commonly Found in Proteins*[a]

R is predominantly hydrocarbon

Glycine (Gly) $H-$

Alanine (Ala) CH_3-

Valine (Val) $(CH_3)_2CH-$

Leucine (Leu) $(CH_3)_2CHCH_2-$

Isoleucine (Ile) $CH_3CH_2CH(CH_3)-$

R contains — OH or other groups

Serine (Ser) $HOCH_2-$

Threonine (Thr) $CH_3CH(OH)-$

Cysteine (Cys) $HSCH_2-$

Methionine (Met) $CH_3-S-CH_2CH_2-$

Asparagine (Asn) $H_2N-\overset{O}{\overset{\|}{C}}-CH_2-$

Glutamine (Gln) $H_2N-\overset{O}{\overset{\|}{C}}-CH_2CH_2-$

R is acidic

Aspartic acid (Asp) $HO-\overset{O}{\overset{\|}{C}}-CH_2-$

Glutamic acid (Glu) $HO-\overset{O}{\overset{\|}{C}}-CH_2CH_2-$

R is basic

Lysine (Lys) $H_2N(CH_2)_4-$

Arginine (Arg) $H_2N-\overset{NH}{\overset{\|}{C}}-NH(CH_2)_3-$

Histidine (His) $HC{=\!=}C-CH_2-$ (imidazole ring with N, NH, and C—H)

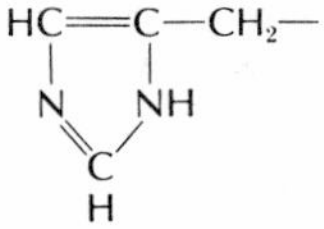

R contains a benzene ring

Phenylalanine (Phe) (benzene)$-CH_2-$

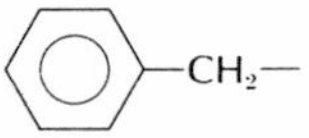

Tyrosine (Tyr) $HO-$(benzene)$-CH_2-$

Tryptophan (Trp) (indole ring)$-CH_2-$

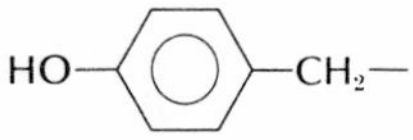

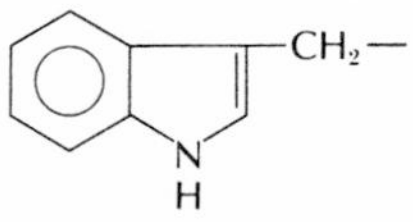

A special case

Proline (Pro) (pyrrolidine ring with H_2C, H_2C, H_2C, $C-COOH$, N, H)

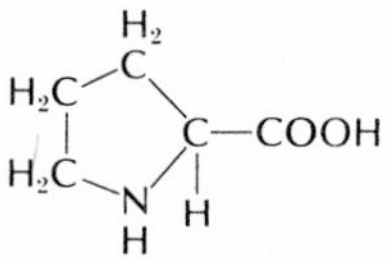

---

[a] The name and conventional symbol of each compound are indicated. The general formula, which applies to all except proline, is

$H_2N-\overset{\overset{H}{|}}{\underset{\underset{R}{|}}{C}}-\overset{O}{\overset{\|}{C}}-OH.$

Of proteins and peptide bonds

The most characteristic groups on any amino acid are those which are common to the entire class, the —NH$_2$ and the —COOH groups. Because of their presence, an amino acid can act as both a base and an acid. Add to this dual functionality the fact that acids and bases tend to react with each other, and you have the stuff that men (and women) are made of—*proteins*. The combination of two amino acids is illustrated below with alanine and valine.

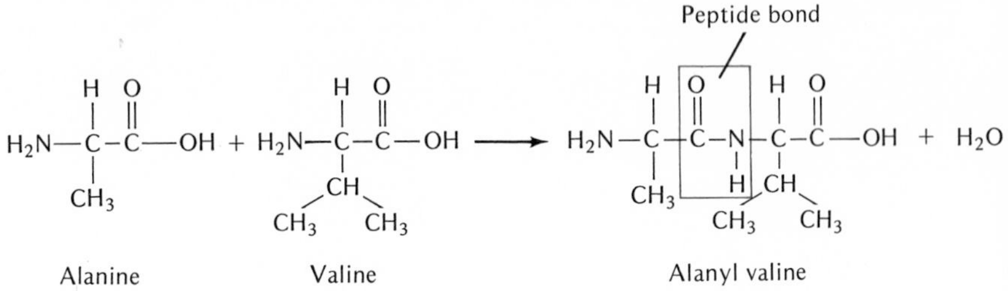

$$\underset{\text{Alanine}}{\text{H}_2\text{N}-\overset{\overset{\text{H}}{|}}{\underset{\underset{\text{CH}_3}{|}}{\text{C}}}-\overset{\overset{\text{O}}{||}}{\text{C}}-\text{OH}} + \underset{\text{Valine}}{\text{H}_2\text{N}-\overset{\overset{\text{H}}{|}}{\underset{\underset{\text{CH}\diagup \text{CH}_3 \diagdown \text{CH}_3}{|}}{\text{C}}}-\overset{\overset{\text{O}}{||}}{\text{C}}-\text{OH}} \longrightarrow \underset{\text{Alanyl valine}}{\text{H}_2\text{N}-\overset{\overset{\text{H}}{|}}{\underset{\underset{\text{CH}_3}{|}}{\text{C}}}-\boxed{\overset{\overset{\text{O}}{||}}{\text{C}}-\overset{\overset{\text{H}}{|}}{\underset{\underset{\text{H}}{|}}{\text{N}}}}-\overset{\overset{\text{H}}{|}}{\underset{\underset{\text{CH}\diagup \text{CH}_3 \diagdown \text{CH}_3}{|}}{\text{C}}}-\overset{\overset{\text{O}}{||}}{\text{C}}-\text{OH}} + \text{H}_2\text{O}$$

Peptide bond

In the course of the reaction, water is eliminated and a dipeptide, called alanyl valine, is formed.

The carbon–nitrogen *peptide bond* identified in the drawing is the linkage which holds amino acids together and makes proteins possible. Often the bond is repeated hundreds of times to form molecules whose weights, expressed on the standard scale, run into the tens of thousands. At one end of such a *polypeptide* chain is a free amino group, at the other an unreacted carboxylic acid. And each *residue* (as the polymerized amino acids are called) contributes its own characteristic R group. Since these side chains carry a variety of functional groups, the properties of proteins can vary widely, depending upon their amino acid composition.

It is interesting that the polymerization of amino acids into peptides and proteins is not a spontaneous process. The reaction needs a thermodynamic push. Just where this assistance came from in the primordial soup or smog is not clear. There are several hypotheses, all of them, needless to say, impossible to prove unequivocably. But somehow or other, proteins got put together; we each provide personal proof of that! Moreover, we internally accomplish the assembly ourselves every day. It is a considerable chemical accomplishment—both the synthesis itself and the discovery of how we do it. More of that later, but this might be the place to mention that proteins have also been synthesized in laboratory glassware. In 1968, two groups of scientists, one at Rockefeller University and the other at the pharmaceutical firm of Merck Sharp and Dohme, independently prepared the enzymatic (catalytic) protein bovine ribonuclease A from its constituent amino acids. That phenomenal feat required about 18 months of actual work plus some 50 years of preparation. A cow does it in a minute or two.

What makes protein synthesis little short of miraculous, whether done in antediluvian ooze, tissue, or test tube, is the incredible number of possible products. Bovine ribonuclease, for example, contains 124 amino acids. Since there are 20 different building blocks to choose from, there could be 20^{124} different proteins of this size. Expressed relative to the more familiar base 10, that's 10^{161}. Arriving at one specific composition out of 10^{161} options is natural selection of mindboggling improbability. Of course, neither a cow nor Bruce Merrifield and his colleagues at Rockefeller operate against the odds which existed in the prebiotic earth. The cow knows what the composition of ribonuclease should be because the knowledge is stored in her nucleic acids. Dr. Merrifield knew because another scientist had determined the sequence of amino acids in a natural sample of the enzyme.

This information is called the *primary structure* of a protein. It represents the identity of the amino acids involved and the order in which they are joined. The amino acid composition of a protein can be obtained with relative ease. In fact, there are analyzers which do the job automatically. The protein is first hydrolyzed—broken down into individual amino acids by the action of concentrated acid. The resulting mixture is run through a cation-exchange column (see Chapter 11). Here the function of the exchanger is not to deionize the solution, but to separate its constituents by a method known as *chromatography*. Under the conditions of the analysis, most of the amino acids are in the following ionic form and hence bear positive charges.

$$^{+}H_3N-\overset{\displaystyle \overset{H}{|}}{\underset{\displaystyle \underset{R}{|}}{C}}-COOH$$

These cations can and do displace some of the Na^{+} ions originally on the exchanger. But the tightness with which the amino acid ions are bound to the column depends upon the nature of the R group. Those compounds that exhibit the weakest attraction for the exchanger are easily washed off and out when an eluting solution is poured through the column. They appear in the first portion of the column effluent. Those amino acid cations with stronger electrostatic affinity for the column material emerge in later fractions. Using the proper eluting solutions, all 20 amino acids can be isolated from a mixture. The instrument measures the concentration of each and neatly displays the results on a graph.

Important as such information is, it is obviously insufficient to fully characterize a protein. For example, we may know that each molecule of ribonuclease contains 12 alanine residues, 4 arginines, 5 aspartic acids, and so on for all the amino acids. But there are thousands of different isomers which would have this identical composition. The transposition of only a single pair of residues would be enough to change the protein. The fact that the enzymatic activity and other properties of ribonuclease are reproducible suggests that there is one specific sequence which characterizes this complex compound.

Finding the order of amino acids which makes any protein what it is, is a painstaking and time-consuming process. Nevertheless, a surprisingly large number have been sequenced. Usually the molecule is first cleaved into several smaller fragments which are then individually investigated and analyzed. The amino acids on the ends of these peptide chains can be determined without too much difficulty. And if overlapping fragments are formed, the puzzle is made easier. But almost always, some stepwise analysis is required. Certain chemical reagents and reactions permit the selective removal of the end residue which bears the free, unreacted amino group. This amino acid is isolated and identified, and the process is then repeated on the newly exposed NH_2-terminal end. The amino acids are then reassembled on paper to yield the original order.

Information of the primary structure of proteins has been a valuable aid in drawing the family tree of the animal kingdom. Certain proteins assume the same function in a wide variety of species. These similar, but often nonidentical substances are said to be *homologous*. Since it is generally accepted that the most closely related species diverged most recently in the evolutionary process, it is reasonable to assume that the amino acid composition and sequence of their homologous proteins should have much in common. And indeed, there is good agreement between this index of kinship and more traditional methods of biological taxonomy. For example, the cytochrome c's of men and monkeys differ by only one amino acid out of 104; but a comparison of the same homologous proteins from men and dogs discloses ten different residues.

These differences between species have evolved as a result of mutations. The instructions for making all the proteins produced by an organism are written in a chemical code and stored in the chromosomes of each cell. Sometimes a molecular mistake occurs and is passed on to subsequent generations. Some of these mutations are favorable and enhance the individual's chance of survival. But often they manifest themselves in hereditary diseases and disorders. The nature of one of these, sickle-cell anemia, is known in molecular detail. Moreover, it provides a graphic demonstration that minor changes in primary structure can have major effects upon protein function.

When an individual with a genetic tendency toward sickle-cell disease is subjected to conditions of high oxygen demand, such as heavy exercise, some red blood cells distort into a rigid sickle or crescent shape. Because these erythrocytes thus lose their normal deformability, they cannot pass through the tiny openings in the spleen and other organs. Some of the cells are destroyed and anemia results. Others can clog organs so badly that their blood supply is seriously reduced.

The property of "sickling" has been traced to the amino acid composition of hemoglobin. Each molecule of this oxygen-carrying pigment protein (molecular weight = 64,500) consists of four iron-containing heme groups and four protein chains. The latter come in two pairs, designated α and β. Sequence determination has demonstrated that in individuals with sickle-cell anemia, the sixth amino acid from the —NH_2 end of the β chain is valine. In normal hemoglobin that position is occupied by glutamic acid. Apparently this very slight substitution is sufficient to cause the abnormal hemoglobin S to polymerize or gel at low oxygen concentrations.

Normally, mutations detrimental to a species are selectively eliminated. The fact that sickle-cell disease is still quite common suggests that the gene which produces it may have been retained because it also imparts some beneficial effect. A clue to what this effect might be is found in a study of the people who have the altered hemoglobin in their blood. In the United States, it is far more common among the black population than the white. In fact, about 10% of Americans of African descent are carriers of sickle-cell trait and, though they do not have the disease themselves, they could pass it on to their children. Moreover, the incidence of hemoglobin S is high in Africa and other tropical and subtropical countries—just the places where malaria is common. The evidence suggests that the correlation is more than coincidental, and several hypotheses have been advanced to explain how sickling erythrocytes might combat malaria. Whatever the ultimate findings will be, one thing is clear: the primary structure of a protein is of great importance in determining its properties.

Information about conformation

Among other things, the amino acid sequence dictates the shape or conformation which is assumed by the molecule. The smallest complete unit of a protein consists of one or more polypeptide chains. In the case of α-chymotrypsin, an enzyme which catalyzes the breaking of peptide bonds, there are three chains. One contains 13 amino acids, another includes 131, and the third has 97 residues. They are covalently joined together by —S—S— or disulfide bonds. These relatively long chains could get hopelessly tangled in a million different ways were it not for the order imposed by nature. As a matter of fact, under fixed conditions of temperature, pH, and so on, a protein molecule always appears to have the same shape. This conformation is described as its *secondary* and *tertiary structure*.

The actual determination of the secondary and tertiary structures of proteins has been achieved through the method of X-ray diffraction. X Rays are bounced off the atoms in a manner already alluded to in Chapter 8. The data consist of a series of spots on a photographic film, spots which signal the impact of the reflected beams. Thanks to certain mathematical relationships, it

FIGURE 15.4.   *The secondary and tertiary structure of α-chymotrypsin, based upon the X-ray diffraction studies of Matthews et al. Nature 214, 652 (1967). The molecule consists of three polypeptide chains, marked A, B, and C and identified by color in the drawing. These chains are held together with disulfide linkages represented by alternating cross-hatched bands. Two specific amino acid residues involved in the enzymatic function of the protein are indicated—histidine 57 and serine 195.*

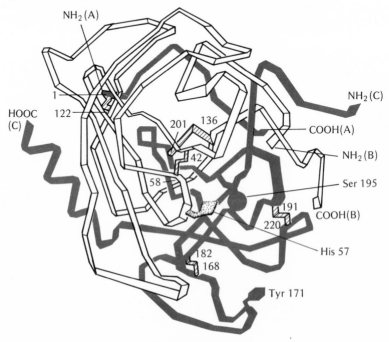

is possible to translate this seeming nonsense into a three-dimensional map which indicates the relative positions of the atoms in a crystal. If the substance is a simple one, such as sodium chloride, the assignment is fairly easy, and structures similar to Figure 9.1 have been known for quite some time. However, determining the crystal structure of a protein like chymotrypsin, with a molecular weight of 25,000, is a formidable task. Thanks to modern high-speed computers, it has been accomplished. Not all 5000 atoms in each chymotrypsin molecule have been located with equal precision; the hydrogen atoms are particularly hard to "see." Nevertheless, the atomic arrangement is known with sufficient certainty to produce the picture of the molecule printed as Figure 15.4. The snarl of tape represents the peptide chain backbone. Individual atoms are not indicated in Figure 15.4, nor are most R groups pictured. However, two residues which are involved in the enzymatic activity of chymotrypsin are identified. They are histidine 57 and serine 195. (The numbering refers to the sequence, starting with the —NH_2 end of the A chain.)

Within the overall conformation is a localized region of *secondary structure.* I refer to the short spiral at the —COOH end of the chain marked C. The regular arrangement of atoms here indicates a segment of α-helix. This structure, pictured in greater detail in Figure 15.5, is the discovery of Linus Pauling, the only person ever to win Nobel Prizes for both Chemistry and Peace. The helix is stabilized by a series of hydrogen bonds which run roughly parallel to its axis. The linkage is always between the —NH group of

one amino acid and the $\diagdown C=O$ of the fourth residue up the chain $\diagup$

from it. This right-handed spiral staircase climbs at a rate of 3.6 amino acids per turn, with a regular repeat distance of 5.4 Å. Judging from the structural determinations which have been made, the α-helix is quite widely distributed in nature. Some proteins contain a much larger fraction of it than chymotrypsin does. Indeed, the peptide chain of hemoglobin is almost all spirally wound. Other proteins exhibit other forms of hydrogen-bonded secondary structure.

The way in which the protein, including its segments of α-helix, is folded into final form is known as its *tertiary structure.* The disulfide bridges marked in Figure 15.4 help stabilize this particular conformation. So do a variety of other intramolecular interactions. Most of these (they are summarized in Figure 15.6) depend upon the nature of the side chains contributed by the various R groups of the constituent amino acids. Certain residues hydrogen-bond to each other. Oppositely charged groups exert mutual electrostatic attraction. And hydrocarbon side chains experience hydrophobic

FIGURE 15.5.   *A polypeptide (protein) chain exhibiting the
α-helical secondary structure proposed by Pauling, Corey, and
Branson, Proc. Natl. Acad. Sci. U. S. **37**, 205 (1951). Atoms of the
various elements involved are designated in different colors, and
the stabilizing hydrogen bonds are represented by colored lines.*

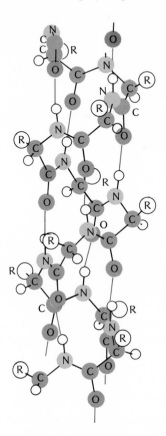

FIGURE 15.6.   *Various methods of intramolecular interaction
which stabilize the secondary and tertiary structure of proteins.*

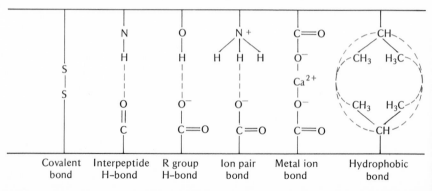

| Covalent bond | Interpeptide H–bond | R group H–bond | Ion pair bond | Metal ion bond | Hydrophobic bond |

interaction. Generally speaking, this latter attractive force is particularly important within the interior of the molecule. The hydrocarbon groups of a globular protein, like α-chymotrypsin, point inward and create a center which is oily in its essential characteristics. The electrically charged side chains, including the acidic and basic groups, are usually found on the outside of the molecule. Here they can exercise their affinity for the aqueous medium which normally surrounds a protein. Thus, in gross structural features globular proteins resemble soap micelles.

Each protein molecule appears to have a specific *native conformation* which exists under physiological conditions. It is generally believed that this particular secondary and tertiary structure is identical with or very similar to that in the solid crystal. For proteins with catalytic capabilities, the native conformation is necessary for enzymatic activity. This is because the geometry of the interaction between the enzyme and its *substrate* (the substance it acts upon) is very critical. As I have already noted, at least two specific amino acids are involved in the catalytic function of chymotrypsin. If the molecule were stretched out in a single chain, histidine 57 and serine 195 would be far removed from each other. In fact, they are on different polypeptide segments. Nevertheless, in the tightly folded native conformation, these residues are in close proximity in sort of a pocket. The substrate, part of a protein, for example, fits into this pocket and is exposed to the chemical action of these two groups. The maximum efficiency of chymotrypsin occurs in nearly neutral solutions. These are the conditions encountered in the small intestine where this enzyme helps digest ingested beefsteak and burgers. But if chymotrypsin is *denatured* by subjecting it to extremes of acidity, alkalinity, or temperature, it loses its catalytic capability. This loss occurs because the protein assumes a different conformation under these new conditions. The molecule unfolds and the serine 195 and histidine 57 become separated from each other. But even the shape of a denatured protein is a consequence of its amino acid sequence.

A further indication that primary structure influences secondary and tertiary structure was provided by synthetic ribonuclease. When the protein was completed, the scientists working on the project did not need to worry about forcing the molecule into its native conformation. The compound took care of that itself, once the correct order of linked amino acids was achieved. Moreover, the synthetic ribonuclease proved its conformational correctness by a demonstration of enzymatic activity.

There are few reactions in the chemistry of life which do not involve one or more enzymes. Over a thousand specific biological catalysts have been isolated. All of them are proteins, but their functions range widely. Some, like ribonuclease A and α-chymotrypsin, promote the breaking of bonds and the digestive disruption of larger molecules into smaller ones. Others catalyze the formation of bonds in the buildup of structural components from food fragments. Still others facilitate the transfer of groups from one molecule to another. And a large class of enzymes are involved in oxidation–reduction reactions. These amazing catalysts are the envy of every synthetic chemist. And a good deal of information has been amassed about the way in which they work.

But I must suppress the temptation to say more about enzymes in particular or proteins in general. This is an area in which I have done a little research, and I could write a hundred additional pages about this fascinating field. However, I realize that not everyone may share my enthusiasm. Nor is this a book about protein chemistry. And so I move reluctantly to the subject of the next section. It is sweet, though not as sweet to me as this.

Sweetness and light

Sugars, or *saccharides*, belong to a larger class of compounds known as *carbohydrates*. The latter name suggests carbon plus water, and so do the formulas of many sugars. For example, in ordinary sucrose, $C_{12}H_{22}O_{11}$, the H:O atomic ratio is 2:1, just that encountered in H_2O. Moreover, anyone who has ever scorched the fudge or burned a marshmallow is well aware that upon heating, sugar loses water to leave a black carbonaceous residue. Saccharides vary in number of carbon, hydrogen, and oxygen atoms present in a single molecule. Indeed, these compounds are often classified on the basis of their size. Since the ending "-ose" signifies a sugar, a triose includes three carbon atoms per molecule, a pentose has five, a hexose six, and so on.

Predictably, the arrangement of atoms in these simple saccharides conforms to a certain structural pattern. Figure 15.7 indicates what it is. Like amino acids, sugars are multifunctional. They all contain several alcoholic hydroxyl groups, normally one less than the number of carbon atoms. A single —OH is thus bonded to each of the carbons except one.* The remaining carbon carries a doubly bonded oxygen. If this carbonyl group is on the end of the molecule, the resulting compound is an aldehyde.

$$\overset{\displaystyle O}{\overset{\displaystyle \|}{}}$$

Sugars containing the —C—H function are called *aldoses* and include glucose, the most ubiquitous sugar of all. Other saccharides,

---

* In deoxyribose, a sugar found in DNA, a hydrogen atom replaces an
hydroxyl group.

FIGURE 15.7.   *The structural formulas of three common sugars, all of them monosaccharide hexoses with the general formula $C_6H_{12}O_6$. D-Glucose and L-glucose are stereoisomeric aldoses; D-fructose is a ketose. The asymmetric, optically active carbon atoms are identified with asterisks.*

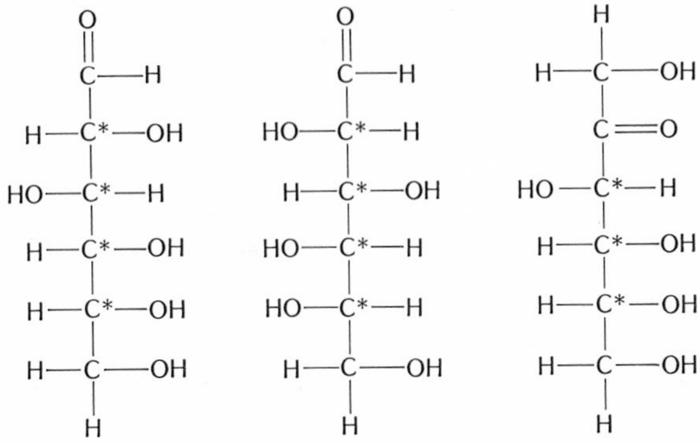

D-Glucose   L-Glucose   D-Fructose

called *ketoses,* have the carbonyl group as the next-to-last carbon.

The origin of the name is apparent, since the R—C—R′ structure with the $\overset{O}{\overset{\|}{}}$ characterizes a ketone. Fructose is a familiar example.

   Sugars resemble amino acids in another important respect: they are optically active. In fact, saccharides generally include several asymmetric carbon atoms. Look at the structure of D-glucose. The four internal carbon atoms each bear four different substituents. Each of these atoms can rotate the plane of polarized light, and the net rotational effect is a composite of contributions from all the asymmetric centers. Furthermore, because there are four carbons capable of assuming alternate atomic arrangements, the total number of possible optical isomers is 2^4 or 16. There is little point in listing all of them. You can imagine the effect of transposing an —H and an —OH on any one of the asymmetric carbon atoms of glucose. It creates a new molecule which again is not superimposable upon its mirror image. As a further example of stereoisomerism, the structure of L-glucose is also pictured. Most natural sugars are in the D-configuration. According to the conventions of carbohydrate chemistry, this means that the arrangement of substituents about the asymmetric carbon farthest from the carbonyl group is the same as that found on the corresponding carbon in D-glucose.

   Under most conditions, the glucose molecule does not have the linear shape suggested by Figure 15.7. Hexoses generally assume the cyclic structures of Figure 15.8. The carbonyl group and the secondary alcoholic group furthest removed from it react with each other to form a linkage which includes an oxygen atom. In glucose, the ring has six members, in fructose it has five. In both cases, the formation of the ring introduces an additional asymmetric center and hence doubles the number of optical isomers.

A pair of simple *monosaccharides,* such as glucose and fructose, can also react with each other to form a double-barreled sugar called a *disaccharide.* The selection of these two specific hexoses was hardly accidental, since together they constitute sucrose, common cane or beet sugar. You have already seen the structure of this molecule as Figure 11.2. Since the sucrose molecule is too large to pass through cell membranes, it must be split into its two constituent monosaccharides before it can be utilized as food. In the body, this hydrolysis is accomplished by an enzyme called sucrase, but acid will catalyze the process *in vitro.** The reaction is called "inversion" because it is attended by a change in optical rotation from positive to negative. "Invert sugar," which frequently appears on labels of candy and jellies, is merely hydrolyzed sucrose. It is slightly sweeter than the disaccharide and has less tendency to crystallize.

Familiar as sucrose may be, it is not nearly as abundant as *cellulose,* a polymer of D-glucose. It has been estimated that over half of the total organic carbon on earth is found in this form. Approximately one hundred billion tons of cellulose are produced every year by photosynthetic green plants. The starting materials are about the simplest imaginable—carbon dioxide and water—and the energy to drive the normally nonspontaneous reaction

FIGURE 15.8. *The α-ring forms of* D-*glucose and* D-*fructose. The model molecules should be visualized with the ring roughly perpendicular to the page, the ring oxygen atom to the back, and the heavy line to the front. The asymmetric, optically active carbon atoms are identified with asterisks.*

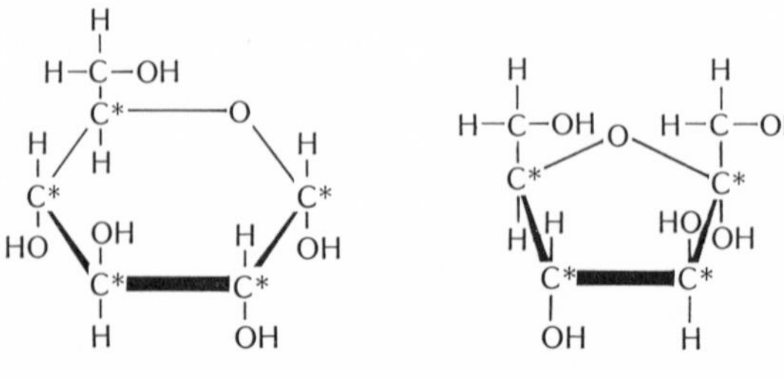

α–D–Glucose   α–D–Fructose

Literally "in glass" and thus opposed to *in vivo,* "in living things."

comes, of course, from sunlight. After the hexose units have been formed, they are linked together to yield the *polysaccharide*. Typically, cellulose molecules contain from 300 to 3000 ᴅ-glucose residues and have relative weights ranging from 50,000 to 500,000.

As cotton (nearly pure cellulose) and wood (about 50% cellulose), this major structural carbohydrate has many obvious uses. Not so well known, but equally important, are its applications as a starting material for the manufacture of paper, fabrics (rayon and acetate), plastics (cellophane and cellulose acetate), and explosives (gun cotton). But there is one direct use of cellulose which human beings have not yet mastered. Notwithstanding King Nebuchadnezzar's peculiar gustatory habits, man cannot live by grass alone. Nor is there much nourishment in chewing toothpicks. To be sure, the glucose units of both grass and toothpicks are full of potential food value, but we don't have the enzymes to take advantage of it. We simply cannot break the ties which bind the hexose residues in cellulose. For that matter, neither can a cow— unless she has help, that is. Fortunately, her kine and kin are assisted by friendly bacteria which inhabit a special stomach called the rumen. While their hostess is contentedly grazing in the green, green grass, these bacteria are busy breaking apart glucose units. The cow's own chemistry takes over from there.*

The bond which bacteria find so appetizing, but the rest of us find indigestible, is the *β-glycosidic* linkage shown in Figure 15.9. The molecular difference between it and the *α-glycosidic* bond (also illustrated) seems almost trivial. But it may literally be the difference between a famine and a feast, between sawdust and mashed potatoes. For the *α*-glycosidic linkage connects ᴅ-glucose units into *starch* and *glycogen*. The former is another plant-produced carbohydrate, but one readily hydrolyzed by the salivary enzyme, amylase. Starch serves plants as a nutritive reservoir and, as you well know, is an important food for animals. Glycogen is a similar polysaccharide which forms an energy store in animal tissue. The two compounds differ primarily in the extent to which the chains of hexose units are branched. Glycogen molecules fan out with many branch points; starch has longer uninterrupted straight segments.

The breakdown of glucose and its precursors, glycogen and starch, is one of the most thoroughly studied and best understood biochemical processes. But before we proceed to that topic, some additional preparation is advisable and some additional compounds are essential. For the degradation of glucose involves not

---

* Even termites require bacterial aid in order to survive on a diet of wood.

FIGURE 15.9.   *Representations of the molecular structure of (a) cellulose and (b) starch and glycogen. In both cases n can be very large, indicating the existence of large molecules with a high degree of polymerization.*

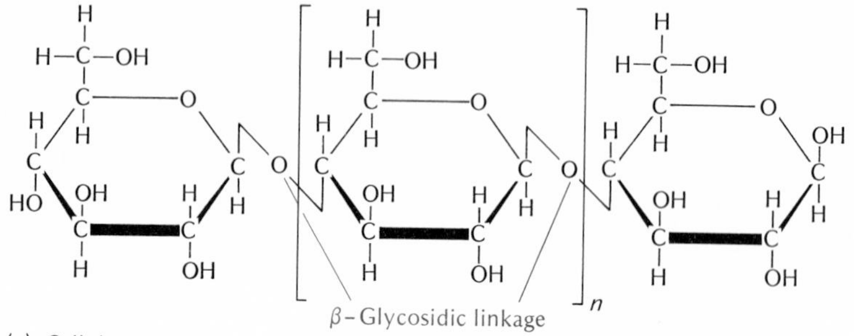

β–Glycosidic linkage

(a) Cellulose

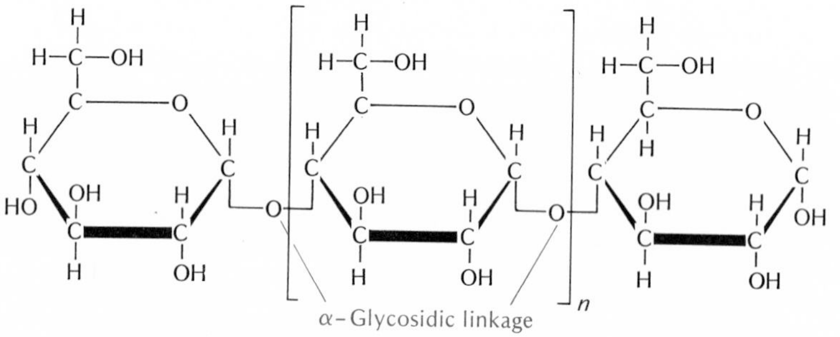

α–Glycosidic linkage

(b) Starch and glycogen

only carbohydrates and proteins, but also purine and pyrimidine bases and their remarkable compounds.

The bases of heredity

The statement that purines and pyrimidines are biochemically basic may be a bad pun but it's good chemistry. These proton-accepting compounds play vital roles in the storage and transfer of information and energy. As the formulas of Figure 15.10 clearly indicate, the *pyrimidines—cytosine, uracil,* and *thymine—*share a common six-membered ring system consisting of four carbon atoms and two nitrogens. These nitrogen atoms are in part respon-sible for the basic properties of the compound. In addition, the cytosine molecule contains an amino group. Note that there are other small but significant structural differences among the three pyrimidines. *Adenine* and *guanine* are both based upon the *purine*

FIGURE 15.10. *Structural formulas of the pyrimidine and purine bases found in nucleotides and nucleic acids. Their common abbreviations are indicated.*

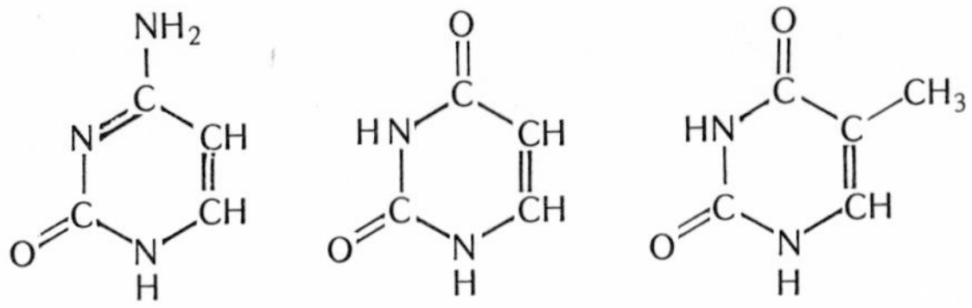

Cytosine (C)   Uracil (U)   Thymine (T)

(a) The pyrimidine bases

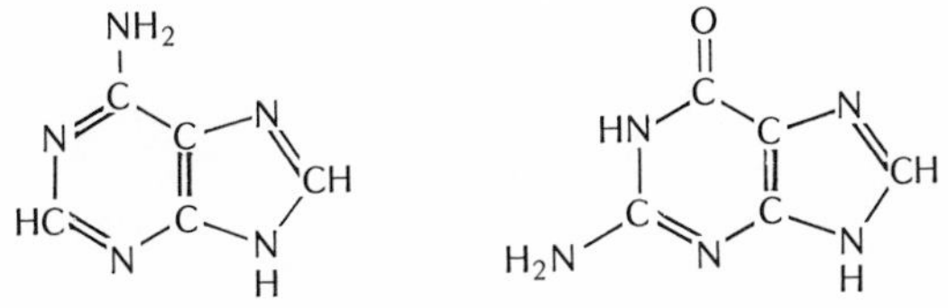

Adenine (A)   Guanine (G)

(b) The purine bases

system of two fused rings. Again, the rings include both carbon and nitrogen atoms.

Cells contain only low concentrations of the free purine and pyrimidine bases. Indeed, these compounds assume their unique biological significance only when chemically coupled with sugars and phosphate groups to form *nucleotides*. The sugar involved in a nucleotide is either D-ribose or 2-deoxy-D-ribose. Both these pentoses and their points of attachment to the pyrimidine and purine rings are illustrated in Figure 15.11. The phosphate group is also bound to the sugar, usually in either of the positions identified in the figure as 3′ or 5′. Of course, the specific structures pictured here are only examples. All five bases form nucleotides with both ribose and deoxyribose.

The four nucleotides containing ribose as the sugar residue and adenine, guanine, cytosine, and uracil bases are the building blocks of *ribonucleic acids*, or *RNA*. *Deoxyribonucleic acids*, or *DNA*, are made of deoxyribose nucleotides based upon adenine, guanine, cytosine, and thymine. Together, the two types of nucleic acids provide the molecular mechanism for storing, copying, transcribing, and translating the genetic information which literally makes us what we are. The extent of our understanding of this process, reflected in what you are about to read, is undoubtedly one of the most remarkable achievements of modern science.

The name "nucleic acid" clearly implies that these macro-

FIGURE 15.11.   *The molecular structures of two representative nucleotides.*

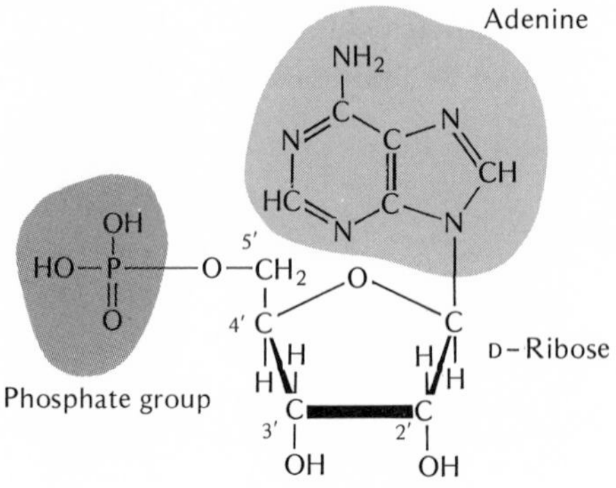

(a) Adenosine 5'–phosphoric acid
     (adenosine monophosphate, AMP)

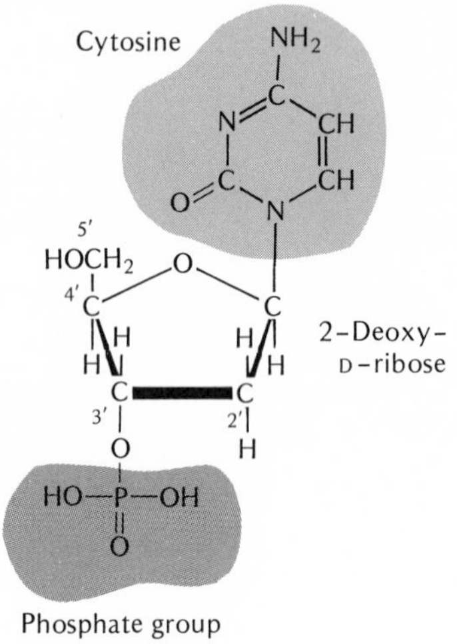

(b) Deoxycytidine 3'–phosphoric acid

molecules are found in the nuclei of cells. And the chromosomal portion of the nucleus has indeed been identified as the source of DNA. But for most cells, the major portion of the RNA is present in the surrounding cytoplasm. Moreover, the total amount of RNA in a cell is frequently five to ten times the DNA concentration.

All nucleic acids exhibit very high molecular weights; experimentally determined values range from 25,000 to several millions. The magnitude of these relative masses indicates that nucleic acids must be polymers. Other evidence shows that they are, in fact, *polynucleotides*. The nucleotides are bonded together to form DNA and RNA much as a string of amino acids constitutes a polypeptide or protein and a chain of monosaccharides makes up a polysaccharide. The covalent bond which links successive subunits is different for each of these three series of biopolymers. In nucleic acids, the phosphate group provides a bridge between the ribose or deoxyribose residues of adjacent nucleotides. The connection is from the 3' carbon of one ring to the 5' carbon of the next, as illustrated in Figure 15.12. Since the sugars in the figure happen to be deoxyribose, the drawing represents DNA. But the same phosphorous-containing linkages also characterize RNA.

The bases in Figure 15.12 were chosen arbitrarily. In nature they are not. Indeed, some of the best early (ca. 1950) evidence for the involvement of nucleic acids in the transmittal of genetic information came as a consequence of quantitative determinations

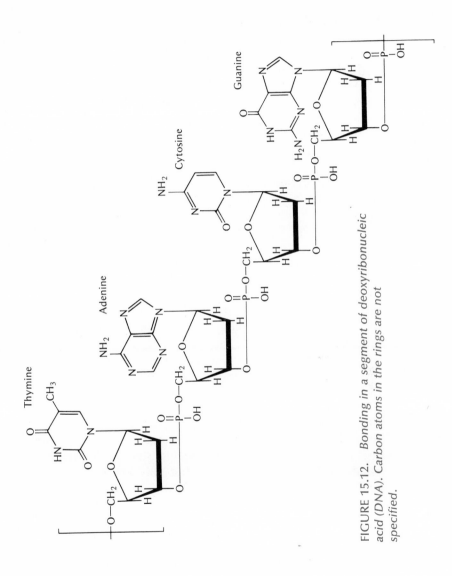

FIGURE 15.12. *Bonding in a segment of deoxyribonucleic acid (DNA). Carbon atoms in the rings are not specified.*

of the bases present in DNA samples. Chargaff and his co-workers were able to show that the relative amounts of the four bases present in a DNA sample are always invariant for all members of the same species and independent of the age, nutritional state, or environment of the organism studied. Closely related species have DNAs of similar base composition, but it is reassuring that man and mold differ considerably in this respect. The species dependence strongly suggests a causal relationship between inherited characteristics and the base distribution in deoxyribonucleic acid. Indeed, DNA is implicated as the actual chemical carrier of chromosomal instructions.

Chargaff's analyses also disclosed that in spite of the differences in base composition among widely divergent species, the DNAs of man, mold, and all other living things obey certain rules. The number of adenine residues present almost always equals the number of thymine residues, and the number of guanine residues in a sample of the acid is essentially identical to the number of cytosine residues. Or, more simply put, A = T and G = C. A correlation of this sort can hardly be coincidental; it must have basis in biochemical fact and function. And the most likely explanation is that the bases of DNA somehow come in pairs of one purine and one pyrimidine. Adenine always appears to be associated with tyrosine, guanine is matched with cytosine.

The discovery of how this pairing is achieved won a Nobel Prize for James Watson, a young American geneticist, and Francis Crick, an English biophysicist. Their chief data were X-ray diffraction patterns obtained by others, and their achievement was to devise a model consistent with the various structural periodicities implied by this evidence. Of course, the proposed molecular arrangement also had to agree with other experimental facts, and here Chargaff's work proved to be of great assistance. If one draws to scale the molecular structures of the purine and pyrimidine bases and pairs them in the manner suggested by Chargaff's rules, the result is Figure 15.13. Note that this arrangement brings into opposition two pair of groups capable of forming hydrogen bonds between adenine and thymine. Three such bonds can form between guanine and cytosine.

Watson and Crick went on to suggest that these *complementary* base pairs are horizontally stacked one on top of the other and uniformly separated by 3.4 Å. This specific value was chosen because it coincides with a periodic repeat distance observed in the X-ray patterns. The same data also yielded a second separation of 34 Å. This the Cambridge scientists identified as the length of one complete helical turn containing ten base pairs. The geometry of the nucleotides suggests an open spiral staircase, with the base pairs as the steps and the deoxyribose residues and phosphate

FIGURE 15.13.   *Paired molecules of thymine and adenine and cytosine and guanine, drawn to approximate scale. Hydrogen bonds are indicated in color.*

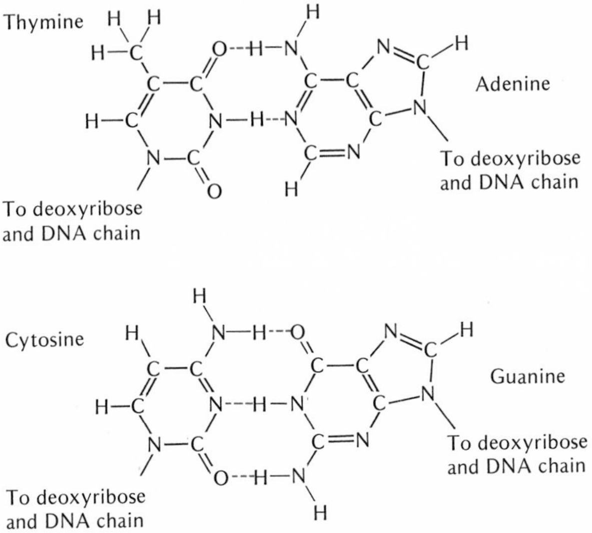

links forming the double helix to which the steps are attached. The helix is double because it consists of two intertwined polynucleotide spirals of complementary bases which are matched and stabilized by hydrogen bonds.

The double-helix model of DNA first appeared in print on April 25, 1953, in Volume 171 of *Nature*. The paper, "Molecular Structure of Nucleic Acids: A Structure for Deoxyribose Nucleic Acids," is brief and written with the customary passionless detachment of contemporary scientific prose. Even the most portentous statement in the communication is delivered with quintessential English understatement.

It has not escaped our notice that the specific pairing we have postulated immediately suggests a possible copying mechanism for the genetic material.

But some 15 years later, James Watson published a far more personal account of the research which culminated in that paper. Appropriately, this very readable book is entitled "The Double Helix." It is warmly recommended, particularly to anyone who still doubts that scientists are flesh and blood with feet of clay.

The copying mechanism mentioned by Watson and Crick has been much investigated and its essential features are understood. Before a cell divides, the DNA double helix which constitutes its chromosomes partially unwinds at a rate approaching 10,000 turns

FIGURE 15.14.   *A schematic sketch of the double helix of deoxyribonucleic acid. Z designates deoxyribose; ● stands for a phosphate group; and Ad, Cyt, Gu, and Thy, respectively, represent adenine, cytosine, guanine, and thymine. Taken from P. Karlson "Introduction to Modern Biochemistry," 3rd edition, page 131, Academic Press, New York, 1968.*

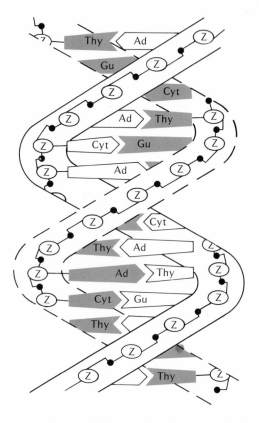

per minute. This results in separated regions of complementary single-stranded DNA. Mononucleotides present in the cell are selectively hydrogen-bonded to these two templates according to the pairing principles enunciated earlier: thymine to adenine, adenine to thymine, cytosine to guanine, and guanine to cytosine. Held in these positions, the monomeric units polymerize under the influence of an enzyme. As the replication process speeds through the cell, adding 90,000 nucleotides every minute, each strand of the original DNA generates a complementary copy of itself. The template chain and its newly formed complement then coil about each other to yield a new double helix. And, since the other single strand separated from the original DNA duplex has likewise produced a partner, there are now two double helices where there once was one. Both of them are identical to their progenitor. As the nucleus splits, one complete set of chromosomes is incorporated into each of the two daughter cells. Thus, the DNA contained in the second generation is identical to that in the first. Indeed, each of the 2×10^{12} cells in a newly born human baby includes all the genetic information which was assembled from parental DNA when the sperm entered the ovum.

It is a prodigious amount of information, expressed in a fantastic length of DNA. A single human cell contains about 2 meters (2 yards) of double helix, made up of roughly 5.5 billion base pairs. Within the cell, this length is coiled and super-coiled into 46 chromosomes, each one essentially a single DNA molecule with a molecular weight of billions. But if all the deoxyribonucleic acid from all the cells of an infant were placed end to end, the resulting double helix could stretch to the sun and back 13 times!

The 5.5 billion base pairs repeated in every cell provide the blueprint for producing one *Homo sapiens,* female, white, blue-eyed, brown-haired, big-boned, etc., etc., etc. Whatever the specifications might be, they are expressed in proteins. The reason for this is clear. Because of their crucial enzymatic functions, proteins, more than any other class of biological compounds, determine the characteristics of every living assemblage of cells. Therefore, the master cryptographers who set out to crack the genetic code concentrated on translating base language into amino acid language. Their goal was to find the symbols which specify the particular amino acids to be incorporated into a protein.

It was obvious at the outset that the code could not be a simple one-to-one relationship of base identity and amino acid identity. Since there are only four different bases in DNA—A, G, C, and T—a one-letter code could represent only four different amino acids. At least 20 symbols are required to encode the 20 common amino acids found in proteins. And so, it seemed very reasonable that the genetic code might consist of words, combinations of bases presumably read in sequence. This would greatly increase the options open for assignment to amino acids, just as combining the members of our 26 letter alphabet yields thousands and thousands of words.

Simple mathematics permits the calculation of just how many different words of given length can be made from an alphabet of known size. One merely raises the number of letters available to a power equal to the number of letters per word. Thus, the four "letters" of the DNA alphabet can be used to form 4^2 or 16 different two-letter "words." Again, the magnitude of the answer indicates that this vocabulary is too limited to provide one unique word for each of the amino acids. In other words, the genetic code cannot consist of pairs of bases. If, however, the four bases are read in groups of three, there would be 4^3 or 64 possible combinations, more than enough to do the cryptic job. It follows logically that the code for amino acids might very well consist of three-base units. Needless to say, more than mathematical logic was required to substantiate that suggestion. But the experimental evidence was obtained, and the existence of a commaless code of three-letter words is now well established.

The next challenge was to find the particular nucleotide word or words equivalent to each amino acid. No Rosetta Stone was available, so M. W. Nirenberg and others made one in a series of elegant and imaginative experiments. Thanks to their research, the dictionary is now complete. If you were to take any four letters of the alphabet (A, G, C, and T if you like) and assemble them in all 64 possible three-letter combinations, you would get a few that make sense (CAT, TAG, ACT) but more that are meaningless (AGC, GCT, CTA). Nature does far better. Sixty-one of the 64 triplet *codons* specify amino acids, and the three "nonsense syllables" are signals to stop the synthesis of a polypeptide chain. Since there are many more code words than amino acids, the code is redundant, that is, certain amino acids are represented by more than one tribasic unit. Leucine, serine, and arginine appear to have the most codon equivalents—six each. On the other hand, tryptophan and methionine are each symbolized by only a single triplet. The code is universally the same in all living things.

The mechanism by which the code is used to direct the synthesis of proteins is known in amazing detail. It is here that ribonucleic acids play their essential functions. As the name implies, *messenger RNA*, or *mRNA*, carries the genetic information from the DNA of the chromosome to the site of protein synthesis, cell components called *ribosomes*. The mRNA molecules are formed on the template of one of the DNA strands and are complementary to that strand. They thus are identical to the other strand of the chromosomal double helix, except that mRNA contains uracil in place of thymine. The long, single-stranded messenger molecules (molecular weights of 25,000 to over 1,000,000) are read during the synthesis process.

Floating around in the cytoplasm near the ribosomes are amino acids and *transfer RNA*, or *tRNA*. The many members of this latter type of ribonucleic acids have the function of transporting the amino acids to the ribosome and the growing peptide chain. All tRNAs have roughly the same molecular weight (23,000 to 30,000). Moreover, each molecule has two regions of particular importance. One is a functional group capable of attaching to an amino acid; the other is a three-base *anticodon* complementary to a code word on messenger RNA. As you might expect, there are specific transfer RNAs for all 20 amino acids and their various code names. Very clever enzymes, also present in the cell, couple the appropriate amino acids to the correspondingly coded tRNAs.

During the actual assembly of a protein molecule, tRNAs bearing their amino acids form hydrogen bonds with the complementary codons of the mRNA. The latter, of course, dictates the desired sequence of amino acids. The process is so organized that the growing end of the polypeptide chain is held adjacent to the next amino acid to be added. This proximity, plus proper enzymatic and energetic involvement, promotes the formation of the peptide bond. In this manner, the protein is sequentially constructed according to the instructions written in purine and pyrimidine bases.

A single human cell contains enough DNA to code for thousands upon thousands of proteins. We can even express the information content of a cell in more familiar units of information storage—books. Since three bases are required for each codon, 5.5×10^9 nucleotide pairs represent 1.85×10^9 bits of information. Now assume that these codons specify letters instead of amino acids. The letters can be assembled into words, just as the amino acids join to form proteins. If the average word consists of five letters, 1.85×10^9 letters correspond to 3.7×10^8 words. This amounts to 7.5×10^5 pages of 500 words each. And if this great stack of paper were bound in books of 500 pages each, it would require almost 1500 volumes to contain the information you carry about in 2 meters of helical thread, invisible to all but electron microscopists. The length of the book of life bearing your name and your individuality testifies that you are an incredibly complex creature. And the miniaturization of this information to the molecular level is truly an awesome aspect of nature—one which puts to shame the most sophisticated computers.

It is a major miracle that so few misplacements of nucleotides occur to cause mutations. As the example of sickle-cell disease clearly indicates, the identity of a single base can literally be a matter of life or death. The two mRNA triplets which code for glutamic acid are GAG and GAA. The four codons for valine include GUG and GUA. It is thus apparent that the substitution of a uracil nucleotide for an adenine can change the molecular blueprint so that a valine is incorporated into the protein instead of a glutamic acid. If this amino acid replacement occurs in the sixth residue of the β chain of hemoglobin, the resulting protein exhibits behavior which leads to sickling red blood cells. Understandably, there is a good deal of scientific interest in the action of chemical mutagenic agents which cause such alterations in base sequence.

Remarkable as the genetic function of purines and pyrimidines may be, it is not the only biological role played by these compounds. Nucleotides serve another purpose which is almost equally important for life. They act as chemical agents for the storage and transmittal of energy as well as information. The properties which make this possible are admirably illustrated by *adenosine tri-phosphate,* or *ATP*. As the name suggests, the compound contains three phosphate groups, linked in a chain and attached to the 5′ carbon of the ribose residue.

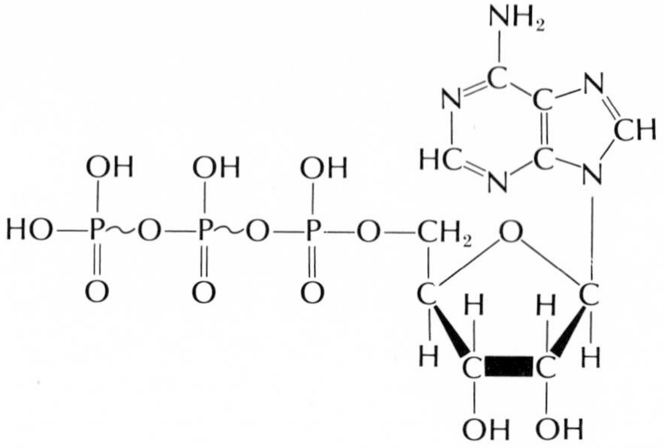

The energy-related function of ATP and similar compounds involving guanine, cytosine, uracil, and thymine are chiefly a consequence of these phosphate groups. The linkages, marked with a squiggle ($\sim$), are so-called "high-energy" bonds. When ATP reacts with water or certain other compounds, the high-energy bond holding the outermost phosphate group in place is broken, a phosphate group is freed, and *adenosine diphosphate (ADP)* is produced. In brief, the reaction is as follows:

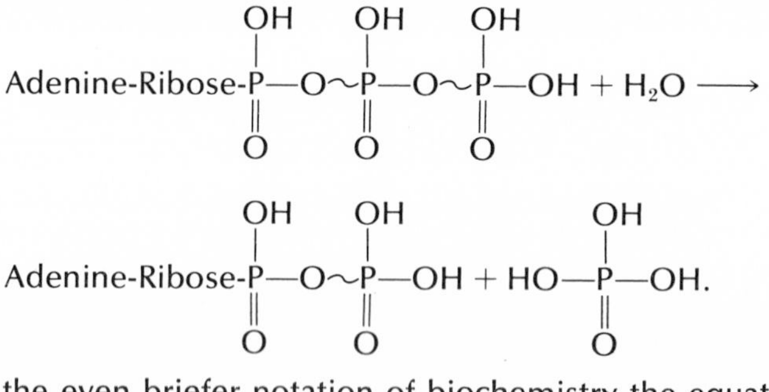

In the even briefer notation of biochemistry the equation requires only a single line:

$$ATP + H_2O \rightarrow ADP + P_i.$$

No matter how you write it, the most important thing about the reaction is the fact that the free energy of the system decreases as it occurs. Under standard conditions, $\Delta G° = -7.3$ kcal per mole of ATP. In other words, 7.3 kcal of energy, previously stored in a high-energy phosphate bond, becomes available to do work when ATP is hydrolyzed to ADP. A free energy change of equal magnitude attends the dephosphorylation of adenosine diphosphate to adenosine monophosphate. But the previous reaction occurs far more frequently in the chemistry of life.

The reaction represented by Equation 15.1 is the immediate source of energy for a wide variety of physiological processes. Many of the things we have our bodies do (or our bodies do without being told) require an input of energy. It obviously takes energy to run, to walk, to think, to exist. Stated in only slightly more thermodynamic terms, the reactions involved in muscle action, in nerve function, and in the synthesis of complex molecules from simpler fragments are in themselves nonspontaneous. They require an energetic push in order to take place. And the push comes from the spontaneous hydrolysis or dephosphorylation of ATP to ADP which is chemically coupled to the reluctant reaction. Thus, the role of ATP resembles that of a battery which expends energy in order to motivate a mechanical King Kong—a highly unlikely event in the absence of some energetic outside (or inside!) intervention.

Of course, a battery cannot go on putting out energy indefinitely; eventually it runs down. Then energy must be added in order to convert the reactants back to the high free energy state which makes a spontaneous change possible. ADP is a partially discharged molecule, but it, too, can be recharged. When a supply of energy is available, reaction 15.1 is run backwards.

$$\text{Energy} + \text{ADP} + \text{P}_i \rightarrow \text{ATP} + \text{H}_2\text{O}$$

The added energy is packed into phosphate bonds where it is again potentially available for subsequent use. The source of energy for this phosphorylation process is the chemical breakdown of food. Obviously, a good deal of energy is released when I subject a peanut butter sandwich to my own gradual internal combustion process. But this energy does not go directly into powering my muscles or my brain cells. It proceeds via the storage and transfer agency of adenosine triphosphate.* The complex and clever way in which it gets there forms the subject of this chapter's final section.

"Better living through chemistry"

Metabolism is a general term applied to biochemical reactions involving nutrients, and as such, it includes most processes carried out in living cells. The degradation of relatively complicated foodstuffs into simpler molecules is a special case of metabolism called *catabolism*. The biological buildup of complex structural com-

---

* It should now be manifestly clear why phosphate is an essential nutrient which, in excess, can cause excessive growth of algae.

ponents from smaller fragments is known as *anabolism*. This section will examine a particular example of catabolism, the reactions of glucose which ultimately yield carbon dioxide, water, and a considerable quantity of energy.

The digestion and degradation of glucose occurs in three distinct series of reactions. The first is called *glycolysis*. In this process, the overall change occurring in the sugar is its conversion to lactic acid. Since each molecule of lactic acid contains three atoms of carbon, two molecules of this compound are formed from each molecule of glucose.

15.2
$$C_6H_{12}O_6 \longrightarrow 2CH_3\overset{\overset{\displaystyle OH}{|}}{\underset{\underset{\displaystyle H}{|}}{C}}\overset{\overset{\displaystyle O}{\|}}{C}-OH$$

Glucose   Lactic acid

In the course of this transformation a moderate amount of energy is released. Specifically, the standard free energy of the system decreases by 47.0 kcal for each mole of glucose which reacts, that is, $\Delta G° = -47.0$ kcal. This free energy is available to do work, but if the degradation were to occur all at once, most of it would probably be lost as heat. In point of fact there are eleven gradual steps between D-glucose and lactic acid, each catalyzed by a specific enzyme. Thus, the energy cascades down in a series of small falls, rather than one abrupt drop.

At two particular points in this chain, the decreasing free energy is salvaged and stored in the high-energy phosphate bonds of adenosine triphosphate. Since 2 moles of ADP are phosphorylated to ATP for every mole of glucose which reacts, 2×7.3 or 14.6 kcal are chemically conserved. The source of this energy is the 47.0 kcal released by the decomposing glucose. The difference of 32.4 kcal is liberated but not concentrated in ATP. Instead, it serves as the driving force for the various individual steps which make up the overall chemical change. The net result of these reactions can be summarized by the equation

$$C_6H_{12}O_6 + 2ADP + 2P_i \longrightarrow 2CH_3\overset{\overset{\displaystyle OH}{|}}{\underset{\underset{\displaystyle H}{|}}{C}}\overset{\overset{\displaystyle O}{\|}}{C}-OH + 2ATP + 2H_2O.$$

ATP is also intimately involved in the control and regulation of glycolysis. Biochemical processes are generally quite efficient, and the various reactions which make up a metabolic series are in delicate interdependent balance. When the body places heavy demand upon the ATP energy reserves, for example during strenuous exercise or deep thought, the concentration of ATP naturally decreases and that of ADP increases. This change in concentration stimulates an enzyme which catalyzes a certain control step in the glycolysis chain. As a result, the rate of conversion of glucose to lactic acid is increased, and the energy evolved in this process is used to promote the transformation of ADP back to ATP.

The same series of reactions just summarized is also involved in the conversion of glucose to ethyl alcohol. The only difference is that the yeasts responsible for alcoholic fermentation contain enzymes which introduce two new terminal steps in place of the final conversion to lactic acid. As a consequence of this chemistry, two molecules of ethanol and two molecules of carbon dioxide are formed from each original sugar molecule.

$$C_6H_{12}O_6 + 2ADP + 2P_i \longrightarrow 2CH_3CH_2OH + 2CO_2 + 2ATP + 2H_2O$$

Note that in the process, ADP is again converted to ATP.

Much more ATP is formed before the degradation of glucose is completed. The conversion of the monosaccharide to lactic acid represents only the first series of steps in a long catabolic ladder which ultimately ends at carbon dioxide and water. The complete change of state experienced by glucose is the same oxidation which occurs when the sugar is burned.

15.3 $$C_6H_{12}O_6 + 6O_2 \longrightarrow 6CO_2 + 6H_2O$$

And, as in the case of combustion, the energy stored in the chemical bonds of glucose is released during the biological breakdown. To be more specific, 680 kcal of free energy are potentially available to do work when 1 mole of $C_6H_{12}O_6$ undergoes reaction 15.3, no matter what particular path is followed.

Only a relatively small fraction of this energy is released during glycolysis. A cursory inspection of Equation 15.2 indicates that the transformation of glucose to lactic acid is not an oxidation at all, but merely a molecular rearrangement. Indeed, oxygen is not even required for this *anaerobic* ("without air") process. The free energy change of −47.0 kcal which attends the reaction is only 7% of the total associated with Equation 15.3. The bulk of the energy change occurs in the oxidation which follows glycolysis. It is a process involving two more series of consecutive reactions.

The first of these is called the *Krebs cycle* after Sir Hans Krebs who did much to illucidate it. The fuel for the cycle is not lactic acid, but pyruvic acid, the compound which just precedes it in the glycolysis sequence. Each molecule of pyruvic acid has both a ketone and a carboxylic acid group and fits the formula,

$$CH_3\overset{O}{\underset{\|}{C}}-\overset{O}{\underset{\|}{C}}-OH.$$ The presence of three carbon atoms correctly suggests that two molecules of pyruvic acid are formed from a

single glucose molecule. Just outside the door to the Krebs cycle, each molecule of pyruvic acid loses a carbon atom as CO_2 and thus becomes converted into an acetyl group,

$$CH_3\overset{\displaystyle O}{\overset{\displaystyle \|}{C}}\text{—.}$$

It is this species which enters the cycle, chemically bound to a carrier called coenzyme A.

The cycle consists of eight steps, each involving an organic acid. Since several of these contain three carboxylic acid groups, the entire process is sometimes known as the *tricarboxylic acid cycle*. It is cyclic because the eight acids regenerate each other and remain at constant overall concentrations. Thus, they are in effect catalysts for the net reaction which is the oxidation of each mole of

$$CH_3\overset{\displaystyle O}{\overset{\displaystyle \|}{C}}\text{—}$$

into 2 moles of CO_2. Other catalysts are also involved; at least one enzyme participates in each step.

The energy released during Krebs cycle oxidation is not directly stored in the phosphate bonds of ATP. Rather it is first chemically concentrated in another compound called nicotinamide adenine dinucleotide. This substance is the first station in the sequence of reactions which completes the degradation of glucose.

The final series of reactions is known as *oxidative phosphorylation*. It includes seven steps, each catalyzed by its own enzyme. In every one of these reactions, a different compound is oxidized. In other words, electrons are passed down the chain of oxidative phosphorylation. And as they are, energy is evolved. At certain specific links in the chain, the energy is used to put a third phosphate group on adenosine diphosphate. And thus, part of the free energy decrease which attends the oxidation of glucose is stored for future use.

We can calculate how much because biochemists have determined that 36 moles of ATP are formed for each mole of reacting $C_6H_{12}O_6$. Under standard conditions, 7.3 kcal are required to convert a single mole of ADP to ATP. It follows that 36×7.3 or 263 kcal of energy are used in charging up 36 moles. All the intermediate steps conveniently cancel out and we can write the net reactions and their overall sum as follows:

$$C_6H_{12}O_6 + 6O_2 \longrightarrow 6CO_2 + 6H_2O \qquad \Delta G^\circ = -680 \text{ kcal}$$

$$36ADP + 36P_i \longrightarrow 36ATP + 36H_2O \qquad \Delta G^\circ = +263$$

$$C_6H_{12}O_6 + 6O_2 + 36ADP + 36P_i \longrightarrow$$
$$6CO_2 + 42H_2O + 36ATP \qquad \Delta G^\circ = -417 \text{ kcal}$$

Clearly, the free energy decrease associated with the oxidation of glucose, 680 kcal, is not all converted to usable form. An amount equal to 417 kcal provides the thermodynamic driving force for the various reactions and is dissipated. But we just calculated that

263 kcal are incorporated into the bonds of ATP. Thus, the overall efficiency of energy recovery from the oxidation of glucose is (263/680) × 100 or 39%. A simplified schematic summary of the entire process is given in Figure 15.15.

FIGURE 15.15. *A simplified flow sheet or schematic chart of carbohydrate catabolism, showing glycolysis, the Krebs cycle, and oxidative phosphorylation. The reactants are enclosed in black boxes, the final products in colored boxes, and some intermediates in colored dashed lines. The overall reaction is seen to be,*

$$C_6H_{12}O_6 + 6O_2 + 36ADP + 36P_i \rightarrow 6CO_2 + 42H_2O + 36ATP.$$

Individual steps are not represented.

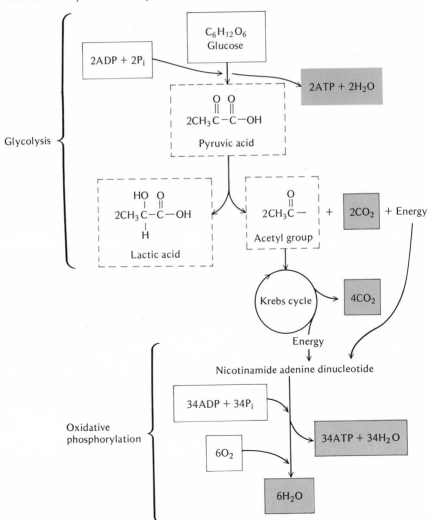

If this were a biochemistry text, I could trace the metabolic fate of the two other important classes of food—proteins and fats. You would see a surprising number of similarities, because the intermediate products obtained from the preliminary digestion of proteins and fats also enter the Krebs cycle and proceed through oxidative phosphorylation. This common mechanism unites various aspects of metabolism in a way which keeps the function and composition of a healthy organism in perfect balance. Moreover, many of the reactions of catabolism can be run backwards to synthesize carbohydrates, proteins, and fats from simpler fragments. To be sure, anabolism is not merely a reversal of catabolism. Some of the steps must differ, and ATP must be converted to ADP to provide the free energy necessary to form new bonds and make more complex molecules. We cannot even begin to consider these processes. But hopefully, this brief account has been enough to convince you of the truly remarkable chemistry going on inside you this very instant. And, if thinking about it has placed a severe drain on your ATP, you may rest assured that it is being quickly and quietly replenished.

For thought and action

15.1   Identify the class (or classes) of organic compounds to which each of the following belong:

(a) 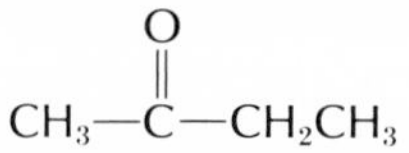 $CH_3-C(=O)-CH_2CH_3$

(b) $(CH_3)_2CHOH$

(c) $C_6H_5-C(=O)-H$

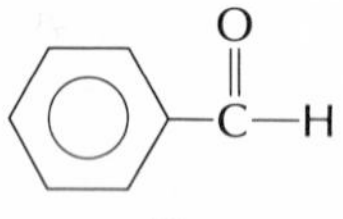

(d) 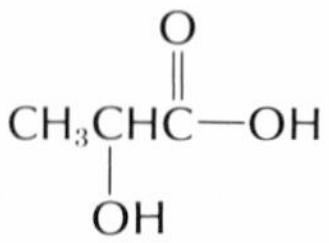 $CH_3CHC(=O)-OH$ with OH

(e) 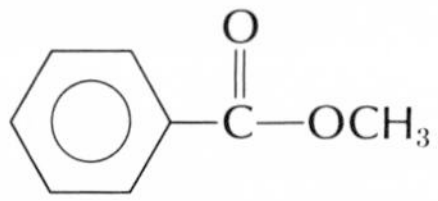 $C_6H_5-C(=O)-OCH_3$

(f) $CH_3-O-C_2H_5$

(g) 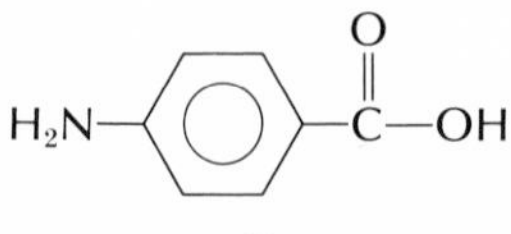 $H_2N-C_6H_4-C(=O)-OH$

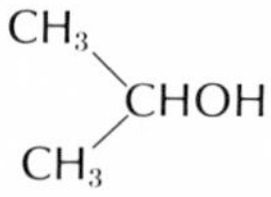

(h) $HOCH_2CHC(=O)-H$ with OH

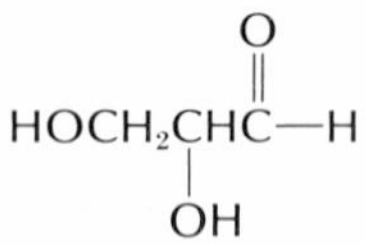

15.2   Pictured below are the structural formulas of some well-known compounds. Identify the functional groups present in each one.

(a)   Aspirin

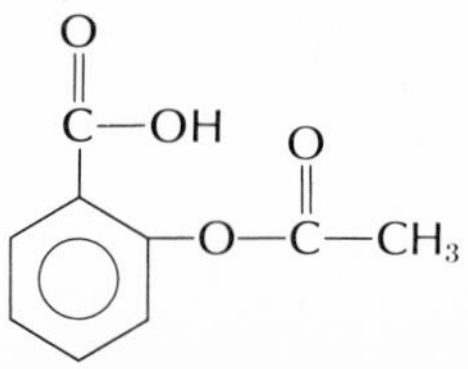

(b) "Permanent" antifreeze
$$\begin{array}{c} CH_2OH \\ | \\ CH_2OH \end{array}$$

(c) Mescaline

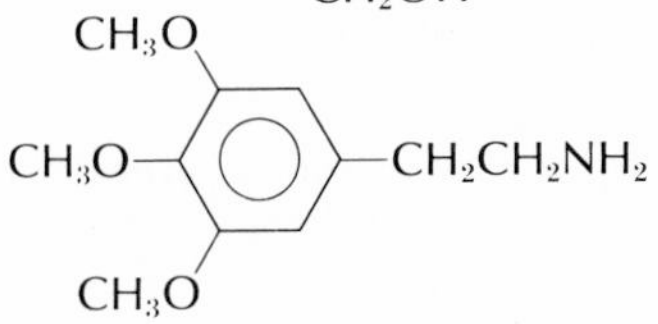

$$CH_3O-\!\!\!\underset{\underset{CH_3O}{|}}{\overset{\overset{CH_3O}{|}}{\bigcirc}}\!\!\!-CH_2CH_2NH_2$$

(d) Banana oil

$$CH_3\overset{\overset{\displaystyle O}{\|}}{C}-OCH_2CH_2CH\overset{\textstyle CH_3}{\underset{\textstyle CH_3}{<}}$$

(e) "Speed"

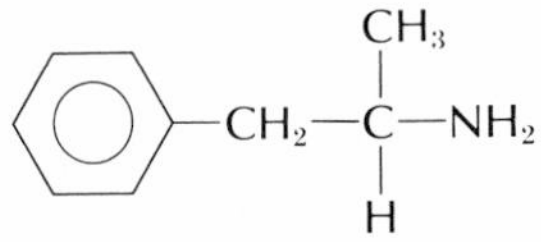

$$\bigcirc\!\!-CH_2-\overset{\overset{\textstyle CH_3}{|}}{\underset{\underset{\textstyle H}{|}}{C}}-NH_2$$

(f) Vanillin (vanilla)

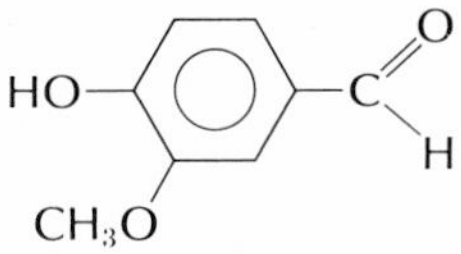

$$HO-\!\!\!\underset{\underset{CH_3O}{}}{\overset{}{\bigcirc}}\!\!\!-\overset{\overset{\displaystyle O}{\nearrow}}{C}\!\!\underset{\textstyle H}{\searrow}$$

(g) 2,4-D

$$Cl-\!\!\!\underset{\underset{Cl}{}}{\overset{}{\bigcirc}}\!\!\!-O-CH_2\overset{\overset{\displaystyle O}{\|}}{C}-OH$$

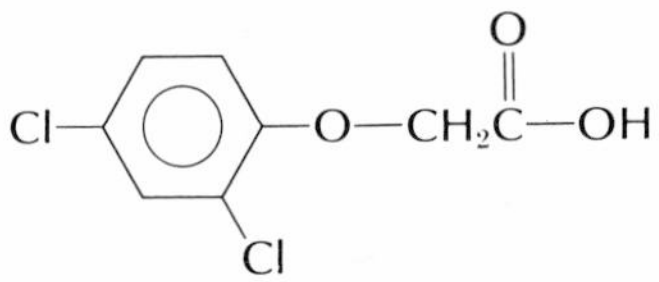

15.3 Draw the structures of all the isomers corresponding to the following formulas. The classes of compounds represented are indicated.
   (a) C_2H_6O alcohol and ether
   (b) C_3H_6O aldehyde and ketone
   (c) C_3H_8O alcohol and ether

15.4 Why is glycerol preferable to glycerine as the name for
$$\begin{array}{ccc} CH_2 & -CH & -CH_2 \\ | & | & | \\ OH & OH & OH \end{array}$$

15.5 Write an equation for the reaction of "wood" alcohol (methanol) and vinegar (acetic acid) and name the product.

15.6 Indicate which of these compounds are optically active.

(a) $CH_3-\overset{\overset{\textstyle H}{|}}{\underset{\underset{\textstyle CH_3}{|}}{C}}-C_2H_5$

(d) $CH_3-\overset{\overset{\displaystyle O}{\|}}{C}-\overset{\overset{\displaystyle O}{\|}}{C}-OH$

(b) $C_2H_5-\overset{\overset{\textstyle H}{|}}{\underset{\underset{\textstyle CH_3}{|}}{C}}-OH$

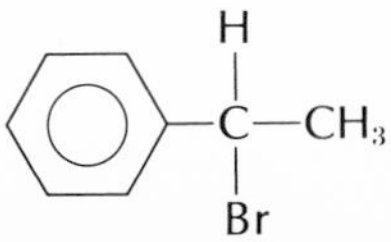

(e) CH_2BrCl

(c) $\bigcirc\!\!-\overset{\overset{\textstyle H}{|}}{\underset{\underset{\textstyle Br}{|}}{C}}-CH_3$

(f) $HOCH_2-\overset{\underset{\underset{\textstyle OH}{|}}{}}{CH}-\overset{\overset{\displaystyle O}{\|}}{C}-H$

15.7   Draw the formula of glycyl serine.

15.8   The first "nylon" (Nylon 6-6) was synthesized in 1935 by Dr. Wallace Carothers of the DuPont Company. The process involved the reaction of adipic acid and hexathylene diamine to form a polymer of alternating acid and amine residues. Write an equation for the reaction and draw the structure of a segment of the product. What similar biochemical reaction have you studied?

$$\underset{\text{Adipic acid}}{HO-\overset{\overset{\displaystyle O}{\|}}{C}(CH_2)_4\overset{\overset{\displaystyle O}{\|}}{C}-OH} \qquad \underset{\text{Hexamethylenediamine}}{H_2N(CH_2)_6NH_2}$$

15.9   Some proteins are said to be "more acidic" than others. What do you suppose this means in terms of amino acid composition?

15.10 It is commonly found that the presence of the amino acid proline interrupts an α-helix and puts a kink in the protein chain. Suggest why.

15.11 How do you suppose scientists measure the molecular weights of proteins, which frequently run into the tens of thousands?

15.12 From your brief study of protein structure and properties offer a molecular explanation for what happens when an egg is cooked.

15.13 Chymotrypsin is an enzyme which catalyzes the breaking of peptide bonds, yet it is itself a protein. What potential difficulties does this situation create?

15.14 Insofar as possible, classify the following sugars as monosaccharides or disaccharides, pentoses or hexoses, aldoses or ketoses.

(a) $C_5H_{10}O_5$

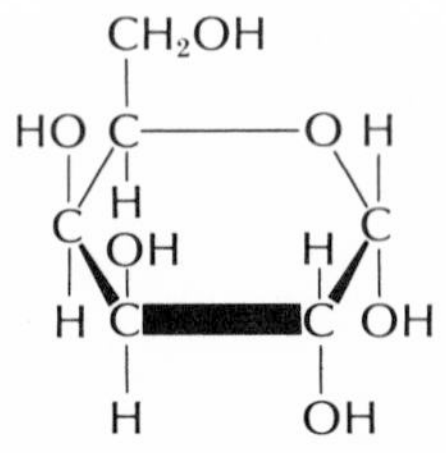

(d) $C_{12}H_{22}O_{11}$

(b)
$$\begin{array}{c} CH_2OH \\ | \\ C=O \\ | \\ H-C-OH \\ | \\ H-C-OH \\ | \\ CH_2OH \end{array}$$

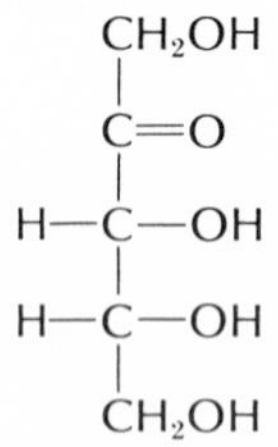

(e)
$$\begin{array}{c} CHO \\ | \\ H-C-OH \\ | \\ HO-C-H \\ | \\ HO-C-H \\ | \\ H-C-OH \\ | \\ CH_2OH \end{array}$$

(c)

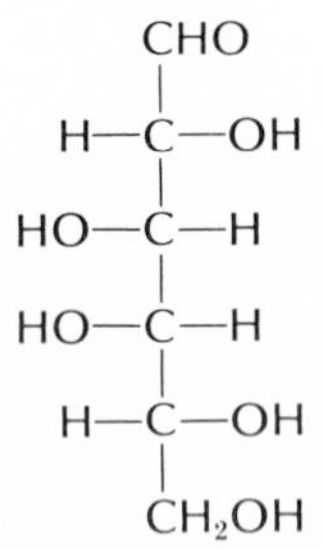

(f)

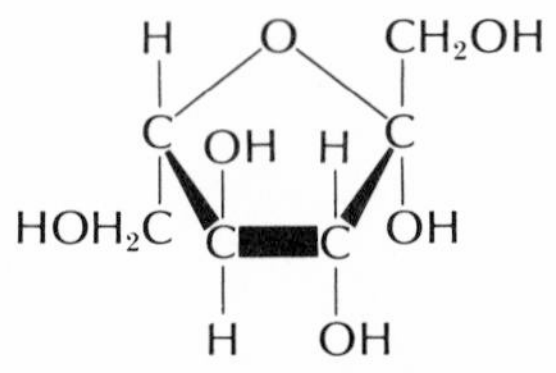

15.15 The general formula for four-carbon aldoses (tetroses) is

$$HOCH_2 \overset{\overset{\displaystyle H}{|}}{\underset{\underset{\displaystyle OH}{|}}{C}} \overset{\overset{\displaystyle H}{|}}{\underset{\underset{\displaystyle OH}{|}}{C}} \overset{\overset{\displaystyle O}{\|}}{C} H$$

(a) How many asymmetric carbon atoms are there in the compound?

(b) How many stereoisomers are possible?

(c) Sketch the stereoisomers in a way which indicates their structure.

15.16 Specify (a) the number of noncyclic stereoisomers and (b) the number of cyclic stereoisomers having the same general formula as ribose. Then specify (c) the number of noncyclic optical isomers and (d) the number of cyclic stereoisomers with the same formula as deoxyribose.

15.17 Why is too much starch about as bad as too much sugar for diabetics who have excess glucose in their blood?

15.18 Cellulose is an important indirect source of food energy for humans, but the conversion to a utilizable form (beefsteak, for instance) is quite inefficient. Suggest more efficient means of utilizing this vast food supply for human consumption.

15.19 Check the astronomical figure on page 467 by calculating the total length of DNA double helix in a newly born baby. Most of the necessary information is in the text, but remember that the distance from the earth to the sun is about 93,000,000 miles.

15.20 Assume the average protein contains 500 amino acid residues. How many protein molecules could be encoded by the 5.5×10^9 base pairs of the DNA in one human cell?

15.21 The DNA in the single circular chromosome of the bacterium, *E. coli,* has a molecular weight of 2,800,000,000 and consists of 0.12 cm of double helix. Calculate the number of base pairs in the chromosome.

15.22 One of Chargaff's findings, cited in the text, was that all members of the same species have DNA with the same base composition. How can this observation be reconciled with the genetic differences among individuals of the same species?

15.23 Each cell contains all the genetic information to fully specify a complete individual. Yet, cells are widely divergent in their characteristics. Thus, brain, blood, liver, and muscle cells all have their own properties and functions. Speculate on the reason for this diversity in spite of the common information found in every cell.

15.24 Experiments with virus DNA indicate that when one or two nucleotides are inserted in or deleted from the normal sequence, a mutation occurs. The amino acid composition of the protein coded for by the modified DNA is different from that normally produced. However, if *three* nucleotides are inserted in or deleted from a single gene, the proper amino acid composition and sequence is reestablished in the protein. Explain the significance of this observation.

15.25 The illucidation of the genetic code involved the use of synthetic messenger RNA of known base composition in protein-synthesizing ribosomal systems. Suggest the design of an experiment or experiments which could be performed to yield such information.

15.26 The redundancy of the genetic code is often associated with variability in the third base of the codon triplet. For example, GUU, GUC, GUA, and GUG in messenger RNA all code for valine. Why is this characteristic advantageous to the organism?

15.27 Our increasing knowledge of the molecular basis of hereditary information may eventually enable us to use "genetic engineering" to produce human beings with specially desired characteristics. Comment on the ethical problems raised by such a prospect. (Aldous Huxley's "Brave New World" and more recent works of science fiction may be relevant to the issue.)

15.28 A soft drink containing glucose is claimed to provide more "quick energy" than sucrose or starch. Is there any biochemical basis to this claim, and if so, what is it?

15.29 A common dosage of glucose administered in intervenous feeding is 100 cc of a 5% glucose solution per hour. This corresponds to 5 g of glucose. Using this information, calculate the following:
   (a) The total number of kilocalories of free energy associated with the complete oxidation of this much $C_6H_{12}O_6$ to H_2O and CO_2
   (b) The number of kilocalories from this total which go to transforming ADP to ATP
   (c) The number of moles of ADP which can be converted to ATP with this much energy.
   Base your computations upon the standard free energies given in the text.

15.30 In working a set of chemistry problems, a student expends 800 kcal of energy driving his brain cells and pushing a pencil. The conversion of ATP to ADP is the immediate source of this energy. Assume this process of energy transfer is completely efficient, that is, all the energy released is used to solve the problem (it really isn't), and determine the following:
   (a) The total number of moles of ATP converted to ADP
   (b) The total number of moles of glucose which must be oxidized to CO_2 and H_2O in order to recharge the ADP produced
   (c) The total weight of glucose represented by your answer to (b).

You may further assume that the standard free energies given in the text apply to this system.

15.31 The various reactions in a metabolic series such as glycolysis are linked in such a manner that there is often one single control point which influences all the steps. What advantages, if any, are associated with this arrangement?

15.32 Certain metabolic intermediates—the acetyl group is one—are involved in the anabolism and catabolism of carbohydrates, proteins, and fats. What value to the organism do you see in this chemical commonality?

Suggested readings

Allison, Anthony C., Sickle Cells in Evolution, *Scientific American* **195**, 87 (1956).

Dickerson, Richard E., The Structure and History of an Ancient Protein, *Scientific American* **226**, 58 (1972).

Dickerson, Richard E., and Geis, Irving, "The Structure and Action of Proteins," Harper, New York, 1969.

Lehninger, Albert L., "Biochemistry," Worth, New York, 1970. An excellent textbook.

Maxwell, Kenneth E., "Chemicals and Life," Dickenson, Belmont, California, 1970.

Merrifield, R. B., The Automatic Synthesis of Proteins, *Scientific American* **218**, 56 (1968).

Neurath, Hans, Protein-digesting Enzymes, *Scientific American* **211**, 68 (1964).

Oparin, A. I., "The Origin of Life" (translated by Sergius Morgulis), Dover, New York, 1953. The book is starting to show its age, but it remains a classic.

Perutz, M. F., The Hemoglobin Molecule, *Scientific American* **211**, 64 (1964).

Schrödinger, Erwin, "What is Life?," Cambridge Univ. Press, London and New York, 1955. The question is asked by a famous physicist.

Watson, James D., "The Double Helix," Signet (New American Library), New York, 1968.

16

Swords and plowshares: the challenge of nuclear energy

A glimpse of the new world

The cryptic message which introduces this chapter was telephoned by a Nobel Prize-winning professor of physics to the president of Harvard University on the afternoon of December 2, 1942. Arthur H. Compton had just come from a squash court under the stands at Stagg Field, the abandoned football field of the University of Chicago. It was no ordinary squash court. Inside that cold cavern were 500 tons of graphite bricks and 50 tons of uranium oxide, arranged in an irregular, roughly spherical pile 28 feet across. There were also instruments and machines, dials and gauges, and about 20 men. They included some of the most illustrious scientists in the world. And they were toasting the arrival of the Italian navigator by sharing a quart of Chianti. When the wine was finished, the bottle was passed once more, and each man solemnly autographed its straw covering.

An unseen observer from another planet —or this one, for that matter—would undoubtedly have concluded that the members of this bizarre party were all mad as hatters. And if he were to have overheard the rest of the conversation between Compton and James Bryant Conant, he would have amassed more evidence of what seemed to be an epidemic of academic insanity.

"What? Already?" asked the excited Conant, a famous chemist in his own right.

"Yes," replied Compton, "the earth was smaller than estimated and he arrived several days earlier than he had expected."

"Were the natives friendly?"

"They were indeed. Everyone landed safe and happy."

In their intentionally unintelligible conversation, Compton and Conant had drawn a parallel between 1492 and 1942. Whether it was done consciously or subconsciously is of little importance. The point is, the analogy was justified. For the strange events of that December day may well rival the landing of Columbus in the magnitude of their impact upon the course of this planet's history. Admittedly, such an evaluation may sound excessive. But how else does one classify the first controlled nuclear fission reaction?

This chapter is about that reaction, the discoveries which led up to it, and the consequences which grew out of it. In presenting this topic I plan to lead you on a narrative, chronological course, from Paris, Cambridge, Berlin, and Copenhagen; back to the squash court in Chicago; to Oak Ridge and Los Alamos; to Hiroshima and Nagasaki; and to the brink of the uncertain future.

There is no need to spend much time in Paris. A quick review of the relevant section of Chapter 6 will be enough to remind you of the cloudy sky which illuminated Becquerel's research and first exposed the phenomenon of natural radioactivity. That was back in 1896. Soon thereafter, Ernest Rutherford determined that two sorts of particles are emitted from the nucleus of radioactive elements. Alpha (α) particles, with relatively low power of penetration, are the nuclei of helium atoms. According to the currently accepted theory of atomic structure, each α particle consists of two protons and two neutrons. This composition accounts for a mass of 4.0 on the standard atomic weight scale and a positive charge equal in magnitude to that of two electrons, but opposite in sign. A β particle, you will recall, is an electron. The γ rays which frequently accompany the emission of these particles are massless and uncharged, part of the electromagnetic spectrum.

The phenomenon of natural radioactivity was intensively investigated during the first decade of this century, culminating in the unprecedented announcement, by Rutherford and Soddy, of the spontaneous transmutation of elements. You are already familiar with the rules which govern the identity of the daughter elements. Moreover, you have also studied the crucial experiment in which Rutherford essentially proved the nuclear nature of the atom by bombarding metal foils with α particles. But foils were not the only targets of Rutherford's submicroscopic bullets. He also fired them at gases.

When he did, he observed a startling transformation. The interaction of a stream of α radiation with nitrogen yields a stable isotope of oxygen. In the symbolic language of nuclear chemistry,

$$^{14}_{7}N + ^{4}_{2}\alpha \rightarrow ^{17}_{8}O + ^{1}_{1}p.$$

During the process in which an α particle is captured by an atom of nitrogen-14, a proton is kicked out of the nucleus and oxygen-17 is formed. Elementary identity is actually altered. Here is surely a case of modern alchemy, and the α particle is the Philosopher's Stone. With the execution of this experiment, man at last succeeded in artificially transmuting one element into another.

Woman was soon to follow suit, but with a significant addition. Irène Curie was, by heredity and environment, admirably equipped to study radioactivity. And, like her mother before her, she chose as a scientific collaborator her physicist husband. Even the element under investigation by Irène and Frédéric Joliot* was a discovery of Marie and Pierre Curie. It was polonium, and the Joliot-Curies were studying the interaction of the α particles it emitted with a variety of other elements.

When the target was boron, an isotope of nitrogen was formed, along with a neutron for each atom involved in the reaction.

$$^{10}_{5}B + ^{4}_{2}\alpha \rightarrow ^{13}_{7}N + ^{1}_{0}n$$

However, the atoms of nitrogen did not retain their elementary identity for long. They in turn underwent a spontaneous radioactive decay. M. and Mme. Joliot-Curie had generated the first artificially produced radioactive isotope. Soon after, they demonstrated that the bombardment of aluminum-27 with α particles yields neutrons and phosphorous-30, a β-emitter. Moreover, similar treatment of magnesium-24 forms a radioactive isotope of silicon with a relative mass of 27. Today, artificial radioactivity is no novelty. But in 1935 it was. And thus it is particularly fitting that Irène Joliot-Curie and her husband followed the footsteps of her illustrious parents in one other respect—the receipt of a Nobel Prize.

Since this pioneering research, hundreds of radioactive isotopes have been artificially produced. Indeed, their number far exceeds the number of naturally occurring radioactive isotopes. The method of generation is still basically that used by the Joliot-Curies: subatomic bullets are fired at atomic targets. But the choice of ammunition is greater today. Neutrons, protons, and

---

* M. and Mme. Joliot preferred to use Joliot-Curie as their surname.

deuterons* are now more frequently used than α particles. More-over, the caliber, complexity, and cost of the artillery used to accelerate these projectiles have risen dramatically. One reason for this latter development is the strong repulsion which exists between particles of like charge. In order for a proton, deuteron, or α particle to actually enter a positive nucleus, the particle must be moving with a tremendous amount of energy. The problem is not acute when the target atoms are relatively light and the nuclear charge is low. Hence, M. and Mme. Joliot-Curie were able to obtain their results without elaborate acceleration equipment. But as the atomic number and nuclear charge increase, so does the repulsive force faced by a positive projectile. Without sufficient energy to overcome this repulsion, the miniscule missiles will be reflected like the slow α particles of Rutherford's famous scattering experiment.

One of the first devices invented to help positive particles penetrate positive nuclei was the *cyclotron*. A schematic sketch of this Nobel-prize winning creation of physicist Ernest O. Lawrence constitutes Figure 16.1. Its principle of operation is described in the caption, but in effect it is sort of an electrically powered slingshot. And, as accelerators go, even large cyclotrons are in the slingshot category. They can accelerate protons to energies of about 250 million electron volts. The really big guns put out particles with energies in the billions of electron volts. I realize that you are probably unfamiliar with an electron volt,† but the magnitude of the numbers which proceed the unit should impress you. They represent formidable amounts of energy—and money! The cost of building one of these mammoth accelerators is not too far below a penny per electron volt. Such costs can only be borne by the public, and the public must ultimately decide if the returns justify the expenditures.

New nuclei: structure and stability

There have been many applications of radioactive isotopes produced in particle accelerators. Some of the most successful have been in medicine. Diagnosis, treatment, and research have all profited so greatly that specific examples in each category will be cited in the last section of this chapter. For the moment, how-

---

* A deuteron is the nucleus of a deuterium atom, i.e., one proton and one neutron. Hence its charge is identical to that of a proton, but its mass is twice as large. It is often represented as ^2_1H.

† It is the amount of energy gained by a proton or electron when it is accelerated through an electric field of 1 volt.

FIGURE 16.1.   *Schematic diagram of a cyclotron. Because of the oscillating high voltage, the hollow evacuated "dees" change polarity very quickly. The timing is such that the positively charged particles produced by the ion source are always attracted across the gap between the electrodes. With each of these jumps the particle accelerates and follows an ever-widening circular path dictated by the influence of the perpendicular magnetic field. Eventually, the negative deflecting electrode is used to focus the beam of accelerated protons, deuterons, or α particles upon the intended target.*

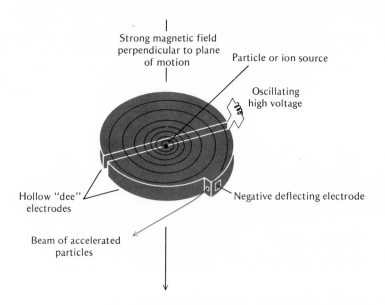

ever, our chief concern will be results of a more theoretical nature—some of them with earth-shattering practical importance.

A study of artificially produced and naturally occurring isotopes has helped our understanding of the structural conditions associated with nuclear stability on the one hand and radioactivity on the other. It is a major miracle that the nucleus stays together at all. The electrostatic repulsion between positively charged protons is fantastic. And yet, it is overcome by other attractive forces. A significant factor appears to be the neutron-to-proton ratio. A quick consultation of the periodic table indicates that for the first 20 elements, the atomic weight is roughly twice the atomic number. To be sure, each tabulated atomic weight is an average over a number of naturally occurring isotopic weights. But this average will most strongly reflect the mass of the most abundant isotope. Moreover, it is reasonable to assume that if a particular isotope occurs in a high relative concentration it is probably non-

FIGURE 16.2.   Two graphs indicating regions of nuclear stability
and instability. (a) A plot of the number of neutrons in the
nucleus vs. the number of protons. Isotopes whose neutron/
proton ratio places them on or near the colored line are stable.
Those falling above the line have an n/p ratio which is too high
for stability and are β-emitters. The n/p ratio for those isotopes
falling below the line is too low, and consequently their nuclei
emit positrons or capture orbital electrons. (b) A plot of the
neutron/proton ratio for stable isotopes vs. atomic number. The
same general considerations apply as in (a).

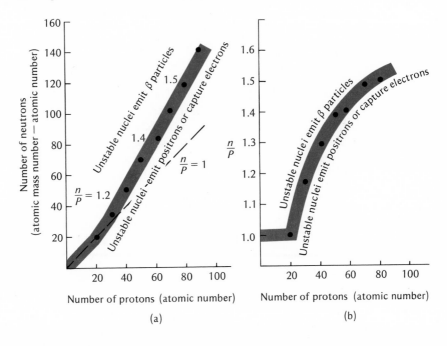

(a)   (b)

radioactive. Thus, we can conclude that among the light elements,
stable nuclei contain equal numbers of neutrons and protons. But
for heavier elements, the average atomic weight is more than
twice the atomic number. Indeed, as Figure 16.2 clearly shows, the
most stable neutron-to-proton ratio (n/p) increases quite smoothly
with increasing atomic number.

Such a trend suggests that deviations from this curve are associated with nuclear instability and radioactivity. The neutron-to-
proton ratio can be either too high or too low. In the former case,
the imbalance is rectified by the loss of a β particle or electron.
This in effect converts a neutron to a proton and hence reduces
the magnitude of n/p. An example of an isotope which undergoes
this decay process is carbon-14. The stable form of carbon (atomic
number 6) has an atomic weight of 12.000 which, incidentally,

serves as the standard for modern atomic weight values. Its neutron-to-proton ratio is exactly 1.00. But the nucleus of an atom of carbon-14 contains eight neutrons in addition to the six protons which are responsible for its elementary identity. Thus, n/p equals 8/6 or 1.33. Emission of a β particle, according to the equation,

$$^{14}_{6}C \rightarrow\ ^{14}_{7}N\ +\ ^{0}_{-1}\beta,$$

changes a neutron to a proton and carbon into nitrogen. The resulting n/p ratio has a stable value of 1.00.

This seems to be an appropriate place for a brief aside on an important application of the above transmutation. The carbon dioxide of the atmosphere contains a steady-state constant concentration of one atom of carbon-14 for every 10^{12} atoms of carbon-12.* Living plants and plant-eating animals incorporate $^{14}_{6}C$ and $^{12}_{6}C$ in this same ratio. But when the organism dies, the process stops. No new $^{14}_{6}C$ is taken up to replace that lost by radioactive decay via β emission. Fortunately, we know the rate of this atomic disintegration. It has a half-life of 5760 years. You recall (from Chapter 6) that this means that starting with any fixed number of $^{14}_{6}C$ atoms, half of them will spontaneously convert to $^{14}_{7}N$ in 5760 years. Thus, the greater the length of time between the death of the organism and the present, the lower its relative concentration of carbon-14. This behavior provides a convenient way of dating carbon-containing artifacts. One merely measures the ratio of $^{14}_{6}C$ to $^{12}_{6}C$ and does a little arithmetic. Charcoal from prehistoric caves, ancient papyri, and suspected art forgeries have all been forced to admit their ages through this technique. Other known natural processes of radioactive decay permit the dating of earth rocks, moon rocks, and other inorganic materials.

Getting back to the main business of nuclear stability, some isotopes have, relatively and figuratively speaking, too few neutrons for comfort. In such cases there are two mechanisms for rectifying the low neutron-to-proton ratio. One involves the capture of an orbital electron by the nucleus. This has the effect of converting a proton to a neutron. In the process, X rays are often liberated. The other way in which a nucleus increases its n/p value and its stability is by emitting a *positron*. You have not yet been formally introduced to this particle. It has the almost-negligible mass of an electron, but bears a positive charge instead of a negative one. Losing a positron converts a proton to a neutron and thus decreases both the atomic number and the nuclear charge by one unit. As a matter of fact, the first "artificial" radio-

---

* The carbon-14 in the atmosphere is produced by the interaction of neutrons from cosmic radiation with nitrogen.

active isotope prepared by the Joliot-Curies, nitrogen-13, is a positron-emitter. The symbol ^0_1e represents a positron, and its emission from $^{13}_7N$ is written

$$^{13}_7N \rightarrow {}^{13}_6C + {}^0_1e.$$

The neutron-to-proton ratio in carbon-13 is not quite 1, but the isotope is stable enough so that it does not undergo radioactive decay.

The introduction of the positron suggests another contribution of modern accelerators. By literally smashing atoms, these fantastic devices have revealed many new fundamental particles among the fragments. As I indicated in an earlier chapter, the center of the atom remains a complex (perhaps too complex) and largely unknown part of the universe. Nevertheless, some progress has been made toward explaining the structure of the nucleus and the nature of the incredibly strong forces which hold the mutually repulsive protons so tightly together.

When the nucleus reaches a certain size, these forces are apparently no longer strong enough to accomplish this feat. Thus, all natural and artificial isotopes of actinium ($_{89}Ac$), thorium ($_{90}Th$), and uranium ($_{92}U$) are unstable. They spontaneously break down to give rise to the three decay series studied by Rutherford, Soddy, and others. As you well recall, these multiple-step transitions involve the emission of much α, β, and γ radiation before they wind up in stable isotopes of lead.

Because of their inherent instability, the heaviest of all naturally occurring elements have been subjected to a particularly big barrage from particle accelerators. The results have been genuinely novel. To date, at least a dozen new transuranium elements have been artificially produced. Immediately after uranium, the terminal point of the "natural" periodic table, come its two planetary partners—neptunium and plutonium. Next are the elements spawned in the cyclotrons of the University of California at Berkeley—americium, californium, and berkelium. Finally, the periodic table ends (for the moment) with a series of elements honoring famous scientists, including Einstein, Fermi, Mendeleev, Nobel, and Lawrence. Elements number 104 and 105 have apparently also made their appearance and have tentatively been given the names kurchatovium (Ku) and hahnium (Ha), respectively.

This uncertainty about the creation of an element may strike you as somewhat strange. But the fact of the matter is that no more than a very few atoms of some of these transuranium elements have ever existed at any one time. And, because of their extreme instability, their existence has typically been very brief. Scientists may catch only a fleeting glimpse of a decaying atom—

a new streak on a photographic film.* Others have been produced in kilogram quantities.

The most notable of these is plutonium. The element was first
detected late in 1940 by a University of California chemist, Glenn
Seaborg, and his associates. The reaction involved the capture of
a deuteron by an atom of uranium-238 to form neptunium-238
plus two neutrons, followed by the β decay of the neptunium. In
equation form, the sequence is

$$\ce{^{238}_{92}U} + \ce{^{2}_{1}H} \rightarrow \ce{^{238}_{93}Np} + 2\ce{^{1}_{0}n}$$
$$\ce{^{238}_{93}Np} \rightarrow \ce{^{238}_{94}Pu} + \ce{^{0}_{-1}\beta}.$$

In addition, the California group prepared another isotope of the
element, plutonium-239, by bombarding uranium-238 with
neutrons.

Later Seaborg was to preside at the birth of several other elements, win a Nobel prize for his contributions, and serve as the
Chairman of the United States Atomic Energy Commission. But
in many respects, his first element was to have the greatest impact
on the course of human events. Indeed, the practical significance
of his discovery was so great that the paper announcing it was
not published until April, 1946. That was almost a year after the
spontaneous disintegration of plutonium-239 reduced Nagasaki to
smoldering rubble.

Fragments and fission

The high drama of the discovery of nuclear fission, the radiochemical reaction responsible for the atomic bomb, is filled with
fateful irony. An amateur psychiatrist might even suggest that it
was a discovery which, subconsciously, no one wished to make.
But such an interpretation badly underestimates the insatiable
need to know which drives most scientists. Nevertheless, the fact
does remain that splitting uranium atoms repeatedly revealed
themselves to some of Europe's finest scientists—to Enrico Fermi
in 1934, to two Swiss scientists in 1936, and to Irène Joliot-Curie
in 1938. And in each case, the observers refused to believe what
they saw. They were looking for new transuranium elements which
should form when uranium atoms are bombarded by neutrons.
And indeed, they did find some of these. But so intent and sharply
focused was their search that it blinded them to other phenomena
and alternative explanations. Or, if the idea that heavy atoms

* Some of the methods of detecting radiation are discussed in the final
section of this chapter.

actually divide into smaller atomic fragments did occur to them, it was quickly dismissed as being too outlandish to entertain.

It is true that Irène Joliot-Curie discovered, in the products formed when neutrons react with uranium, an element with many of the properties of lanthanum. But lanthanum has an atomic number of 57, that of uranium is 92. There was no known mechanism for such transformation. Just to make sure, Mme. Joliot-Curie carried out a series of chemical tests originally devised by her mother. They appeared to prove conclusively that the element in question was not lanthanum. It was somewhat more likely that it might be number 89, actinium. Since actinium and lanthanum are in the same periodic family, they do have similar properties. But here again the experimental results were negative. So Irène Joliot-Curie logically concluded that a new transuranium element had been formed by the bombardment of uranium with neutrons.

The proposed new element failed to fit the scheme devised by three scientists at the Kaiser Wilhelm Institute in Berlin—Lise Meitner, Otto Hahn, and Fritz Strassmann. And so they repeated the French work. The outcome was similar. Again there was evidence for a relatively lightweight element among the reaction products. Only this time it was barium, atomic number 56. Perhaps it was because Hahn and Strassmann were primarily trained as chemists that they found their chemical results so convincing. They were less committed to the prevailing paradigms of physics than Dr. Joliot-Curie. After their analyses repeatedly revealed the presence of barium and not its heavier relative, radium, the German researchers finally accepted the evidence as correct. Even then, however, the contradiction between chemistry and physics was enough to make their report cautious and qualified. These characteristics are obvious in the following words which Hahn and Strassmann wrote in *Die Naturwissenschaften* for January 6, 1939.

We have come to the conclusion our 'radium isotopes'
have the properties of barium; as chemists we must
properly say that the new substances behave not as
radium, but as barium . . . . As nuclear chemists, closely
allied to physics, we cannot commit ourselves to making
this leap, so contradictory to all the phenomena previously
observed in nuclear physics.

Obvious in its absence from that epoch-making paper was the name Lise Meitner. The reason was the antisemitism of the Third Reich. Dr. Meitner was Jewish. After the German invasion of her native Austria in March, 1938, she was dismissed from her position at the Institute. With the aid of friends she was able to escape to Holland, Denmark, and finally Sweden. It was there, in a small resort town, that Lise Meitner first learned of the splitting of the atom. Although her colleagues in Berlin had been unable to prevent her dismissal, they did keep her informed of the progress of the research she helped launch. Naturally, Hahn wrote her a detailed letter describing the results of his most recent study. Immediately Miss Meitner grasped the significance of the discovery. It was the unbelievable but unavoidable conclusion that under neutron bombardment, a uranium atom breaks into two lighter atoms of roughly equal size. The unknown element in Irène Joliot-Curie's experiment must have been lanthanum after all!

By good fortune, Lise Meitner had a young visitor with whom she could discuss this startling news. It was her physicist nephew, Otto Frisch, up from Neils Bohr's Institute for the Christmas holidays. At first incredulous, but finally convinced by his aunt's persuasive arguments, Frisch hurried back to Copenhagen to tell Bohr the news and then to do the test which proved unambiguously that the uranium atom was indeed splitting. On January 15, 1939, he once more repeated the familiar experiment. But this time he looked at the energy which attended the process. On a relative scale it was gigantic: gram for gram, thirty million times the explosive energy of TNT! The result was exactly what he and Lise Meitner had predicted.

At once Frisch telephoned his aunt, and the two composed a letter to the editor of *Nature*. It occupies only one page and one additional paragraph in the February 11, 1939, issue of that weekly. But in spite of its brevity, the note on "Disintegration of Uranium by Neutrons: A New Type of Nuclear Reaction" was (and is) an announcement of immense magnitude. Under the influence of the neutrons, uranium atoms were dividing like cells. Indeed, the analogy was so apt that Frisch applied a term long used to denote this biological process—*fission*. The word and the nuclear phenomenon it describes have become an intrinsic part of modern existence.

The news spread like wildfire. Dozens of physicists rushed to duplicate the experiment and investigate other aspects of nuclear fission. On March 3, less than a month after the appearance of the Meitner-Frisch paper, flashes of light on a cathode ray tube at Columbia University signaled yet another breakthrough. To Leo Szilard and Walter Zinn, the flashes meant that each splitting uranium atom was yielding two or three neutrons along with the elementary fission fragments. But in those streaks of light Szilard also saw a dreadful spectre. "That night," he wrote later, "there was little doubt in my mind the world was headed for grief."

Szilard had both physical and political reasons for his pessimistic premonition. The former have to do with his discovery that neutrons not only trigger the fission of uranium nuclei, they are also

FIGURE 16.3. *Representation of a nuclear fission chain reaction in uranium-235. Note that two or three neutrons are released for each uranium atom which splits into smaller fragments. These neutrons can in turn trigger additional fission reactions.*

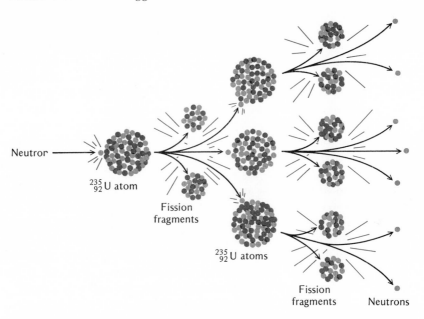

produced by the reaction. Moreover, they are produced in greater numbers than they are consumed. Hence, there is a net gain of one or two neutrons for every atomic disintegration. These in turn can trigger additional fission reactions. And quickly—incredibly quickly—the reaction can fan out through a block of uranium, smashing atoms as it spreads. Fissioning atoms become linked to each other in a growing chain which can race through 15 kilograms of uranium in a fraction of a second. And at every step of this *chain reaction* a little bit of matter disappears, only to manifest itself as a fantastic amount of energy.

Matter and energy, the two most fundamental concepts of physical science, had been united by Albert Einstein in 1905. The great theoretical genius joined them with a trivially simple equation. Now mankind was experiencing theory metamorphosed into fact, matter transformed into energy. Mathematically, Einstein's famous equation is easily understood. It states that $E = mc^2$, where E is the energy released when a mass of matter, m, is fully converted into energy. But the enormity of this energy defies comprehension. The reason is the term c^2; for c represents the speed of light, 3.0×10^{10} cm/sec. Thus, c^2 is 9.0×10^{20} cm²/sec². And 1.00 g of matter is equivalent to $1.00 \text{ g} \times 9.0 \times 10^{20}$ cm²/sec² or 9.0×10^{20} ergs. This is enough energy to raise a weight of 1 million tons 6 miles, enough to turn 30 thousand tons of water into steam, enough to yield the explosive force of 30 thousand tons of TNT.

Einstein's formula is perfectly general, and hence applies even to ordinary chemical reactions. Any change which liberates energy should also be characterized by a decrease in mass. But the energy changes associated with chemical processes are, relatively speaking, very small. And, as a problem at the end of this chapter illustrates, the change in mass is so infinitesimal as to be undetectable by even the most sensitive methods. Thus, the Law of Conservation of Mass and Matter accurately and adequately describes all chemical reactions. In the case of nuclear fission, however, mass differences between products and reactants can be determined by mass spectrometry. True, the decrease is small, both on an absolute and a relative basis. Only about one-tenth of one percent of the matter involved is actually converted into energy. For every gram of uranium-235, just 1 milligram (0.001 g) is "destroyed." Yet, according to Einstein's equation, this amounts to a total energy release of 9×10^{17} ergs or 2×10 kcal. In stark contrast, the chemical energy released by a gram of exploding TNT is 0.66 kcal. It follows that it would require 3×10^7 g or 33 tons of TNT to equal the explosive energy of just a single gram of disintegrating uranium atoms.

The origin of this "atomic" energy is, strictly speaking, the nucleus. One way to better understand its release is to examine in more detail the fission of uranium. Note first of all that the reaction involves a particular isotope of element number 92. By far the most abundant isotope of uranium has an atomic weight of about 238. But for every 140 atoms of $^{238}_{92}U$, there is one of $^{235}_{92}U$. It is this slightly lighter atomic species which is susceptible to splitting and the self-propagation of the process. I could write numerous equations which would correctly represent this fragmentation, because a wide variety of products have been identified. Three reactions are given below, including the one which yielded the barium found by Hahn and Strassmann and another which produced the lanthanum missed by Joliot-Curie. Note too that more neutrons are produced than are required to initiate the original fission, spores for the mushroom growth of the reaction.

$$^{235}_{92}U + ^{1}_{0}n \begin{cases} ^{146}_{57}La + ^{87}_{35}Br + 3^{1}_{0}n \\ ^{141}_{56}Ba + ^{92}_{36}Kr + 3^{1}_{0}n \\ ^{144}_{55}Cs + ^{90}_{37}Rb + 2^{1}_{0}n \end{cases}$$

It is not obvious from these equations that matter disappears in the fission process. The superscripts of the products all add up to 236, the combined mass of one atom of uranium-235 and a neutron. But these are only approximate atomic weights—really *mass numbers*, that is, the total number of neutrons and protons involved. More accurate determinations indicate that the mass in fact decreases when atoms split. The reason sounds a bit peculiar at first hearing: neutrons and protons weigh more in heavy nuclei than in atoms of intermediate mass. When unstable heavy atoms divide to form lighter, more stable ones, the average mass of a nuclear particle in effect decreases. Thus, there is matter left over which becomes converted into energy.

Many of the fragments are also unstable, typically having a neutron-to-proton ratio which is too high. And so, to correct this imbalance, they give off β particles, frequently accompanied by γ rays. The emission accounts for part of the radioactive hazard associated with nuclear bombs and reactors. For example, $^{90}_{38}Sr$, an isotope with a half-life of 28 years, is a common fission product. Because of its chemical similarity to calcium, the strontium-90 becomes incorporated into milk and hence into bones. A secondary effect is the creation of new radioactive isotopes when originally stable atoms capture some of the neutrons evolved in the chain reaction. Thus, radioactive tritium, $^{3}_{1}H$, can form in the cooling water of a nuclear power plant. Similarly, the dust sucked up by a nuclear explosion can become radioactive.

Leo Szilard no doubt knew that all these implications were inherent in the experiment which first demonstrated that uranium-235 can sustain a chain reaction. But on March 3, 1939, there were also other, more obvious indications that "the world was headed for grief." Spain was torn by civil war. Fighting between Japan and China was in its second year. And German troops were massing on the Czechoslovakian border, preparing for an invasion. Six months later another Nazi invasion, that of Poland, triggered World War II.

"I am become Death, the shatterer of worlds"

In a strife-ridden world, the discovery of nuclear fission assumed towering significance. The energy unleashed by splitting atoms provided the potential for a bomb orders of magnitude more destructive than any weapon previously known to man. And the nation or nations which could first build such a bomb would have an incalculable military advantage. But the vast majority of military and political leaders were unaware of the crucial experiments done in late 1938 and early 1939. Indeed, few besides professional scientists understood nuclear energy and its enormous potential.

A small number of European-born physicists were convinced that it was absolutely essential to proceed as rapidly as possible with the investigation of nuclear fission and its applications. Leo Szilard,* Edward Teller,* and Eugene Wigner were Hungarian by birth; Victor Weisskopf was Austrian; and Enrico Fermi was Italian. All were well aware that the fragmentation of the uranium nucleus had first been recognized in Berlin. Moreover, they knew, from bitter personal experience, the wickedness of Hitler's aspirations. They had come to the United States, refugees from Nazi antisemitism, a policy which ironically cost Germany some of its best scientific brains.

After several unsuccessful attempts at following official channels, all of which ended tangled in red tape, the five physicists decided to go directly to the top—President Franklin D. Roosevelt. In order to dramatically gain the attention of the President they enlisted the aid of the world's most prestigious scientist, the shy, gentle theoretician who first formulated the equation which had now been proved in practice. On August 2, 1939, Albert Einstein signed his famous letter to Franklin Roosevelt. It began as follows:

Some recent work by E. Fermi and L. Szilard, which has been communicated to me in manuscript, leads me to expect that the element uranium may be turned into a new and important source of energy in the immediate future. Certain aspects of the situation which has arisen seem to call for watchfulness and, if necessary, quick action on the part of the Administration. I believe therefore that it is my duty to bring to your attention the following facts and recommendations.†

The letter goes on to make several suggestions for maintaining "permanent contact between the Administration and the group of physicists working on chain reactions in America."

President Roosevelt is said to have reacted to this warning with the words, "This requires action!" But the action was slow in coming. It was not until late in 1941, over two years after the writing and receipt of the Einstein letter, that the decision to build the bomb was finally made. It was a decision that launched a 2 billion dollar project which still ranks, even in this day of lunar

---

* After the War, Teller was instrumental in the development of the hydrogen (fusion) bomb and became an outspoken advocate of a strong national defense posture based upon a large arsenal of nuclear weapons. Szilard, on the other hand, was an early supporter of nuclear disarmament.
† Quoted by William L. Laurence in "Men and Atoms," pp. 57–58, Simon and Schuster, New York, 1946, 1959, 1962.

landings, as the most astounding achievement of science and technology in history.

One of the items of prime importance was to determine whether a nuclear chain reaction could be sustained and controlled. It was this task which brought the twenty-odd scientists to the squash court at Stagg Field. The first functional atomic "pile" which they created there was literally that. The fuel inside it was uranium-235. But the graphite blocks which made up the bulk of the reactor had an important function, too. The carbon served to slow down the neutrons emitted by the fragmenting $^{235}_{92}U$ atoms to a speed of about 1 mile per second. Normally, these neutrons travel much faster and are absorbed by the uranium-238 which makes up 99.3% of the element. Consequently, a chain reaction cannot be sustained in the naturally occurring mixture of uranium. The neutrons are swallowed up before they can trigger many additional fission reactions. However, $^{238}_{92}U$ has a much lower affinity for slow neutrons, and under these conditions it does not interfere with the propagation of fission. Branching occurs as described earlier.

In order to control the spread of this reaction, the Chicago pile also included cadmium and boron rods. When these were inserted into the uranium/carbon heap they absorbed sufficient neutrons to prevent a chain from forming. But as these control rods were slowly withdrawn, under Fermi's watchful eye, the number of neutrons absorbed decreased and the flux of slow neutrons increased. Finally, on that fateful December 2nd, the wild clicking of the neutron detectors signaled that a new Promethean fire had been kindled. The pile had gone "critical." The Italian navigator had landed in the New World!

As Fermi and his co-workers signed their names to the straw-covered *fiasco*, they must have been keenly aware of the significance of their successful experiment. The heap of black blocks before them was the prototype of many nuclear reactors which would tap a vast new source of energy. The basic characteristics of all these "atomic piles" are essentially the same. Fissionable material spontaneously emits neutrons which are slowed by moderating elements to the point where they will initiate additional neutron-producing nuclear disintegrations without being absorbed by other atomic species. The process is kept in check by cadmium and boron control rods which regulate the neutron flow and hence the rate of fission. The energy which is literally created from the attendant loss of matter is expressed as heat. The remaining step is basically that found in any fossil fuel power plant—the transformation of heat into work. The experiment had unequivocally proved that nuclear fission can be controlled and used as a practical source of power.

FIGURE 16.4.   *A schematic design of a nuclear power plant. The heat flow is indicated by large colored arrows. In some nuclear reactors, the primary coolant also serves as the moderator.*

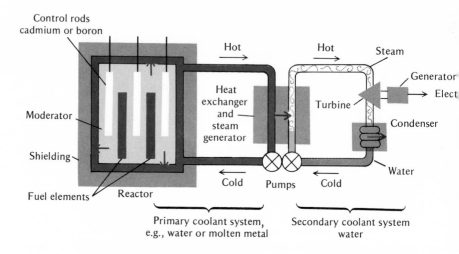

But on December 2, 1942, the production of controlled nuclear energy was not uppermost in the minds of the men in the squash court. For the experiment also proved that an uncontrolled chain reaction was also possible—a reaction which could release its incredible energy in a flashing fraction of a second.

Although the results of Fermi's test were clear, many technical difficulties had to be overcome before a bomb could be built. Without the presence of the moderating graphite, a chain reaction cannot occur in a block of ordinary uranium. The U-238 gobbles up the neutrons as fast as they are formed by the fragmenting U-235. In order to prevent fission from fizzling out, most of the interfering heavier isotope must be removed. This presents a separatory problem of colossal magnitude. Chemically, the two isotopes of uranium are identical. Thus, no reactions can discriminate between the two. Moreover, their physical properties are almost identical, making it extremely difficult to fractionate uranium-235 and uranium-238 by any means whatsoever.

Out of the several suggested procedures, two were selected for the most intensive investigation. One made use of differences in the charge-to-mass ratio for ions of the two isotopes. Because $^{235}_{92}U^+$ is a little lighter than $^{238}_{92}U^+$, the former ion has a slightly higher q/m value. You will recall from our discussion of J. J. Thompson's work on the electron, that the path of a charged particle in an electric and/or magnetic field depends upon this very ratio. Therefore, the two isotopic ions should follow slightly different trajectories in a mass spectrometer. Indeed, the exact isotopic masses and the natural abundances of the two species of uranium were determined in just such a device. It should be possible, Lawrence and others argued, to use the same sort of instrument as a means of separation.

This electromagnetic method proved incredibly difficult. But so was the other procedure, gaseous diffusion. The basic concept has already been outlined (in Chapter 10). To recapitulate, uranium was reacted with fluorine to form a gaseous compound, uranium hexafluoride. The fact that the molecular weight of $^{235}_{92}UF_6$ is slightly lower than that of $^{238}_{92}UF_6$ means that at any fixed temperature, the former, lighter gas will diffuse more rapidly than the latter. By repeating the diffusion process through thousands of consecutive pinholes, a considerable enrichment of uranium-235 can be achieved. In fact, both methods of separation were employed by engineers and scientists at the atomic energy installation at Oak Ridge, Tennessee. The yield was sufficient $^{235}_{92}U$ to form a *critical* mass, roughly 15 kg or 33 lb. When this much uranium-235 is brought together in a compact lump, it spontaneously undergoes uncontrolled and uncontrollable fission. And so it did over Hiroshima on August 6, 1945.

But uranium-235 is not the only fissionable isotope. Evidence amassed in the early 1940s indicated that plutonium-239, an element already mentioned, will also sustain a chain reaction. Moreover, by a quirk of nature, this very isotope is formed from uranium-238 in a nuclear pile. The heavy, plentiful isotope of uranium captures neutrons and, after several intervening steps, yields plutonium-239.

$$^{238}_{92}U + ^1_0n \rightarrow ^{239}_{94}Pu + 2^{\ 0}_{-1}\beta$$

Thus, the success of the squash-court pile also indicated that it was possible to create another deadly nuclear explosive from ordinary uranium.

To achieve this end, huge "breeder" reactors were built at Hanford on the Columbia River in Washington. Meanwhile, Seaborg's colleagues were doing a complete chemical characterization of plutonium, using microgram quantities. Many of the tests were conducted on the stage of a microscope. Nevertheless, the chemistry was accurate. And the process devised on these minute quantities were successfully scaled up a billion-fold to treat the spent fuel slugs from the Hanford reactors. Because of the elementary differences between plutonium and uranium, the former substance could be chemically separated from the latter with relative ease. But the extreme radioactivity of $^{239}_{94}Pu$ made it necessary to conduct all operations by remote control from behind heavy shielding.

As the purified plutonium was obtained it was transferred to what was once a private boys' school on a remote New Mexico mesa. The Los Alamos Ranch School had been purchased by the United States Government in late 1942. Shortly thereafter, some of the greatest luminaries of chemistry and physics began to move into its buildings and the uncomfortable temporary structures hurriedly added. For this was the site of the final critical stage in what had come to be called "the Manhattan Project." To the Los Alamos laboratory was given the responsibility for designing, testing, and assembling the bomb. At its head was J. Robert Oppenheimer—brilliant, brooding, charismatic, above all, enigmatic.

The work on the mesa went on at a feverish pace. The supply of uranium-235 from Oak Ridge was so scant that no plans were made to test the element in an explosive device. It would all go into a bomb named "Little Boy." But there were serious questions about the workability of the plutonium bomb, and a test was deemed essential.

On June 30, 1945, the final supply of plutonium necessary to complete the critical mass of about 5 kg arrived at Los Alamos by automobile. It was incorporated into the complex bomb design, and on the morning of July 12, a small convoy transported the device to the designated test site—a place called "Trinity." Oppenheimer had selected the name, inspired by the opening lines of a holy sonnet by John Donne:

Batter my heart, three-personed God; for, you
As yet but knock, breathe, shine, and seek to mend . . .

There was irony, too, in other names. For Trinity lay in a valley called *Jornada del Muerto*, ringed by the peaks of the *Sangre de Christo* range—the "Journey of Death" and the "Blood of Christ."

Carefully, methodically, the final assembly took place. On Saturday, July 14, the 10,000-pound "gadget" was raised to the top of a 100-foot steel tower. All that day and the next, the scientists at the site rechecked the firing mechanism, tested their instruments, and nervously watched the clouds gather. The test was scheduled for 4:00 a.m., Monday morning, July 16. But at 2:00, wind, rain, and lightning raked the desert. The test was postponed an hour, then another half hour. The tension in the shelters grew with each passing moment. Under the tower itself, a handful of men waited. At least one slept.

Then, at 4:00 a.m. the rain stopped; half an hour later the wind had died down and the clouds were breaking up. The decision was made to go ahead with the test at 5:30. At about 5:00, the party at the tower completed the final arming of the bomb and closed the switches which connected the bomb to the electronic firing mechanism in a shelter 6 miles away. Then they climbed into a jeep and quickly drove those 6 miles. Each second of that last half hour before the dawn of the Atomic Age must have seemed interminable. Men 10 and 20 miles away donned heavy welders' goggles and peered into the darkness, lost in their own thoughts.

One of them was William L. Laurence, Science Editor of the *New York Times*. His eloquent words capture the awesome, terrible beauty of that fateful moment: 5:29:45 a.m., July 16, 1945.

And just at that instant there rose as if from the bowels of the earth a light not of this world, the light of many suns in one. It was a sunrise such as the world had never seen, a great green supersun climbing in a fraction of a second to a height of more than eight thousand feet, rising ever higher until it touched the clouds, lighting up earth and sky all around with a dazzling luminosity.

Up it went, a great ball of fire about a mile in diameter, changing colors as it kept shooting upward, from deep purple to orange, expanding, growing bigger, rising as it expanded, an elemental force freed from its bonds after being chained for billions of years. For a fleeting instant the color was unearthly green, such as one sees only in the corona of the sun during a total eclipse.

On that moment hung eternity. Time stood still. Space contracted to a pin point. It was as though the earth had opened and the skies had split. One felt as though he had been privileged to witness the Birth of the World.*

Less than a month later, Laurence was privileged to witness the Death of Nagasaki.

Of course, the aim of the entire Manhattan Project had always been to build a bomb. In spite of security and compartmentalization, essentially all the scientists and engineers involved knew that. The chief original motivation for the project was the fear that the Nazis would acquire and employ atomic weapons. But with the surrender of Germany on May 8, 1945, this threat was removed. Nevertheless, the war in the Pacific was still claiming thousands of casualties each week. As the New Mexico test neared, it became more and more likely that the bomb would be completed before the surrender of Japan. And so, the nation's political and military leaders, ultimately President Harry Truman, were faced with the question of whether or not to use the bomb, and, if the answer was yes, how best to employ it.

* Quoted, with permission, from William L. Laurence, "Men and Atoms," pp. 116, 117, Simon and Schuster, New York, 1946, 1959, 1962.

FIGURE 16.5.   *The fire ball of the first manmade atomic explosion, Almogordo, New Mexico, July 16, 1945. Reproduced from M. Russell Wehr and James A. Richards, Jr., "Physics of the Atom," 2nd ed., Fig. 10.7, Addison-Wesley, 1967. Courtesy of P. M. S. Blackett and D. S. Lees, Imperial College of Science and Technology, London.*

On May 31 and June 1, 1945, the "Interim Committee," chaired by Secretary of War Henry Stimson, met to discuss the latter issue only. One of the most momentous decisions of this century, the decision to drop the atomic bomb, had more or less made itself by the sheer inertia of mobilized momentum. With the consultation and agreement of a Scientific Advisory Panel, the Interim Committee recommended to the President that the bomb should be used against a Japanese military site surrounded by residential or other fairly fragile buildings, as quickly as possible and with no prior warning.

In the first two weeks of June there was an ineffectual flurry of activity from scientists who had been willing to build a bomb, but who did not wish to see it used. The report of the committee chaired by physicist James Frank and a petition circulated by Szilard called for a demonstration of the bomb in an uninhabited area of Japan before actual employment against a military or civilian target. But the Interim Committee had already considered and rejected this alternative. Such a demonstration, they argued, would not speak eloquently enough to force quick surrender. Besides, there were only two bombs ready—"Little Boy" (the uranium bomb) and "Fat Man" (the plutonium bomb). To eliminate the necessity of an invasion of Japan, an undertaking estimated to cost 3 million Japanese and American lives, these bombs had to be used for maximum effect. And a well-heralded dud would only weaken the American bargaining position.

On June 14, the Scientific Advisory Panel met to review the decision made two weeks earlier by the Interim Committee. They noted the division among scientists, but their conclusion reinforced that already arrived at.

The opinions of our scientific colleagues on the initial use of these weapons are not unanimous; they range from the proposal of a purely technical demonstration to that of the military application best designed to induce surrender . . . We can propose no technical demonstration likely to bring an end to the war; we see no acceptable alternative to direct military use.

Significantly, the Panel added another paragraph, one pregnant with implications for science and society.

With regard to these general aspects of atomic energy, it is clear that we, as scientific men, have no proprietary rights. It is true that we are among the few citizens who have had occasion to give thoughtful consideration to these problems during the past few years. We have, however, no claim to special competence in solving the political, social and military problems which are presented by the advent of atomic power.*

On August 6, 1945, a uranium fission bomb destroyed 60% of the Japanese city of Hiroshima, killing 78,000 men, women and children and injuring 37,000 more. Three days later, a plutonium fission bomb was dropped on Nagasaki. There were 100,000 casualties. On August 14, Japan sued for peace. The war was over.

Stealing the sun

There is a medium-sized yellow star situated on the edge of the Milky Way. The inhabitants of a blue-green planet about 93,000,000 miles from the star call it the sun. Every second, for billions of years, that fiery ball has been sending out into the cold blackness of space 7×10^{22} kcal of energy. This energy literally comes from the body of the sun in an atomic fire which consumes 5 million tons of matter each second. Nevertheless, the mass of our star is so great (2.2×10^{27} tons) that measured on a human scale, its supply of matter and energy seems infinite.

Astrophysicists believe that the principal net reaction responsible for the sun's energy is the *fusion* of four hydrogen atoms to yield a helium atom.

---

* Quoted by William L. Laurence in "Men and Atoms," p. 141, Simon and Schuster, New York, 1946, 1959, 1962.

$$4^1_1H \rightarrow {}^4_2He + 2^0_1e$$

You will note that in the process two positrons are emitted for each atom of helium formed. So are 4.0×10^{-5} ergs. This energy change may seem small, but remember it arises from only four hydrogen atoms. At this rate, a single gram of fusing hydrogen generates 6.0×10^{18} ergs or 1.4×10^8 kcal, the energetic equivalent of almost 20 tons of burning coal. On a gram for gram basis, this is about eight times the energy of fissioning uranium-235.

The ultimate origin of the energy in fusion and fission is fundamentally the same, the conversion of matter to energy. In the former case, the combined mass of the products—helium atoms and positrons—is less than the mass of the hydrogen atoms which are the reactants. There is one significant difference, however: the fusion of ordinary hydrogen nuclei only occurs at temperatures of 20 million °C or higher. Therefore the possibility of duplicating this *thermonuclear* reaction on earth long seemed an academic impossibility.

But the successful fission of heavy nuclei provided a match to light this solar fire. For the temperature associated with a uranium explosion approaches 100 million degrees. Therefore, it should be possible to surround a critical mass of ${}^{235}_{92}U$ with hydrogen and thus create a monster bomb. For practical reasons, ${}^1_1H$, the ordinary isotope of hydrogen which constitutes 99.985% of the element, is not suitable for this purpose. However, the heavier isotopes of hydrogen—deuterium, ${}^2_1H$, and tritium, ${}^3_1H$, can be used as fusion fuel. In practice, a solid compound called lithium deuteride is employed. The lithium is a specific rare isotope containing three neutrons and three protons in each nucleus. When one of these atoms absorbs a neutron, produced by a fission bomb, it is converted into an atom of tritium and one of helium.

$$^6_3Li + {}^1_0n \rightarrow {}^3_1H + {}^4_2He$$

The deuterium is, of course, already present in the lithium deuteride. The two isotopes of hydrogen then fuse in the following energy-producing reaction.

$$^3_1H + {}^2_1H \rightarrow {}^4_2He + {}^1_0n$$

This design was employed in the first hydrogen-bomb test at Eniwetok Atoll on November 1, 1952. The energy released was equal to that of 5 million tons (5 megatons) of TNT. It was 250 times the energy of the Trinity test, 2.5 times the energy of the total explosives dropped by all the belligerents in World War II. It is just possible that the horror potentially present in that ball of fire has done more to prevent another major war than all the diplomats and all the moralists in history.

Unfortunately, the controlled, constructive use of nuclear fusion is far more difficult than its uncontrolled employment as a means of destruction. There are two major difficulties, both associated with the fantastic temperatures required for this thermonuclear reaction. The first problem is how to achieve 50 million degrees or more without a fission explosion; the second is how to contain the fire once it is kindled. The most heat-resistant materials on earth vaporize instantly at temperatures above 6000°C. The situation is thus akin to finding a container for a universal solvent. The answer may be magnetism, not matter. Ions of deuterium can be squeezed into a narrow beam by strong magnetic fields. The resulting charged gas, or *plasma,* can then be heated to incredibly high temperatures by the passage of electric current. *If* the temperature can reach the level necessary for thermonuclear reaction, and *if* the beam of plasma can be kept away from the walls of the magnetic bottle, the controlled fusion of deuterium nuclei may yet be achieved. Because deuterium is far more plentiful than uranium, such a breakthrough would provide a vast, almost limitless new source of power. A single cubic kilometer of sea water contains enough deuterium to yield the energetic equivalent of 2 trillion barrels of oil, roughly the world's total oil reserve. Considering the vastness of the oceans, the amount of readily accessible deuterium is sufficient to supply ten times the world's current energy consumption for several billion years. But the *ifs* represent prodigious technical problems which must be mastered before mankind can create an earth-bound sun.

Peaceful atoms

Much of this chapter has dealt with destruction. To attempt to disguise the fact that science and engineering have been intimately involved with warfare would be dishonest—transparently dishonest. It is a fact which science and society must face squarely. But on the other hand, to end this chapter and this book with a cry of *mea culpa* would neglect the great good which can come to this troubled world through nuclear science in particular and science in general. For atomic energy and radioactivity have been and can be used to enhance the quality of life upon earth.

Ironically, but not surprisingly, the very properties of nature which permit the production of bombs and make radiation a significant health hazard also lead to power plants and new ways of saving lives. The application of fission and possibly fusion as sources of controlled energy have already been discussed. There is no need to go into additional detail. But it is worth noting that nuclear power plants are not unmixed blessings. The emission from such installations must be carefully monitored and controlled. Moreover, the disposal of spent fuel elements and other radioactive wastes is an item of major concern. It is a long-term problem because the half-lives of some isotopes keep them dangerously active for hundreds of years. At present the most promising solution seems to be sealing the long-lived emitters in nonporous ceramic materials and burying them in abandoned salt mines.

Of course the reason for this concern is the potentially harmful physiological and genetic effects which can result from exposure to radiation. The biological damage caused by α, β, and γ rays (and X rays and cosmic rays as well) arises from the interaction of high-energy particles and photons with tissue. As a consequence of such collisions, electrons are often knocked out of atoms in the target material. In other words, ions are formed. During the ionization, chemical bonds are sometimes broken and molecular rearrangements can occur. Such changes alter the delicate chemical balance of a cell. Furthermore, the water which forms the bulk of most living things is another likely point of attack. Through a series of radiation-induced steps, H_2O molecules react to yield H_2O_2, hydrogen peroxide. The latter compound is a strong oxidizing agent which can seriously interfere with the normal function of a cell.

White blood cells, or leukocytes appear to be particularly susceptible, and an early symptom of radiation disease is a decrease in white blood cell count. Because these cells are an essential part of the body's defense mechanism, lowering their number greatly increases susceptibility to infection. Radiation also causes changes in the bone marrow; and since a function of the marrow is the production of red blood cells, anemia results. Exposure to higher doses gives rise to burns and ulcers which heal only with extreme difficulty. Loss of hair, nausea, vomiting, and hemorrhaging may all be symptomatic of more severe internal disorders. Moreover, the cumulative buildup of radiation damage over many years may lead to cancer, particularly leukemia.

In other cases, prolonged exposure may produce changes in the DNA of sperm or ova. These mutations are then expressed as birth defects in the next generation. Fortunately, the number of human beings who have been highly irradiated is relatively small. But as a consequence, statistical studies of the mutation rate among their offspring are limited. It is particularly frustrating that our knowledge of the genetic effects of a lifetime of low level exposure is so inadequate. Apparently there is no safe threshold. All radiation, however slight, can produce potentially deleterious mutations. To be sure, none of us can escape the daily bombardment of cosmic rays and natural geological radiation. But we can minimize the amount of additional artificial radiation to which we

FIGURE 16.6.   A Geiger-Müller tube. An entering emission
particle ionizes an argon atom which conducts electricity
between the oppositely charged electrodes.

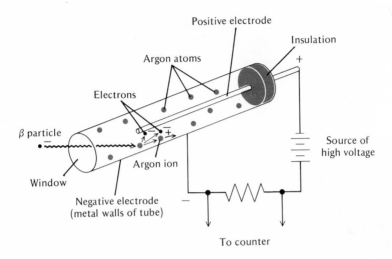

are subjected. This, of course, was a strong motivating force behind the Nuclear Test Ban Treaty of 1963. Furthermore, similar
considerations are involved in establishing radiation emission
standards for nuclear power plants and exposure limits for the
scientific and technical uses of radioisotopes.

All this implies that there are means of measuring the intensity
of radiation. Generally speaking, they are based upon the same
ionizing property which destroys cells. Perhaps the best known
detector, the Geiger-Müller counter, is pictured in Figure 16.6. As
you can see, it is a tube filled with gas, often argon, at a low
pressure. Through the center of the cylinder runs a wire which is
connected to a source of high voltage. The metal wall of the tube
acts as the other electrode. Because there is a large difference in
electrical potential between the wire and the wall, there is a
tendency for current to flow. However, argon atoms are neutral
in their net charge, and hence cannot conduct electricity. Thus,
the circuit is broken. But then α or β particles enter through the
thin window, they knock electrons out of the argon atoms. The
resulting ions and electrons give the gas the property of electrical
conductivity, just as the ions in a solution of an electrolyte such as
sodium chloride carry current. The circuit is completed and electricity flows from the electrode of higher potential to that of
lower potential. Each of these atomic events is detected and
amplified by an electronic device which literally counts the number of individual ionizations occurring.

FIGURE 16.7.   *α-Particle tracks in a cloud chamber. Photo by Los Alamos Scientific Laboratory.*

Other detectors work on the same fundamental phenomenon. In a *scintillation counter,* each particle or packet of energy absorbed induces a flash of light in a chemical called a scintillator. These flashes are counted by a photosensitive cell. In a *cloud chamber,* droplets of dew condense on the ions formed in the path of the radiation. Thus, the trajectory of the ray is made visible. Similarly, a string of bubbles signals the passage of a high-energy photon or particle through a *bubble chamber.* All these devices have been of great importance in radiochemistry and nuclear physics.

They have, among other things, led to the establishment of units for expressing the intensity of radioactive emission or absorption. A *curie* is defined as the number of disintegrations which occur in 1 gram of radium in 1 second, 3.7×10^{10}. A more useful unit for measuring total radiation dosage is the *roentgen,* the amount of radiant energy which produces a certain arbitrarily fixed number of ions in 1 cubic centimeter of air under specified conditions. Unlike the curie, the roentgen does not treat all particles or photons equally. The roentgen rating of highly energetic radiation is greater than that of low-energy emission. Moreover, the roentgen is independent of time. It can therefore be used to measure the cumulative effect of radiation. According to present standards, a full body absorption of 0.1 roentgen per day is the maximum safe limit. Significantly, this value is lower than that originally established, and further reductions are possible. It is estimated that exposure to 500 roentgens over a brief period of time is probably lethal to most human beings. A similar unit is the *rad* (radiation absorbed dose). It corresponds to the absorption of 100 ergs of radiant energy per gram of absorber.

Not surprisingly, curies, roentgens, and rads are frequently employed in the health sciences. With careful control, modern medicine has turned a potentially fatal phenomenon into a means of saving lives. Obviously, radioactive emission is no respecter of cells. It destroys both healthy and diseased tissue. However, malignant cells, because they grow at a faster rate than normal ones, are more susceptible to radiation damage. Certain tumors can be destroyed or arrested by exposure to a focused beam of particles from radium or cobalt-60. Other cancers can be treated by the internal administration of radioactive isotopes. The most convenient is probably cancer of the thyroid. This gland, located in the throat, selectively concentrates iodine. Indeed, the reason for adding potassium iodide to table salt is to prevent an iodine deficiency which leads to goiter, an enlarged thyroid. In the case of malignant goiter, the patient drinks a dilute solution containing radioactive iodine, I-131 or I-128, probably again present as potassium iodide. The element preferentially accumulates in the thyroid where it can combat the cancer without significantly harming the rest of the body.

Other radioactive isotopes serve as diagnostic messengers. An example is the use of radioactive compounds to detect certain circulatory and coronary disorders. A small amount of the substance, perhaps salt containing radioactive sodium, is injected into the bloodstream. As it moves through the body, it emits its telltale signals. The level of the radiation is not high enough to cause damage. But when the radioactivity builds up in a certain region of the body, it is good evidence of an impediment to normal circulation. Various medicinal or surgical treatments may then be indicated.

Biomedical research has also profited greatly from the introduction of radioisotopes. The tortuous pathways of body chemistry can be followed by feeding compounds containing radioactive carbon-14 or oxygen-18. The distribution of the radioactivity in urine, feces, exhaled carbon dioxide, and among the various tissues of the experimental animal provide invaluable insight into metabolic mechanisms and molecular metamorphoses. Coupled with the sensitive techniques of modern analysis, radiotracer research can often reconstruct the specific biochemical events which befall an injected or ingested food additive or drug. And similar applications of isotopes have helped illucidate the steps in relatively simple reactions and the complex chemical chain in photosynthesis. Surely information of this sort is fascinating in its own right; but more than that, its implications for a biologically better life are legion.

Thanks to the techniques of radiochemistry and the identifying emission of radioisotopes, incredibly minute concentrations of certain elements can be detected. In fact, it is even possible to make stable isotopes temporarily radioactive by bombarding them with neutrons. The newly born but rapidly decaying nuclei then signal their presence and their concentration with escaping radiation. For certain metals, a sample as small as 10^{-10} g is detectable. Such sensitivity suggests that the method, called *neutron activation analysis*, has great potential in measuring and monitoring certain environmental pollutants, particularly heavy metals like cadmium, mercury, and lead.

Even hydrogen bomb explosions are not without their peaceful applications. They could be used to dig new harbors and canals, release oil trapped in tar sands and shale, crush low-grade ore deposits before mining, and create underground water reservoirs or even heat reservoirs. Of course, these projects would all require extreme caution. But, for that matter, so do all uses of nuclear energy. The dangers and difficulties are great. But so are the advantages which can accrue to us if we beat the nuclear sword of Damocles which threatens us daily into the sharpest and brightest of plowshares.

Epilogue

The dilemma posed by nuclear energy is by no means unique. It only emphasizes with particular eloquence the ironical issue always raised when man discovers more about his world. For the logical, harmonious, beautiful truths of science are also power— power for good or evil. Society chooses which. Printing presses can promote the irradication of illiteracy or poison minds with propaganda; radio and television can educate millions or brainwash them; computers can make living easier or more inhumane; wonder drugs and new surgical techniques can increase the fullness of life or prolong existence to the point where it becomes intolerable; and the creations of chemistry can serve the needs and comforts of people everywhere or suffocate them in effluent offal.

One thing is clear—science alone will neither destroy nor save this planet and its inhabitants. That is a decision for all of mankind. The challenge has never been so great, and there has never been so much at stake. On the one hand is the nightmare of the world ending with a nuclear bang; on the other is the last whimper of a starving child. Faced with these spectres, the temptation to flee into an artificial euphoria or fall into a deep melancholia is undeniably compelling. But the times demand more than naïve optimism or dark despair. Rather it is your responsibility and mine to face our perplexing problems with honesty, intelligence, and concern for the human condition. I wish us strength, courage, and, above all, success in our common and universal cause.

For thought and action

16.1 Listed below are the atomic numbers of certain elements. Using Figure 16.2 (and *not* the periodic table), predict the number of neutrons present in the most stable nucleus of each of these elements. Also estimate the atomic weight of the most plentiful isotope.
(a) 15   (b) 25   (c) 45   (d) 65   (e) 82

16.2 For each of these unstable isotopes, predict the mode of radioactive decay (for example, β emission) and write a radiochemical equation representing the process.
(a) ^3_1H (b) $^{10}_6C$ (c) $^{90}_{38}Sr$ (d) $^{58}_{27}Co$ (e) $^{60}_{27}Co$

16.3 The $^{14}_6C/^{12}_6C$ ratio in a bit of charcoal from a cave is one-fourth that found in a freshly chopped piece of wood. How long ago did the tree that yielded the charcoal die?

16.4 The combustion of 1.00 g of glucose liberates 3.74 kcal of heat. Use Einstein's formula to calculate the mass change associated with the evolution of this much energy.

16.5 Calculate the quantity of energy (in ergs and calories) which could be obtained by the *complete* conversion of a charcoal briquet (mass = 25 g) into energy.

16.6 A statement made in the text indicates that 9.0×10^{20} ergs is enough energy to raise a weight of 1 million tons, 6 miles. Check the accuracy of these estimates. (*Hint:* Recall a similar calculation in the first section of Chapter 12.)

16.7 It is estimated that in 1973, the energy consumption in the United States will be 1.8×10^{16} kcal.
(a) Calculate the weight of fissioning uranium-235 required to supply that quantity. Note that the reaction does not involve the *complete* conversion of U-235 to energy, but only about 0.1% of the mass. Further details are given in the text.
(b) Also calculate the weight of coal required to supply the same amount of energy. The combustion of 1 g of coal yields about 8 kcal.

16.8 Complete the following equations representing various reactions in uranium fission:
(a) $^{235}_{92}U + ^1_0n \rightarrow ^{140}_{54}Xe + \underline{\hspace{1cm}} + 2^1_0n$
(b) $^{235}_{92}U + ^1_0n \rightarrow \underline{\hspace{1cm}} + ^{103}_{42}Mo + 2^1_0n$

16.9   Suppose each fissioning uranium-235 nucleus releases, on the average, two neutrons, each of which in turn causes another atom to split with a similar evolution of neutrons. Calculate the total number of atoms split in two, three, four, and five "generations," starting with a single initiating neutron and plenty of uranium.

16.10   The fusion reaction involved in the hydrogen bomb is expressed by the equation

$$_1^2H + {}_1^3H \rightarrow {}_2^4He + {}_0^1n.$$

Accurate determinations yield the following nuclear masses:

$$_1^2H = 2.01355, \quad {}_1^3H = 3.01550, \quad {}_2^4He = 4.00150, \quad {}_0^1n = 1.00867$$

Each of these values represents the mass of the nucleus of the indicated atom relative to the carbon-12 standard. Thus, 1 mole of $_1^2H$ nuclei weighs 2.01355 g, and so on. Use these data to calculate the energy released (in ergs and calories) during the fusion of 1 mole of $_1^2H$ and 1 mole of $_1^3H$.

16.11   Figures quoted in the body of the text indicate that 1 g of fusing hydrogen generates 1.4×10^8 kcal of energy.
   (a)   Calculate the energy (in kcal) which would be liberated if all the hydrogen atoms in a glassful of water (250 g H_2O) underwent fusion.
   (b)   The heat of vaporization of water is 540 cal/g. How many grams, kilograms, and tons of water at its boiling point could be converted into steam by the amount of energy calculated in part (a)?

16.12   A recent Supreme Court ruling held that states cannot establish more stringent standards for radioactive emission than those set by the United States Atomic Energy Commission. Debate the pros and cons of that decision.

16.13   Obviously, much of the nuclear research done just before and during World War II was classified. Under what conditions, if any, should classified research be permitted at a university? Defend your answer.

16.14   Comment on the statement by the Scientific Advisory Panel which is quoted on p. 503. Was it a cop-out?

16.15   Defend or refute the following proposition: "Scientists should refuse to work on research which has a good chance of yielding information, materials, or processes which could ultimately be used for harmful or destructive purposes."

16.16   This would be an appropriate time to reconsider your answers to the questions at the end of Chapter 1. Question 1.3 is particularly pertinent to this final chapter.

16.17 Reread the lines from Poe's "Sonnet—To Science" quoted in Question 1.4. If your conception of science has changed as a result of this course, specify the nature of the change.

Suggested readings

Bulletin of the Atomic Scientists: Science and Public Affairs, a monthly magazine which, as its subtitle suggests, is concerned with the impact of science upon the larger society.

Compton, Arthur H., "Atomic Quest," Oxford Univ. Press, New York, 1956.

Groueff, Stephane, "Manhattan Project: The Untold Story of the Making of the Atomic Bomb," Little, Brown, Boston, Massachusetts, 1967.

Hersey, John, "Hiroshima," Bantam Books, New York, 1946.

Lamont, Lansing, "Day of Trinity," Atheneum, New York, 1965.

Laurence, William L., "Men and Atoms," Simon and Schuster, New York, 1946, 1959, 1962.

Pauling, Linus, "No More War!" Dodd, Mead, New York, 1958.

Teller, Edward, and Latter, A. L., "Our Nuclear Future," Criterion Books, New York, 1958.

Appendix 1   Playing the numbers

The power of exponents

Exponential notation provides a compact and convenient way of writing very large and very small numbers. And, since chemistry often involves both of these extremes, this book manages to save a good deal of paper and ink by employing the exponential convention. The idea is to make use of positive or negative powers of 10. The following tabulation should remind you of the meaning of positive exponents.

$10^0 = 1.$
$10^1 = 10.$
$10^2 = 10 \times 10 = 100.$
$10^3 = 10 \times 10 \times 10 = 1000.$
$10^4 = 10 \times 10 \times 10 \times 10 = 10,000.$

You will observe that the exponent indicates the number of 10's multiplied together to give the desired value. It is also apparent that the exponent equals the number of zeros between the 1 and the decimal point. Thus, 10^6 could be written out as 1 followed by six 0's, that is 1,000,000. Similarly, 10^8 is equal to 100,000,000.

On the other hand, a negative exponent means a reciprocal. Thus, one can write

$$10^{-1} = \frac{1}{10^1} = \frac{1}{10} = 0.1$$

$$10^{-2} = \frac{1}{10^2} = \frac{1}{10 \times 10} = \frac{1}{100} = 0.01$$

$$10^{-3} = \frac{1}{10^3} = \frac{1}{10 \times 10 \times 10} = \frac{1}{1000} = 0.001$$

Once again, the absolute numerical value of the exponent indicates the number of 10's multiplied together. But remember that this product appears in the denominator to yield a fraction less than 1. Also note that the exponent is one more than the number of 0's between the decimal point and the 1. It follows that 10^{-6} can be written as a decimal point followed by five 0's and then a 1: 0.000001. Working the opposite way, it is apparent that 0.00000001 with its seven 0's must be identical to 10^{-8}.

Of course most of the quantities and constants of chemistry are not simple whole-number powers of 10. Therefore, they must be written as a number multiplied by 10 raised to some appropriate integral exponent. Consider, for example, Avogadro's number. In its extended form, it almost takes up an entire line: 602,300,000,000,000,000,000,000. But this number can be broken down, or factored, so that the result is the mathematical product of a smaller number and a whole-number power of 10. It thus becomes $6.023 \times 100,000,000,000,000,000,000,000$. Counting the zeros discloses 23 of them, so the second part of the number can be conveniently expressed as 10^{23}. Therefore, in exponential notation, Avogadro's number is 6.023×10^{23}. Of course, it would be equally correct to write it as 60.23×10^{22}, 602.3×10^{21}, 0.6023×10^{24}, or any other equivalent form. However, the common convention is to express the number in such a way that only one digit appears to the left of the decimal point, that is, in the exponential manner originally employed.

The reverse process, the transformation of the exponential form of a number into its extended version, can be illustrated with a particular value of the gas constant, $R = 8.314 \times 10^7$ ergs/(mole °K). Introducing the extended form of 10^7, $R = 8.314 \times 10,000,000$ ergs/(mole °K) = 83,140,000 ergs/(mole °K).

Negative exponents are useful in writing very small numbers, such as the mass of a hydrogen atom,

0.0000000000000000000001674 g.

This number corresponds to

1.674 × 0.0000000000000000000000001 g.

In exponential shorthand, the second part of the product is simply 10^{-24} since the number contains 23 zeros to the right of the decimal point. Therefore, the mass of a hydrogen atom can be written

1.674×10^{-24} g.

Similarly, the charge on an electron, 1.6021×10^{-19} coulomb, can be strung out with 18 zeros between the decimal point and the first non-zero digit,

0.00000000000000000016021 coulomb.

One advantage of exponential notation is the fact that multiplication and division are simplified. To multiply two powers of 10, one merely adds exponents. Thus,

$$10^2 \times 10^3 = (10 \times 10)(10 \times 10 \times 10) = 10^{(2+3)} = 10^5.$$

If one (or both) of the exponents happen to be negative, the same procedure is followed, remembering that the algebraic rules of adding require that signs be considered.

$$10^5 \times 10^{-3} = (10 \times 10 \times 10 \times 10 \times 10)\frac{1}{(10 \times 10 \times 10)}$$
$$= 10 \times 10 = 100$$
$$= 10^{[5+(-3)]} = 10^2$$

Similarly,

$$10^{-3} \times 10^{-2} = 10^{[-3+(-2)]} = 10^{-5}.$$

To divide powers of 10, one subtracts exponents. For example,

$$10^5 \div 10^3 = (10 \times 10 \times 10 \times 10 \times 10) \div$$
$$(10 \times 10 \times 10) = 10 \times 10 = 100$$
$$= 10^{(5-3)} = 10^2.$$

You will observe that this problem is, in effect, identical to $10^5 \times 10^{-3}$. Again, it is necessary to watch signs carefully, as the four following examples illustrate:

$$10^2 \div 10^5 = 10^{(2-5)} \quad = 10^{-3}$$
$$10^{-5} \div 10^2 = 10^{(-5-2)} \quad = 10^{-7}$$
$$10^5 \div 10^{-2} = 10^{[5-(-2)]} = 10^{(5+2)} = 10^7$$
$$10^{-5} \div 10^{-2} = 10^{[-5-(-2)]} = 10^{(-5+2)} = 10^{-3}.$$

When the numbers involved in a multiplication or a division are not simply 10 raised to some whole-number exponent, the appropriate mathematical operations must also be carried out with the numbers which proceed the exponential multipliers. For example, one can calculate the gram atomic weight of hydrogen by multiplying the mass of a single atom by Avogadro's number.

$$1.674 \times 10^{-24} \frac{g}{atom} \times 6.023 \times 10^{23} \frac{atoms}{GAW} =$$
$$1.674 \times 6.023 \times 10^{-24} \times 10^{23} \frac{g}{GAW} =$$
$$10.08 \times 10^{(-24+23)} = 10.08 \times 10^{-1} = 1.008 \frac{g}{GAW}$$

The division of exponential numbers is illustrated by an example from Chapter 3. The mass of a single atom of copper is obtained by dividing the gram atomic weight of the element by the number of atoms in one gram atomic weight, that is, Avogadro's number.

$$\frac{68.5 \text{ g/GAW}}{6.023 \times 10^{23} \text{ atoms/GAW}} = \frac{10.5 \text{ g/atom}}{10^{23}}$$

But $1/10^{23}$ is identical to 10^{-23}, and therefore, the mass of one atom of copper is 10.5×10^{-23} g or 1.05×10^{-22} g.

Clearing the logjam

Logarithms (or logs) are really exponents—in the case of common logarithms, exponents of the base 10. Thus, log $1000 = $ log $10^3 = 3$, and log $0.01 = $ log $10^{-2} = -2$. Conversely, 1000 is the antilogarithm of 3, that is, the number whose logarithm is 3. Similarly, antilog $-2 = 0.01$. Quite obviously, any exact multiple or submultiple of 10 will have a logarithm which is a whole number. But *any* number can be written as 10 raised to the appropriate exponent. And, aside from the special case already mentioned, this exponent or logarithm will *not* be a whole number. For example, $5.0 = 10^{0.699}$, that is, log $5.0 = 0.699$. Note that the exponent is between 0 and 1. This is ·consistent with the fact that 5.0 lies between 1 (or 10^0) and 10 (or 10^1).

Tables of logarithms are listings of these exponents. Since logs are employed in only one part of this text, the optional section on pH, there is little point in including an extensive table. The following should be sufficient.

Number	Logarithm
1.0	0.000
1.5	.176
2.0	.301
2.5	.398
3.0	.477
3.5	.544
4.0	.602
4.5	.653
5.0	.699
6.0	.778
7.0	.845
8.0	.903
9.0	.954
10.0	1.000

You will probably recall that this same tabulation can be used to obtain the logarithm for any number larger than 10 or smaller than 1. One merely expresses the number as the product of a number between 1 and 10, and 10 raised to the appropriate whole-number exponent. When the number is written in this form, its logarithm is easily specified.

$$\log 5.0 = 0.699$$
$$\log 50. = \log(5.0 \times 10^1) = \log 5.0 + \log 10$$
$$= 0.699 + 1.000 = 1.699$$
$$\log 500. = \log(5.0 \times 10^2) = \log 5.0 + \log 10^2$$
$$= 0.699 + 2.000 = 2.699$$
$$\log 0.05 = \log(5.0 \times 10^{-2}) = \log 5.0 + \log 10^{-2}$$
$$= 0.699 - 2.000 = -1.301$$

Logarithms are a convenient time-saver in solving problems, particularly those which involve large numbers. Because of the properties of exponents, numbers can be multiplied by adding their logarithms and divided by subtracting their logarithms. To obtain the answer, one then looks up the antilogarithm of the result. Readers desiring more detailed information on the use of logarithms or a log table expressed to more places are referred to mathematics books or compilations of mathematical tables and formulas.

The significance of figures

One of the great strengths of the physical sciences is the fact that scientists often know how good (or how bad) their data are. And, since these data are commonly expressed in numerical form with appropriate units appended, the numerical values become indications of experimental precision. Understandably, the way in which this is done is a topic of considerable importance in the laboratory science of chemistry.

Consider, for example, a chemist who seeks to determine the density (mass per unit volume) of a liquid. He begins by placing some of the liquid in an empty cylinder with volume graduations marked on its side. Using a simple, rough balance which weighs to the nearest tenth of a gram, he finds that the mass of the cylinder and its contents is 46.2 g. Assuming that the balance is operating correctly, this reading indicates that the actual weight of the object is closer to 46.2 g than to 46.1 or 46.3 g. We say that the mass is expressed to three *significant figures*. It would be futile to claim any greater precision for this balance or its readings. Thus, it would be misleading to write 46.20 g instead of 46.2 g. The two numbers are *not* identical. The former value implies that the mass is known to the nearest 0.01 g. But a mass of 46.2 g could in fact be 46.24 or 46.17, or any other number which rounds off to 46.2.

One obviously cannot improve the precision of a weighing by arbitrarily adding zeros to the answer. That can only be achieved by using a more sensitive balance. Therefore, the chemist places his graduated cylinder on an analytical balance capable of weighing to the nearest tenth of a milligram, that is, to the nearest 0.0001 g. He finds that this balance reads 46.1907 g, a value which has six significant figures. The chemist also notes that the level of liquid in the cylinder indicates a volume of 8.5 ml, expressed to the highest precision obtainable from the graduations.

He then empties the cylinder, washes and dries it thoroughly, and weighs it. This time only the rough balance is convenient, and he finds that it indicates a mass of 36.8 g. He now must subtract this mass from the initial mass of the cylinder plus the liquid in order to obtain the mass of the liquid itself. Note, however, that there is no point in carrying along all six significant figures in the combined mass of the cylinder and the liquid. The results of a chemical calculation can be no more precise than the least precise piece of data included. And here that dubious distinction goes to the mass of the empty cylinder. To write

$$
\begin{array}{lr}
\text{mass of cylinder plus liquid} & = 46.1907 \text{ g} \\
-\text{ mass of cylinder} & = 36.8 \quad\ \text{ g} \\
\hline
\text{mass of liquid} & = \ 9.3907 \text{ g}
\end{array}
$$

creates an air of precision which is completely unjustified. It may seem less impressive, but it is far more honest to express the two original masses to three significant figures and their difference to two:

$$
\begin{array}{lr}
\text{mass of cylinder plus liquid} & = 46.2 \text{ g} \\
-\text{ mass of cylinder} & = 36.8 \text{ g} \\
\hline
\text{mass of liquid} & = \ 9.4 \text{ g}
\end{array}
$$

Of course, if the chemist really wished to know the mass of the liquid to five significant figures, he could weigh the empty cylinder to the nearest tenth of a milligram on a more precise balance and subtract this answer.

mass of cylinder plus liquid = 46.1907 g
— mass of cylinder = 36.8145 g
mass of liquid = 9.3762 g

But in this particular experiment there is little point in doing so. Recall that the desired property is the density, and in order to get it, one divides a known mass of substance by the known volume it occupies. Here the volume of the liquid is known only to a precision reflected in the two significant figures of 8.5 ml. Therefore, a more precise mass is of no value in the calculation. The correct computation is simply

$$\text{density} = \frac{9.4 \text{ g}}{8.5 \text{ ml}} = 1.1 \text{ g/ml.}$$

Note one other thing. A chemist is not allowed to "improve" the precision of his data by mathematical manipulations (aside from certain legitimate operations such as averaging many results). To be sure, one could go on dividing 9.4 g by 8.5 ml from now until Friday, but it would be a colossal waste of time. Indeed, the result, 1.1058823529 . . . g/ml would incorrectly imply a degree of precision totally incompatible with the experiment. For this case, the density can be justifiably expressed to only two significant figures.

In preparing the examples and problems for this book, I have attempted to keep the principles of significant figures well in mind. Obviously, such attention to indicators of experimental precision is even more important in the laboratory. Nevertheless, you are urged to exhibit similar care in working problems. To aid you, this final observation might be useful. For large numbers written out in extended form, ambiguities sometime arise over the number of significant figures intended. For example, the precision with which 86,400 g is expressed is not clear from this writing. One or both of the two zeros could be significant figures with real experimental meaning; or, they could be included only to fix the decimal point. A way around this uncertainty is to use exponential notation. Thus, 8.64×10^4 g has three significant figures, 8.640×10^4 g has four, and 8.6400×10^4 g has five. Negative powers of 10 are also useful for reminding readers that zeros to the right of the decimal point and to the left of the first nonzero digit are not significant figures. For example, the first two zeros to the right of the decimal point in 0.0020 g only serve to locate the decimal. But the last zero does indicate the precision of the measurement. Again, the fact that this is a number expressed to two significant figures is apparent in the form 2.0×10^{-3} g.

Appendix 2   Measure for measure

The metric system: from micro to mega

The metric system was a child of the French Revolution and the spirit of reason and optimism which ushered in the age. Ironically, it was the same age, but not the same spirit, which brought the founder of modern chemistry to the guillotine. The great Lavoisier was himself a contributor to the study of weights and measures submitted to the French National Assembly by the Paris Academy of Sciences in 1791. The project was an answer to the worldwide mess of wildly divergent units, some of them based on such irreproducible and improbable standards as the length of barley corns and royal appendages. The new system was to be rationally based upon the properties of the earth and common substances, thus permitting, at least in principle, the experimental redetermination of the units anywhere. There were only to be a few of these units, and, insofar as possible, they were to be logically related to one another. Moreover, the system was designed to be decimal in nature, thus greatly simplifying conversions and calculations. No wonder, then, that Talleyrand saw in the metric system an "enterprise whose result should belong some day to the whole world." In the nearly two centuries that have elapsed, such universal acceptance has nearly been attained. Of all the major industrial nations of the world, only the United States still remains conspicuously outside the metric community.

The basic unit of the system is the meter, originally intended to be one ten-millionth (10^{-7}) of the distance along a meridian following the curvature of the earth from the north pole to the equator by way of Paris. In other words, the circumference of the earth was set at 4×10^7 meters. The actual determination was done by a group of surveyors, working in France and Spain, who carefully measured the distance of arc subtended by a known angle. The calculated meter (just over a yard) was then transferred to a bar of platinum which thus became the official standard of length. The bar is still in the International Bureau of Weights and Measures near Paris, but it no longer defines the meter. As you have read or will read in Chapter 6, the new standard is the wavelength of light emitted by krypton-86. This length can be measured to within one part in 10^8, a much higher degree of precision than that obtainable from a metallic bar subject to expansion and contraction as the temperature changes.

Once the meter was adopted as the fundamental unit of length, it of course became possible to specify a unit of area (the square meter) and a unit of volume (the cubic meter). Moreover, the meter also became the basis for the standard unit of mass, the kilogram. This connection was achieved via the properties of water. The kilogram was defined as the mass of one cubic decimeter of distilled water at the temperature of its maximum density (4°C). This volume corresponds to a cube, one-tenth of a meter on each side. It is also known as a liter and is approximately equal to 1 quart. A kilogram weighs roughly 2.2 pounds.

The metric system involves many multiples and submultiples of these basic units of length and mass. But in all cases, the multipliers are simple powers of 10. Therefore, the exponential notation discussed in Appendix 1 is admirably suited for use here. Moreover, the values of the multipliers are indicated by prefixes. The most commonly used are tabulated below.

Multiplier	Prefix	Abbreviation	Multiplier	Prefix	Abbreviation
10^1	deka-	dk	10^{-1}	deci-	d
10^2	hecto-	h	10^{-2}	centi-	c
10^3	kilo-	k	10^{-3}	milli-	m
10^6	mega-	M	10^{-6}	micro-	μ
10^9	giga-	G	10^{-9}	nano-	n

The use of these prefixes and the ease with which one changes metric units by moving the decimal point can be illustrated with the volume of water used to define the kilogram (kg). Recall that it corresponds to a cube 0.1 or 10^{-1} meter (m) on a side. As the table indicates, the multiplier, 10^{-1}, is designated by the prefix *deci-*. Therefore,

1 decimeter (dm) = 10^{-1} m.

Cube both sides of this identity, and it follows that

1 dm^3 = $(10^{-1}$ m$)^3$ = 10^{-3} m^3,

which is read, "1 cubic decimeter equals 10^{-3} cubic meter." But in chemistry, volume is more commonly expressed in cubic centimeters (cm^3 or cc). Because *centi-* stands for 10^{-2},

1 centimeter (cm) = 10^{-2} m = 10^{-1} dm.

In other words, there are 100 centimeters in each meter, and 10 centimeters in each decimeter. We can make use of this latter equivalence to calculate the number of cubic centimeters in a cubic decimeter. Merely start with the identity,

10 cm = 1 dm,

and cube both sides of the equation to get from one dimension into three

(10 cm)^3 = (1 dm)^3
10^3 cm^3 = 1 dm^3.

As the numbers indicate, 1000 cubic centimeters equal 1 cubic decimeter.

Recall, now, that 1 liter was identified as equal in volume to 1 dm^3. Since *milli-* means 10^{-3},

1 milliliter (ml) $= 10^{-3}$ liter.

Of course the same ratio is maintained if we multiply both sides of the equation by 10^3:

10^3 ml $= 1$ liter.

And, since 1 liter and 1 dm^3 are identically equal, 1 cm^3 and 1 ml are also equal. We can therefore observe that by definition, 1 dm^3, 1000 cm^3, 1 liter, and 1000 ml represent exactly the same volume. Moreover, this volume of distilled water weighs exactly 1 kg at 4°C. From the table of prefixes, it is apparent that there are 1000 of the widely used 1 gram (g) units in 1 kg.

10^3 g $= 1$ kg

And, since 1 ml is the thousandth part of a liter, exactly 1 ml of water must weigh exactly 1 g. In other words, the maximum density of liquid water is exactly 1 gm/ml or 1 g/cm^3.

The frequency with which the word "exactly" appears in the last few paragraphs indicates that the equalities expressed there are matters of definition, not experiment. Therefore, there are no limits of precision upon statements such as

10^3 milligrams (mg) $= 1$ g $= 10^{-3}$ kg   or

10^3 millimeters (mm) $= 10^2$ cm $= 1$ m $= 10^{-3}$ kilometers (km).

And because the numbers are exact, the concept of significant figures (Appendix 1) does not apply. No limits to precision are introduced by a decimal conversion from one metric unit of length or mass to another. On the other hand, the relationship between metric units and other units is seldom expressed in exact powers of 10 and often involves experimentally determined values. Consequently, such translations are slightly more complicated—enough so to warrant the section which follows.

There'll be some changes made

The conversion factors and constants of nature tabulated at the end of this appendix are frequently employed by chemists and students of chemistry. They are collected here to facilitate easy access. The meaning and use of the constants are discussed in the text and there is no point in repeating this information. However, your attention is directed to the fact that these are experimentally determined quantities. The number of significant figures specified in the table does not necessarily indicate the maximum precision with which the value of any particular constant is known. The values cited correspond to those used in the text and represent sufficient precision for use in most calculations.

With a few exceptions, the conversion factors or unit equivalencies listed below are also experimentally obtained values. Several relationships, for instance, that between the angstrom and the centimeter, are exactly defined. But the number of centimeters which correspond to one inch is based upon an actual comparison of length. Similarly, most of the other equivalencies reported have also been determined by experiment. In most cases, the values are known to higher precision than is reflected in the three or four significant figures reported. But here again, the precision is adequate for most of the data found in the problems.

For many of you, the use of these conversion factors will present nothing new, and no particular difficulties. But some readers might find a brief review helpful. Besides, you may not have met the beautiful simplicity of the conversion factor method. To use it, one treats the expression relating two units as an equation—which is clearly what it is. Suppose, for illustration's sake, that I wish to calculate the number of centimeters in a foot. Without even thinking, I begin with an equation,

$$1 \text{ ft} = 12 \text{ in.}$$

This expression of equality exhibits all the ordinary properties of an equation, and therefore, I can divide both sides of it by "1 ft." The resulting ratio has a value of unity.

$$\frac{12 \text{ in.}}{1 \text{ ft}} = 1$$

Note the importance of writing down the units. It is patently obvious that 12/1 does *not* equal 1. But 12 in. and 1 ft are identical; they represent the same length. Consequently, *their* ratio does equal 1.

Now I recall from somewhere that I can multiply anything by 1 and still have that same anything—its value does not change. So, with this justification, I multiply:

$$1 \text{ ft} \times \frac{12 \text{ in.}}{1 \text{ ft}} = 12 \text{ in.}$$

You see that I have canceled the "ft" in the numerator and denominator (units cancel just as easily as numbers) and am left with 12 in. Admittedly, this is not exactly news. You and I have known for quite a while that the way to find the number of inches in any number of feet is to multiply the number of feet by 12, or, to be a good deal more exact, by 12 in./ft.

This ratio is a conversion factor, and we have been using it, and others like it, for years, whether or not we've thought about the process in those terms. But sometimes, it does help to think a bit about conversion factors, particularly if the units involved are a little unfamiliar. I again illustrate the procedure with the second step of the determination of the number of centimeters in one foot. Now I need to know the number of centimeters in an inch, and the table provides the necessary equation,

$$1 \text{ in.} = 2.54 \text{ cm.}$$

Once more, I rearrange this into a conversion factor which, as always, is equal to 1:

$$\frac{2.54 \text{ cm}}{1 \text{ in.}} = 1.$$

The ratio simply states the fact that there are 2.54 centimeters per inch. Therefore, to determine the number of centimeters in 12 inches, I multiply 12 inches by this factor:

$$12 \text{ in.} \times \frac{2.54 \text{ cm}}{1 \text{ in.}} = 30.48 \text{ cm.}$$

This time the "inch" units cancel, and I am left with 30.5 cm (to three significant figures). The length is still equal to 12 in.; I have done nothing to change its magnitude. However, I have succeeded in changing the units in which this length is expressed. In fact, I could have made the transformation from feet to centimeters by combining the two steps into one:

$$1 \text{ ft} \times \frac{12 \text{ in.}}{1 \text{ ft}} \times \frac{2.54 \text{ cm}}{1 \text{ in.}} = 30.5 \text{ cm.}$$

Exactly the same approach can be used on any conversion problem, including those easy ones involving only metric units. First determine what the change in units is to be, and then select the appropriate equation or equations which relate the units. Set each of these up as a ratio or conversion factor in such a way that the unwanted units will cancel when the original measurement is multiplied by the factor or factors. Thus, to calculate the number of inches in a meter stick, use the following sequence of steps:

$$1 \text{ m} \times \frac{100 \text{ cm}}{1 \text{ m}} \times \frac{1 \text{ in.}}{2.54 \text{ cm}} = 39.4 \text{ in.}$$

Writing the names of the units as you go along may seem like a nuisance. But this practice immediately provides an indication of whether or not your conversion is properly set up. If you do not cancel everything but your desired units, you are obviously using conversion factors incorrectly. Hopefully, that will be an infrequent occurrence.

How hot is hot?

The relationship between the Fahrenheit and the Centigrade or Celsius temperature scales deserves a few words because it does not quite fit the conversion factor formula. The two scales are commonly compared at the normal freezing and boiling points of pure water. The former frigid event occurs at 32°F or 0°C. And water boils at a temperature called 212°F or 100°C, depending upon your nationality or the extent of your scientific training. These two points are sufficient to permit the derivation of a general formula for the conversion of degrees Fahrenheit into degrees Centigrade,

$$°C = \frac{5}{9} [°F - 32].$$

And, with a little rearrangement, the equation can be turned around to accomplish the opposite transformation, a useful capability if you ever encounter a European who complains that the temperature is 37.8° in the shade.

$$°F = \frac{9}{5} \times °C + 32 = 1.8 \times °C + 32$$

The relationship between the two temperature scales can be schematically summed up by these two thermometers.

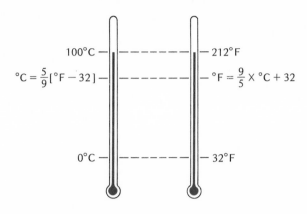

Conversion factors

Ångstrom	$1 \text{ Å} = 1 \times 10^{-8} \text{ cm} = 1 \times 10^{-1} \text{ nm}$
Atmosphere	$1 \text{ atm} = 760 \text{ mm Hg}$
	$1 \text{ atm} = 1034 \text{ g/cm}^2$
Calorie	$1 \text{ cal} = 4.18 \text{ joules}$
	$= 4.18 \times 10^7 \text{ ergs}$
Centimeter	$1 \text{ cm} = 0.394 \text{ in.}$
Degrees Celsius (Centigrade)	$°C = \frac{5}{9}[°F - 32]$
Degrees Fahrenheit	$°F = \frac{9}{5} \times °C + 32$
Degrees Kelvin (Absolute)	$°K = °C + 273$
Erg	$1 \text{ erg} = 1 \times 10^{-7} \text{ joule}$
	$1 \text{ erg} = 2.39 \times 10^{-8} \text{ cal}$
Faraday	$1 \text{ } \mathfrak{F} = 96,500 \text{ coulombs}$
	$1 \text{ } \mathfrak{F} = 6.023 \times 10^{23} \text{ electrons}$
Inch	$1 \text{ in.} = 2.54 \text{ cm}$
Joule	$1 \text{ j} = 1 \times 10^7 \text{ ergs}$
	$1 \text{ j} = 0.239 \text{ cal}$
Kilocalorie	$1 \text{ kcal} = 4180 \text{ joules}$
	$1 \text{ kcal} = 41.3 \text{ liter atm}$
Kilogram	$1 \text{ kg} = 2.21 \text{ lb}$
Kilometer	$1 \text{ km} = 0.621 \text{ mile}$
Liter	$1 \text{ liter} = 1.057 \text{ qt}$
Liter atmosphere	$1 \text{ liter atm} = 24.2 \text{ cal}$
Meter	$1 \text{ m} = 39.4 \text{ in.}$
Pound	$1 \text{ lb} = 454 \text{ g}$
Quart	$1 \text{ qt} = 0.946 \text{ liter}$
Mile	$1 \text{ mile} = 1.609 \text{ km}$

Constants

Acceleration due to gravity	$g = 981$ cm/sec^2
Avogadro's number	$N = 6.023 \times 10^{23}$
Electronic charge	$e = 1.6021 \times 10^{-19}$ coulomb
	$e = 4.80 \times 10^{-10}$ electrostatic unit (esu)
Electronic mass	$m = 9.1091 \times 10^{-28}$ g
Gas constant	$R = 0.0821$ liter atm/(mole °K)
	$R = 82.1$ ml atm/(mole °K)
	$R = 62.3$ liter mm Hg/(mole °K)
	$R = 8.31 \times 10^7$ ergs/(mole °K)
	$R = 8.31$ joules/(mole °K)
	$R = 1.99$ cal/(mole °K)
Planck's constant	$h = 6.625 \times 10^{-27}$ erg sec
Pi	$\pi = 3.1417$
Speed of light	$c = 2.998 \times 10^{10}$ cm/sec

Appendix 3 What's in a name?

Chapter 4 of this book contains a lengthy example of how the formulas of chemical compounds are experimentally determined. But it does not discuss the rules for naming compounds. This appendix is a brief summary of some of the more important of these conventions.

Inorganic compounds

Many inorganic compounds consist of one metallic element and one nonmetal. The names given to these compounds reflect the identity of their constituent elements. The name of the metal comes first, just as the symbol of the metal comes first in the formula. The nonmetal is named last, but with a slight modification. Instead of merely writing the name of the element, its normal ending is replaced with the suffix -*ide*. Thus, the compound of sodium and fluorine, NaF, is called sodium fluoride. Similarly, magnesium and oxygen form magnesium oxide, MgO.

In Chapter 9, these specific compounds are identified as ionic in nature. They are made up of positively charged cations, formed when atoms of metallic elements lose electrons, and negatively charged anions which arise when atoms of nonmetals acquire electrons. The ions are named in the manner just described. A sodium atom minus an electron is represented by Na^+ and is known as the sodium ion, Mg^{2+} is the magnesium ion, and Al^{3+} is the aluminum ion. On the other hand, F^- is referred to as the fluoride ion, Cl^- is the chloride ion, O^{2-} is the oxide ion, and S^{2-} is the sulfide ion.

These same rules of nomenclature apply even if the compound in question does not actually exist in ionic form. For example, $AlCl_3$ is not an ionic substance (its bonding is chiefly covalent), but it nevertheless is called aluminum chloride. And even compounds of two nonmetallic elements are named according to this convention. Thus, the gaseous compound of hydrogen and chlorine, HCl, is known as hydrogen chloride, H_2S is hydrogen sulfide, and H_2O is hydrogen oxide (though it is seldom called that).

Since it is fairly common for pairs of nonmetals to combine in more than one atomic ratio, it is necessary to be able to distinguish these differences with names as well as formulas. For instance, it would obviously not do to call both compounds of carbon and oxygen, carbon oxide. Potential ambiguities such as this are usually resolved by making use of Greek prefixes: *di-* for two, *tri-* for three, *tetra-* for four, *penta-*

for five, and so on. Thus, CO, the compound in which the carbon: oxygen atomic ratio is 1:1 is called carbon monoxide. But CO_2 is carbon dioxide. Similar considerations apply to the five oxides of nitrogen listed in Table 3.1.

N_2O	dinitrogen oxide or nitrous oxide
NO	nitrogen oxide or nitric oxide
N_2O_3	dinitrogen trioxide
NO_2	nitrogen dioxide
N_2O_5	dinitrogen pentoxide

You will note from this tabulation that some widely used common names do not conform to this convention. The names nitrous oxide and nitric oxide reflect another method of nomenclature sometimes also used with metals which exist in more than one valence state. The -*ous* ending identifies the lower oxidation number or ionic charge; the -*ic* suffix represents the higher oxidation number. According to this rule, Cu^+ is the cuprous ion, Cu^{2+} is the cupric ion. The chloride of the former ion, CuCl, is consequently called cuprous chloride, while $CuCl_2$ is cupric chloride. It is currently more common to designate the oxidation state of the metal with a Roman numeral in parentheses. Thus, Cu^+ is known as the copper(I) ion, and CuCl becomes copper(I) chloride. Similarly, Cu^{2+} is the copper(II) ion and $CuCl_2$ is copper(II) chloride.

Many inorganic compounds involve more than just two elements. Typically, these are arranged in polyatomic ions which have, of course, been given names. For example, NH_4^+ is the ammonium ion, a cation with properties similar to those of the alkali metal ions. The compound, NH_4Cl, ammonium chloride, thus resembles sodium chloride and other salts of the Group 1A elements. Some of the more common of the oxygen-containing polyatomic anions are listed below. Certain of these oxyions appear, along with their structures, in Table 9.1.

NO_3^-	nitrate	PO_4^{3-}	phosphate
NO_2^-	nitrite	CO_3^{2-}	carbonate
SO_4^{2-}	sulfate	HCO_3^-	bicarbonate or hydrogen carbonate
SO_3^{2-}	sulfite	OH^-	hydroxyl

You will observe that in this tabulation two oxyions are listed for both nitrogen and sulfur. Note their formulas. Whenever such a situation arises, that is, whenever one element forms two or more oxyions, the ions always differ in the number of oxygen atoms which they contain. Moreover, this difference is recognized in the ending of the name given to the ion. When there are only two ionic species, the -*ate* suffix identifies the ion with the relatively larger number of oxygen atoms, for example, the nitrate ion, NO_3^-. The -*ite* ending is applied to the ion which includes the smaller number of oxygen atoms, the nitrite ion, NO_2^-, for instance. If there are more than two distinguishable polyatomic ions involving the same ions, additional means of semantic distinction are introduced in the form of prefixes. We will forego that introduction and instead consider briefly the compounds of the polyatomic anions.

All these ions exist in chemical combination with either positive metallic ions or hydrogen ions. In the former case, the resulting compounds are named according to the "cation anion" convention. Thus, KNO_3 is potassium nitrate, $MgSO_4$ is magnesium sulfate, and $CaCO_3$ is calcium carbonate. Although the latter substance happens to be insoluble in water, many of these compounds do dissolve and dissociate into their constituent ions. In other words, they are electrolytes. A somewhat special case is the class of compounds containing metal cations and hydroxyl anions. These hydroxides—NaOH, (sodium hydroxide) is a familiar example—impart basic or alkaline properties to solutions.

Of course, an H^+ and an OH^- form HOH or H_2O, but the association of most negative oxyions with hydrogen ions yields acids. Frequently, this association is fairly weak. The H^+ ions are free enough to give characteristic acidic properties to the compounds and their aqueous solutions (see Chapter 11). These oxyacids are named after the anions, but another modification occurs at the ends of the name. Acids containing the *-ate* ion end in *-ic*. Thus, HNO_3 is nitric acid and H_2SO_4 is sulfuric acid. But if the polyatomic ion present is an *-ite* ion, the corresponding acid has *-ous* as a suffix. Accordingly, HNO_2 is nitrous acid and H_2SO_3 is sulfurous acid. Oxygen-free acids arising from aqueous solutions of binary compounds of hydrogen and other nonmetal also end in *-ic*, but they are preceded by the prefix *hydro-*. For example, a water solution of hydrogen chloride gas, HCl, is called hydrochloric acid.

Organic compounds

As you might expect, the plethora of carbon-containing compounds and the many options introduced by isomerism make the naming of organic compounds exceedingly complex. The task of translating a three-dimensional structure into a few words is a formidable one. And a structural formula is undoubtedly subject to less potential misinterpretation than a name. But names are convenient, and for that reason, international conferences have been convened to discuss and devise uniform systems of nomenclature. Such conventions (about conventions!) have undoubtedly promoted the communication of chemists the world over by creating a more-or-less universal language. But there still remain numerous synonyms in common usage.

In this very brief survey, I can consider only the commonest of organic compounds. Nor will I attempt to strictly adhere to the international rules. My aim is merely to set down a few of the conventions used in naming certain of the compounds mentioned in Chapter 9, Chapter 15, and elsewhere in the text. For more comprehensive information you must consult sources like the "Handbook of Chemistry and Physics," Robert C. Weast, editor, published every year in a new edition by The Chemical Rubber Co., Cleveland.

The number of carbon atoms in a single molecule of the compound in question is designated by a prefix which takes on various endings, depending upon the chemical nature of the substance. The first ten of these roots are included in the following list.

Carbons	Prefix	Carbons	Prefix
1	*meth-*	6	*hex-*
2	*eth-*	7	*hept-*
3	*prop-*	8	*oct-*
4	*but-*	9	*non-*
5	*pent-*	10	*dec-*

For alkanes, saturated hydrocarbons with the general formula C_nH_{2n+2}, the appropriate ending is *-ane*. Thus, the simplest member of this series, CH_4, is methane; C_2H_6, or CH_3CH_3, is ethane, and so on. If the hydrocarbon contains a single double bond, the compound is an alkene and conforms to the general formula C_nH_{2n}. The accepted suffix is *-ene*, so C_2H_4 or CH_2=CH_2 should be ethene. Instead it goes by its traditional name, ethylene. Propene or propylene is C_3H_6 or CH_3CH=CH_2. The ending *-yne* identifies an alkyne, a hydrocarbon with one triple bond and the general formula C_nH_{2n-2}. Thus, C_3H_4, or CH_3C≡CH, is named propyne. But the common name, acetylene, is retained for C_2H_2, or CH≡CH, the most familiar alkyne.

To name compounds containing various functional groups, the root words are converted into adjectives by adding the suffix *-yl*. This ending identifies a hydrocarbon group or radical, not a molecule. In effect, these radicals are the corresponding alkanes minus one hydrogen atom. For example, CH_3— is the methyl group or radical, C_2H_5— is the ethyl group, and so on. This adjective modifies the general name of the class of compounds, just as the radical modifies the functional group charac-

teristic of the class. Thus, CH_3OH is methyl alcohol, $CH_3\overset{\displaystyle O}{\overset{\displaystyle \|}{C}}C_2H_5$ is methyl ethyl ketone, C_2H_5Cl is ethyl chloride, and CH_3NH_2 is methylamine.

The endings employed for organic acids and related compounds are the same as those used for comparable inorganic substances. The suffix *-ic* indicates an acid and *-ate* identifies the ion formed from the acid by the loss of a hydrogen ion. For example, $CH_3\overset{\displaystyle O}{\overset{\displaystyle \|}{C}}OH$ is acetic acid and $CH_3\overset{\displaystyle O}{\overset{\displaystyle \|}{C}}O^-$ is the acetate ion. Compounds including this ion bear its name, so $CH_3\overset{\displaystyle O}{\overset{\displaystyle \|}{C}}ONa$ is sodium acetate. Similarly, $CH_3\overset{\displaystyle O}{\overset{\displaystyle \|}{C}}OCH_3$ is called methyl acetate, although esters do not dissociate into ions. The one-carbon acid, $H\overset{\displaystyle O}{\overset{\displaystyle \|}{C}}OH$, is named formic acid, after the Latin word for ant, a source of the substance. But the other organic acids and ions follow the pattern based upon the previous list of prefixes. Thus, $C_2H_5\overset{\displaystyle O}{\overset{\displaystyle \|}{C}}OH$ is propionic acid and $C_2H_5\overset{\overset{\displaystyle O}{\|}}{C}O^-$ is the propionate ion. Aldehydes are named after the corresponding acid, so $CH_3\overset{\overset{\displaystyle O}{\|}}{C}H$ is acetaldehyde.

Obviously, there are some glaring omissions in this cursory account of organic nomenclature. If you have already read Chapter 9, you are well aware that specifying the number of atoms in a molecule is often not enough to unambiguously identify a compound. The way in which the atoms are arranged is also a significant variable which influences the properties and, hence, the identity of a substance. For example, the atoms of butane, C_4H_{10}, can be bound together in two distinct molecular structures or isomers. Similarly, differences in the branching of the carbon skeleton give rise to three isomers of pentane, five hexanes, and so on. For the same reason, propyl, butyl, and all higher hydrocarbon radicals exist in more than one structural form. As a consequence, any compound containing these groups can exhibit a variety of atomic arrangements. Moreover, the placement of a double or triple bond is yet another structural variable which influences the number of isomeric alkenes and alkynes and their derivatives.

Geometrical isomerism about double bonds, substitutional isomerism in benzene rings, and optical isomerism arising from asymmetrically substituted carbon atoms require still more rules. Unfortunately, a summary of the system used to bring order to the complexities of structural isomerism cannot be justified in a book of this sort. I think it is more important that you acquire a general understanding of isomerism through a careful study of the examples in Chapters 9 and 15.

Appendix 4 For thought and action: Answers to problems

Chapter 1

The questions posed in this chapter (and in many of the others) have no pat right or wrong answers. That makes them challenging and, I hope, interesting. I will not presume to offer definitive answers where there may be none. But I believe it is vitally important that we all think—and think hard—about some of these issues.

Chapter 2

2.1   By and large, I think alchemy was a protoscience because it attempted to explain nature by a reasonably coherent system.

2.2   The inertness of gold, as exemplified by its reluctance to react with oxygen, means that it can exist in pure metallic form without undergoing chemical combination with the elements around it. On the other hand, because sodium reacts so readily with oxygen and other elements, it is never found in an uncombined state.

2.3   Phlogiston Theory:

Calx of copper + Carbon → Copper + Fixed air

Carbon, which is flammable, transfers its phlogiston to the calx of copper which, as a result, is converted into the calcinable metal.

Oxygen Theory:

Copper oxide + Carbon → Copper + Carbon dioxide

Carbon reacts with the oxygen in the copper oxide, forming an oxide of carbon (carbon dioxide) and liberating the element copper.

2.4   Priestley:

Iron → Calx of iron + Phlogiston

Lavoisier:

Iron + Oxygen → Iron oxide

2.5   Lead oxide, tin oxide, silver oxide

2.6   Try burning sulfur in the absence of air. Then burn it in the presence of air and investigate what you get. Are any new substances formed? Is the amount of oxygen reduced? Is nitrogen still present in the same amount?

2.7   The source of the sulfur dioxide is the sulfur which occurs naturally in many deposits of coal and petroleum. Some solutions are suggested in Chapter 11.

2.8   (a) Compound   (b) element   (c) compound
    (d) mixture   (e) mixture   (f) element   (g) compound
    (h) element   (i) mixture   (j) compound

2.9   Weighing hot and cold objects, and objects in the light and in the dark; trying to form specific compounds containing heat and light. (Chapter 12 includes a discussion of Count Rumford's experiments on the weight of caloric.)

Chapter 3

3.1 H_2O_2; molecular weight = 34 (approximately)
    $H_2 + O_2 \rightarrow H_2O_2$

3.2   Atomic weight of nitrogen = 4.68 (assuming the atomic weight of hydrogen = 1.00)
    $N_2 + 3H_2 \rightarrow 2NH_3$

3.3   GEW = 7.0 g

3.4 $N_2 + 2O_2 \rightarrow 2NO_2$
    1 vol + 2 vol → 2 vol

3.6   Compound A: $\dfrac{\text{wt iron}}{\text{wt oxygen}} = 3.50$

         Compound B: $\dfrac{\text{wt iron}}{\text{wt oxygen}} = 2.33$

    $\dfrac{3.50}{2.33} = \dfrac{3}{2}$

3.7   Grain equivalent weight = 95 grains; HgO

3.8   (a) 6.25   (b) 112.5

3.9   (a)–(c) 6.023×10^{23} atoms

3.10   (a) 5.98×10^{24} atoms   (b) 9.48×10^{22} atoms
    (c) 2.91×10^{22} atoms

3.11   (a) 13.9 moles   (b) 8.37×10^{24} molecules   (c) 1.67×10^{25} atoms of hydrogen, 8.37×10^{24} atoms of oxygen

3.12 3.27×10^{-22} g

3.13 9.27×10^{-14} g

3.14   207

3.15   (a) 16   (b) 70.7   Chlorine usually exists as diatomic molecules.

3.16   (a) Approximate atomic weight = 97   (b) GEW = 15.9 g
    (c) Exact atomic weight = 95.4

3.17   Atomic weight of carbon = 12.0; atomic weight of oxygen = 16.0; oxide A is CO (carbon monoxide); oxide B is CO_2 (carbon dioxide).

3.18 Using modern atomic weights, "olefiant gas" is CH_4 and "carbonic oxide" is CO_2.

Chapter 4

4.1 (a) 108.0   (b) 154.0   (c) 32.0   (d) 342.0   (e) 211.0
(All molecular weights based on atomic weights expressed to the nearest tenth of a gram.)

4.2 (a) 25.9% N, 74.1% O
(b) 7.8% C, 92.2% Cl
(c) 37.5% C, 12.5% H, 50.0% O
(d) 42.1% C, 6.4% H, 51.5% O
(e) 62.6% C, 8.1% H, 22.7% O, 6.6% N

4.3 (a) N_2O (b) NaCl   (c) H_2SO_4 (d) CH_4N_2O

4.5 $2HgO \rightarrow 2Hg + O_2$

4.6 HO is the simplest formula.

4.7 CH is the simplest formula; C_6H_6 is the true molecular formula.

$$C_6H_6 + \frac{15}{2}O_2 \rightarrow 6CO_2 + 3H_2O$$

or

$$2C_6H_6 + 15O_2 \rightarrow 12CO_2 + 6H_2O$$

4.8 Not all of the oxygen in the H_2O and CO_2 formed comes from the compound. Some of it comes from the atmosphere in which the sample is burned.

4.9 C_2H_6O or C_2H_5OH

4.10 (a) N_2 $+ 3H_2$ $\rightarrow 2NH_3$
(b) $2HCl$ $+ Ba(OH)_2 \rightarrow BaCl_2$ $+ H_2O$
(c) $2HNO_3 + Zn$ $\rightarrow Zn(NO_3)_2 + H_2$
(d) $C_6H_{12}O_6 + 6O_2$ $\rightarrow 6CO_2$ $+ 6H_2O$

4.11 3.09 g CO_2; 1.42 g H_2O

4.12 (a) $C_3H_8 + 5O_2 \rightarrow 3CO_2 + 4H_2O$
(b) 3 moles CO_2, 1.5 moles CO_2, 15 moles CO_2
(c) 8 moles H_2O, 4.82×10^{24} molecules
(d) 3.64 g O_2. If less than this amount of oxygen were present, the combustion would be incomplete and some CO would probably be formed.

4.13 (a) 76.7 g salicylic acid and 33.3 g acetic acid
(b) 309 tablets

4.14 (a) 0.70 ton of iron   (b) 0.70 ton of iron   (c) 0.62 ton of iron   (d) In parts (a) and (b) there is an excess of coke; iron oxide is present in the limiting equivalent amount; in (c) the limiting reactant is the carbon.

4.15 2.73×10^{26} atoms

4.16 80 lb of sulfur

4.17 (a) 64.1% lead   (b) 1.36 kg of lead

4.18 67,100

Chapter 5

5.1 $CaO + CO_2(g) \rightarrow CaCO_3(s)$

5.2 $MgCO_3(s) \rightarrow MgO(s) + CO_2(g)$

5.3 $CaCO_3$ plus acid liberates CO_2 but CaO plus acid does not yield this, or any other gas. The equations for these reactions are, respectively.

$$CaCO_3 + 2H^+ \rightarrow Ca^{2+} + H_2O + CO_2(g)$$

and

$$CaO \quad + 2H^+ \rightarrow Ca^{2+} + H_2O$$

(See Chapters 9 and 11 for more information on ions and acids.)

5.4   (a) LiOH   (b) NaBr   (c) CsCl   (d) BaO   (e) $SrCO_3$

5.5   (a) Potassium oxide   (b) barium chloride   (c) rubidium carbonate   (d) sodium iodide   (e) barium hydroxide

5.6   (a) $2Na + 2H_2O \rightarrow 2NaOH + H_2$
      (b) $4K + O_2 \quad \rightarrow 2K_2O$
      (c) $Ca + 2H_2O \rightarrow Ca(OH)_2 + H_2$
      (d) $2BaO \quad \rightarrow 2Ba + O_2$

5.7   Sodium reacts violently with water.

5.8   You can't believe everything you read in the papers!

5.9   The magnesium is precipitated as solid $Mg(OH)_2$ by adding a soluble hydroxide to the seawater. The element is subsequently isolated from its hydroxide by electrolysis.

5.10   This is a tough one! Most other explanations, for example, differences in isotopic distribution, are not very plausible.

5.11   The smaller helium atoms diffuse more easily and more rapidly through the pores of the balloon. (See Chapter 10.)

5.12   It prevents the filament from burning up, as it would in an oxygen-containing atmosphere.

5.13   Inertness. The metal should not rust and the gas should not react with its contents.

5.14   On the basis of periodic properties, Mendeleev could tell with considerable certainty that some elements were missing from the body of his table. Because *all* the rare gases were undiscovered, he had no way of knowing any of them were absent. Our current knowledge of atomic structure (Chapter 8) makes it very unlikely that any elements have been omitted from the body of the periodic table. However, there may be some elements at the end of the table which are not yet discovered, or better said, not yet made. All the elements beyond uranium are man-made. (See Chapter 16.)

5.15   Atomic weight = 23; density = 0.70 g/cm³;
      melting point = 122°C; boiling point = 1044°C

5.16   (a) Metal   (b) nonmetal   (c) nonmetal   (d) metal
      (e) metal

5.17   (a) 4   (b) 6   (c) 5   (d) 3   (e) 7   (f) 1
      (g) 8   (h) 2

Chapter 6

6.1   (a) 4.04720 × 10⁻⁵ cm   (b) 404.720 nm
        (c) 1.59 × 10⁻⁵ in.   (d) 7.408 × 10¹⁴ sec⁻¹

6.2   500 sec or 8.33 min

6.3   4.06 × 10¹³ km or 2.52 × 10¹³ miles

6.4   Remember, matter must interact with energy in some fashion.

6.5   1.098 × 10²⁷ electrons

6.6   9.33 × 10⁻² faraday or 5.62 × 10²² electrons

6.7   The biologically harmful effects of radiation. (See Chapter 16.)

6.8   Metallic objects absorb X rays and thus show up in X-ray photographs.

6.9   The electrical charge and bulk of an α particle are greater than the comparable properties of a β particle.

6.10   (a) 0.50 mg   (b) 0.25 mg   (c) 0.125 mg

6.11   49.2 years

6.12   (a) Helium-3, ³He   (b) radon-222, ²²²Rn
        (c) radium-224, ²²⁴Ra   (d) uranium-234, ²³⁴U

6.13   (a) ⁴⁰Ca   (b) ²¹⁸Po   (c) ²³⁸U   (d) β

6.14   (a) ¹⁴C → ¹⁴N + β
        (b) ⁹⁰Sr → ⁹⁰Y + β
        (c) ²³⁸U → ²³⁴Th + ⁴α

6.15   Strontium is chemically similar to calcium and consequently it is also deposited in bone.

6.16   Because the identity of the parents was known, but the identity of the offspring was, at first, uncertain.

6.17   By making use of the constancy of radioactive decay rates. (More details are given in Chapter 16.)

Chapter 7

7.1   Coleridge's definition of beauty, which introduces Chapter 5, might bear on this question.

7.2   (a) 4   (b) 1   (c) 2   (d) 3

7.3   Good luck!

7.4   How about relevance—to experiment, that is?

7.5   (a) 6 electrons, 6 protons,   ? neutrons
        (b) 6 electrons, 6 protons,   6 neutrons
        (c) 11 electrons, 11 protons,   12 neutrons
        (d) 9 electrons, 9 protons,   ? neutrons
        (e) 82 electrons, 82 protons, 125 neutrons
        (f) 88 electrons, 88 protons, 138 neutrons

7.6   ¹⁶O: 8 electrons, 8 protons, 8 neutrons
        ¹⁷O: 8 electrons, 8 protons, 9 neutrons
        ¹⁸O: 8 electrons, 8 protons, 10 neutrons

7.7   HH, 2; HD, 3; HT, 4; DD, 4; DT, 5; TT, 6

7.8   (a) $_2^3$He   (b) $_{87}^{219}$Fr   (c) $_{-1}^{0}\beta$ (d) $_{84}^{210}$Po

7.9   By their color

7.10 Red, yellow, white, blue

7.11 (a) 3.03×10^{-12} erg   (b) red

7.12 3×10^{17} photons

7.13 Maybe because it changed their world, or at least their theories about it.

7.14 Emission spectra (see text for details)

7.15 8.49×10^{-8} cm or 8.49° Å

7.16 4.57×10^{14} sec^{-1}; 6.17×10^{14} sec^{-1}; 6.91×10^{14} sec^{-1}

Chapter 8

8.1

l	m_l	m_s	l	m_l	m_s
0	0	$\pm\frac{1}{2}$	3	-3	$\pm\frac{1}{2}$
1	-1	$\pm\frac{1}{2}$	3	-2	$\pm\frac{1}{2}$
1	0	$\pm\frac{1}{2}$	3	-1	$\pm\frac{1}{2}$
1	1	$\pm\frac{1}{2}$	3	0	$\pm\frac{1}{2}$
2	-2	$\pm\frac{1}{2}$	3	1	$\pm\frac{1}{2}$
2	-1	$\pm\frac{1}{2}$	3	2	$\pm\frac{1}{2}$
2	0	$\pm\frac{1}{2}$	3	3	$\pm\frac{1}{2}$
2	1	$\pm\frac{1}{2}$			
2	2	$\pm\frac{1}{2}$		32 sets	

8.2 (a) O.K.   (b) $m_l = -1$ is inconsistent with $l = 0$
(c) O.K.   (d) l has an incorrect negative sign
(e) O.K.   (f) m_s should be either $+\frac{1}{2}$ or $-\frac{1}{2}$.

8.3 (a) There are two electrons with parallel spin in the 2s orbital.
(b) There is only one electron in the 1s orbital.
(c) O.K.
(d) There are three electrons in the 2s orbital.
(e) Electrons in singly occupied p orbitals are usually represented with parallel spins.
(f) This is tricky. It appears as though the 2s orbital is underpopulated, that is, it should contain one of the electrons now in a 2p orbital. In Chapter 9 you will learn that such an electron configuration does exist in something called an sp^3 hybrid.

8.4 Atomic number 7: $1s^2 2s^2 2p^3$

1s   2s   2p
[↑↓] [↑↓] [↑|↑|↑]

Atomic number 15: $1s^2 2s^2 2p^6 3s^2 3p^3$

1s   2s   2p   3s   3p
[↑↓] [↑↓] [↑↓|↑↓|↑↓] [↑↓] [↑|↑|↑]

Atomic number 33: $1s^2 2s^2 2p^6 3s^2 3p^6 4s^2 3d^{10} 4p^3$

1s   2s   2p   3s   3p   4s   3d   4p
[↑↓] [↑↓] [↑↓|↑↓|↑↓] [↑↓] [↑↓|↑↓|↑↓] [↑↓] [↑↓|↑↓|↑↓|↑↓|↑↓] [↑|↑|↑]

They are members of the same elementary family.

8.5   (a) Ca: $1s^22s^22p^63s^23p^64s^2$
      (b) Fe: $1s^22s^22p^63s^23p^64s^23d^6$
      (c) Sn: $1s^22s^22p^63s^23p^64s^23d^{10}4p^65s^24d^{10}5p^2$
      (d) Xe: $1s^22s^22p^63s^23p^64s^23d^{10}4p^65s^24d^{10}5p^6$
      (e) Hg: $1s^22s^22p^63s^23p^64s^23d^{10}4p^65s^24d^{10}5p^66s^24f^{14}5d^{10}$
      (f) Lu: $1s^22s^22p^63s^23p^64s^23d^{10}4p^65s^24d^{10}5p^66s^24f^{14}5d^1$
      (g) U: $1s^22s^22p^63s^23p^64s^23d^{10}4p^65s^24d^{10}5p^66s^24f^{14}5d^{10}6p^67s^26d5f^3$

8.6   (a) 0   (b) 2   (c) 5   (d) 0   (e) 1

8.7

n	l	m_l	m_s
1	0	0	$\pm\frac{1}{2}$
2	0	0	$\pm\frac{1}{2}$
2	1	-1	$\pm\frac{1}{2}$
2	1	0	$\pm\frac{1}{2}$
2	1	1	$\pm\frac{1}{2}$

The four $2p$ electrons of oxygen can be represented by six different sets of quantum numbers.

8.8   Configurations such as $1s^23s$ correspond to excited electronic states.

8.9   One would probably expect element 104 to fall under hafnium (Hf), exhibit similar properties, and have the electronic configuration: $[\text{Rn}]\ 7s^25f^{14}6d^2$. Note that kurchatovium (Ku) is placed accordingly.

8.10 (a) $m_l = -4, -3, -2, -1, 0, 1, 2, 3, 4$
      (b) 9 orbitals   (c) 18 electrons   (d) $n = 5$

8.11

n	l	m_l	m_s
1	1	-1	$\pm\frac{1}{2}$
		0	$\pm\frac{1}{2}$
		1	$\pm\frac{1}{2}$
2	1	-1	$\pm\frac{1}{2}$
		0	$\pm\frac{1}{2}$
		1	$\pm\frac{1}{2}$
2	2	-2	$\pm\frac{1}{2}$
		-1	$\pm\frac{1}{2}$
		0	$\pm\frac{1}{2}$
		1	$\pm\frac{1}{2}$
		2	$\pm\frac{1}{2}$

1	2	3	4	5	6												6
7	8	9	10	11	12	13	14	15	16	17	18	19	20	21	22		

8.12 See text.

8.13 After all, quantum mechanics is pretty arty stuff—and abstract, at that.

8.14 6.63×10^{-18} g cm/sec

8.15 Don't let your subject know you're watching him!

8.16 How about mind and body for a starter?

8.17 Incidentally, what is really REAL?

Chapter 9

9.1   (a) Br^-: $1s^2 2s^2 2p^6 3s^2 3p^6 4s^2 3d^{10} 4p^6$ 36 electrons, 35 protons
    (b) Al^{3+}: $1s^2 2s^2 2p^6$ 10 electrons, 13 protons
    (c) Cu^+: $1s^2 2s^2 2p^6 3s^2 3p^6 3d^{10}$ 28 electrons, 29 protons

9.2 P^{3-} (o.n. $= -3$), S^{2-} (o.n. $= -2$), Cl^- (o.n. $= -1$),
    K^+ (o.n. $= +1$), Ca^{2+} (o.n. $= +2$), Sc^{3+} (o.n. $= +3$)

9.3   (a) Be^{2+}: $1s^2$
    (b) Rb^+: $1s^2 2s^2 2p^6 3s^2 3p^6 4s^2 3d^{10} 4p^6$
    (c) I^-: $1s^2 2s^2 2p^6 3s^2 3p^6 4s^2 3d^{10} 4p^6 5s^2 4d^{10} 5p^6$

9.4   (a) KI (b) $AgBr$ (c) $BaSO_4$ (d) ZnO

9.5   (a) K^+, Br^- (b) NH_4^+, Cl^- (c) Pb^{2+}, $2NO_3^-$
    (d) Ca^{2+}, $2OH^-$ (e) $3Na^+$, PO_4^{3-}

9.6   1 faraday: 39.1 g potassium, 35.5 g chlorine
    0.1 faraday: 3.91 g potassium, 3.55 g chlorine

9.7   Reduction: $2Al^{3+}$ $+ 6e^- \rightarrow$ $2Al$
                    (o.n. $= +3$)   (o.n. $= 0$)
    Oxidation: $3O^{2-}$ $\rightarrow$ $\frac{3}{2}O_2$ $+ 6e^-$
                  (o.n. $= -2$)   (o.n. $= 0$)
    Al^{3+} is reduced and O^{2-} is oxidized.

9.8   Oxidizing agent: Cu^{2+} or CuO
    Reducing agent: C
    Element oxidized: C (o.n. $0 \rightarrow +4$)
    Element reduced: Cu (o.n. $+2 \rightarrow 0$)

9.9   (a) o.n. $= +2$,   o.n. $= +4$
    (b) $SnCl_2$:   62.6% Sn,   37.4% Cl
        $SnCl_4$:   45.5% Sn,   54.5% Cl
    (c) $SnCl_2$: $\dfrac{wt\ Sn}{wt\ Cl} = 1.67$; $SnCl_4$: $\dfrac{wt\ Sn}{wt\ Cl} = 0.835$;
          $\dfrac{1.67}{0.835} = \dfrac{2}{1}$

9.10   Oxide A: FeO, o.n. $= +2$
    Oxide B: Fe_2O_3, o.n. $= +3$

9.11   (a) Covalent   (b) ionic   (c) covalent   (d) covalent
    (e) ionic

9.12 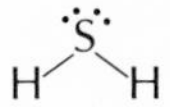   bent

9.13   tetrahedral
                                 sp^3 hybrid

9.14   Be sp hybrid
    C sp hybrid 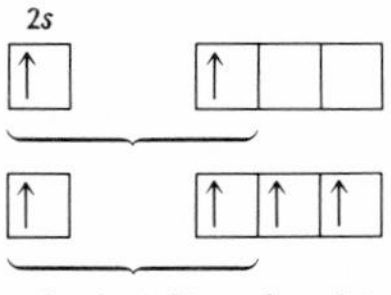
                     hybridized orbitals

9.15 $\ddot{O} :: C :: \ddot{O}$ no permanent dipole

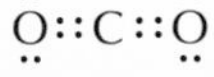

9.16

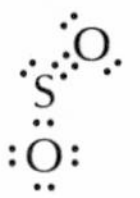

9.17   The results suggest electron resonance.

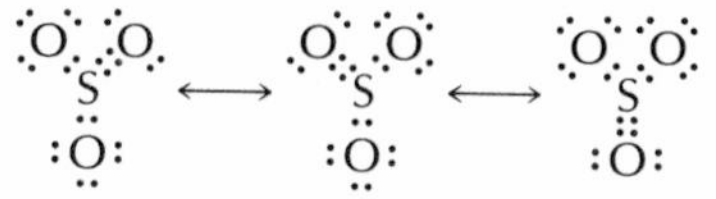

By inference (and experiment) SO_2 also exhibits resonance.

9.18 NH_3 is bent into a pyramidal shape by a pair of unbonded electrons on the nitrogen atom, which is negative relative to the hydrogen atoms. The boron atom in BF_3 carries only the three pair of bonding electrons and hence the molecule is planar and symmetrical.

9.19   (a) Permanent dipole   (b) no permanent dipole
(c) no permanent dipole   (d) no permanent dipole
(e) no permanent dipole   (f) permanent dipole

9.20

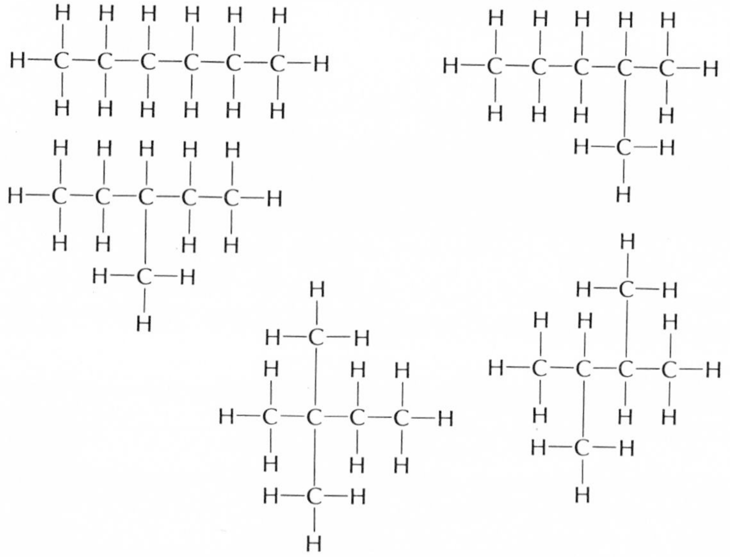

9.21

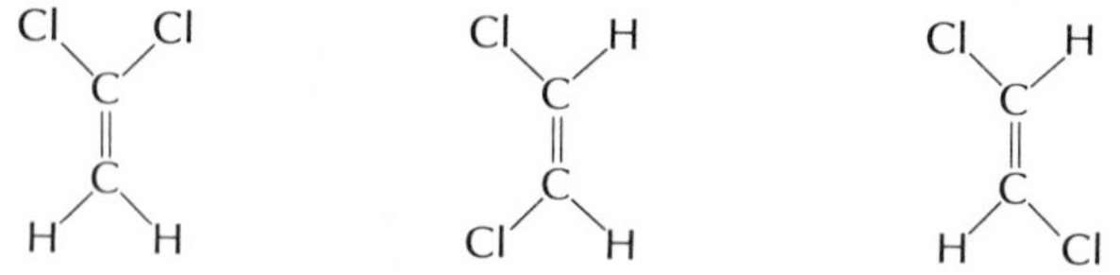

gem   *cis*   *trans*

9.22

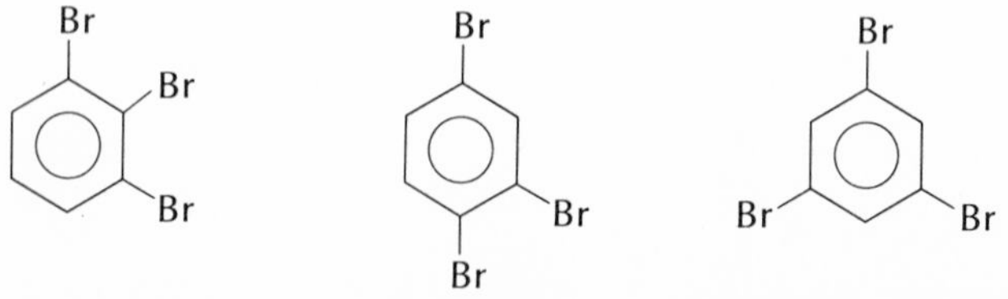

9.23   Remember that the σ bond is distributed in perfect circular symmetry around the line joining the two atoms.

Chapter 10

10.1   681 mm Hg or 0.896 atm

10.4   Atmospheric pressure will only support a column of water 33.9 ft high.

10.5   1.13 liters

10.6   1.73 liters. In fact, the pressure probably increases.

10.7   484°K or 211°C

10.8   22.4 liters

10.9 3.88×10^{-2} moles of air; 2.34×10^{22} molecules of air; 4.89×10^{21} molecules of oxygen.

10.10   309 atm

10.11   69.9 g

10.12 $335°K = 62°C = 144°F$

10.13   84.3

10.14   (a) 1.40×10^{-20} atm   (b) 1.48×10^5 cm/sec or 3.31×10^3 mi/hr (3310 miles/hour)

10.15   (a) 4.22×10^{23} molecules/cm^3 (b) 1.32×10^9 atm
(c) 6.98×10^7 cm/sec or 1.56×10^6 mi/hr (1,560,000 miles/hour)

10.16   The answers to Problem 10.14 are much more likely correct. The interior of the sun has very little in common with a perfect gas.

10.17 $v(^{235}UF_6)/v(^{238}UF_6) = 1.004$ $E(^{235}UF_6)/E(^{238}UF_6) = 1.000$

10.18 1.15×10^{-3} mole

10.19 2.59×10^5 cal or 259 kcal

10.20   (a) H_2O, C_2H_5OH, $C_2H_5OC_2H_5$
(b) $C_2H_5OC_2H_5$, C_2H_5OH, H_2O
(c) $C_2H_5OC_2H_5$, C_2H_5OH, H_2O
(d) H_2O, C_2H_5OH, $C_2H_5OC_2H_5$

10.22   (a) Dispersion   (b) covalent   (c) metallic   (d) ionic
(e) dispersion, dipolar, hydrogen-bonded   (f) dispersion
(g) dispersion   (h) dispersion, dipolar

10.23   (a) Dispersion forces increase with increasing molecular size and weight.
(b) The high-pressure treatment is involved.
(c) Electrostatic forces are stronger than dispersion and dipolar forces and hydrogen bonds.
(d) H_2O molecules are more tightly packed in liquid water than in ice.
(e) Remember the heat of evaporation!
(f) It's a matter of electrons, and how they bond (or don't bond)—covalent forces vs. dispersion forces.
(g) The water is dirty.

10.24   How about "extra" electrons?

Chapter 11

11.1   (a) 33.7%   (b) 6.13 M (c) 8.48 M
11.2   4.31 × 10⁻² M; higher
11.3   4.76 × 10⁻⁵ mole/liter
11.4   342 g sucrose + 1000 g distilled water. The sugar concentration is higher in the 1.00 M solution.
11.5   (a) 1.39 m (b) 7.01 × 10⁻² M
11.6   (a) Water   (b) water   (c) benzene   (d) soluble in both
11.7   (a) CH₃OH and HOH are quite similar and hydrogen-bond together.
        (b) The charged end helps.
        (c) Those covalent bonds are too strong to break.
        (d) Ions do the carrying.
11.8   Heat flows from the beer to the solution.
11.9   (a) Once-dissolved gas.
        (b) Remember that the solubility of air in water decreases with increasing temperature.
11.10   Undoubtedly you have discovered the solution!
11.11   Freezing point = −32.0°C; ethyl alcohol is volatile.
11.12   (a) 21.5 m (b) 57.1%
11.13   Methyl alcohol is more volatile than water and tends to boil away too easily at engine temperatures.
11.14   (a) 0.53 m (b) 100.55°C   (c) −1.97°C
11.15   (c), (a), (b)
11.16   (a) 5.00 × 10⁻⁴ M (b) 25,000
11.17   The key is osmotic pressure. The vascular system osmotically steals water from the rest of the body.
11.18   Remember that Cl⁻ ions go along with the H⁺ ions in HCl, and Na⁺ ions accompany the OH⁻ ions in NaOH.
11.19   Between 0.019 and 0.19 mg DDT/day; between 6.9 and 69 mg DDT/year.
11.20   4.00 × 10¹⁷ molecules CO
11.21   (a) 11.4 g PO₄³⁻   (b) 10.1 g CaO
11.22   (a) 1.0 × 10⁻² M (acidic)   (b) 3.2 × 10⁻⁶ M (acidic)
        (c) 4.0 × 10⁻⁸ M (slightly basic)   (d) 1.6 × 10⁻³ M (acidic)
11.23   (a) 2.0   (b) 11.0   (c) 7.8   (d) 6.6

Chapter 12

12.1   "Because that's the way things are" isn't a bad answer.
12.2   (a) and (b) Perhaps the speaker never heard of the First Law of Thermodynamics.   (c) Someone doesn't know the difference between temperature and heat.
12.3   (a) 5600 cal   (b) 9.2°C
12.4   While you're at it, you might disprove the existence of the ether.
12.5   All that work heats it up.
12.6   2.13 × 10⁵ cm (about 1.3 miles!)

12.7 $Q_1 = \Delta E_1 = +3000$ cal; $Q_2 = \Delta E_2 = -2000$ cal; $Q_{total} = \Delta E_{total} = +1000$ cal

12.8   (a) $\Delta E = +1000$ cal, $W = -1440$ cal, $Q = -440$ cal
        (b) Some energy (-440 cal worth) was lost from the system, probably as heat radiated to the surroundings.

12.9   (a) $Q = -3000$ kcal, $W = +2000$ kcal, $\Delta E = -5000$ kcal, for a net energy change of -1000 kcal.
        (b) The athlete obviously extracted some energy from the reserves chemically stored in his body.
        (c) He'd waste away by consuming himself.

12.10   (a) Exothermic   (b) endothermic   (c) exothermic
        (d) endothermic   (e) endothermic   (f) exothermic

12.11   (b), (e), (a), (f), (d), (c)

12.12 $\Delta H^\circ = -2.99 \times 10^4$ kcal

12.13   (a) $\Delta H^\circ = -530.7$ kcal/mole. Heat is evolved.
        (b) 2.58×10^4 kcal of heat can be obtained.

12.14   Isooctane: 1.10×10^{-3} cents/kcal; propane: 1.83×10^{-3} cents/kcal

12.15   (a) $\Delta H^\circ = -326.7$ kcal
        (b) 249 kcal are evolved ($Q = -249$ kcal)

12.16   (a) 93.5 kcal   (b) 16.3 min

12.17 2.30×10^{12} kg   or 2.53×10^9 tons of glucose

12.18   (a) $\Delta H^\circ = -1.10 \times 10^6$ kcal   (b) $\Delta H^\circ = -2.49 \times 10^7$ kcal

12.19   (a) $\Delta H^\circ = Q_p = -212.8$ kcal   (b) 150 kcal
        (c) 11.3 g CH_4

12.20 $\Delta H^\circ = +110.7$ kcal; heat is absorbed

12.21 $\Delta H^\circ_{f,n-C_4H_{10}(g)} = -29.8$ kcal

12.22 $\Delta H^\circ_{f,NO(g)}$ is positive.

12.23   See text. Also note that enthalpy changes are easier to measure than energy changes.

12.24   Where does the energy come from?

Chapter 13

13.4   (a) 10.0 liter atm, 242 cal, 1010 joules
        (b) $Q = +242$ cal or 1010 joules, $\Delta E = 0$, $\Delta S = +1.37$ cal/°K

13.5   Path-dependent function: heat, work; state-dependent function: enthalpy, entropy

13.6 $\epsilon = 0.236$; remember the pressure cooker?

13.7   Nuclear power plants tend to operate at lower efficiency.

13.8   A heat source which is clean in the home may be pretty dirty at the power plant.

13.11   The text should help on this one.

13.13   (c),   (e),   (d),   (b),   (a)

13.16   (b),   (e),   (a),   (f),   (d),   (c)

13.17 $+26.0$ cal/°K

13.18   (a) The evaporation of shaving lotion   (b) the burning of a match   (d) the dissolving of sugar in coffee   (d) the freezing of water.

13.19   The chemist hasn't read this chapter and doesn't know that $\Delta G = \Delta H - T\Delta S$.

13.20   (a) $\Delta G = -260$ cal, water boils
      (b) $\Delta G = +260$ cal, steam condenses

13.21   (a) 353°K or 80°C   (b) $\Delta G = +1300$ cal. The tendency for benzene vapor to condense is greater than the tendency for liquid benzene to vaporize.

Chapter 14

14.1   The oxidation of magnesium is spontaneous under standard conditions. Its $\Delta G°$ is -136.1 kcal/mole. The reverse reaction, the decomposition of magnesium oxide to elementary magnesium and oxygen is nonspontaneous, and has an associated $\Delta G°$ of $+136.1$ kcal/mole.

14.2   The fact that Fe_2O_3 has a smaller negative standard free energy of formation than does Al_2O_3 explains this behavior.

14.3   Aluminum oxide forms a tight film which protects the underlying metal from the action of oxygen. Rust (iron oxide) easily flakes off, exposing more metal to air.

14.4   You are thermodynamically well advised to invest in something else; $\Delta G°$ is positive for the proposed reaction.

14.5   (a) 94.3 kcal or 3.94×10^5 joules   (b) 127 g of carbon
      (c) We live in a nonideal, irreversible world. Some energy gets "lost" as heat.

14.6 $\Delta H° = -47.5$ kcal, $\Delta S° = +0.67$ cal/°K

14.7   (a) $\Delta G° = -21.9$ kcal   (b) $\Delta H° = -42.3$ kcal
      (c) $\Delta S° = -68.5$ cal/°K. Note the volume decrease.

14.8   ATP forms a very efficient means of internal energy transfer. More details are available in Chapter 15.

14.9   The enthalpy of graphite is lower than that of diamond, and the entropy of graphite is greater than that of diamond. Hence, heat is evolved ($\Delta H° < 0$) and disorder increases ($\Delta S° > 0$) when the atoms of diamond are rearranged into graphite. Both factors favor the spontaneity of the process.

14.10   (a) Diamonds were formed under nonstandard conditions of high temperature and pressure where ΔG is negative.
      (b) "It takes time, man. The reaction needs a shove."

14.11   (a) $\Delta E = 21$ kcal   (b) ΔH will be more negative than ΔE
      (c) $\Delta H° = -22$ kcal   (d) negative, because the volume of gases decreases

14.12   "It's all thermodynamics, Junior"

14.13   (a) $K = \dfrac{[CO_2][H_2O]^2}{[CH_4][O_2]^2}$

      (b) $K = \dfrac{[HI]^2}{[H_2][I_2]}$

      (c) $K = \dfrac{[CO][Cl_2]}{[COCl_2]}$

14.14　Some of the N_2 and H_2 will react to form NH_3 because the original nonequilibrium mixture contains, chemically speaking, too much N_2 and H_2 and too little NH_3.

14.15　(a) $K = \dfrac{[CO_2][H_2]}{[CO][H_2O]}$

(b) $[CO_2] = [H_2] = 0.075$ moles/liter
$[CO] = [H_2O] = 0.025$ moles/liter

(c) $K = 9.0$

14.16　(a) $K = \dfrac{[SO_3]^2}{[SO_2]^2[O_2]}$　(b) $\Delta G° = -33.4$ kcal

(c) K will be greater than 1.

14.17　6.8×10^{18}

14.18　$\Delta H°$ is negative; $\Delta H° = -47.0$ kcal

14.19　(a) Some NO_2 would dissociate into NO and O_2.

(b) Some NO and O_2 would react to form NO_2.

(c) Some NO_2 would dissociate into NO and O_2.

14.20　The trout would experience about a 50% increase in the rate of his internal chemistry.

14.21　$NO(g) + \tfrac{1}{2}O_2(g) \rightarrow NO_2(g)$

$\underline{SO_2(g) + NO_2(g) \rightarrow SO_3(g) + NO(g)}$

$SO_2(g) + \tfrac{1}{2}O_2(g) \rightarrow SO_3(g)$

14.22　By surface catalysis

14.23　(a) $\Delta G° = -123.0$ kcal　(b) yes　(c) $\Delta H° = -135.4$ kcal

(d) More CO_2 would form at tailpipe temperatures, so put the catalyst there.

14.24　Remember that a large value of K means the position of equilibrium is far to the right; the CO binds very tightly to the hemoglobin.

14.25　Fortunately, you react slowly.

Chapter 15

15.1　(a) Ketone　(b) alcohol　(c) aldehyde　(d) alcohol, acid
(e) ester　(f) ether　(g) amine, acid　(h) alcohol, aldehyde

15.2　(a) Acid, ester　(b) alcohol　(c) ether, amine　(d) ester
(e) amine　(f) ether, aldehyde, phenol (you probably called the —OH group alcoholic)　(g) ether, acid

15.3　(a) CH_3CH_2OH,　CH_3—O—CH_3

(b) $CH_3CH_2\overset{\displaystyle O}{\overset{\|}{C}}$—H,　CH_3—$\overset{\displaystyle O}{\overset{\|}{C}}$—$CH_3$

(c) $CH_3CH_2CH_2OH$,　$\overset{CH_3}{\underset{CH_3}{>}}CHOH$,　CH_3CH_2—O—CH_3

15.4　The -ol ending signifies an alcohol.

15.5

$$CH_3\overset{\displaystyle O}{\overset{\|}{C}}\!-\!OH + HOCH_3 \longrightarrow CH_3\overset{\displaystyle O}{\overset{\|}{C}}\!-\!OCH_3$$

acetic acid   methyl   methyl acetate
alcohol

15.6   (b),   (c),   (f)

15.7

$$H_2N\!-\!\underset{\underset{\displaystyle H}{|}}{\overset{\overset{\displaystyle H}{|}}{C}}\!-\!\overset{\displaystyle O}{\overset{\|}{C}}\!-\!N\!-\!\underset{\underset{\displaystyle CH_2}{|}}{\overset{\overset{\displaystyle H}{|}}{C}}\!-\!\overset{\displaystyle O}{\overset{\|}{C}}\!-\!OH$$

with $-N-$ having H below, $\underset{\underset{\displaystyle OH}{|}}{CH_2}$

15.8

$$n HO\!-\!\overset{\displaystyle O}{\overset{\|}{C}}\!-\!(CH_2)_4\!-\!\overset{\displaystyle O}{\overset{\|}{C}}\!-\!OH + n H_2N\!-\!(CH_2)_6\!-\!NH_2 \longrightarrow$$

$$\cdots\!\!\left[\!-\!\underset{\underset{\displaystyle H}{|}}{N}\!-\!(CH_2)_6\!-\!\underset{\underset{\displaystyle H}{|}}{N}\!-\!\overset{\displaystyle O}{\overset{\|}{C}}\!-\!(CH_2)_4\!-\!\overset{\displaystyle O}{\overset{\|}{C}}\!-\!\underset{\underset{\displaystyle H}{|}}{N}\!-\!(CH_2)_6\!-\!\underset{\underset{\displaystyle H}{|}}{N}\!-\!\overset{\displaystyle O}{\overset{\|}{C}}\!-\!(CH_2)_4\!-\right.$$

Note the similarity to the reaction of amino acids to form proteins.

15.9   Amino acids with extra acid groups on their side chains (for example, glutamic acid) introduce acidic properties into proteins, since these groups do not react to form peptide bonds.

15.10   Proline is an atypical amino acid. Because of its ring structure, it doesn't fit into an α-helix arrangement.

15.11   You will recall osmotic pressure measurements from Chapter 11 and Problem 11.16. Other methods make use of high-speed centrifuges, the scattering of light, and chromatographic columns packed with material of known porosity.

15.12   The exact mechanism is not known, but it most likely involves changes in tertiary structure and the aggregation or association of protein molecules.

15.13   It can "eat" itself—or, less colloquially, it can catalyze its own hydrolysis.

15.14   (a) Monosaccharide, pentose   (b) monosaccharide, pentose, ketose,   (c) monosaccharide, hexose, aldose   (d) disaccharide (two hexoses)   (e) monosaccharide, hexose, aldose (f) monosaccharide, hexose, ketose

15.15   (a) 2   (b) 4

$$\begin{array}{cccc}
CHO & CHO & CHO & CHO \\
H\!-\!C\!-\!OH & HO\!-\!C\!-\!H & HO\!-\!C\!-\!H & H\!-\!C\!-\!OH \\
H\!-\!C\!-\!OH & HO\!-\!C\!-\!H & H\!-\!C\!-\!OH & HO\!-\!C\!-\!H \\
CH_2OH & CH_2OH & CH_2OH & CH_2OH
\end{array}$$

15.16   (a) 8   (b) 16   (c) 4   (d) 8

15.17   The starch is broken down into glucose molecules.

15.18   By imitating cows, and letting bacteria (or maybe chemicals) do some of our digesting.

15.20 3.7×10^6 (that's 3,700,000) proteins. (This calculation assumes only one strand is read and neglects base triplets used as start and stop signs.)

15.21 3.5×10^6 base pairs

15.22   The genetic similarities among members of a single species are far greater than their differences.

15.23   This question raises one of the most interesting unknowns in modern biochemistry: the mechanism by which the vast majority of genetic information in a cell remains suppressed, while only a minute fraction is actually utilized. Significant advances in our understanding of the process have been made recently, but the embryonic development of each fantastically complicated, functioning human being is still awe-inspiring.

15.24   The experiment provides evidence that the genetic code is written nucleotide triplets.

15.25   Carry out protein synthesis using messenger RNA of known base composition and attempt to relate this information to the amino acid composition of the protein formed.

15.26   It makes mistakes less serious; indeed many changes in base identity are never observed in the protein because the meaning of the code remains unaltered.

15.28   The necessary step of breaking down the sucrose or starch to monosaccharides is skipped if only glucose is present.

15.29   (a) 18.9 kcal   (b) 7.4 kcal   (c) just over 1 mole (6.023×10^{23} molecules)

15.30   (a) 110 moles ATP   (b) 3.06 moles glucose   (c) 550 g glucose

15.31   Single-point control simplifies the regulation of body chemistry. Moreover, it promotes economy, rather like shutting down an entire assembly line with a single switch.

15.32   This chemical commonality makes for highly efficient utilization of a wide variety foods for various purposes.

Chapter 16

16.1   (a) 15 neutrons, atomic weight $= 30$ (b) 27, 52   (c) 60, 105   (d) 94, 159   (e) 123, 205

16.2   (a) β emission: $^3_1\text{H} \rightarrow \, ^3_2\text{He} + \, ^0_{-1}\beta$

        (b) positron emission: $^{10}_6\text{C} \rightarrow \, ^{10}_5\text{B} + \, ^0_1\text{e}$

        (c) β emission: $^{90}_{38}\text{Sr} \rightarrow \, ^{90}_{39}\text{Y} + \, ^0_{-1}\beta$

        (d) positron emission: $^{58}_{27}\text{Co} \rightarrow \, ^{58}_{26}\text{Fe} + \, ^0_1\text{e}$

        or electron capture: $^{58}_{27}\text{Co} + \, ^0_{-1}\text{e} \rightarrow \, ^{58}_{26}\text{Fe}$

        (e) β emission: $^{60}_{27}\text{Co} \rightarrow \, ^{60}_{28}\text{Ni} + \, ^0_{-1}\beta$

16.3   11,520 years ago

16.4 1.74×10^{-10} g

16.5 2.25×10^{22} ergs; 5.38×10^{14} cal

16.7   (a) 9.0×10^8 g (about 1000 tons) of U-235

(b) 2.3×10^{15} g (about 2.5 billion tons) of coal

16.8   (a) $^{94}_{38}Sr$ (b) $^{131}_{50}Sn$

16.9   2 generations, 3 atoms; 3 generations, 7 atoms;
4 generations, 15 atoms; 5 generations, 31 atoms

16.10 1.70×10^{19} ergs, 4.07×10^{11} cal

16.11   (a) 3.89×10^9 kcal   (b) 7.20×10^9 g, 7.20×10^6 kg,
7.93×10^3 tons

Index

Numbers set in boldface type identify pages on which terms are defined, laws are stated, or formulas of compounds are introduced.